Jiaqi Zhu, Jiecai Han

Infrared Antireflective and Protective Coatings

Also of Interest

Series: Advanced Composites.
J. Paulo Davim (Ed.)
ISSN 2192-8983
Published titles in this series:
Vol. 6: Wood Composites (2017) Ed. by Aguilera, Alfredo/Davim, J. Paulo
Vol. 5: Ceramic Matrix Composites (2016) Ed. by Davim, J. Paulo
Vol. 4: Machinability of Fibre-Reinforced Plastics (2015)
Ed. by Davim, J. Paulo
Vol. 3: Metal Matrix Composites (2014) Ed. by Davim, J. Paulo
Vol. 2: Biomedical Composites (2013) Ed. by Davim, J. Paulo
Vol. 1: Nanocomposites (2013) Ed. by Davim, J. Paulo/
Charitidis, Constantinos A.

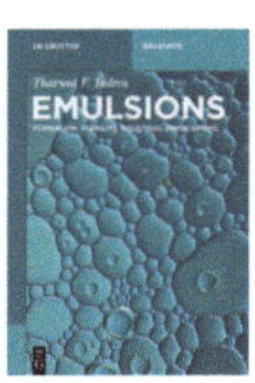

Emulsions
Formation, Stability, Industrial Applications
Tadros, 2016
ISBN 978-3-11-045217-4, e-ISBN 978-3-11-045224-2

Metals and Alloys Industrial Applications
Benvenuto, 2016
ISBN 978-3-11-040784-6, e-ISBN 978-3-11-044185-7

Thermophysical Properties of Multicomponent Liquid Alloys
Brillo 2016
ISBN 978-3-11-046684-3, e-ISBN 978-3-11-046899-1

Jiaqi Zhu, Jiecai Han

Infrared Antireflective and Protective Coatings

國防工業出版社
National Defense Industry Press

DE GRUYTER

Authors
Prof. Jiaqi Zhu
Harbin Institute of Technology
Shool of Astronotics
Yikuang Road 2
150080 Harbin, Nangang District
China
zhujq@hit.edu.cn

Prof. Jiecai Han
Harbin Institute of Technology
Shool of Astronotics
Yikuang Road 2
150080 Harbin, Nangang District
China
hanjc@hit.edu.cn

ISBN 978-3-11-048809-8
e-ISBN (PDF) 978-3-11-048951-4
e-ISBN (EPUB) 978-3-11-048819-7

Library of Congress Cataloging-in-Publication Data
A CIP catalog record for this book has been applied for at the Library of Congress.

Bibliographic information published by the Deutsche Nationalbibliothek
The Deutsche Nationalbibliothek lists this publication in the Deutsche Nationalbibliografie; detailed bibliographic data are available on the Internet at http://dnb.dnb.de.

Cover image: Carpe89 / iStock / Getty Images
Typesetting: PTP-Berlin, Protago-TEX-Production GmbH, Berlin
Printing and binding: CPI books GmbH, Leck
♾ Printed on acid-free paper
Printed in Germany

www.degruyter.com

Foreword I

Infrared transmitting materials placed at the front of the infrared detectors ensure the high-efficiency transmission of target infrared radiation and protect the optical imaging/detecting system, and they are typical structure-function integrated materials. Infrared transmitting materials are divided into two categories: infrared window substrates and infrared antireflective/protective coatings. In order to enhance the survivability of aircraft and realize "efficient detection, rapid withdrawal, and precise striking", infrared transmitting materials must possess excellent service capabilities in extreme environments and meet various application demands including multiband spectral detection. With the rapid growth of related military demands in recent years, the research and application of infrared transmitting materials have made great progress. It is gradually becoming a rather independent discipline with increasingly expanding prospects.

Since the establishment of the National Key Laboratory of Special Environment Composite Material Technology upon its approval in 2000, the realization of high-temperature-resistant, light, multifunctional, composite materials for use in extreme environments has always been the primary research direction of this lab, with infrared transmitting materials, in particular, as a priority. After continuous effort over a decade, specialized equipment for single-crystal growth and coating was developed, and breakthroughs were made in key technologies such as homogeneous deposition on large flat surfaces/semispherical surfaces. The single-crystal growth of sapphire as well as antireflection and protection technology of infrared windows/domes won the national technological invention awards. They have been applied to various key models with mass production, contributing to the tackling of critical challenges in national defence technology.

The authors of this book have been undertaking research on infrared antireflective/protective coatings for a long time. They have done tremendous work on the associated fundamental theory, coating system design, preparation techniques, structure characterization, performance modulation, and application development. Based on their research over a decade and by integrating the latest research progress in infrared antireflection/protection technology, they have authored this systematic and comprehensive book for readers: *Infrared Antireflective and Protective Coatings*. The publication of this book is highly meaningful and indicates that research on infrared antireflection/protection technology in China has entered a new level.

Harbin, December 20, 2014

https://doi.org/10.1515/9783110489514-001

Foreword II

Infrared antireflective/protective coating materials play a crucial rule from the development to the practical applications of infrared precision guidance and detection systems, attracting great attention from strong military powers worldwide. In order to meet the increasingly rigorous technical requirements of infrared precision guidance and detection systems, the research and industrialization of the infrared antireflective/protective coating that directly faces the service environment have been rapidly advanced.

This book systematically introduces the preparation, structure, performance, characterization, and applications of typical infrared antireflective/protective coating materials, encompassing multiple cutting-edge fields in infrared antireflective/protective coating material research. Among the ten chapters of the book, the chapters on the basic knowledge of infrared transmitting materials discuss infrared physics, materials chemistry, and materials processing; the chapters on infrared antireflective/protective coatings include optical design, physics & chemistry at surfaces & interfaces, thin-film preparation techniques, and service environment effects. At the end of the 20th century, SPIE Press released the monograph of Daniel C. Harris, *Materials for Infrared Windows and Domes: Properties and Performance*. In 2007, Huaizhi Yu, from the General Research Institute for Nonferrous Metals, compiled *Infrared Optical Materials*. The authors of the present book have long been engaged in technical work related to infrared antireflective/protective coating materials and have accumulated rich experience. Apart from their latest research results, research progress in this field attributed to other researchers around the world is also included in this book, making it a comprehensive academic publication specialized in infrared antireflective/protective coating materials.

In recent years, the research and development of infrared materials have been active in China, and industrialization has gained initial momentum in terms of scale. I believe that the publication of this book will not only offer guidance for the research, manufacture, and education of infrared transmitting materials and infrared antireflective/protective coatings, but also have great practical significance in promoting discipline construction and industrial development of infrared optical materials in China.

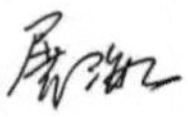

December 24, 2014

https://doi.org/10.1515/9783110489514-002

Preface

Infrared windows or domes mounted on the front of aircraft-borne infrared searching/tracking systems and missile-borne infrared precision guidance systems must not only ensure the efficient transmission of target signals, but also protect interior parts, making them critical structure-function integrated system components. With vast and extensive applications, their development has shown a trend towards multi-waveband, multi-function, and large windows. Infrared window materials can be divided into two categories: window substrates and infrared antireflective/protective coatings. In order to enhance the survivability of future aircraft and realize "efficient detection, rapid withdrawal, and precise striking," infrared transmitting materials must possess excellent service capabilities in extreme environments and meet various application demands including antireflection/protection, multi-band spectral detection, and radar stealth. To satisfy the urgent demands of our country in recent years, research on infrared antireflective/protective coating material technology has made rapid progress with remarkable achievements.

Service in extreme environments requires the functional film to have high surface hardness, excellent stability, and good attachability. Multi-spectral optical transmission demands extremely low absorption in the visible band and the far-infrared band, adjustability of optical performance parameters, and optical antireflection in particular bands; radar stealth requires a relatively low area resistance. Therefore, the mechanical, optical, electronic, and thermal performance of functional coatings have to be well coordinated to satisfy the service requirements. The development of comprehensively high-performance infrared antireflective/protective coating materials has been a challenge that draws much attention in the international research community. Based on different application environments and window substrates, various antireflective/protective coating material systems, including diamond-like carbon, germanium carbide, boron phosphide, oxides, and nitrides, have been developed. They are able to meet different application needs and demonstrate exciting application prospects.

Studies and applications of infrared antireflective/protective coating materials worldwide have made notable achievements with an expanding group of scientific and technological personnel engaged in research. However, thus far, no monograph that comprehensively and systematically discusses infrared antireflective/protective coating materials has ever been published. In the published books on infrared optical materials, infrared antireflective/protective coating materials only account for few sections, which can hardly satisfy the needs of professionals to know the status quo of technology development in this area. In order to promote the development of infrared antireflective/protective coating material technology, we attempt to present the basic characteristics, fabrication techniques, micro-scale structures, optical properties, me-

https://doi.org/10.1515/9783110489514-003

chanical properties, and development & applications of typical infrared antireflective/protective coating materials from a comprehensive and systematic perspective.

Based on the results of research over a decade achieved by the authors' research groups, we also try our best to include detailed primary achievements attributed to international researchers working in the field of infrared antireflective/protective coatings in their latest research. We endeavour to make the concepts clear and readily understandable, as well as to reflect the state-of-art technology of this subject area.

This book is divided into ten chapters. Their main content and authors are listed here: Chapter 1, foundations of infrared-transparent materials (Qiuling Yang, Jiaqi Zhu); Chapter 2, service environment of infrared antireflective and protective coatings (Tianyu Ma, Jiaqi Zhu); Chapter 3, optical design of infrared antireflective and protective film systems (Bing Dai, Pei Lei, Jiaqi Zhu); Chapter 4, preparation methods of infrared antireflective and protective coatings (Lei Yang, Shuai Guo, Jiaqi Zhu); Chapter 5, amorphous diamond films (Jiaqi Zhu, Jiecai Han); Chapter 6, germanium carbide films (Chunzhu Jiang, Jiaqi Zhu, Jiecai Han); Chapter 7, boron phosphide films (Weixia Shen, Jiaqi Zhu, Jiecai Han); Chapter 8, alumina films (Xiaopeng Zhang, Jiaqi Zhu); Chapter 9, yttrium oxide films (Pei Lei, Jiaqi Zhu); Chapter 10, infrared transparent conductive oxide thin films (Lei Yang, Jiaqi Zhu, Jiecai Han).

Infrared antireflective/protective films are a type of emerging structure-function integrated coating materials, the study of which involves multiple disciplines including solid-state physics, thin-film optics, material science, and plasma technology. For practical application, even more disciplines such as engineering thermal physics and impact mechanics need to be considered simultaneously. Therefore, the associated theories are deep, extensive, and relatively new, with a distinct cross-disciplinary feature. Limited by our knowledge and strength, mistakes/omissions are almost unavoidable in the book. Corrections and suggestions from readers are warmly welcome.

We would like to express our sincere gratitude to the National Defence Industry Press for their full support in the course of publication. We would like to thank Mr. Shanyi Du and Mr. Hailing Tu, both well-known material scientists and members of the Chinese Academy of Engineering, for the forewords they wrote for this book. This book also contains the research work of PhD students Wei Gao, Xiao Han, Manlin Tan, Aiping Liu, Li Niu, Chunzhu Jiang, Xing Liu, and Yuanchun Liu, who are greatly appreciated.

Authors
February 2015

Contents

Foreword I —— v

Foreword II —— vii

Preface —— ix

1 Foundation of infrared-transparent materials —— 1
1.1 Overview —— 1
1.2 Infrared signal on the battlefield —— 1
1.2.1 Electromagnetic spectrum and atmospheric window —— 1
1.2.2 Blackbody radiation —— 2
1.2.3 Infrared signal transmission through infrared-transparent materials —— 6
1.3 Typical infrared-transparent substrate materials —— 8
1.3.1 Classes of materials —— 9
1.3.2 Single-crystal materials —— 9
1.3.3 Polycrystalline materials and ceramics —— 20
1.3.4 Optical glass —— 28
1.4 Infrared antireflective and protective coatings —— 32
1.4.1 Common optical thin-film materials and their main characteristics —— 32
1.4.2 Surface superstructures —— 34
1.4.3 Transparent conducting thin-films —— 35
1.5 Optical properties of infrared antireflective and protective coatings —— 37
1.5.1 Optical constants —— 37
1.5.2 Transmittance —— 39
1.5.3 Emittance —— 41
1.5.4 Transmission wave range of common infrared materials —— 43
1.6 Mechanical and thermal properties of infrared antireflective and protective coatings —— 45
1.6.1 Hardness and elastic modulus —— 45
1.6.2 Coating–substrate adhesion —— 48
1.6.3 Thermal conductivity and thermal expansion coefficient —— 49

2 Service environment of infrared antireflective and protective coatings —— 55
2.1 Overview —— 55
2.2 Aerodynamic heat/strength environment and aerodynamic heat/strength failures —— 56

2.2.1 Basic forms of aerodynamic heat/strength failure of infrared windows — 56
2.2.2 The development of infrared window materials — 57
2.2.3 Design method for infrared window functional films — 61
2.2.4 Ground failure tests of infrared window — 67
2.2.5 Numerical simulation of failure of infrared window — 71
2.3 Analysis of typical aerodynamic heat/strength environment — 73
2.3.1 Calculation model of supersonic aircraft outer flow field — 74
2.3.2 Verification of supersonic aircraft outer flow-field model — 81
2.3.3 Calculation of blunted cone outer flow field — 83
2.3.4 Calculation of the pointed cone outer flow field — 86
2.4 Rain erosion — 91
2.4.1 Rain erosion and influential factors — 92
2.4.2 Test method and facilities of rain erosion — 95
2.4.3 Evaluation index of rain erosion resistance — 98
2.4.4 Protection from rain erosion — 99
2.5 Sand erosion — 100
2.5.1 Sand erosion damage behavior — 100
2.5.2 Factors influencing sand erosion — 101
2.5.3 Introduction of sand erosion test — 103

3 Optical design of infrared antireflective and protective coating system — 113
3.1 Overview — 113
3.2 Basic theories of coating design — 113
3.2.1 Mathematical method — 113
3.2.2 Monolayered uniform dielectric film — 117
3.2.3 Multilayered uniform dielectric film — 123
3.3 Oblique incidence theory and coating design — 126
3.3.1 Depolarization design of incident medium, coating and substrate combinations — 132
3.3.2 Quarter-wave stack depolarization design — 133
3.4 Discussion on the parametric variations of coating systems — 138
3.4.1 Influence of incident angle — 138
3.4.2 Influence of thin-film optical thickness — 140
3.4.3 Influence of thin-film refractive index — 145
3.4.4 Influence of thin-film physical thickness — 146

4 Preparation methods of infrared antireflective and protective coatings — 147
4.1 Overview — 147
4.2 Magnetron sputtering — 147

4.3 Thermal evaporation deposition — 150
4.3.1 Vacuum condition of the evaporation process — 151
4.3.2 Evaporation condition in the deposition process — 152
4.4 Plasma-enhanced chemical vapor deposition — 152
4.5 Filtered cathodic vacuum arc deposition — 154
4.6 Large-area filtered cathodic vacuum arc uniform deposition — 157
4.7 Large-size plane uniform magnetron sputtering — 158
4.8 Heavy-calibre spherical uniform magnetron sputtering — 164

5 Amorphous diamond films — 167
5.1 Overview — 167
5.2 Determination of deposition energy for rich sp^3 hybridization — 168
5.2.1 Mechanical properties of thin films — 169
5.2.2 Raman analysis — 170
5.2.3 X-ray photoelectron spectrum analysis — 174
5.2.4 Electron energy loss spectrum analysis — 178
5.3 Deposition mechanism of amorphous diamond films — 179
5.3.1 Cross-section layered density distribution of thin films — 180
5.3.2 Structural model of amorphous diamond films — 185
5.3.3 Surface morphology of thin films — 188
5.3.4 Surface composition of thin films — 191
5.4 Process control rules of optical parameters of thin films — 192
5.4.1 Basic theory — 192
5.4.2 Refractive index — 194
5.4.3 Extinction coefficient — 197
5.4.4 Optical band gap — 198
5.5 Effects of thin-film thickness and stress — 200
5.5.1 Mechanical properties of amorphous diamond thin films with different thickness — 200
5.5.2 Raman characterization of amorphous diamond films in different thicknesses — 202
5.5.3 Effect of thin-film stress — 204
5.6 Thermal stability — 206
5.6.1 Thermal stability of thin films in air — 207
5.6.2 Thermal stability of thin films in vacuum — 212
5.7 Adhesion with infrared window materials — 217
5.7.1 Interface properties of amorphous diamond thin films with zinc sulphide and magnesium fluoride — 218
5.7.2 Interface properties of amorphous diamond films with germanium — 220

5.8 Structure and stress analysis of multilayered amorphous diamond films — 225
5.8.1 Stress theory of multilayered films — 225
5.8.2 Microstructure of multilayered amorphous diamond films — 227
5.8.3 Stress analysis of multilayered amorphous diamond films — 231
5.9 Mechanical properties of multilayered amorphous diamond films — 233
5.9.1 Hardness and Young's modulus — 234
5.9.2 Fracture properties — 236
5.9.3 Scratch resistance and adhesion performance — 240
5.9.4 Effect of sublayer thickness on the mechanical properties of multilayered films — 244

6 Germanium-carbide film — 263
6.1 Overview — 263
6.2 Deposition rate of germanium-carbide film prepared by magnetron co-sputtering — 264
6.2.1 Preparation of germanium-carbide film by magnetron co-sputtering — 264
6.2.2 Effect of power on deposition rate — 265
6.2.3 Effect of substrate temperature on deposition rate — 266
6.3 Surface morphology, crystal structure and composition — 267
6.3.1 Surface morphology — 267
6.3.2 Crystal structure — 269
6.3.3 Composition analysis — 270
6.4 Bonding structure and bonding mechanism — 272
6.4.1 FTIR spectra of germanium-carbide thin films — 272
6.4.2 Raman spectra of germanium-carbide thin films — 275
6.4.3 X-ray photoelectron spectra of germanium-carbide films — 281
6.4.4 Bonding mechanism and rules — 285
6.5 Film density — 286
6.6 Optical properties of thin films — 290
6.6.1 Visible optical properties — 290
6.6.2 Infrared optical properties — 296
6.7 Electrical properties of thin films — 302
6.8 Mechanical properties of thin films — 305
6.8.1 Hardness and Young's modulus — 305
6.8.2 Residual stress — 307
6.9 Thermal stability — 313
6.9.1 Analysis of bonding structure — 313
6.9.2 Change of hardness — 316
6.10 Germanium-carbide composite films — 317

6.10.1 Germanium carbide/germanium-carbide composite films — **317**
6.10.2 Amorphous diamond films/germanium-carbide composite films — **320**

7 Boron phosphide thin films — 325
7.1 Overview — **325**
7.2 Research progress on the structure and properties of boron phosphide thin films — **326**
7.2.1 Structural features of boron phosphide thin films — **326**
7.2.2 Performance features of boron phosphide thin films — **329**
7.2.3 Composite films — **331**
7.3 Structural analysis of boron phosphide thin films by reactive magnetron sputtering — **332**
7.3.1 Process design of reactive magnetron sputtering — **332**
7.3.2 Composition analysis — **333**
7.3.3 Surface morphology — **335**
7.3.4 Crystal structure and bonding states — **338**
7.3.5 Growth mechanism — **344**
7.4 Thermal stability of boron phosphide thin films — **345**
7.4.1 XRD analysis — **346**
7.4.2 Raman analysis — **347**
7.5 Mechanical and optical properties of boron phosphide thin films — **349**
7.5.1 Mechanical properties — **349**
7.5.2 Optical properties — **353**
7.6 Corrosion resistance and wear resistance of boron phosphide thin films — **355**
7.6.1 Corrosion resistance — **355**
7.6.2 Wear resistance — **357**

8 Alumina thin films — 365
8.1 Overview — **365**
8.2 Research status of alumina thin films — **365**
8.2.1 Basic properties — **365**
8.2.2 Research status of crystalline alumina — **366**
8.3 Preparation methods of alumina thin films — **369**
8.3.1 Deposition process of conventional magnetron sputtering — **369**
8.3.2 Deposition process of resputtering assisted magnetron sputtering — **371**
8.4 Growth rate, composition, and chemical bonding — **373**
8.4.1 Growth rate — **373**
8.4.2 O/Al composition ratio — **374**
8.4.3 Chemical bonding — **375**

8.5 Morphology and structure analysis — 376
8.5.1 Surface morphology — 376
8.5.2 Crystal structure — 377
8.6 Low-temperature crystallization by resputtering assisted deposition — 378
8.6.1 Crystal structure — 378
8.6.2 Mechanical properties — 380
8.6.3 Optical properties — 382

9 Yttrium oxide thin films — 387
9.1 Overview — 387
9.2 Crystal structure control of yttrium oxide thin films — 388
9.2.1 Effect of oxygen partial pressure — 389
9.2.2 Effect of substrate temperature — 391
9.2.3 Effect of substrate bias — 393
9.2.4 Effect of sputtering power — 395
9.3 Performance control of yttrium oxide film — 397
9.3.1 Optical properties of yttrium oxide thin films — 398
9.3.2 Mechanical properties of yttrium oxide thin films — 405
9.3.3 Surface wettability of yttrium oxide thin films — 409
9.4 Growth pattern and deposition rate — 417
9.4.1 Reactive hysteresis loop and growth pattern — 418
9.4.2 Control of reactive hysteresis loop — 420
9.4.3 Growth pattern and deposition rate — 422
9.5 Adhesion of yttrium oxide thin film and zinc-sulphide substrate — 423
9.5.1 XPS composition analysis of interface region — 424
9.5.2 Interfacial adhesive strength — 434

10 Infrared transparent conductive oxide thin film — 439
10.1 Overview — 439
10.2 Films prepared by plasma bombardment magnetron sputtering — 440
10.2.1 Microstructure of In_2O_3 and In_2O_3:Sn films — 440
10.2.2 Composition and chemical bonding of In_2O_3 and In_2O_3:Sn films — 448
10.2.3 Photoelectric performance of In_2O_3 and In_2O_3:Sn films — 455
10.3 Structure and performance of Y_2O_3:Ru — 461
10.3.1 Surface morphology of ruthenium-doped yttrium oxide film — 461
10.3.2 Crystal structure of ruthenium-doped yttrium oxide film — 462
10.3.3 Composition and chemical bonding of ruthenium-doped yttrium oxide film — 464
10.3.4 Electrical and optical properties of ruthenium-doped yttrium oxide film — 472

10.4 Mechanism analysis and application of infrared transparent conduction — **474**
10.4.1 Mechanism analysis of infrared transparent conduction — **474**
10.4.2 Zinc sulphide window far-infrared transparent conductive oxide thin film — **477**

Index — 481

1 Foundation of infrared-transparent materials

1.1 Overview

Infrared technology, including infrared imaging, infrared guidance, and infrared countermeasures, is a very important strategic and tactical measure for modern warfare and plays an important role in the military. The infrared window and fairing are key parts of infrared technology. The functional film on the surface of the infrared window and fairing often needs to function in harsh environments and withstand thermal shock, seawater corrosion, and impact from wind, sand, rain, and snow. In particular, the recent development of stealth target and antimissile technology has further demanded higher requirements for infrared materials. Therefore, research on infrared-permeable materials has gained increasingly widespread attention from various countries and has become a hot topic in the field of materials science. This chapter will introduce the basic properties and typical applications of infrared-permeable materials and provide a brief introduction on the basic properties and typical applications of antireflective and protective coatings.

1.2 Infrared signal on the battlefield

1.2.1 Electromagnetic spectrum and atmospheric window

In nature, infrared radiation will be produced whenever the temperature of an object exceeds absolute zero. This type of infrared radiation is essentially thermal radiation. Infrared radiation is a type of electromagnetic wave; hence, thermal radiation is also known as electromagnetic radiation. Electromagnetic radiation can be characterized by its frequency ν and wavelength λ. An electromagnetic wave has undulatory property, and its velocity of propagation c can be expressed as

$$c = \lambda \nu, \tag{1.1}$$

where c is the speed of light ($c = 2.997\,924\,58 \times 10^8$ m/s).

Electromagnetic radiation also has quantum-mechanical properties and exists in the form of photons. The energy and frequency of photons satisfy

$$E = h\nu = \frac{hc}{\lambda}, \tag{1.2}$$

where h is the Planck constant ($h = 6.626 \times 10^{-34}$ J·s).

It can be seen from equation (1.2) that photons with a higher frequency will have a shorter wavelength and higher energy.

https://doi.org/10.1515/9783110489514-004

Tab. 1.1: Electromagnetic spectrum.

Radiation type	Wavelength (m)	Frequency (Hz)
Cosmic rays	$< 10^{-12}$	$> 3 \times 10^{20}$
γ-rays	10^{-12}–10^{-11}	3×10^{20}–3×10^{19}
X-rays	10^{-11}–10^{-8}	3×10^{19}–3×10^{16}
Ultraviolet	10^{-8}–3.80×10^{-7}	3×10^{16}–7.89×10^{14}
Visible light	3.80×10^{-7}–7.80×10^{-7}	7.89×10^{14}–3.84×10^{14}
Violet light	3.80×10^{-7}–4.30×10^{-7}	
Blue light	4.30×10^{-7}–4.80×10^{-7}	
Green light	4.80×10^{-7}–5.30×10^{-7}	
Yellow light	5.30×10^{-7}–5.80×10^{-7}	
Orange light	5.80×10^{-7}–6.20×10^{-7}	
Red light	6.20×10^{-7}–7.80×10^{-7}	
Infrared	7.80×10^{-7}–10^{-3}	3.84×10^{14}–3×10^{11}
Near-infrared	7.80×10^{-7}–3×10^{-6}	
Mid-infrared	3×10^{-6}–5×10^{-6}	
Far-infrared	8×10^{-6}–1.4×10^{-5}	
Microwave	10^{-3}–10^{-1}	3×10^{11}–3×10^{9}
Radio wave	$> 10^{-1}$	$< 3 \times 10^{9}$

Tab. 1.1 lists the names of radiation with different wavelengths in the electromagnetic spectrum as well as the corresponding wavelength and frequency. It can be observed that the infrared wavelength is usually at the micrometer level (10^{-6} m), whereas visible light is at the nanometer level (10^{-9} m).

When electromagnetic waves penetrate the atmosphere, various types of gases and aerosols in the atmosphere will absorb and scatter the electromagnetic radiation, which causes a gradual decrease in the intensity of the electromagnetic waves during propagation. The spectral bands where the reflection, absorption, and scattering effects are weak and the transmittance of electromagnetic waves is high are called atmospheric windows. Mainly, gases such as H_2O, CO_2, and O_3 absorb electromagnetic radiation in the atmosphere. Fig. 1.1 shows the typical infrared transmittance spectrum of the atmosphere. It can be seen that the two important infrared sensing regions of atmospheric windows are the mid-wavelength infrared (MWIR) window at 3–5 μm and the long-wavelength infrared (LWIR) window at 8–14 μm, where the MWIR window is divided into two parts by the cut-off point at 4.3 μm, owing to the strong absorption effect of CO_2 on the electromagnetic waves in the atmosphere.

1.2.2 Blackbody radiation

Thermal radiation is the phenomenon where any object will radiate energy at a certain temperature. When an object radiates energy, its energy is constantly consumed, and its temperature decreases until it reaches thermal equilibrium. Thus, objects will

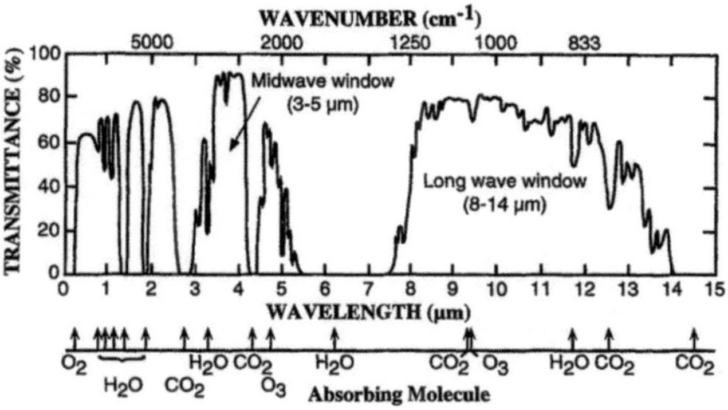

Fig. 1.1: Atmospheric infrared transmittance spectrum.

constantly absorb energy as they radiate energy. In 1859, Kirchhoff pointed out the relationship between the energy radiation and the absorption of objects, i.e. for all objects, at a given temperature, the ratio of the emissive power e_λ and absorption power a_λ is not related to the properties of the object; it is equal to the emissive power of a perfect blackbody as follows:

$$\frac{e_{\lambda 1}}{a_{\lambda 1}} = \frac{e_{\lambda 2}}{a_{\lambda 2}} = e_{\lambda 0} = f(\lambda, T), \tag{1.3}$$

where $e_{\lambda 0}$ is the emissive power of a perfect blackbody.

Equation (1.3) is the mathematical expression of Kirchhoff's law. From the equation, it can be concluded that different objects have different emissive powers and absorption powers; a higher emissive power results in a higher absorption power. The ratio is constant for all objects, and it is a function of the temperature and wavelength. A perfect blackbody is an object that absorbs all of the energy projected to its surface. The absorption power of a blackbody $a_\lambda = 1$. A perfect blackbody is an idealized model that does not exist in reality. The absorption power of actual objects is $a_\lambda < 1$.

In 1900, Planck proposed the quantum theory of blackbody radiation. Planck assumed that the energy of blackbody radiation is discrete, and that it radiates and absorbs energy in the form of photons. The expression describing the spectral radiant exitance of a blackbody is as follows:

$$M_\upsilon = \frac{2\pi h\upsilon^3}{c^2}\frac{1}{e^{h\upsilon/kT} - 1} \tag{1.4}$$

or

$$M_\lambda = 2hc^2\lambda^{-5}\frac{1}{e^{hc/\lambda kT} - 1}, \tag{1.5}$$

where k is the Boltzmann constant ($k = 1.380\,65 \times 10^{-23}$ J/K).

Fig. 1.2 shows the Planck distribution curves of blackbody radiation for different temperatures calculated according to equation (1.5).

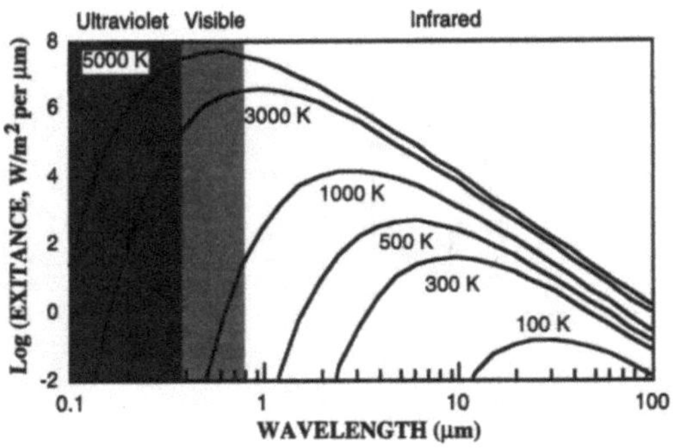

Fig. 1.2: Planck distribution curves of blackbody radiation.

The following conclusions can be obtained from Fig. 1.2:

(1) at an arbitrary wavelength, the spectral radiance of the blackbody increases as the temperature increases;
(2) the spectral radiance changes continuously with wavelength, with only one peak for each curve;
(3) two curves for different temperatures do not intersect;
(4) as the temperature increases, the peak wavelength of the curve $\lambda_{\max}$ moves toward short wavelengths;
(5) the area under each curve is equal to σT^4.

The expression for the full radiant exitance M of a blackbody can be obtained by integrating the wavelength from 0 to ∞ using equation (1.5):

$$M = \int_0^\infty M \, d\lambda = \int_0^\infty 2\pi h c^2 \lambda^{-5} \frac{d\lambda}{e^{hc/\lambda kT} - 1}. \tag{1.6}$$

After deduction, the following can be obtained:

$$M = \frac{2\pi^5 k^4}{15c^2 h^3} T^4.$$

Assuming

$$\sigma = \frac{2\pi^5 k^4}{15c^2 h^3},$$

then

$$M = \sigma T^4, \tag{1.7}$$

where σ is the Stefan–Boltzmann constant ($\sigma = 5.67051 \times 10^{-8}\ \mathrm{W \cdot m^{-2} \cdot K^{-4}}$).

Equation (1.7) is known as the Stefan–Boltzmann law. It indicates that the full radiant exitance of a blackbody is directly proportional to the fourth power of temperature. If the temperature of blackbody is doubled, the total radiant energy of the blackbody per unit area per unit time will increase by $2^4 = 16$ times.

When integrating the wavelength from λ_1 to λ_2 using equation (1.5), the energy radiated by a blackbody within any given wavelength range can be obtained as follows:

$$M_{\lambda_1-\lambda_2} = \int_{\lambda_1}^{\lambda_2} M \, d\lambda. \tag{1.8}$$

According to equation (1.8), the energy radiated by a blackbody per unit area for the wavelength ranges of 3–5 µm, 8–10 µm, and 8–14 µm at temperatures of 300 K, 600 K, and 2000 K can be obtained. The results are summarized in Tab. 1.2. It can be observed from the table that at room temperature, the energy radiated by a blackbody in the LWIR range is much higher than that in the MWIR range. At 600 K, the energy radiated by the blackbody in these two wavebands is approximately the same. However, at 2000 K, the energy radiated by the blackbody in the MWIR range is higher than that in the LWIR range.

Tab. 1.2: Blackbody radiation.

Temperature (K)	Radiance (W/m^2)		
	3–5 µm	8–10 µm	8–14 µm
300	5.9	61.1	172.6
600	1719	957	1937
2000	1.60×10^5	1.07×10^4	1.86×10^4

Taking the derivative of the wavelength in equation (1.6) and assuming $dM/d\lambda = 0$, the expression of the relationship between the wavelength corresponding to the peak radiant exitance of the blackbody λ_{max} and the temperature can be obtained as follows:

$$\lambda_{max} T = C, \tag{1.9}$$

where C is a constant ($C = 2.878 \times 10^{-3}$ µm/K).

Equation (1.9) is known as Wien's displacement law. This law states that the maximum blackbody wavelength λ_{max} is inversely proportional to the temperature of the blackbody. When the temperature of an object is 300 K, the maximum wavelength can be obtained according to Wien's displacement law (1.9) as $\lambda_{max} = 9.6$ µm. When the temperature is 800 K, $\lambda_{max} = 3.6$ µm. Therefore it can be concluded that the maximum emission wavelength of an object at room temperature is located in the far-infrared region. However, the temperature of gas expelled by a jet engine is 500 °C, and its maximum emission wavelength is in the MWIR region.

The peak radiant exitance of the blackbody spectrum can be obtained by substituting the $\lambda_{\max}T$ value in Wien's displacement law into equation (1.5):

$$M_{\lambda_{\max}} = BT^5, \tag{1.10}$$

where B is a constant ($B = 1.2876 \times 10^{-11}\ \mathrm{W \cdot m^{-2} \cdot \mu m \cdot K^{-5}}$).

Equation (1.10) is known as the Wien maximum radiance law. It states that the spectral radiant exitance of a blackbody is proportional to the fifth power of the temperature.

1.2.3 Infrared signal transmission through infrared-transparent materials

The modulation transfer function (MTF) is commonly applied in the quantitative description of image degradation in optical imaging systems. A larger MTF value indicates better imaging quality.

Fig. 1.3 (a) shows a typical grating pattern describing the imaging capability of optical systems. In Fig. 1.3 (a), the black bars with fixed intervals represent the object, and the square wave on the right represents the transmitted image. The topmost panel in Fig. 1.3 (b) shows that the edge brightness of the image is lower than that of the original object. This is because in an optical imaging system where diffraction is limited as far as possible, the diffraction at the edge of optical elements can still cause image blurring. In a real optical system, not only will diffraction cause blurring in the images of objects, the spatial frequency of objects will also affect image clarity. Spatial frequency refers to the number of lines per millimetre. As the spatial frequency increases, the image resolution will decrease. As shown in Fig. 1.3 (b), when the lines in the object are more closely spaced, i.e. when the spatial frequency increases, the difference between the maximum and minimum brightness will decrease, thus reducing image clarity. As can be seen from Fig. 1.3 (c), when the original object contains sinusoidal changes, the image will also present sinusoidal changes, but with reduced brightness.

With reference to Fig. 1.3 (b), the modulation of the grating pattern can be defined as

$$\text{modulation} = \frac{\text{maximum brightness} - \text{minimum brightness}}{\text{maximum brightness} + \text{minimum brightness}}. \tag{1.11}$$

The value of modulation ranges from 0 to 1; a larger value implies greater contrast. When the modulation is 0, there is no contrast within the image and object, whereas when the modulation depth is 1, the image is a perfect reproduction of the object.

Fig. 1.4 demonstrates the relationship between the modulation transfer function (MTF) and the spatial frequency. As can be seen from the sine wave in Fig. 1.4 (b), the modulation of images will increase as the spatial frequency of the objects increases. The curve in Fig. 1.4 (a) represents MTF, which shows the variation in modulation as the spatial frequency changes in both ideal and real optical systems. As the spatial

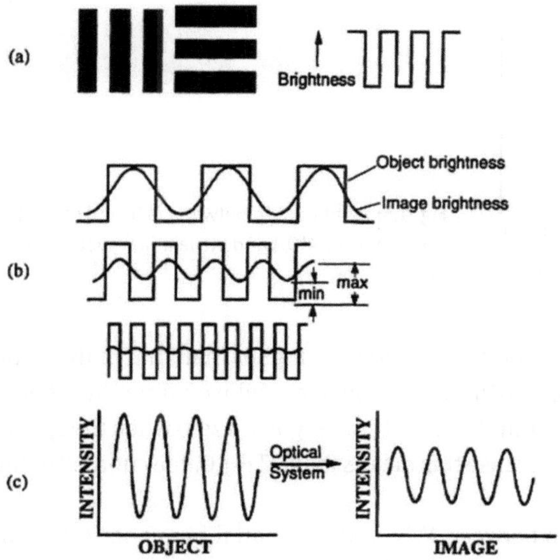

Fig. 1.3: Imaging of bar and sinusoidal patterns for an infrared window.

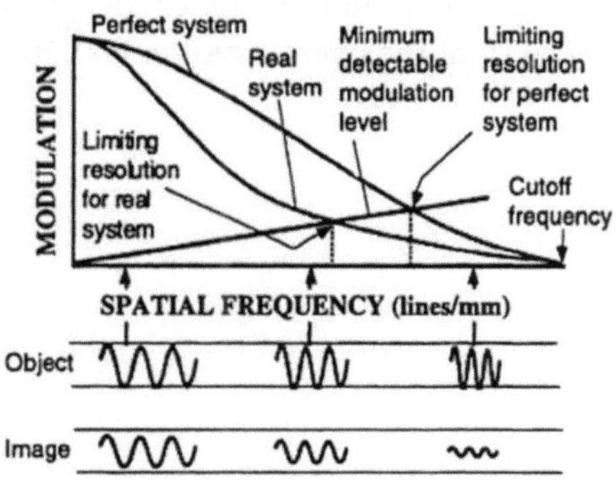

Fig. 1.4: Relationship between MTF and spatial frequency.

frequency increases, the MTF value decreases. The spatial frequency increases to a certain value, known as the limiting resolution, for which MTF = 0. Beyond this point, signals cannot be transmitted by the system.

The MTF of a window can be determined by measuring the MTF of an optical system with and without the window, for which the ratio (MTF with window)/(MTF without window) represents the MTF of the window. For an optical system with relatively high requirements on the imaging quality, MTF = 0.95 will imply image blurring, and might be considered unacceptable. Hence, the higher the imaging quality of the system, the larger the MTF value should be to avoid image degradation. Fig. 1.5 shows the relationship between the MTF and the spatial frequency for aluminium oxynitride (AlON) and yttrium oxide (yttria) windows. As shown in Fig. 1.5, as the spatial fre-

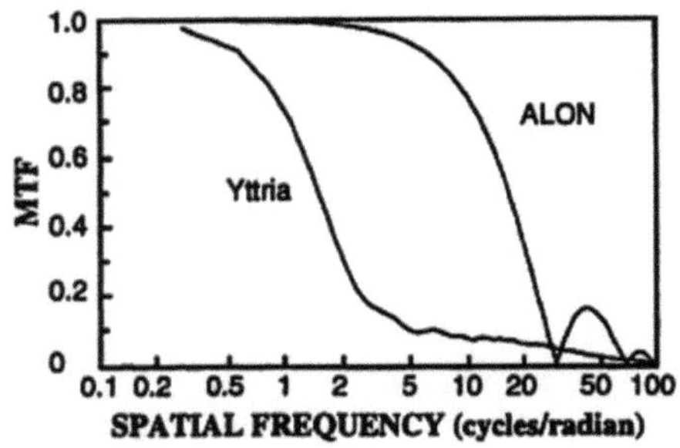

Fig. 1.5: Relationship between MTF and spatial frequency for AlON and yttria at a wavelength of 0.633 µm.

quency increases, the MTF value of the AlON window is higher than that of the yttria window, which indicates that using AlON as window material would produce better imaging results than using yttria would. The AlON curves goes to zero when the spatial frequency is 30–70 cycles/radian. Because this means that MTF = 0, the contrast of the window here is 0.

Fig. 1.6 shows the relationship between MTF and the spatial frequency for a gallium phosphide dome 76 mm in diameter, manufactured by Raytheon Systems Company. The upper curve represents the relationship curve between MTF and the spatial frequency of an optical system without the dome. MTF is multiplicative. Hence, if a window has MTF = 0.80 at a particular spatial frequency, and the optical detection system has MTF = 0.60 at the same frequency, MTF of the overall system is $0.80 \times 0.60 = 0.48$.

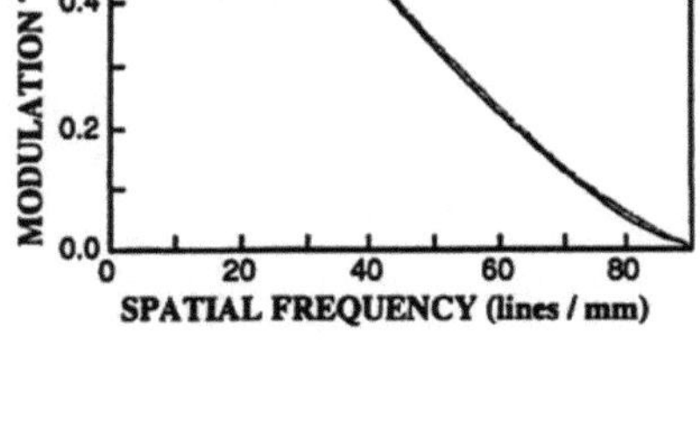

Fig. 1.6: Relationship between MTF and spatial frequency with and without a dome.

1.3 Typical infrared-transparent substrate materials

Infrared-transparent materials are mainly employed to manufacture lenses, prisms, windows, domes, and other optical components. This section will introduce a few typical infrared-transparent substrate materials and briefly describe their performance.

1.3.1 Classes of materials

Materials for infrared-transparent windows can be divided into three major classes: single-crystal, polycrystalline, and optical glass. Specific examples for each class are listed in Tab. 1.3.

Tab. 1.3: Classification of infrared-transparent substrate materials.

Class of material	Typical examples
Single-crystal	Sapphire (Al_2O_3), Si, Ge, GaAs, CaF_2, LiF, KBr, NaCl
Polycrystalline	MgF_2, ZnS, MgO, ZnSe, CdTe, Y_2O_3, spinel ($MgAl_2O_4$)
Optical glass	Calcium aluminate glass, germanate glass, fluoride glass, arsenic trisulfide glass, chalcogenide glass

1.3.2 Single-crystal materials

Currently, infrared optical components are predominantly produced using single-crystal materials. Compared with other types of crystalline materials, single-crystal materials have more diverse ranges of refraction and dispersion which can fulfil the requirements of different applications. Furthermore, several of the crystals have a high melting point, and hence they are materials with good thermal stability. Single-crystals include two major classes: semiconductor crystals and ionic crystals.

1.3.2.1 Semiconductor crystals

Semiconductor crystals constitute a class of infrared optical material that has been applied widely in recent years. Semiconductor crystals involved in extensive research and applications include germanium (Ge), silicon (Si), and diamond. Their physical properties are listed in Tab. 1.4. The optical performance of semiconductor crystals is mainly determined by their intrinsic, extrinsic, and free-carrier absorptions. Hence, they are relatively sensitive to temperature. Semiconductor crystals are primarily utilized as lenses and other optical components. Due to their high refractive index and large reflection losses, the semiconductor crystals need antireflective coating in practical applications.

Germanium (Ge)

Germanium is chemically stable, insoluble in water, and opaque to visible light. Germanium crystals constitute the fundamental material for fabrication of infrared optical windows, lenses, and other optical components. The atomic arrangement of Ge crystals is the same as for diamonds, i.e. a face-centered cubic lattice structure,

Tab. 1.4: Physical properties of some semiconductor crystals.

Material	Cut-off wavelength (μm)	Refraction (4.3 μm)	Knoop hardness	Melting point (°C)	Density ($g \cdot cm^{-3}$)
Germanium (Ge)	25	4.02	800	940	5.33
Silicon (Si)	15	3.42	1150	1420	2.33
Diamond	30	2.4	8820	>3500	3.51
Indium antimonide (InSb)	16	3.99	223	523	5.78
Gallium arsenide (GaAs)	18	3.34 (3.5 μm)	750	1238	5.31
Gallium antimonide (GaSb)	2–4	3.70	448	705	5.62
Cadmium sulfide (CdS)	16	2.26	121	1500	4.82
Cadmium selenide (CdSe)	25	2.4	71	>1350	5.81
Cadmium telluride (CdTe)	30	2.56	54	1045	5.85

where each unit cell contains 4 metal atoms. In single-crystal Ge, the absorption cross-section of photons by electrons is 1/20 that by electron holes; hence, n-type single-crystal Ge has excellent infrared transmission, and the majority of materials applied in infrared optics are n-type single-crystals. Fig. 1.7 shows the infrared transmission characteristics of single-crystal Ge. Fig. 1.8 shows the infrared transmission characteristics of single-crystal Ge for different temperatures. As shown in Fig. 1.8, the transmittance of Ge decreases significantly as the temperature increases, which is the result of greater free-carrier absorption. In addition, Ge is a brittle material with poor impact resistance. Therefore, it is not suitable for use under higher temperatures.

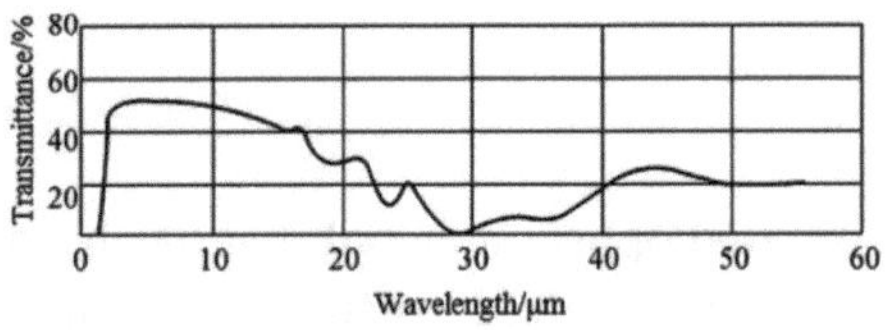

Fig. 1.7: Infrared transmission characteristics of Ge (sample thickness: 2 mm).

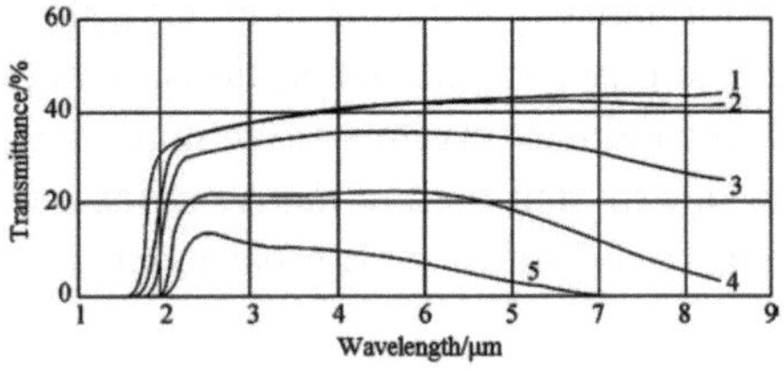

Fig. 1.8: Relationship between infrared transmission of Ge and temperature (sample thickness: 2 mm; resistivity: 30 Ω·cm): (1) 25 °C; (2) 100 °C; (3) 200 °C; (4) 250 °C; (5) 300 °C.

As single-crystal Ge has a relatively high refractive index ($n = 4$) and a surface reflection loss of up to 53 %, it requires high-intensity infrared antireflective and protective coatings prior to application. On one hand, the coating will be able to reduce the surface reflection losses and improve the transmittance of optical elements; on the other hand, it will also protect the elements, and improve their resistance to erosion by rain

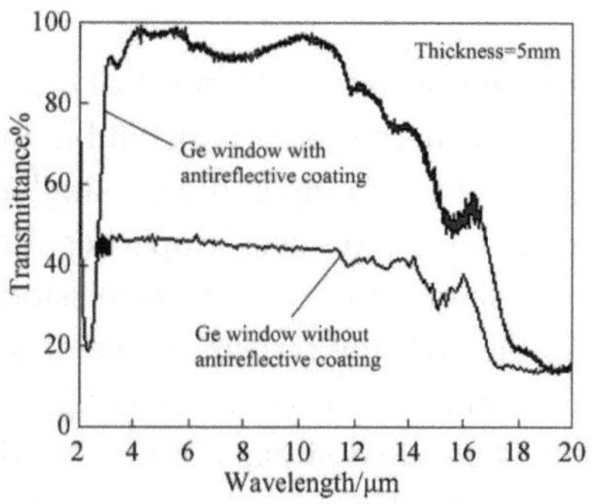

Fig. 1.9: Infrared transmission characteristics of Ge window before and after antireflective coating.

and sand. Fig. 1.9 shows the optical transmittance curves of a Ge window (manufactured by Thorlabs Inc., USA) before and after coating. As can be seen, the Ge window coated with an antireflective film shows a significant increase in transmittance, with the maximum transmittance exceeding 98 %.

In infrared optics, Ge is mainly applied to fabricate infrared optical lenses and infrared optical windows for protecting infrared optical lenses in infrared detection systems. More than 60 % of mid-to-low-end infrared optical lenses are produced by employing Ge materials, while 50 % of high-end infrared optical lenses are produced using single-crystal Ge. Therefore, Ge is very widely utilized in infrared lenses, which are commonly used in military and civilian applications. Fig. 1.10 lists the lenses fabricated using single-crystal Ge materials. Military applications of infrared windows mainly include aircrafts and armored fighting vehicles, in particular helicopters, which require large quantities. Infrared lenses are used in both military and civilian applications: military applications mainly involve airborne, shipboard, roadbed, vehicle-mounted, helmet-mounted, and handheld infrared lenses, as shown in Fig. 1.10 (a). Civilian applications of infrared windows include various surveillance thermal imagers and automotive night-vision systems. Civilian use of infrared lenses mainly involves applications for security, civilian vehicles, civilian vessels, civilian aircrafts, and police use, as shown in Fig. 1.10 (b).

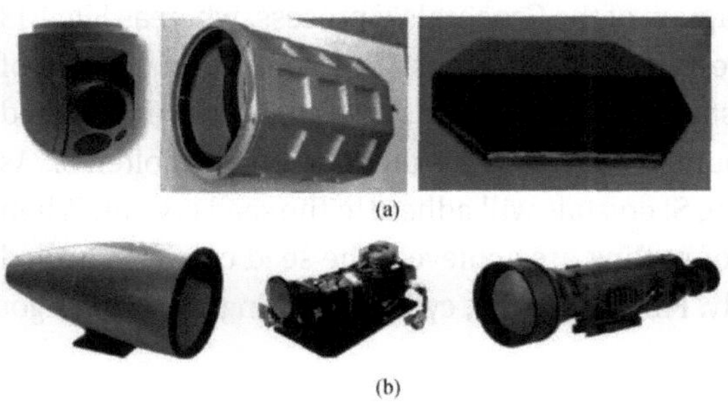

Fig. 1.10: Infrared lenses fabricated using single-crystal Ge materials: (a) military applications; (b) civilian use.

Silicon (Si)

The properties of Si are similar to those of Ge, because it is also a semiconductor crystal that is chemically stable and insoluble in water. Furthermore, Si is insoluble in most acids, and also has a diamond cubic crystal structure. Compared with single-crystal Ge, Si has a higher level of hardness, and better mechanical properties for impact resistance. Fig. 1.11 shows the relationship between the infrared transmission characteristics of Si and temperature. As shown in Fig. 1.11, the transmittance of Si decreases as the temperature increases, but the magnitude of the decrease is significantly smaller than that for single-crystal Ge. This is because Si has a larger band gap and lower free-carrier absorption, causing temperature changes to have a smaller influence on single-crystal Si; hence, Si can be used at higher temperatures. The application of antireflective coating results in a significant increase in the Si transmittance, achieving a maximum transmittance for mid-wavelength infrared of up to 95 %. Therefore, in addition to windows and lenses, single-crystal Si can also be chosen as optical materials for domes.

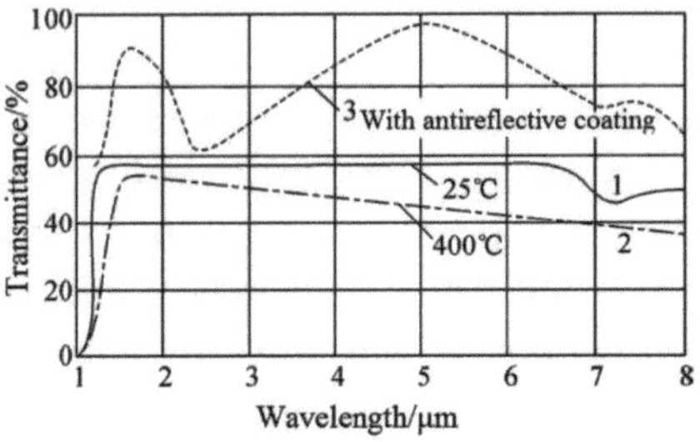

Fig. 1.11: Relationship between infrared transmission characteristics of Si and temperature.

Single-crystal Si is predominantly fabricated using the Czochralski process of crystal growth. The Czochralski process can produce crystals of larger sizes, and the crystals will not be affected by mechanical constraints throughout the process of crystal growth. In addition, the crystal shape and structural integrity can easily be controlled. In addition, the single-crystal products grown with this method have good optical performance. Fig. 1.12 is a schematic diagram of the Czochralski process, whereas Fig. 1.13 is a schematic diagram of the corresponding apparatus. The fabrication process of single-crystal Si starts with polycrystalline Si and dopants that are heated and melted in a quartz crucible. Then, a rod-mounted seed crystal is dipped into the molten Si. As the surrounding solution cools down, Si crystals will adhere to the seed crystal. When the required temperature and rate of pulling are achieved, the seed crystal is pulled upwards and rotated simultaneously. Finally, a large, cylindrical, single-crystal ingot can be obtained at the end of the rod.

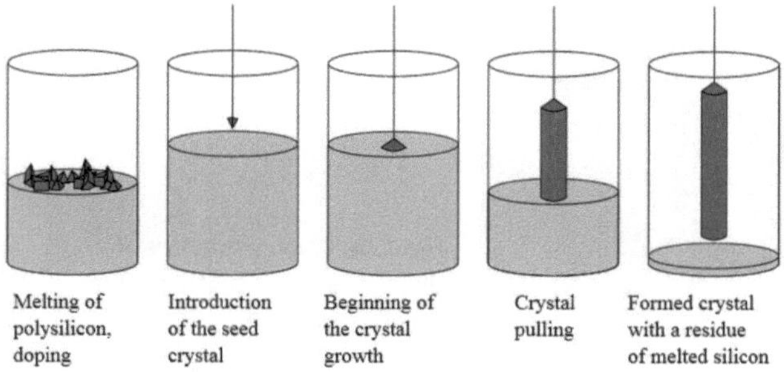

Fig. 1.12: Schematic diagram of the Czochralski process.

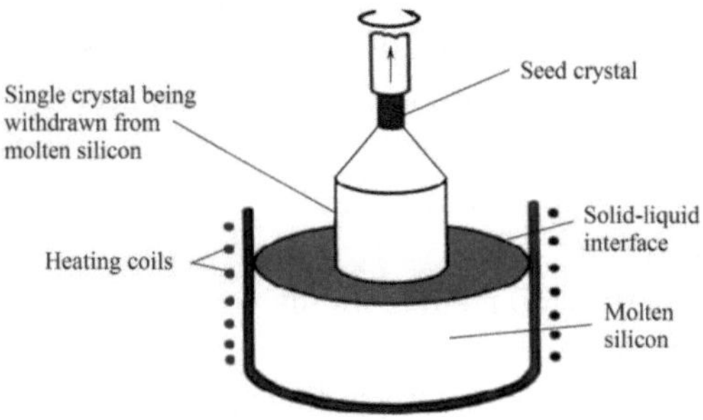

Fig. 1.13: Schematic diagram of the apparatus for the Czochralski process.

Diamond

Diamond is currently the hardest material known to people in nature, and it has excellent mechanical properties. Diamond is chemically stable, acid- and alkali-resistant, and has a broad infrared transmission spectrum, which enables multispectral transmission for ultraviolet, visible, and infrared light. Diamond is currently the ideal material for windows in missiles, rockets, and aircrafts.

Diamond is a face-centred cubic crystal. Fig. 1.14 shows its crystal structure and real single-crystal diamonds. Each carbon atom in the crystal forms covalent bonds with 4 neighboring carbon atoms through sp^3 orbital hybridization, and 4 adjacent carbon atoms form a regular tetrahedron. The color of diamonds varies greatly, ranging from colorless to black. They can be transparent, translucent, or opaque. Many diamonds are yellow, which is caused by the presence of impurities. Due to the extremely high refractive index and strong dispersion of diamonds, they will often appear to be very colourful when exposed to sunlight.

Due to the unique physical and chemical properties of diamond, it is an ideal infrared optical material. The synthesis of diamonds through chemical vapor deposition (CVD) has received widespread attention. Typical polycrystalline diamonds are composed of randomly oriented diamond grains and relatively weak grain boundaries containing sp^2 carbon. However, the optical, electrical, and mechanical performances

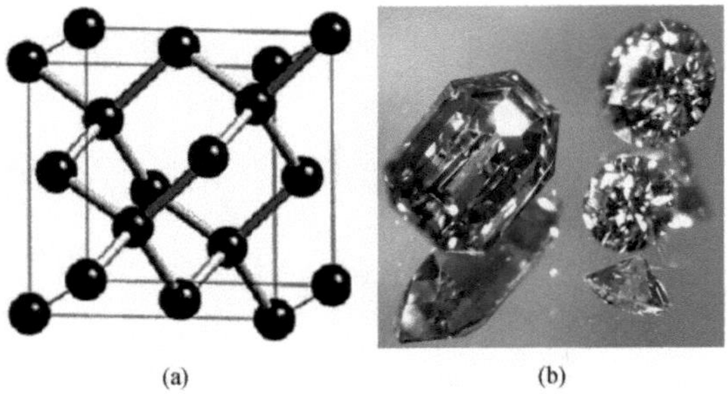

Fig. 1.14: Crystal structure of diamonds: (a) crystal structure; (b) real diamonds.

of synthetic diamonds are much poorer than those of natural single-crystal diamonds. The high-pressure high-temperature (HPHT) method is currently the main method adopted to synthesize single-crystal diamonds. It has also been commercialized. Nevertheless, diamond synthesis by HPHT imposes substantial restrictions on diamond size and quality. Furthermore, the cost-effectiveness of HPHT for the production of single-crystal diamonds is still in need of further improvement. Currently, CVD diamonds are employed for coating and are relatively thin. The transmittance of high-quality CVD diamonds is already very similar to that of natural type IIa diamonds, and their conductivity might even be higher.

Fig. 1.15 shows a schematic diagram of the CVD method for diamond production. CVD is performed to produce diamond thin-films on a substrate, mainly using H_2 and methane as reaction gases. Then, various energy sources, including microwave (MW), radio frequency (RF), laser induction (LI), direct current (DC), hot filaments, and chemical activation (CA), are provided to activate the reaction. The amount of gas, the temperature, and other parameters are also set. This will allow the continuous growth of the diamond film on the surface of the substrate. During this process, the hydrogen (or oxygen) atoms are the most important component of the gas-phase reactions, whereas the mixture of methane and other hydrocarbons only serves as the carbon source. The primary function of hydrogen (or oxygen) is to terminate dangling carbon bonds on the diamond surface or core layer. Furthermore, hydrogen atoms can react with neutral hydrocarbons, and generate reactive groups (e.g. CH_2). Active methylene groups can react with the exposed carbon to produce graphite with three sp^2 hybridized bonds (a-C) or tetrahedral amorphous carbon with sp^3 hybridized bonds (ta-C). Another function of the hydrogen gas is to prevent the production of graphite; this is because atomic hydrogen cleaves the sp^2 bonds in graphite much quicker than the sp^3-bonded carbon in diamond. Hydrocarbons added for hydrocracking can also inhibit the aggregation of polymers or macrocyclic structures on its surface.

The absorption positions for CVD diamonds and natural type IIa diamonds are consistent, with a higher absorption at the 3–5 μm wavelength range. However, for infrared optical materials, the optical properties at 8–12 μm are more important. Therefore, CVD diamonds at 8–12 μm is the ideal material for windows and domes. Natural type IIa diamonds have high thermal shock resistance and the thermal properties

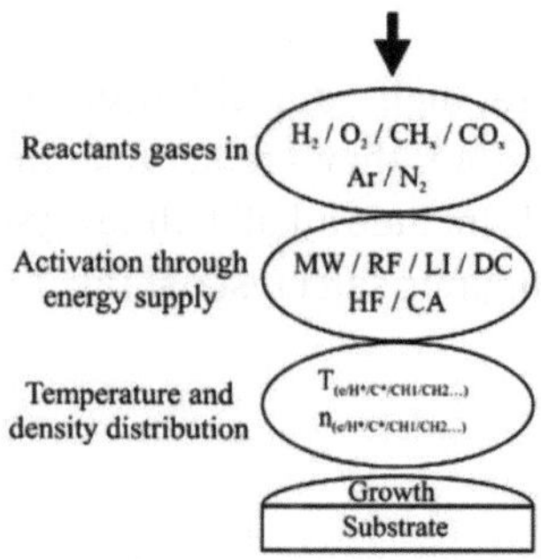

Fig. 1.15: Schematic diagram of the production method of CVD diamonds.

of high-quality CVD diamonds are now approaching those of natural diamonds. Diamonds have excellent chemical stability under room temperature. However, when exposed to air and high temperatures, diamonds are prone to the occurrence of graphitization. In addition, diamonds can also react with certain metals, such as Fe, Co, Ni, and Mn. Therefore, oxidation-resistant coatings are needed when diamond is used for windows and domes. Currently, the main oxidation-resistant coating used on the surface of CVD diamond is aluminium nitride (AlN) film. After coating, the transmittance of the material will essentially remain unchanged at 1000 °C.

The refractive index of diamond is 2.39, and the single-surface reflection loss is 16.9 %. Hence, antireflective coating is needed on the surface when it is employed for windows and domes. Commonly used antireflective coating materials include: ytterbium oxide (Yb_2O_3), yttrium oxide (Y_2O_3), and zirconium oxide (ZrO_2). Prior to the antireflective coating, the diamond surface should receive etching treatment to remove nondiamond carbon, polishing debris, and other impurities. Then, the diamond is activated using microwaves and the surface hydrogenated in a hydrogen plasma atmosphere. Before the application of RF magnetron sputtering for the antireflective coating, sputter cleaning should be performed on the diamond surface in order to prevent oxidation of the graphite. Fig. 1.16 shows the transmittance of diamond surfaces coated with Y_2O_3 and Yb_2O_3 antireflective layers. After the application of Y_2O_3 and Yb_2O_3 antireflective coatings, the transmittance of the diamond surfaces increases by 10 % and 12 %, respectively, thus demonstrating a significant antireflective effect.

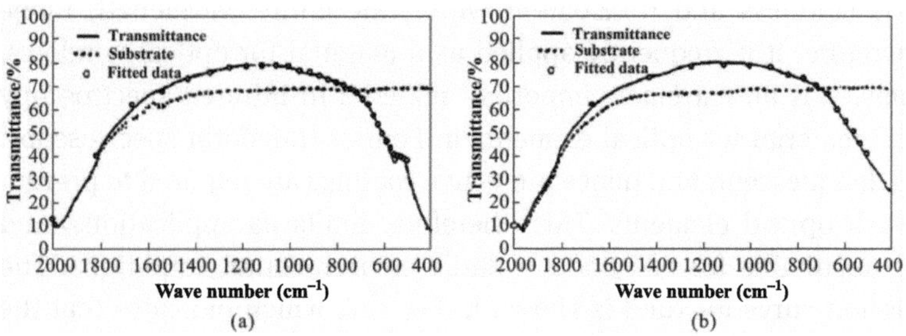

Fig. 1.16: Transmittance of diamond with Y_2O_3 (a) and Yb_2O_3 (b) antireflective coating.

1.3.2.2 Ionic crystals

Ionic crystals have very wide infrared transmission spectra, high transmittance, and are minimally influenced by temperature changes. Therefore, they are suitable for locations with major environmental changes. Ionic crystals include alkali halide crystals, alkaline earth halide crystals, oxide crystals, and inorganic salt crystals. Tab. 1.5 lists the physical properties of a few typical ionic crystals.

Tab. 1.5: Physical properties of typical ionic crystals.

Type	Material	Cut-off wavelength (μm)	Refractive index / 4.3 μm	Hardness (Knoop)	Melting point (°C)	Density ($g \cdot cm^{-3}$)
Alkali halide	NaCl	25	4.02	800	940	5.33
	LiF	15	3.42	1150	1420	2.33
	CsI	30	2.4	8820	>3500	3.51
Alkaline earth halide	MgF_2	16	3.99	223	523	5.78
	CaF_2	18	3.34	750	1238	5.31
Oxide	Al_2O_3	2–4	3.70	448	705	5.62
	Fused silica	16	2.26	121		4.82
	MgO	25	2.4	71	>1350	5.81
Inorganic salt	$SrTiO_3$	30	2.56	54	1045	5.85

Ionic crystals have relatively high transmittance, a long wavelength limit, and are also easily cultivated into single-crystal materials of large sizes. However, they are extremely prone to deliquescence, and require an outer protective coating prior to application. In addition, they have poor mechanical strength, low hardness, and low melting point, and are hence not suitable for field use. Currently, ionic crystals are mainly utilized to manufacture prisms and windows for infrared spectrometers.

Sodium chloride (NaCl)

NaCl is a typical alkali halide, which is a commonly used broadband and optically transparent material. It has a low refractive index and relatively low reflection losses. Furthermore, it is colorless and transparent within the entire wavelength range (0.25–16 μm). Therefore, it is frequently applied as a material for optical windows, lenses, and prisms. It is an extremely important material in infrared spectroscopy and also a crucial material for optical elements in Fourier transform spectroscopy. However, NaCl is deliquescent, and hence protective coatings are required to prevent the erosion of such optical elements. This, therefore, limits its applications, and it is particularly unsuitable for use under volatile environmental conditions. The infrared transmission curve for NaCl is shown in Fig. 1.17, which indicates that the transmittance of NaCl at 5–16 μm could be more than 90 %.

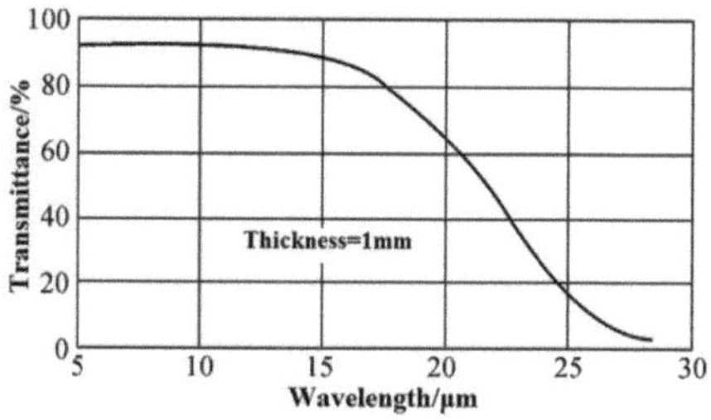

Fig. 1.17: Infrared transmission characteristics of NaCl.

NaCl has a cubic close-packed crystal structure. Fig. 1.18 (a) shows its crystal structure, and Fig. 1.18 (b) shows a real NaCl crystal. Within the NaCl unit cell, each ion is surrounded by six other ions, where the larger chloride ions are arranged in a close-packed cubic array, while the smaller sodium ions fill the octahedral voids between the chloride ions.

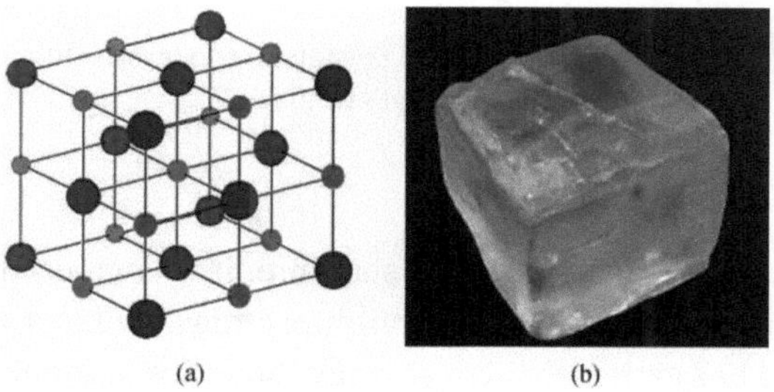

Fig. 1.18: NaCl crystal: (a) structure and (b) real object.

Magnesium fluoride (MgF_2)

MgF_2 is an alkaline earth halide crystal. It has high transmittance and low reflection losses, and thus does not require antireflective coating. Compared with the alkali halides, MgF_2 has superior mechanical strength and hardness, as well as lower solubility, and is virtually insoluble in water. The infrared transmission curve for MgF_2 is shown in Fig. 1.19, which indicates that the transmittance of MgF_2 could be more than 95 % at 2–6 μm.

MgF_2 has a tetragonal crystal system, its crystal structure is shown in Fig. 1.20 (a). MgF_2 crystals have the following characteristics: very high transmittance between the vacuum ultraviolet and infrared (0.11–7.5 μm) wavelength range; resistant to impact, thermal fluctuations and radiation; and excellent chemical stability. Hence, it can be employed in optical prism lenses, wedge prisms, windows, and related optical systems. Because MgF_2 has the properties of tetragonal birefringent crystals, it can also be used in optical communications. Therefore, MgF_2 has been widely used as a material in optical windows between vacuum ultraviolet and infrared wavelengths, e.g. as ultraviolet laser windows. Some MgF_2 windows are shown in Fig. 1.20 (b). Furthermore, due to the birefringent effect of MgF_2 crystals, they can also serve as polarizing elements.

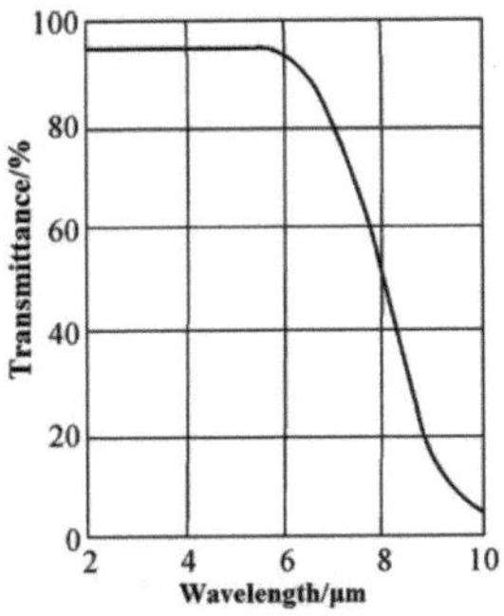

Fig. 1.19: Infrared transmission characteristics of MgF_2.

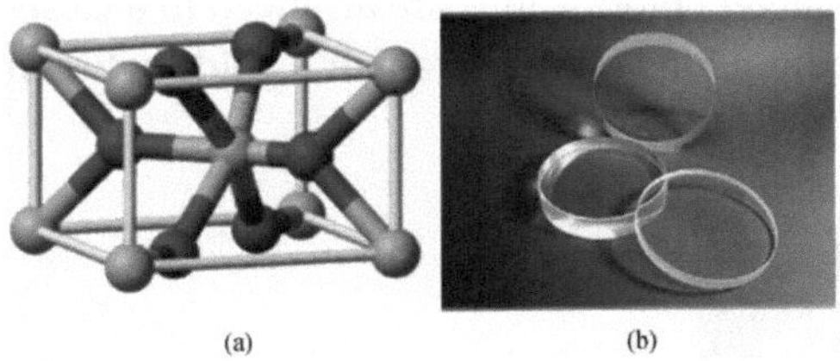

Fig. 1.20: MgF_2 crystal: (a) crystal structure of MgF_2; (b) MgF_2 windows.

Sapphire

Single-crystal aluminium oxide (α-Al_2O_3) is also known as sapphire. It is composed of the covalent bonding of three oxygen atoms and two aluminium atoms, and has a hexagonal crystal structure. Sapphire has extremely high chemical stability, is generally insoluble in water, and is resistant to acid and alkali corrosion. Sapphire crystals have a very high level of hardness, second only to the hardest, i.e. diamond. It has excellent transparency, thermal conductivity and electrical insulation, good mechanical performance, and is resistant to wear and wind erosion. Therefore, it is an exceptional material for infrared windows and domes. Fig. 1.21 show the infrared transmission curve for sapphire. Nevertheless, despite its desirable mechanical and optical properties, growing large single-crystal sapphires is costly and time-consuming, which has limited its applications to a certain extent.

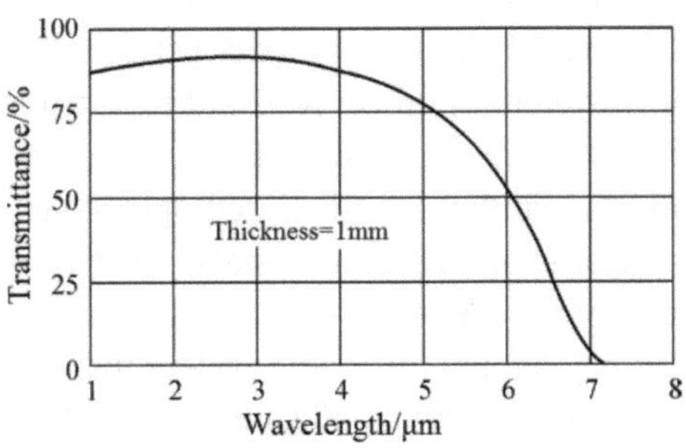

Fig. 1.21: Infrared transmission characteristics of sapphire.

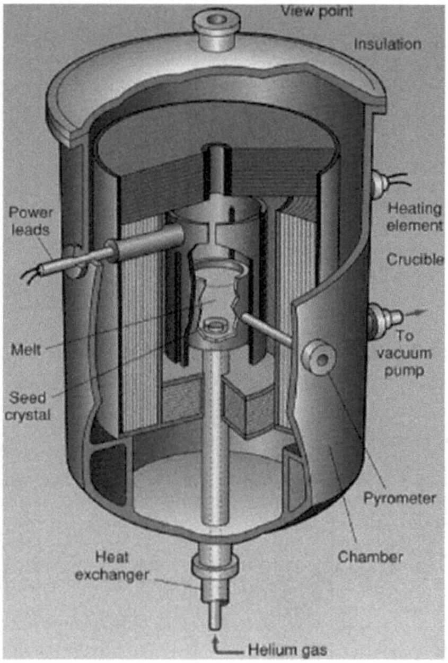

Fig. 1.22: Schematic diagram of HEM furnace.

The heat exchanger method (HEM) is one of the most important techniques for the fabrication of large high-quality single-crystal sapphires. Fig. 1.22 shows a schematic diagram of this method. The production of single-crystal sapphires starts with Al_2O_3 raw material and a sapphire seed crystal which are heated to above 2040 °C in a molybdenum crucible and melted. The temperature of the furnace is then reduced, and the cooling capacity of the cold finger condenser is enhanced in order to obtain the sapphire seed. The sapphire seed is obtained by cooling using helium gas in the cold finger. The entire solidification process requires a few days. After that, annealing and cooling is performed at 2040 °C to remove residual impurities, ultimately producing a large sapphire crystal. Different methods of heat exchange coexist within the sapphire in the crystal growth furnace, such as thermal conduction, thermal convection, thermal radiation between parallel solid surfaces, phase transition, and internal radiation within the melt and crystal. Within the heat exchange furnace, the precise control of the heat transfer is particularly crucial for the production of perfect sapphire crystals. This is because the crystal is relatively sensitive to temperature during the growth process. Hence, precise control of the heat transfer is needed during crystal growth. Fig. 1.23 shows a sapphire crystal, being 34 cm in diameter and having a mass of 65 kg, grown using HEM by Khattak et al.

Sapphire is an ideal window material for reconnaissance, surveillance, and remote-sensing, due to its excellent resistance to wind and sand, as well as its optimal optical and mechanical properties. In addition, sapphire is resistant to thermal shock, and thus is also a good option for missile windows and domes. Images of

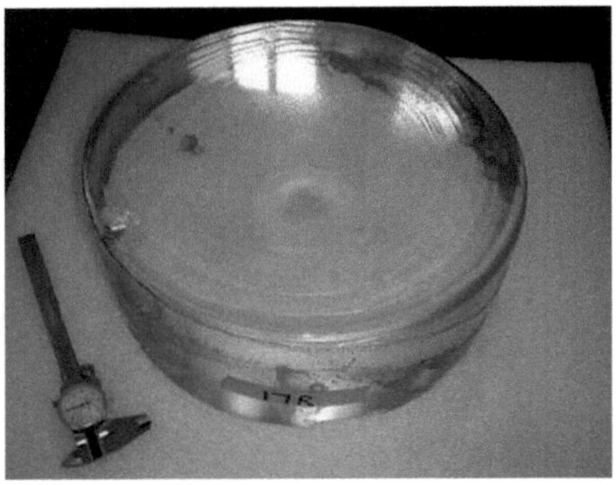

Fig. 1.23: Sapphire grown using HEM.

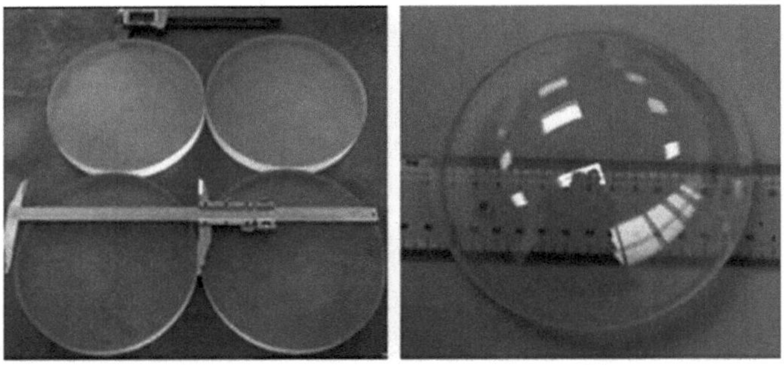

(a) (b)

Fig. 1.24: Sapphire windows (a) and dome (b).

sapphire windows and domes are shown in Fig. 1.24. Furthermore, due to its good crystallinity, high-temperature stability, and relatively low costs, sapphires are often applied as substrate wafers, laser matrices, and optical elements in microelectronics.

1.3.3 Polycrystalline materials and ceramics

With the increasingly widespread application of infrared technology in aerospace industry and the military, there is an urgent need for infrared materials that are resistant to high temperatures and thermal shock. Evidently, normal single-crystal materials are not suitable for such working environments. This, therefore, has stimulated the development of hot-pressed polycrystals and ceramics.

1.3.3.1 Hot-pressed polycrystalline materials

The hot-pressing technology emerged in the early 1950s and 1960s, and is currently one of the most important techniques for the fabrication of polycrystalline infrared optical materials. Tab. 1.6 lists the properties of a few commonly used hot-pressed

Tab. 1.6: Properties of hot-pressed polycrystalline materials.

Material (hot-pressed)	Transmission wave range (μm)	Refractive index / 5 μm	Knoop hardness	Melting point (°C)	Density ($g \cdot cm^{-3}$)	Solubility ($g/100\,g\ H_2O$)
MgF_2	0.45–9.5	1.34	576	1396	3.18	0
ZnS	0.57–15	4.09	354	1020	4.09	0
CaF_2	0.2–12	1.37	200	–	3.18	Trace amounts
ZnSe	0.48–22	2.4	150	–	5.27	0
CdTe	2.0–30	2.7	40	–	5.85	0

polycrystalline materials. Among them, MgF_2 and ZnS are the most frequently used as infrared optical materials.

Hot-pressed polycrystalline MgF_2 is produced by applying high temperature and high pressure to high-purity magnesium fluoride powder with uniform particle distribution and appropriate particle sizes. Hot-pressed MgF_2 has excellent polarization effects and infrared transmission properties, particularly at infrared wavelengths of 1–7 μm, where transmittance is more than 90 %, thus making it a good infrared optical material. It also has numerous advantages, such as high mechanical strength, strong thermal shock resistance, resistance to chemical corrosion, and isotropy. Therefore, it can be utilized widely for infrared domes and windows of precision-guided missiles, and civilian infrared detectors. Furthermore, it has a very low refractive index ($n = 1.34$) and low reflection losses. Hence, it is employed for lenses or windows without antireflective coating, and is also frequently used as an antireflective coating.

The properties of hot-pressed polycrystalline ZnS are similar to those of hot-pressed polycrystalline MgF_2. Hot-pressed polycrystalline ZnS is a ceramic material produced by applying high temperature and high pressure moulding to zinc sulfide powder. It has good transmittance for infrared in the 8–11.5 μm wave range, while also having high mechanical strength, strong thermal shock resistance, and resistance to chemical corrosion. Therefore, it is the preferred material for long-wavelength infrared precision-controlled missile domes and other windows. Due to its low refractive index and reflection losses, it is also frequently used as antireflective coating.

1.3.3.2 Transparent ceramics

The transparent ceramics make up a new class of advanced functional materials. Due to their unique mechanical properties (characteristic strength, impact resistance, and hardness) and transparency, they have been applied widely in military and civilian departments. Some transparent ceramics that are currently under research include: aluminium oxynitride (AlON), submicron alumina (Al_2O_3), sapphire (Al_2O_3), spinel ($MgAl_2O_4$), and zinc sulfide (ZnS) ceramics. The properties of the above transparent ceramics are listed in Tab. 1.7.

Tab. 1.7: Properties of transparent ceramics.

Material	Density ($g \cdot cm^{-3}$)	Hardness (GPa)	Elastic modulus (GPa)	Torsional strength (MPa)	K_{Ic} ($MPa \cdot m^{-2}$)
AlON	3.67	13.8 ± 0.3	315	228	2.4 ± 0.11
Spinel	3.58	12.1 ± 0.2	277	241	1.7 ± 0.06
Submicron alumina	3.986	20–23	390	600	3.5
Sapphire	3.984	16–17	386	350	3.5
ZnS	4.08	2.5	75	103	0.8–0.9

Fig. 1.25 describes the infrared transmission characteristics and transmittance wave ranges of the above transparent ceramics. As can be seen in the figure, the wavelength ranges of aluminium oxynitride (AlON), submicron alumina (Al_2O_3), sapphire (Al_2O_3), and spinel ($MgAl_2O_4$) fall mainly within the mid-infrared wave range (0.9–5 μm), and the transmittances are all above 85 %. Zinc sulfide (ZnS) has a broader wavelength range (2–11 μm), and its transmittance is up to 70 %.

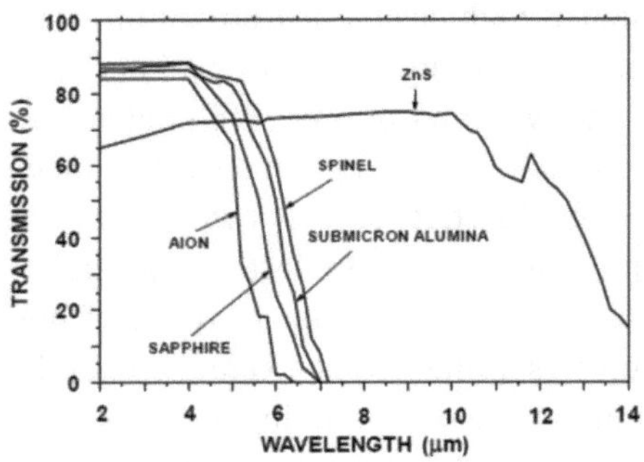

Fig. 1.25: Infrared transmission curves of transparent ceramics.

In 1959, Yamaguchi and Yanagida discovered that heating Al_2O_3 in a N_2 atmosphere produced a spinel phase, which was the first research report on AlON. Subsequent research found that AlON did exist in this system, and its structure could be described as containing a 5 % mass fraction of cubic γ-Al_2O_3 (spinel) structure. Following this discovery, the Raytheon Company and several research groups began to conduct investigations on the application of AlON ceramics. In addition to the good mechanical, thermal, and chemical properties of AlON transparent ceramics, it also has good transparency within the 0.2–5.0 μm wavelength range (this includes the ultraviolet, visible light, and infrared regions), and its theoretical transmittance could reach 85 %. Furthermore, it also has good optical and mechanical isotropy. Therefore, AlON transparent ceramics has a wide range of application prospects in military defence and numerous commercial fields. The fabrication of AlON is technologically mature, and it is an infrared material with extensive applications. For example, MgF_2 and other materials have insufficient resistance to high temperature and erosion, hence can only be used as domes for single-mode infrared-guided, low-speed missiles with a flight speed lower than Mach 2. AlON, single-crystal sapphire, and polycrystalline spinel have been recognized as three ideal candidates suitable for infrared windows of missiles above Mach 3. Comparatively speaking, the overall performance of single-crystal sapphire with regards to optical, mechanical, and thermal properties is the best, but its cost is also the highest. Transparent polycrystalline spinel has excellent optical properties as well, in addition to relatively low costs, but its mechanical properties are not ideal. In contrast, AlON has the best performance/cost ratio out of the three main thermally resistant infrared materials, thus making it the preferred material for high-Mach missiles, and next-generation transparent armour. The USA has already listed this material as a key development material for 21st century national defence. Currently, AlON has extensive applications, mainly as materials for domes, transparent armour, infrared windows, and hyper-hemispherical domes in military aircraft and missiles, laser windows, lenses in military aircraft, semiconductor processing, and scanner windows.

Methods for synthesizing AlON includes the following: carbothermal reduction of aluminium nitride, plasma-arc melting of aluminium oxide and aluminium nitride, reaction sintering of aluminium oxide and aluminium nitride, reaction sintering of aluminium oxide and boron nitride, and microwave-assisted synthesis of aluminium oxide and aluminium nitride. Due to considerations of the process simplicity, raw materials, and production of minimal pollutants, AlON is primarily synthesized using the following formula:

$$9Al_2O_3 + 5AlN \rightarrow Al_{23}O_{27}N_5. \tag{1.12}$$

The production methods for AlON can be categorized into one- and two-step methods. The one-step method involves directly performing reaction sintering with Al_2O_3 and AlN powders of equal composition as raw material to produce AlON ceramics. In other words, dense AlON is produced from Al_2O_3 and AlN in a single reaction stage, where the composition and moulding are both formed during the one-step reaction sintering. In the two-step method, the Al_2O_3 and AlN powders are used as raw materials to synthesize AlON powder, which is then sintered to form AlON ceramics. The entire process of the two-step method can be divided into two stages, where the first step involves moulding the reaction mixture after heat treatment at 1750 °C (i.e. the formation of AlON powder), and the second step involves the AlON densification process. Both the one- and two-step methods have their benefits and drawbacks. A benefit of the one-step method is the simple production process. Some drawbacks are that the solid-state reaction during the production process is difficult to control, the localized inhomogeneous composition of Al_2O_3 and AlN will cause nonuniform morphology in AlON grains during the moulding process, and obstruction of the densification during sintering, thereby resulting in AlON ceramics with low density and poor optical performance. Some benefits of the two-step method:

(1) compared with directly sintering of a mixture of Al_2O_3 and AlN powders, using synthesized high-purity AlON powder can effectively reduce the sintering temperature and time, while also facilitating the formation of uniform AlON grains, which ensures a high transmittance in the ceramics;
(2) it facilitates the effective and homogenous addition of sintering aids, thereby reducing the sintering temperature and time, as well as producing ceramics with good density and optical properties. Drawback: complex production process and high costs of fabricating AlON ceramics.

Polycrystalline aluminium oxide (Al_2O_3) was the first transparent ceramic that was successfully developed, and its performance is very similar to that of sapphire. It has excellent optical properties, with good transmittance in the visible light and infrared regions, high level of hardness, and good resistance to wear and thermal shock. Hence, it can be used as an infrared optical material for windows and domes. Aside from the properties mentioned above, Al_2O_3 also has low density, which enables its use as lightweight bulletproof armor. In fact, its mass efficiency is superior even to boron carbide/silicon carbide materials. Among the transparent ceramics listed above, polycrystalline alumina with submicron grains and near theoretical density is the most difficult to achieve. Using advanced defect-free processing techniques, sintered polycrystalline alumina can be fabricated with particle sizes less than 1 μm and a density approaching the theoretical value (99.95 %) which will possess the abovementioned properties.

When light passes through polycrystalline alumina, not only will it be absorbed, it will also be scattered at microporous defects, and birefringence at the gaps between adjacent crystals will lead to reflection losses, as shown in Fig. 1.26 (a). The effect of thickness on the "real" in-line transmittance (RIT) is fairly significant; a greater level of thickness will lead to a smaller RIT value, and lower transmittance. Thus, the material will present a translucent state. When the thickness is 1 mm, the RIT of dense sintered alumina with particle size 0.5 μm is 60 %. As the thickness increases, the losses would also increase; when the thickness exceeds 1 mm, RIT will be 30 %. Therefore, compared with other isotropic transparent materials, the effect of thickness on sintered submicron alumina is much more significant, as shown in Fig. 1.26 (b). The maximum RIT of 1-mm-thick submicron alumina in the visible light region is only 70 %–75 %. Evidently, this level of thickness is not safe for effective ballistics, and this has therefore limited its application as lightweight high-strength armor. However, polycrystalline alumina is a new material for windows and domes. Due to its high resistance to high temperatures and corrosion, it is extremely suitable as a material for the lamp housing of high-pressure sodium lamps. The European Starlight Project has stated that metal-halide lamps with transparent and high-strength lamp housings would be able to replace energy-wasting halogen lamps on a large scale.

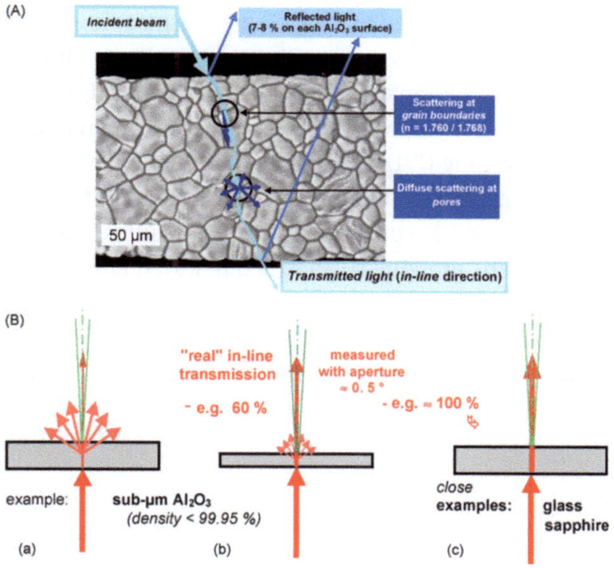

Fig. 1.26: Light transmission through submicron alumina (the actual RIT < 100 due to scattering and absorption).

1.3.3.3 CVD ZnS and CVD ZnSe

ZnS and ZnSe are widely employed polycrystalline, infrared optical materials, which have good transmittance and broad infrared regions. In 1960, the American Fastman Kodak Company fabricated polycrystalline ZnS and ZnSe materials through hot pressing. Because ZnS and ZnSe have high absorption and scattering in the visible light and infrared regions, other production techniques were needed. In 1970, the American Raytheon Company first applied the CVD technique to produce ZnS and ZnSe, which led to breakthroughs in the fabrication of large CVD ZnS and CVD ZnSe. Compared to other processes, CVD has the advantage of enabling simpler and more precise control of the product composition and purity. Thus, it has been extensively applied. Furthermore, the products obtained using this technique showed high levels of physical and chemical integrity. CVD ZnS and CVD ZnSe have unique thermal, mechanical, and optical properties. They are one of the most important long-wavelength infrared transparent ceramics worldwide, and are extensively applied in the military industry.

CVD ZnS is widely used within the transmission wavelength range of 3–10 µm, and is a semiconductor ceramic with a wide band gap. It is mainly used in electroluminescent devices and flat panel displays, thin-film photovoltaic cells, photocatalytic infrared windows, etc.

CVD ZnS is obtained through the direct reaction between zinc vapor and hydrogen sulfide gas. The reaction formula can be expressed as follows:

$$Zn(g) + H_2S(g) \rightarrow ZnS(g) + H_2(g). \tag{1.13}$$

Specific details on the deposition reaction can be seen in Fig. 1.27 (a). At 600–750 °C, ZnS obtained through the direct reaction between Zn and H_2S is deposited directly onto the reactor wall. The crystals of ZnS produced using CVD are anisotropic and the growth is faster in the direction parallel to the substrate than under the conditions of normal mechanism, thus producing a columnar structure. Furthermore, H_2 is produced as a by-product during the CVD growth process, which causes trace amounts of H atoms or H_2S molecules in the resulting CVD ZnS, thereby producing ZnH particles. Due to the absorption effect of ZnH bonds, a reduction in the transmittance of ZnS can be observed in the visible light and 3–5 μm regions. In 1979, Willingham and Pappis obtained ZnS through hot isostatic pressing (HIP) which removed ZnH, allowing ZnS to achieve excellent transmittance in the visible light and infrared regions. The ZnS samples shown in Fig. 1.27 (b) and (c) were both fabricated by an Indian research group led by Roy Johnson. The CVD ZnS sample was yellow, whereas the ZnS sample treated with HIP became white.

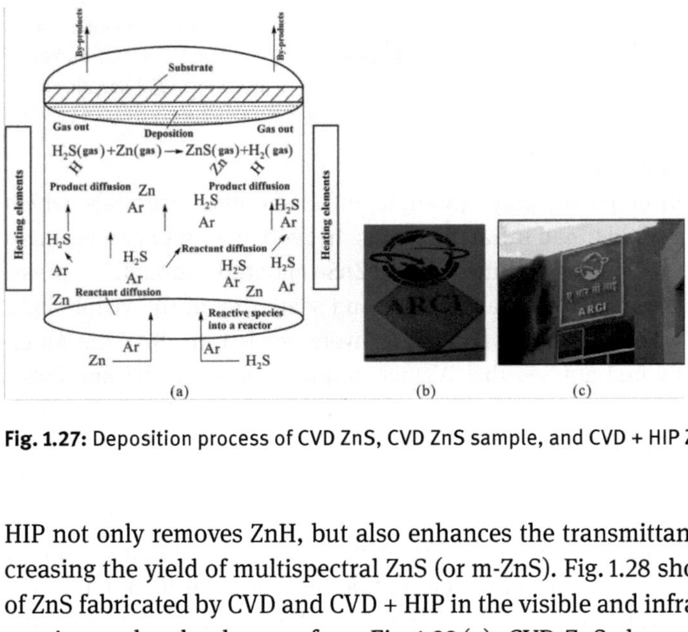

Fig. 1.27: Deposition process of CVD ZnS, CVD ZnS sample, and CVD + HIP ZnS sample.

HIP not only removes ZnH, but also enhances the transmittance of ZnS, thereby increasing the yield of multispectral ZnS (or m-ZnS). Fig. 1.28 shows the transmittance of ZnS fabricated by CVD and CVD + HIP in the visible and infrared regions.

As can be clearly seen from Fig. 1.28 (a), CVD ZnS shows absorption at 6.5 μm, transmittance above 60 % within the 3–5 μm wavelength range, and transmittance above 70 % for 7–10 μm. The strong absorption by ZnH at 6.5 μm is a typical characteristic of CVD ZnS. After treatment by HIP, the transmittance of the resulting CVD + HIP ZnS for the entire wave range of 3–10 μm is above 70 %, indicating that ZnH has been removed. Fig. 1.28 (b) shows that the transmittance at the visible light region for CVD ZnS is essentially 0, but increases to 50–70 % after HIP. Research has indicated that the underlying reason for the improvement of ZnS transmittance and removal of ZnH by HIP is that HIP altered the orientation and size of the ZnS grains, while also removing excess pores.

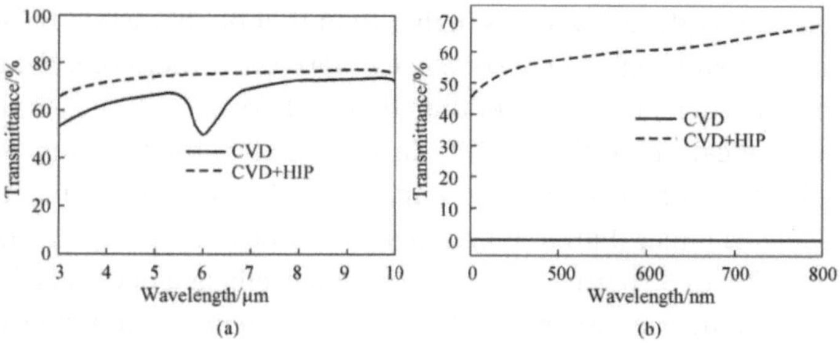

Fig. 1.28: Infrared transmission spectrum of CVD and CVD + HIP ZnS samples: (a) infrared region; (b) visible light region.

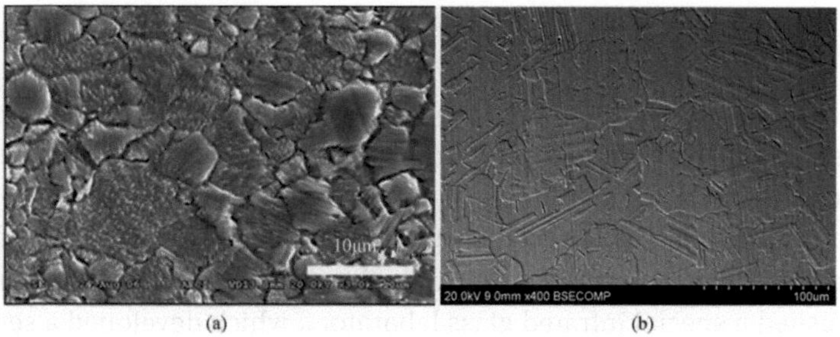

Fig. 1.29: Microstructure of CVD ZnS and CVD + HIP ZnS samples.

Fig. 1.29 (a) and (b) show the scanning electron micrographs of CVD ZnS and CVD + HIP ZnS samples, respectively. As can be seen from the micrograph of the CVD ZnS sample, the columnar crystal structures with maximum diameters of 5–10 μm were dispersed along the direction parallel to the substrate, and lamellar structures were observed along the grain boundaries. Furthermore, Fig. 1.29 (a) also shows small gaps, which were formed by the absorption of H_2 by ZnS during CVD. The CVD + HIP ZnS sample obtained after HIP shows a significant increase in grain sizes, which are between 50 and 100 μm, as shown in Fig. 1.29 (b). In addition, due to the recrystallization of ZnS, the conventional grain boundaries and lamellar structure disappeared as well. The most likely mechanism leading to these changes might be the plastic deformation of the densification process in HIP, which caused morphological changes, such as the formation of crystal twinning.

The production apparatus for CVD ZnSe is similar to that for CVD ZnS. In concrete terms, CVD ZnSe is obtained through the direct reaction between Zn vapors and hydrogen selenide (H_2Se), which can be expressed as follows:

$$Zn(g) + H_2Se(g) \rightarrow ZnSe(g) + H_2(g)\ (730\text{–}825\ °C). \tag{1.14}$$

However, the deposition temperature of ZnSe is higher than that of ZnS, while H_2Se has poorer stability than H_2S and begins to decompose at 160 °C. Furthermore, H_2Se is a highly toxic gas, and its toxicity is far greater than that of H_2S gas.

The inhomogeneous optical properties of CVD ZnSe are due to the presence of defects, which commonly include inclusions, clouding, and patterns. Of these, inclusions are macroscopic defects larger than 0.05 mm in size and possible components of inclusions include ZnSe, Zn, or Se particles. The absorption coefficient of CVD ZnSe is very small and its transmittance shows minimal changes with temperature; hence it can withstand the thermal effects of lasers on windows. However, hot spots are created at the sites of inclusions when ZnSe is applied as laser windows, and window temperature will increase with excessive inclusion density.

1.3.4 Optical glass

Infrared optical glass was discovered around World War II, and was developed in response to the military need for infrared night vision, detection, tracking, and navigation. Research on infrared glass was successively initiated by the USA Bureau of Standards, Columbia University, Eastman Kodak, and other companies. Schott AG in Germany even established a special infrared glass laboratory, which developed a series of infrared optical glasses. Russia also carried out vigorous research work in this area. Optical glass has several advantages over other infrared optical materials, such as good optical homogeneity, which allows casting of parts with various shapes and sizes; different refractive indices for different glasses, which can then be made into compound lenses; relatively high mechanical strength; high surface hardness and stability; less prone to deliquescence; abundant raw materials for glass fabrication; and low costs. The disadvantages include relatively short transmission wavelength for the majority of optical glasses and relatively low softening point.

Optical glasses can mainly be divided into two classes: oxide glass and chalcogenide glass. Tab. 1.8 lists the main properties of a few common optical glasses.

1.3.4.1 Oxide glass

Oxide glasses can be further divided into two classes: silicate and nonsilicate.

Conventional oxide glasses (silicate glass composed of SiO_2, B_2O_3, P_2O_5, PbO, etc.) are able to transmit visible light up to 3 µm (infrared). Replacing SiO_2 with GeO_2, TeO_2, TeO_3, Sb_2O_3, Al_2O_3, Ga_2O_3, Bi_2O_3, La_2O_3, TiO_2, etc. will produce aluminate, antimonite, tellurite, gallate, titanate, and other glasses. The transmission of long-wave infrared in this type of glass is significantly enhanced. We will focus on calcium aluminate glass in this section. The limit of long-wave infrared transmission of glass mainly composed of calcium aluminate is 6 µm. BS39B and BS37A, which are currently extensively employed, are calcium aluminate glasses. The transmittances of BS39B and

Tab. 1.8: Properties of common optical glasses.

Material	Composition	Transmission wave range (μm)	Refractive index	Softening point (°C)	Hardness (Knoop)	Thermal expansion coefficient (10^{-6}/°C^{-1})
Oxide glasses						
Silicate glass	SiO_2-B_2O_3-P_2O_5-PbO	0.3–3	1.5	700	300–600	4–10
BS37A aluminate glass	SiO-BaO-MgO-Al_2O_3	0.3–5	1.5	≈ 700		9.3
BS39B aluminate glass	CaO-BaO-MgO-Al_2O_3	0.3–5.5	1.6	≈ 700		9.7
Gallate glass	SrO-GaO-Mg-BaO- Ga_2O_3-Al_2O_3	0.3–6.65		670		
Tellurite glass	BaO-ZnO-TeO_3	0.3–6	2	320		
Chalcogenide glasses						
Arsenic trioxide glass	As_4S_6	1–11	2.41	210	109	1.6
Arsenic selenide glass	$As_{38.7}Se_{61.3}$	1–15	2.79	202	114	1.7
Glass No. 20	$Ge_{33}As_{12}S_{55}$	1–16	2.49	474	171	3.1
Germanium-antimony-selenium glass	$Ge_{28}Sb_{12}Se_{60}$	1–15	2.62	326	150	3.0
Germanium-arsenic-tellurium glass	$Ge_{10}As_{20}Te_{70}$	2–20	3.55	178	111	
Silicon-germanium-arsenic-tellurium glass	$Si_{15}Ge_{10}As_{20}Te_{50}$	2–12.5	3.06	320	179	
Germanium-selenium-gallium glass	$Ge_{35}Se_{60}Ga_{50}$	1–15	2.5	372		
Germanium-selenium-mercury glass	$Ge_{35}Se_{60}Hg_5$	1–16	2.5	365		
Arsenic-germanium-selenium-tellurium glass	$As_{50}Ge_{20}Se_{20}Te_{10}$	1–13	2.51	195	94	1.1

Fig. 1.30: Su-30MK fighter aircraft.

BS37A are similar to that of sapphire, while also maintaining good thermomechanical properties at 800 °C. Fig. 1.30 shows a Su-30 fighter aircraft manufactured by the Sukhoi Company, in which the glass cockpit was made using calcium aluminate glass.

Gallate glass has similar properties to BS37A, but with a slightly lower annealing temperature of approximately 670 °C, and long-wave transmission limit of 6.65 µm. Tellurite glass has a similar transmission wavelength, but its annealing temperature is 320 °C, and the temperature during use is generally lower than 250 °C. Tellurite glass is suitable for windows of InSb and PbS detectors. The addition of lanthanum oxide (La_2O_3), cerium oxide (CeO_2), and neodymium oxide (Nd_2O_3) to phosphate glass can improve its infrared transmission and reduce the absorption coefficient.

1.3.4.2 Chalcogenide glass

Chalcogenide glass refers to glass mainly containing elements in group VIA of the period table, namely S, Se, and Te, with the addition of fixed quantities of other metallic elements.

In 1870, Schulz–Selack first discovered that the sulfur element alone could form a transparent glass and then fabricated arsenic sulfide and arsenic selenide glasses. This was the first published report on chalcogenide glasses. However, chalcogenide glass did not receive much attention at the time, and did not have practical applications. It was not until 1950, when Frerichs et al. from the USA reexamined As_2S_3 glass and postulated that this optical material could be used in infrared systems, which led the way to an upsurge in research on chalcogenide glasses. In the 1950s, successive studies on various types of chalcogenide glass systems, as well as their optical, electrical, and semiconductor properties were published. Arsenic sulfide glass was mainly used as a window material for infrared wavelengths of 3–5 µm. At the same time, as the concept of thermal imaging began to emerge, researchers started to investigate and explore chalcogenide glasses that can transmit long-wave infrared. At the end of the 1950s, Jerger et al. from the US Air Force, the Ioffe Institute from Russia, and Nielsen et al. from the UK, successively developed selenide- and telluride-based chalcogenide glasses which could transmit wavelengths of 8–12 µm and longer. During the early stages, chalcogenide glasses were mainly utilized in national defence and infrared weapons systems. After the 1970s, especially with the dramatic improvements in laser technology, the research and application of chalcogenide glasses began to develop rapidly. Substantial progress was made in numerous fields, including phase-change memory, xerography, holography, information optics, and fibre-optic communication.

Fig. 1.31 compares the optical transmittance of S-, Se-, and Te-based chalcogenide glasses, quartz glass, and fluoride glass. Chalcogenide glass has very high long-wavelength cut-offs. At a sample thickness of 2 mm, the cut-off of S-based glass is above 10 µm, Se-based glass nearly 20 µm, and for certain Te-based glasses it could be up to 30 µm. This is because, compared with oxide glass, chalcogenide glass has higher density, weaker bond strength, smaller band gap (generally 1–3 eV, thus giving a higher infrared cut-off wavelength, > 12 µm), and its wavelength range covers three atmospheric windows.

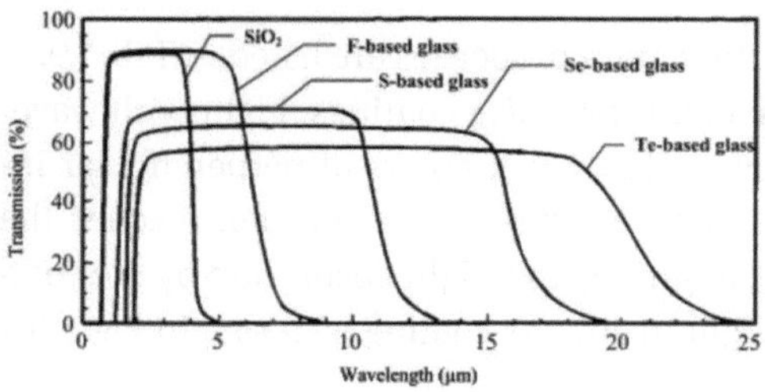

Fig. 1.31: Comparison of optical transmittance among three types of chalcogenide glass, quartz glass, and fluoride glass.

The advantages of chalcogenide glass compared with other infrared optical materials are as follows: minimal changes in the refractive index with temperature, making it an excellent athermal infrared material; relatively low refractive index (2.0–3.0), with refractive index and dispersion at long wavelengths comparable to ZnSe, making it an excellent achromatic infrared material; precision moulding technology can be employed to fabricate infrared optical elements, with low processing costs.

Currently, chalcogenide glasses are generally fabricated through vacuum and die casting, which can easily lead to segregation, bubbles, and other defects. Furthermore, the production yield of large-scale high-quality chalcogenide glass is relative low because oxidation during the production process could lead to the degradation of infrared performance; the majority of constituent elements in chalcogenides are toxic and explosive; and there are difficulties in melting and quenching. The recent development in mechanical alloying (MA) has provided another pathway for the fabrication of large high-quality chalcogenide materials. Combining MA under high vacuum with HIP sintering will seal the entire fabrication process, thereby reducing the dissipation of toxins, while also avoiding oxidation and explosions during the production process. In addition, MA can also produce ultra-fine and homogenous elemental compositions, which provide the necessary conditions for dense HIP sintering.

1.4 Infrared antireflective and protective coatings

Infrared antireflective coatings are extensively applied on windows, in order to reduce the reflection at specific wavelengths or wavelength ranges, enhance window transmittance, and obtain the ideal transmission wavelength ranges. Protective coatings are resistant to erosion by wind, sand, and rain, while also preventing scratches, thereby improving the mechanical performance of materials. Infrared antireflective and protective coatings are extremely important to optical elements, particularly for the practical applications of windows and domes. This section will briefly introduce the materials for widely used infrared antireflective and protective coatings.

1.4.1 Common optical thin-film materials and their main characteristics

Common optical thin-film materials and their main properties are listed in Tab. 1.9.

Fig. 1.32 shows the working principle of antireflective coatings. Light with wavelength λ_0 travelling through air (refractive index = n_0), enters at perpendicular incidence into an antireflective coating (refractive index = n_1) and then reaches the substrate material (refractive index = n_2), assuming that light absorption by the coating is negligible. The wavelength of the light inside the coating is λ_0/n_1. In Fig. 1.32,

Tab. 1.9: Properties of common optical thin-film materials.

Material	Melting point (°C)	Wavelength range (μm)	Refractive index	Chemical stability	Coating method
CaF_2	1360	0.15–12	1.36–1.42	Hygroscopic	Evaporation at 1100 °C, 1.33×10^{-2} Pa
MgF_2	1390	0.2–10	1.38	Stable	Evaporation at 1540 °C, 1.33×10^{-2} Pa
YF_3	1387	0.35–12	1.37	Stable	Evaporation
ThF_4	1110	0.2–15	1.52	Stable	Evaporation at 750 °C, 1.33×10^{-2} Pa
SiO_2	1750	0.2–4.5	1.46	Stable	Evaporation at 1025 °C
Al_2O_3	2050	0.2–7	1.62	Stable	Evaporation at 1550 °C
MgO	2800	0.23–8	1.68	Unstable	Electron beam at 1500 °C
Y_2O_3	2464	0.25–11	1.9	Stable, Hard	Electron beam at 2000 °C
HfO_2	2812	0.6–12	1.9	Stable, Hard	Electron beam at 2500 °C
CeO_2	2600	0.4–12	2.1–2.2	Stable, Hard	Electron beam at 2310 °C
ZnS	1830	0.35–14	2.25	Stable	Electron beam
ZnSe	1520	0.5–14	2.4	Stable	Electron beam
As_2Se_3	Sublimation	0.8–18	2.79	Stable	Electron beam
GaP	1467	0.5–11	2.90	Hard film	Electron beam
Si	1410	1.1–10	3.5	Stable	Evaporation
Ge	950	1.8–100	4.0	Stable	Evaporation

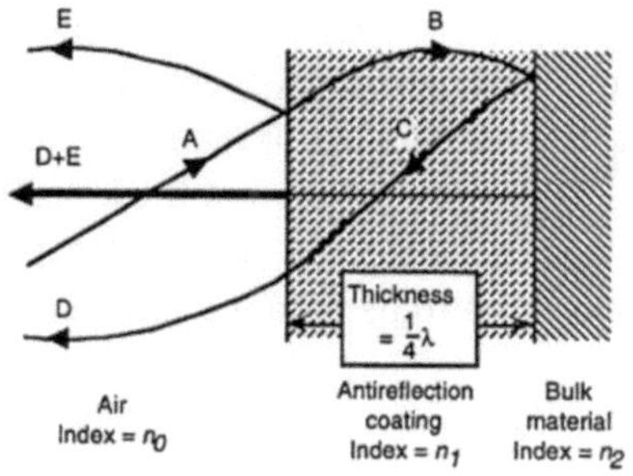

Fig. 1.32: Working principle of $\lambda/4$ antireflective coating.

curve A represents the incident ray; when light wave is transmitted to the surface of the coating, reflection E will occur, and some part will enter the coating (B). When light wave B reaches the surface of the substrate material, reflection light C and transmitted light will occur (transmitted light is not shown in the figure). Finally, light wave D will be transmitted from the antireflective coating (light wave C is ignored). The magnitude of light waves D and E will be identical but they will have opposite phases. Therefore, if they can completely cancel each other out, the net reflection of the coated material will be 0, which is the working principle of antireflective coatings. To ensure that the magnitudes of light waves D and E are identical, the refractive index of the material should satisfy

$$n_1 = \sqrt{n_0 n_2}, \tag{1.15}$$

where n_1 is the refractive index of the thin film, n_0 is the refractive index of air, and n_2 is the refractive index of the substrate material

Therefore, in order to obtain light waves D and E with opposite phases (phase difference of 180°), the thickness of the antireflective coating should be $\lambda/4$ the wavelength of the light wave in the coating (i.e. λ_0/n_1). As long as these two conditions are met, the reflection of light of wavelength λ_0 will approach 0.

Fig. 1.33 shows the changes in transmittance properties of Si coated with mid-wave infrared antireflective thin-film. It can be seen from the figure that the maximum transmittance of Si with antireflective coating reaches 95 % near 4.5 µm, indicating a significant improvement in the transmittance of Si.

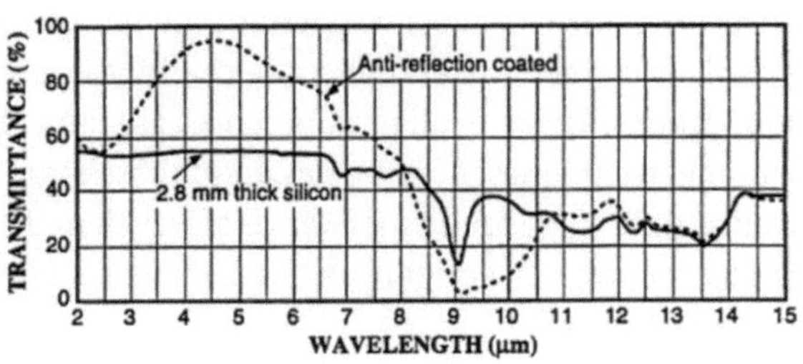

Fig. 1.33: Infrared transmission characteristics of Si with and without antireflective coating.

Fig. 1.34 (a) shows the relationship between the refractive index and thickness of multilayer coatings. When using homogenous multi-layer coatings, increasing the number of layers will cause the refractive index to increasingly approximate continuous gradation. Fig. 1.34 (b) shows the transmittance curve of a 2 mm thick ZnSe window with graded refractive index. The multi-layer antireflective coating greatly increases the transmittance of the ZnSe window, and the transmittance essentially remains above 90 %.

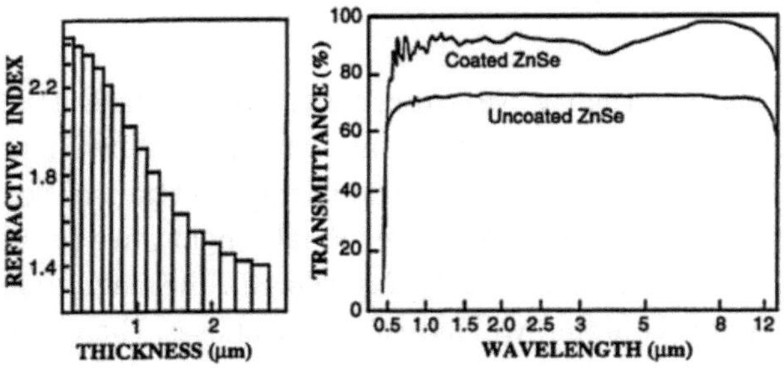

Fig. 1.34: (a) Relationship between refractive index and thickness of multi-layer coating; (b) transmittance curve of 2 mm ZnSe window.

1.4.2 Surface superstructures

The majority of antireflection is achieved through the antireflective coatings described above. However, there are several limitations to the application of antireflective coatings in reducing reflection. For example, the refractive index of the optimal coating material needs to fulfil the $\lambda/4$ condition mentioned above, but it is difficult for the vast majority of infrared optical materials to satisfy this condition. In infrared optical systems, the most commonly used infrared wavelength ranges are 3–5 μm and 8–12 μm, or multiple wavelength ranges. To this end, multilayer coatings, even up to several dozen layers, might be needed. However, not only will this result in extremely complicated processes, it will also cause problems in the thermal expansion coefficient and adhesion between the multiple coating layers and the substrate, as well as the problem of high-energy proton bombardment.

By introducing a surface structure with regular arrays of small protuberances, generally known as a "moth eye" structure, one can effectively reduce the refractive index of air and the coating material. This can replace antireflective coatings in numerous infrared and visible light applications. These small protuberances are formed by etching the surface using lithographic techniques, and the design of these structures can reduce reflection losses to less than 0.5 % for a broad range of wavelengths (3–12 μm). Therefore, there are several advantages to designing a 'moth eye' structure compared with multilayer antireflective coating.

Fig. 1.35: SEM micrographs of "moth eye" structure on diamond thin-film surface.

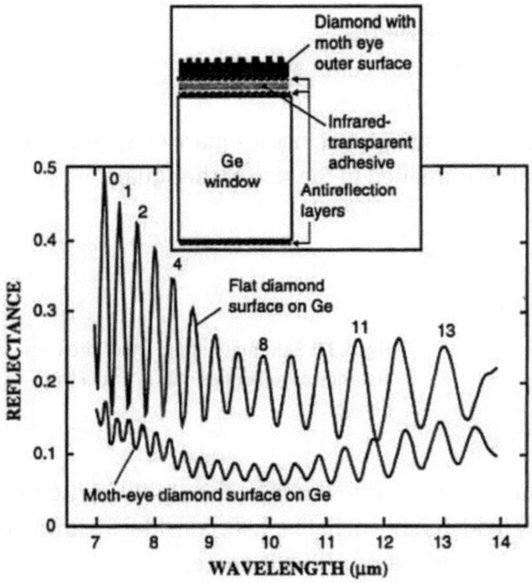

Fig. 1.36: Reflectance spectrum of Ge window with "moth-eye" diamond coating and flat diamond coating.

Fig. 1.35 shows the SEM micrographs of diamond surface with the "moth eye" structure. Fig. 1.36 shows the reflectance spectrum of Ge windows with "moth eye" diamond coating and flat diamond coating. As can be seen from Fig. 1.36, the Ge window with "moth eye" diamond coating had significantly lower refractive index compared with that with normal diamond coating, as well as significantly higher transmittance.

1.4.3 Transparent conducting thin-films

The surfaces of windows and domes might require a layer of conductive coating in order to provide shielding from electromagnetic interference. Such coatings should be transparent in the visible light and infrared regions, but opaque to microwaves and radio waves.

For general bulk materials, as shown in Fig. 1.37 (a), the resistance R can be expressed as

$$R = \frac{\rho L}{A}, \tag{1.16}$$

where ρ is the electrical resistivity, A the cross-sectional area, and L the sample length.

Equation (1.16) shows that as the length of the bulk material increases and the cross-sectional area decreases, resistance will increase. Now considering the resistance of a square conductive film with length L and thickness h, as shown in Fig. 1.37 (b). By referring to the equation for the resistance of bulk materials, one can obtain the resistance of the conductive film, which is also known as the sheet resistance or surface resistance, R_S, and expressed as

$$R_S = \frac{\rho L}{Lh} = \frac{\rho}{h}. \tag{1.17}$$

It can be seen from equation (1.17) that the resistance across a square cross-section of the thin film is a constant (ρ/h), which is unrelated to the size of the square.

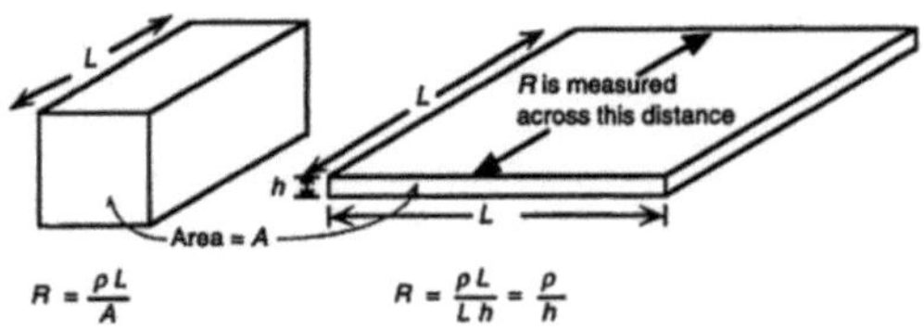

Fig. 1.37: Schematic diagram of the resistance of bulk material (a) and a thin film (b).

Shielding effectiveness (SE) is a parameter applied to describe the ability of the thin film to shield from the radio frequency field. As shown in Fig. 1.38, consider an infrared window with thickness d and dielectric constant ε, which is coated by a conductive thin film with thickness h and sheet resistance R_S. Assuming that the incident radiant power is P_0 and the transmitted power P, the frequency transmittance T and shielding effect SE should satisfy the following equation:

$$T = \frac{P}{P_0} = 10^{-\mathrm{SE}/10}, \tag{1.18}$$

where the unit of SE is dB. It can be seen from equation (1.18) that when SE is 10, the frequency transmittance is reduced to 10 %. When SE is 30, the frequency transmittance is reduced to 0.1 %.

Generally speaking, the reflection and absorption of incident light will strengthen shielding. SE of electromagnetic radiation with perpendicular incidence can be expressed using the two equations below:

$$\mathrm{SE}(\lambda/2) = 20\lg\left(1 + \frac{188.5}{R_S}\right), \tag{1.19}$$

$$\mathrm{SE}(\lambda/4) = 20\lg\left(\frac{1+\varepsilon}{2\sqrt{\varepsilon}} + \frac{188.5}{R_S\sqrt{\varepsilon}}\right), \tag{1.20}$$

where λ is the wavelength of the incident radiation.

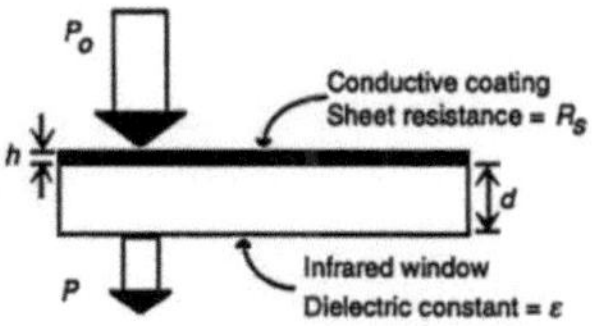

Fig. 1.38: Conductive coating on an infrared window.

As can be seen from the equation above, greater coating conductivity and smaller sheet resistance imply higher SE. Equation (1.19) is applicable to window thicknesses $d = \lambda/2, \lambda, 3\lambda/2, 2\lambda, \ldots$. Equation (1.20) is applicable to window thicknesses $d = \lambda/4$, $3\lambda/4, 5\lambda/4, 7\lambda/4, \ldots$. In order to establish equations (1.19) and (1.20), a further condition needs to be satisfied: the thickness of the conductive thin-film h needs to be smaller than the skin depth δ. For a conductive coating of thickness h and sheet resistance R_S, the skin depth is δ:

$$\delta = \sqrt{\frac{2R_S h}{\omega\mu}}, \tag{1.21}$$

where ω is the angular frequency of the radiation ($\omega = 2\pi f$), μ the magnetic permeability of the coating, and $R_S h$ the resistivity of the coating. For nonmagnetic materials, μ equals the magnetic permeability of free space, i.e. $\mu = 4\pi \times 10^{-7}$ H/m.

1.5 Optical properties of infrared antireflective and protective coatings

1.5.1 Optical constants

Refraction will occur when light travels between two different transmission media, as shown in Fig. 1.39. The angle of incidence θ_1 and the angle of refraction θ_2 satisfy Snell's law:

$$n_1 \sin\theta_1 = n_2 \sin\theta_2 . \tag{1.22}$$

The refractive index n of the material is related to the transmission speed of light in the media, which is expressed as follows:

$$c = c_0/n, \tag{1.23}$$

where c is the transmission speed of light through the medium, and c_0 the speed of light in vacuum. In other words, when light passes through a quartz window with a refractive index of 1.5, its transmission speed is $1/1.5 \approx 67\,\%$ of the speed of light in vacuum. During the transmission process, its frequency v does not change, but since $c = \lambda v$, its wavelength λ decreases to 67 % of that in vacuum.

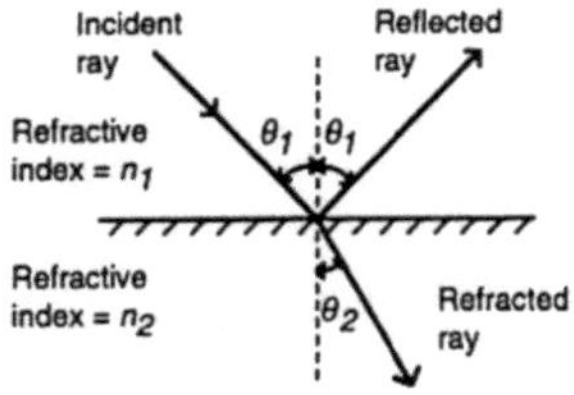

Fig. 1.39: Refraction of light travelling through two different media.

For a majority of materials, the Sellmeier equation below can be used to accurately describe the changes in refractive index with wavelength:

$$n^2 - 1 = \sum_i \frac{A_i \lambda^2}{\lambda^2 - \lambda_i^2} = \frac{A_1 \lambda^2}{\lambda^2 - \lambda_1^2} + \frac{A_2 \lambda^2}{\lambda^2 - \lambda_2^2} + \cdots, \tag{1.24}$$

where n is the refractive index, λ the wavelength of the material, and A_i and λ_i are the expandable empirical constants measured from the materials. Under typical circumstances, only two or three terms are needed to achieve the required accuracy.

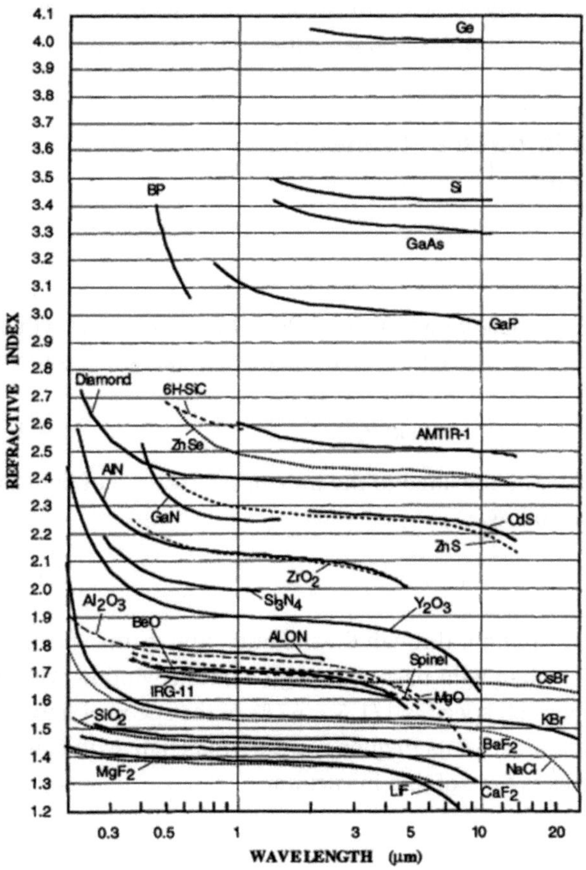

Fig. 1.40: Plots of the relationships between refractive index and wavelength for common infrared optical materials.

Fig. 1.40 shows the plots of the relationship between the refractive index and wavelength for common infrared optical materials. As shown in the figure, the refractive indices of infrared optical materials tend to decrease gradually with increasing wavelength. The refractive indices of most oxides and fluorides (containing the element O or F) fall between 1.3 and 1.8. The refractive indices of sulfides and selenides (containing the elements S or Se) are above 2. The refractive indices of silicon or germanium compounds are even higher, approaching 4. This is because heavy atoms have more polarizable electron clouds, which results in a higher refractive index.

1.5.2 Transmittance

In practical applications of infrared windows, if we consider the perpendicular incidence of light through the window into air, and ignore absorption by the window, then the single-surface reflectance R can be expressed as follows:

$$R = \left(\frac{1-n}{1+n}\right)^2. \tag{1.25}$$

The total reflectance r for both window surfaces is

$$r = \frac{2R}{1+R} = 1 - \frac{2n}{1+n^2}. \tag{1.26}$$

It can be seen from equation (1.26) that in an infrared optical window polished on both surfaces, the total reflectance $r > R$. Reflection losses are determined by the refractive index of the material only. If absorption and scatter are negligible, then the theoretical transmittance t will be

$$t = 1 - r = \frac{2n}{1+n^2}. \tag{1.27}$$

Fig. 1.41 shows the position on the transmittance vs refractive index curve (equation (1.27)) of a few infrared optical materials. One can see that the transmittance of MgF_2 at 4 μm is up to 96 %, whereas that of the Ge window is only 47 %. This is caused by the high refractive index of Ge, which lead to high reflectance and therefore low transmittance. Therefore, for materials with high refractive indices such as Ge, antireflective coatings are necessary in order to ensure that sufficient transmittance is attained for practical applications.

If one considers the absorption and reflection of light by the material, and modifies the transmittance t accordingly, the following expression will be obtained:

$$t = \frac{(1-R)^2 e^{-\alpha b}}{1 - R^2 e^{-2\alpha b}}, \tag{1.28}$$

where α is the absorption coefficient, b the sample thickness, and R the single-surface reflectance.

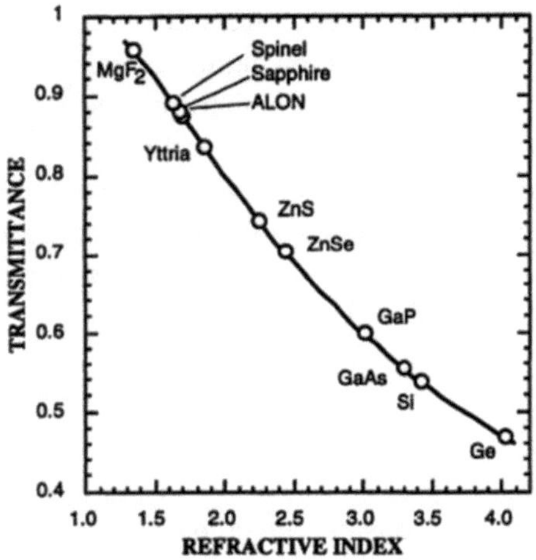

Fig. 1.41: Positions of some infrared window materials with respect to their refractive index (at 4 μm) and transmittance (equation (1.27)).

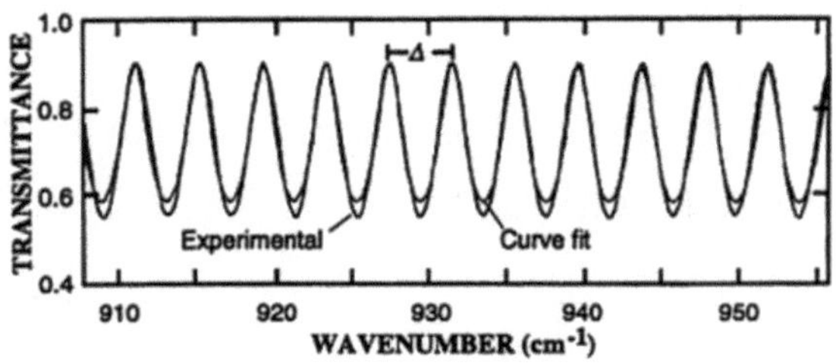

Fig. 1.42: Transmittance fluctuations of a polished diamond surface.

Because polishing is required for all window materials prior to application, an oscillating transmission spectrum might occasionally emerge from parallel surfaces, as shown in Fig. 1.42. Such fluctuations are known as the etalon effect, because an etalon device consists of two parallel plates, and its transmittance is determined by the interference of light waves bouncing between the two plates. The transmittance t of a window exhibiting the etalon effect can be expressed as follows:

$$t = \frac{-(1-R)^2 \mathrm{e}^{-\alpha b}}{1 + R^2 \mathrm{e}^{-2\alpha b} - 2R\mathrm{e}^{-\alpha b}\cos\phi}, \tag{1.29}$$

where $\phi = 4\pi nb/\lambda_0$, n the refractive index, λ_0 the wavelength of radiation in vacuum, and $\lambda = \lambda_0/n$.

The change in the distance between the peaks Δ and temperature T satisfy the following function:

$$\frac{\mathrm{d}\Delta}{\mathrm{d}T} = -\Delta\left[\frac{1}{n}\frac{\mathrm{d}n}{\mathrm{d}T} + \frac{1}{b}\frac{\mathrm{d}b}{\mathrm{d}T}\right], \tag{1.30}$$

where n is the refractive index, b the window thickness, and $(1/b)(\mathrm{d}b/\mathrm{d}T)$ the thermal expansion coefficient, usually denoted by α. If the absorption coefficient of the material α is known and independent, then the change in refractive index $(\mathrm{d}n/\mathrm{d}T)$ can be reduced by decreasing the distance between the peaks Δ.

Generally speaking, as temperature increases, the absorption edge of the infrared window material will shift towards shorter wavelengths. Fig. 1.43 shows the relationship between transmittance and temperature for ZnS and sapphire. We can conclude that the absorption edge of ZnS and sapphire shifts towards shorter wavelengths as the temperature increases.

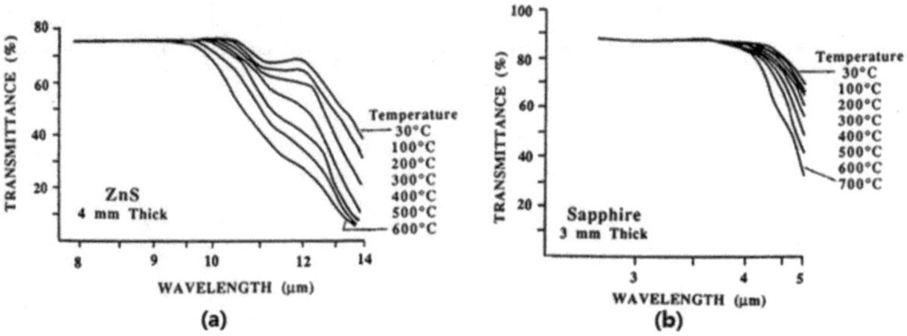

Fig. 1.43: Relationship between transmittance and wavelength in ZnS (a) and sapphire (b).

1.5.3 Emittance

The emittance of the infrared optical materials for windows and domes is extremely important in the selection of a suitable material. There are two main methods for selection: the first is to select materials with low emittance, given that they satisfy the required optical and mechanical properties; the second is to reduce emittance of the selected material through cooling.

The emittance perpendicular to the surface is as follows:

$$\varepsilon = 1 - \frac{(1-R)(1-e^{-\alpha b})}{Re^{-\alpha b}}, \tag{1.31}$$

where R is the single-surface reflectance, α is the absorption coefficient, and b is the sample thickness.

The single-surface reflectance is air is given by

$$R = \frac{(1-n)^2 + k^2}{(1+n)^2 + k^2}. \tag{1.32}$$

Equation (1.31) shows that the emittance and absorption satisfy the following relationship: at a given wavelength, strong absorption implies strong emittance.

The main limiting factor in the usage of infrared windows under high temperature is the emission of light from the window and not the absorption. At excessively high temperatures, the signals emitted by the window might be stronger than the signals observed, especially when the window is nearer to the detector than the observed object, which will cause blurring in the radiation of the observed signals. Furthermore, when the signals of the window or dome are sufficiently strong, the detector will be saturated by photons and become unresponsive to incoming signals.

Fig. 1.44 shows the emittances of sapphire, AlON, spinel, and Y_2O_3, respectively. To facilitate comparison, Fig. 1.45 shows the emittance of these materials at 700 K. The emittance of sapphire increases from 3.2 % to 6.6 %, whereas that of AlON increases from 9.7 % to 14.5 %. The emittance of Y_2O_3 is relatively low (approximately 2.2 %) compared with the other materials, and the effect of temperature increase is relatively small. Therefore, Y_2O_3 is suitable for applications under high temperatures.

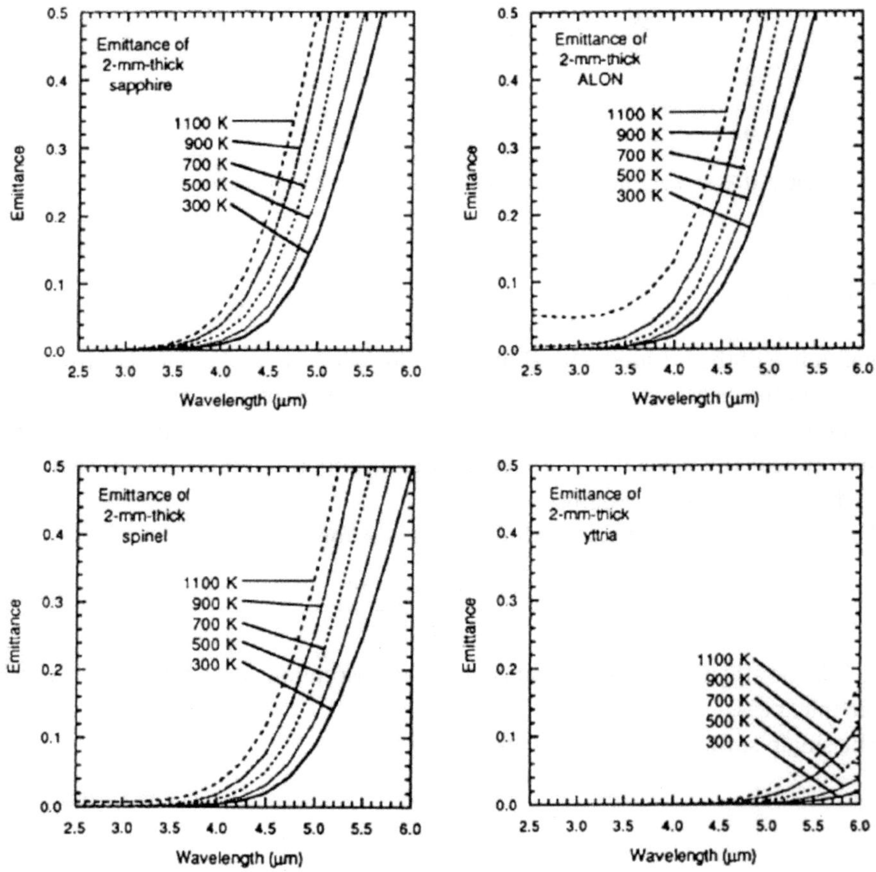

Fig. 1.44: Plots of the relationship between emittance and temperature for sapphire (a), AlON (b), spinel (c), and Y_2O_3 (d).

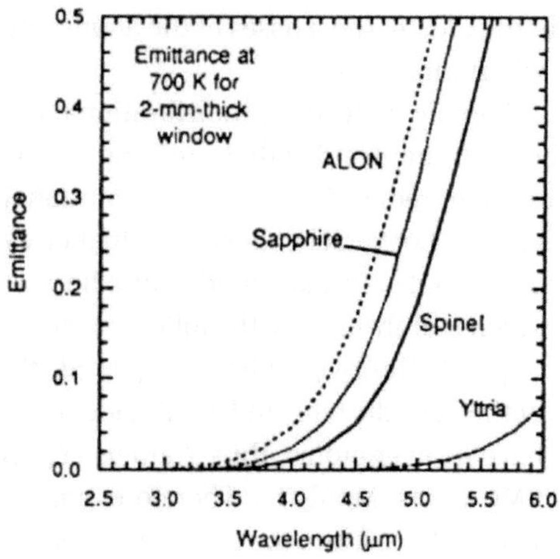

Fig. 1.45: Comparison of emittance at 700 K for sapphire, AlON, spinel, and Y_2O_3.

1.5.4 Transmission wave range of common infrared materials

Fig. 1.46 shows a simple model of the vibration of a diatomic molecule.

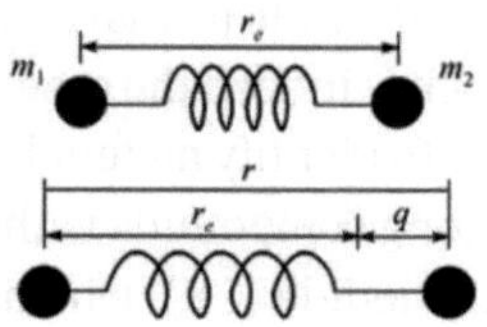

Fig. 1.46: Model of the vibration of a diatomic molecule.

In the figure, m_1 and m_2 are the masses of the two atoms, and k is the spring constant. When the distance between the diatomic molecule is stretched from its equilibrium length r_e to length $r = r_e + q$, the restoring force F will be proportional to the displacement q, which satisfies the following function:

$$F = -kq. \tag{1.33}$$

The stronger the chemical bond within the diatomic molecule is, the larger is the value of the spring constant k.

The vibrational energy levels E can be expressed as

$$E = \left(\upsilon + \frac{1}{2}\right)h\omega, \tag{1.34}$$

where υ is the vibrational quantum number ($\upsilon = 0, 1, 2, 3, \ldots$), h Planck's constant, and ω given by

$$\omega = \frac{1}{2\pi}\sqrt{\frac{k}{\mu}}, \tag{1.35}$$

$$\mu = \frac{m_1 m_2}{m_1 + m_2}. \tag{1.36}$$

One can see from equations (1.34)–(1.36) that the greater the masses of the atoms, the larger the reduce mass, and the smaller the vibrational energy.

With regards to crystalline and gaseous diatomic molecules, the vibrational energy increases with increasing interatomic bond strength and with decreasing atomic masses. Atoms with larger masses will have greater reduced mass, lower vibrational energies, and therefore longer cut-off wavelengths. Conversely, those with higher vibrational energies will have shorter cut-off wavelengths. Weakly bonded and heavier atoms will have lower vibrational energies, which mainly include the following materials: yttrium oxide (Y_2O_3), sapphire (Al_2O_3), spinel ($MgAl_2O_4$), AlON ($9Al_2O_3 \cdot 5AlN$), ZnS, ZnSe, etc. As shown in the partial view of the periodic table in Fig. 1.47, elements at the lower positions in the periodic table have heavier atoms, thus Y atoms (Y_2O_3) are heavier than Al atoms (Al_2O_3, AlON) and Mg atoms ($MgAl_2O_4$). Therefore, among these materials, the vibrational energy of Y_2O_3 is the lowest and the cut-off wavelength is the longest. As for oxides, oxygen (O) is too light, and its bonds are too strong, thus resulting in vibrational frequencies that are too high for transmission in the long wavelength region. Sulfur (S) and selenium (Se) are below oxygen in the periodic table, hence they can form compounds which allow transmission in the long wavelength region. Phosphorus (P) is to the left of sulfur, and arsenic (As) is to the left of selenium. Arsenic is heavier than P. Therefore, one can predict that gallium arsenide (GaAs) has a longer cut-off wavelength than gallium phosphorus (GaP), which is indeed the case: the cut-off wavelength of GaAs is 15 μm, while that of GaP is 11 μm. To identify materials with longer cut-off wavelengths, we will only need to search among compounds with heavier atoms near the bottom of the periodic table, e.g. caesium iodide (CsI), which has a cut-off wavelength of 80 μm. However, this crystal has relatively weak chemical bonds due to its heavy atoms, and the mechanical stability of such infrared materials is poorer than those composed of lighter atoms.

Li	Be				B	C	N	O	F	Ne
Na	Mg				Al	Si	P	S	Cl	Ar
K	Ca	Sc	...	Zn	Ga	Ge	As	Se	Br	Kr
Rb	Sr	Y	...	Cd	In	Sn	Sb	Te	I	Xe
Cs	Ba	La	...	Hg	Tl	Pb	Bi	Po	At	Rn

Fig. 1.47: Partial view of the periodic table.

Fig. 1.48 lists the transmission wavelength ranges of common infrared optical materials. As can be seen, infrared optical materials containing heavy atoms have longer cut-off wavelengths. Most oxides do not transmit in the long-wave infrared region, whereas compounds containing S and Se, which are heavier than O, are generally able to transmit in the long wavelength infrared region.

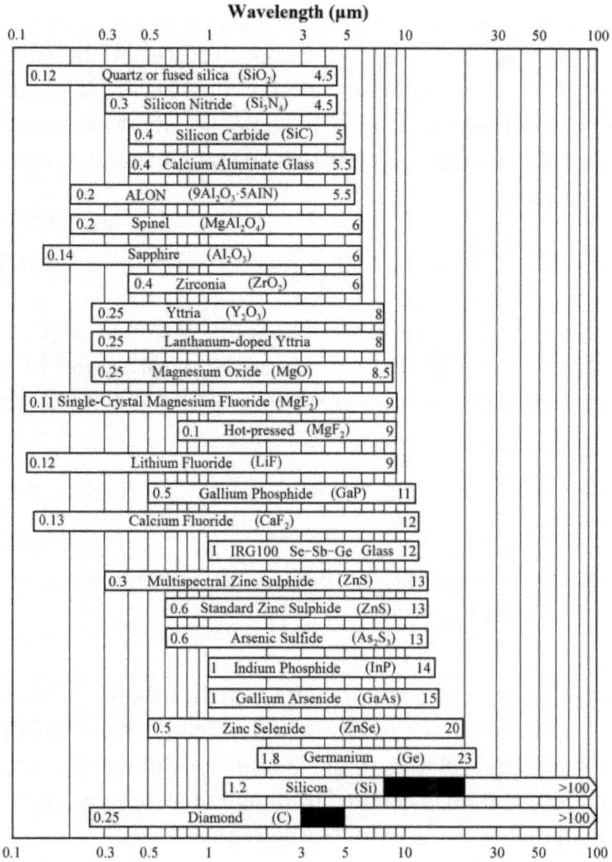

Fig. 1.48: Transmission wave ranges of common optical materials

1.6 Mechanical and thermal properties of infrared antireflective and protective coatings

This section briefly summarizes the parameters of mechanical and thermal properties of common infrared optical materials.

1.6.1 Hardness and elastic modulus

1.6.1.1 Hardness

Hardness is a key parameter that characterizes the mechanical properties of infrared optical materials. It reflects the ability of materials to resist indentation caused by localized pressure. Different hardness values will be obtained by applying different measurement methods in the same material. Infrared optical materials are mainly measured using the Vickers and Knoop hardness values.

Fig. 1.49 is a schematic diagram of the Vickers indenter and indentation. The indenter is a square pyramid with an angle of 136° between opposite faces. An indenter load of 10–100 N is applied and then removed after a period of time, leaving a square indentation pattern on the surface of the material. The diagonal length of the square indentation is denoted by d, and Vickers hardness (HV) of the material can be calculated through the following equation:

$$\mathrm{HV} = \frac{2P\sin(\theta/2)}{d^2} = 1.854\frac{P}{d^2}, \tag{1.37}$$

where P is the indenter load (N), θ the pyramid angle, and d the diagonal length of the indentation (mm). The units used for Vickers hardness are N/m^2, MPa, and GPa. Harder materials will have shallower indentations, and thus smaller d values.

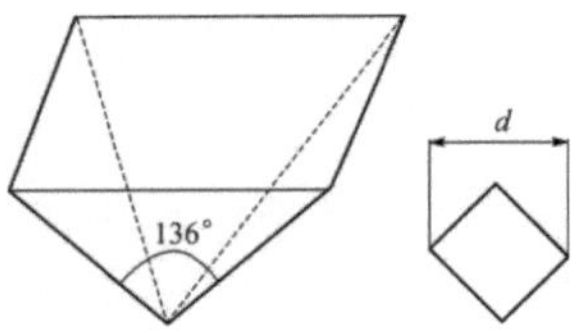

Fig. 1.49: Schematic diagram of Vickers hardness indenter and indentation pattern.

Fig. 1.50 is a schematic diagram of a Knoop hardness indenter and indentation pattern. Unlike the Vickers indenter, the pyramidal angles of the Knoop indenter are 137°30′ and 130°, and a load of 2–40 N is applied to obtain a diamond-shaped indentation. The diagonal length of the diamond indentation is denoted by d, and the Knoop hardness (HK) of the material can be calculated from the following equation:

$$\mathrm{HK} = \frac{P}{cd^2} = \frac{P}{0.0702d^2} = 14.229\frac{P}{d^2}, \tag{1.38}$$

where P is the indenter load (N) and c the indenter constant,

$$c = \frac{1}{2}\cos\left(\frac{172.5°}{2}\right)\tan\left(\frac{130°}{2}\right) = 0.0702;$$

the unit for Knoop hardness is kg/mm^2.

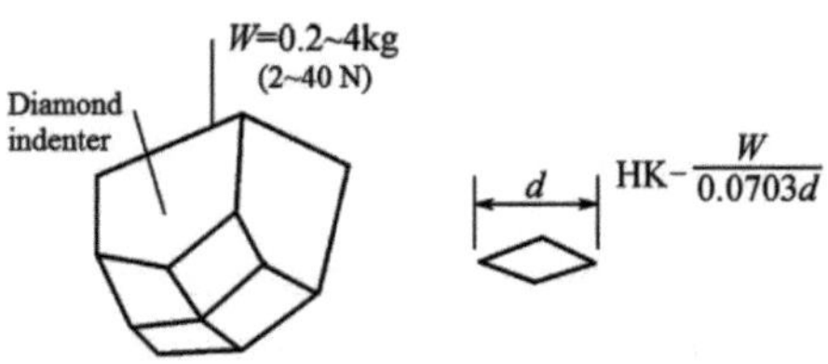

Fig. 1.50: Schematic diagram of Knoop hardness indenter and indentation pattern.

1.6.1.2 Elastic modulus

Consider a cylindrical optical ceramic material with length L and cross-sectional area S which is subject to a tension F. The tension exerted per unit area is known as stress, denoted by σ, $\sigma = F/S$. Subject to the tension F, the cylinder length L is stretched to $L + \Delta L$, and the increase per unit length is known as strain, denoted by ε, $\varepsilon = \Delta L/L$.

Fig. 1.51 shows the reactions of brittle and elastic materials when tension is applied. In brittle materials, when mild tension is applied, strain will be proportional to stress. When the stress exceeds the strength of the material, it will fracture. The point at which the material fractures when a certain level of tension is applied, is when its strength is equal to the strain. The slope of the stress-strain curve represents the elastic modulus E (or Young's modulus) of the material, given by

$$E = \frac{\sigma}{\varepsilon}. \tag{1.39}$$

One can see from equation (1.39) that the physical interpretation of the elastic modulus is the tensile stress required for the relative displacement per unit length of each particle within the material. Evidently, a larger elastic modulus implies that greater stress is required, and thus the stiffness of the material is higher. Ductile materials can be stretched more easily than brittle materials can.

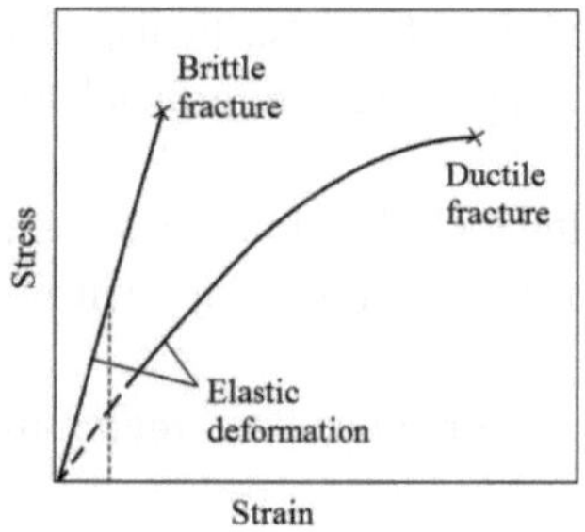

Fig. 1.51: Relationship between tensile stress and strain in brittle and ductile materials.

Consider a cylindrical ceramic solid with length l and diameter d. When a tensile stress σ is applied, its length will increase to $l + \Delta l$ and its diameter decrease to $d - \Delta d$. The Poisson ratio is defined as

$$\upsilon = \frac{\Delta d/d}{\Delta l/l}. \tag{1.40}$$

The Poisson ratio represents the ratio between the transverse rate of change, and the axial rate of change in a material under tensile stress. It is a constant for each type of material, and generally is about 0.2–0.3 for most infrared ceramics.

1.6.2 Coating–substrate adhesion

The foremost requirement in the application of coating materials is sufficient interfacial adhesion between the coating and substrate materials. The level of coating–substrate adhesion is a key indicator in the evaluation of coating quality. It is also a fundamental prerequisite to ensure that the coating fulfils the mechanical, physical, and chemical properties required during application. Chalker et al. pointed out that ideal tests of coating–substrate interfacial adhesion should include

(1) a good physical model that reasonably reflects the separation of the coating from the substrate material, and
(2) the ability to accurately measure the relevant mechanical parameters.

The scratch test is the most sophisticated and widely used among all existing tests. It involves scratching the sample surface with an indenter at a fixed speed, and simultaneously applying a normal force which is increasing in a stepwise or continuous manner until the coating is detached. The minimum force needed to remove the coating from the substrate (critical load L_C) is regarded as a measure of the coating–substrate interfacial adhesion. The majority of coatings can be tested using this method, which has good reproducibility and simple operation. However, the scratch test still has several problems:

(1) different methods used in coating detachment will produce different critical loads, and hence the test results cannot truly represent the magnitude of interfacial adhesion;
(2) numerous external factors can influence the test results, including loading speed, scratching speed, substrate properties, coating properties, and testing environment;
(3) the scratch test reflects the adhesion properties only within a localized region in the coating–substrate system.

Furthermore, the scratch test is destructive to the material, which might limit its practical applications.

The indentation test is currently the most promising method for application in production and experiments. It can be performed using conventional hardness testers, and is nondestructive. In addition, a mathematical model can easily be constructed for this test, thereby allowing targeted failure criteria to be established based on coating fracture or detachment methods. One of the advantages of the indentation test compared to the scratch test is that its interfacial adhesion coefficients P_{cr} or K_{li} are not sensitive to substrate hardness. Furthermore, in addition to testing the coating–substrate adhesion, the indentation test also allows us to measure the coating hardness, elastic modulus, fracture toughness, and other mechanical properties. Cross-section nanoindentation (CSN) has also been extensively applied in the testing of interfacial adhesion. Sánchez et al. applied CSN to measure the interfacial adhesion of

SiO_2 and SiN multilayer coatings deposited on a silicon substrate. The indenter was used to crack the interface, then the crack length at the interface and load-tip displacement curves were used to measure interfacial adhesion.

The underlying principle of the pull-off test is to apply a pulling force in the direction perpendicular to the coating–substrate interface. The load is gradually increased, and the tensile force required for coating–substrate detachment is regarded as the interfacial adhesion strength. This method was first proposed by Agrawal and Raj, who employed it to test metal-ceramic interfacial adhesion. During the experiment, a crack was formed along a direction perpendicular to the load. The relationship between the maximum crack density and interfacial shear strength τ is given by

$$\tau = \frac{\pi \sigma_b h}{\delta_{max}}, \tag{1.41}$$

where σ_b is the tensile strength (fracture strength) of the coating, h the crack depth in the coating, and δ_{max} the maximum spacing between cracks in the coating.

The bend test usually involves making a V-shaped notch in a fixed region of the sample to produce a crack. The length of the V-shaped notch and indenter tip radius are key factors influencing the test results. The samples are usually bonded together using an adhesive during testing, which increases the testing difficulty. Furthermore, this method cannot be used to test coating–substrate interfaces with extremely strong bonds. The statistical dispersion of the results is also very large, which implies that multiple tests are required, and the results need to be analyzed and processed using Weibull statistical theory.

There are currently several methods for the evaluation of interfacial adhesion, such as the scratch test, indentation test, pull-off test, bend test, torsion test, and ultrasonic test. To summarize, all of them involve using a certain method to apply a critical load (L_c) that is just enough to destroy the coating–substrate interface and using it to represent adhesion strength. However, the critical load, which is used to determine adhesion strength, is measured under stationary loading conditions. Hence, it is unable to accurately reflect the behavior of the coating–substrate interface under dynamic loading conditions, and it is also influenced by multiple factors such as the discrepancy in coating–substrate hardness, and coating thickness. Therefore, there are certain limitations to its application.

1.6.3 Thermal conductivity and thermal expansion coefficient

1.6.3.1 Thermal conductivity

When there is a temperature difference between the two ends of a rectangular solid, heat will naturally be transferred from the region with high temperature T_2 to the region with low temperature T_1. In an isotropic material, the heat DQ conducted across the solid material per unit time can be given by

$$\Delta Q = -\lambda \frac{dT}{dx} \Delta s, \tag{1.42}$$

where dT/dx is the temperature gradient along the longitudinal direction, Δs the cross-sectional area of the material, and λ the thermal conductivity (W/m·K).

The change in temperature with time at a certain point per unit cross-section can be obtained using the heat equation as follows:

$$\frac{\partial T}{\partial t} = \frac{\lambda}{\rho c_P} \frac{\partial^2 T}{\partial^2 x} = a \frac{\partial^2 T}{\partial^2 x}, \tag{1.43}$$

where ρ is the density of the material, c_P the heat capacity at constant pressure, and $a = \lambda/\rho c_P$ the thermal diffusivity.

As can be inferred from equation (1.43), the rate of temperature change across the material is proportional to the thermal diffusivity and the temperature gradient. As the thermal diffusivity a increases, the rate of temperature change will be faster. Thermal diffusivity reflects the rate of temperature diffusion across the material.

Most infrared optical materials have high electrical resistance or are electrical insulators. Thermal conductivity λ can be expressed as follows:

$$\lambda = \frac{1}{2} c_V \bar{\upsilon} l = \frac{1}{2} c_V \sqrt{\frac{E}{\rho}} l, \tag{1.44}$$

where c_V is the heat capacity at constant volume, $\bar{\upsilon} = \sqrt{E/\rho}$ the mean velocity of acoustic waves (mean phonon velocity), l the mean free path of phonons, E the elastic modulus, and ρ the density of the solid.

Greater thermal conductivity implies faster heat transfer across a solid material.

1.6.3.2 Thermal expansion coefficient

Most materials will expand when heated. When the temperature of an object with length l is increased by ΔT, the increase in length will be $l + \Delta l$, where the relative increase in length $\Delta l/l$ and the increase in temperature ΔT will satisfy

$$\frac{\Delta l}{l} = \alpha \Delta T, \tag{1.45}$$

where α is the thermal expansion coefficient.

In general, the thermal expansion coefficient is not a constant, but is related to temperature, whereby the thermal expansion coefficient will increase as temperature increases. However, the thermal expansion coefficients for a majority of infrared optical materials are very small, on the order of 10^{-6}. Therefore, it can be roughly considered as a constant.

Thermal expansion refers to the expansion of solids due to the increased vibration of the crystal lattice when the material is heated. The heat capacity of solids indicates the amount of heat required to raise one unit mass of solid material by one unit of

Tab. 1.10: Optical, mechanical, and thermal properties of common infrared optical materials.

Material	Transmission wave range (μm)	Refractive index (3.10 μm)	Melting point (°C)	Knoop hardness ($kg \cdot mm^{-2}$)	Elastic modulus (GPa)	Thermal conductivity (W/·mK)	Thermal expansion coefficient ($10^{-6}\,K^{-1}$)
Hot-pressed MgF_2	0.7–9	1.4	1255	640	114	14.7	10.41
Quartz	0.12–4.5	1.5427		820	73	1.6	0.42
MgO	0.25–8.5	1.6920	2800	640	332	59	10.5
Al_2O_3	0.14–6	1.7122	2050	1600–2200	344	34	5.3
Y_2O_3	0.25–8	1.9253	2464	720	170	13.5	6.6
AlON	0.2–8	1.790	2158	1950	317	12.6	5.8
Spinel	0.2–5.5	1.698	2135	1600	193	14.6	5.6
CVD ZnS	0.2–6	2.2002	1830	250	74	19	7.0
m-ZnS	0.6–13	2.2008	1830	160	88	27	7.0
ZnSe	0.35–13	2.4065	1520	150	70	16	7.6
GaP	0.5–20	2.9	1467	840	103	110	5.3
GaAs	0.5–11	3.2769	1238	750	83	55	5.7
Ge	1.8–23	4.0032	937	850	103	59	6.1
Si	1.2–5.8	3.426	1420	1150	131	163	2.6
SiC	0.4–5	2.576 (1.15 μm)	2600	2540	465	200	1.9
Type a diamond	0.25–35–100	2.376	3770	9000	1050	2000	0.8

temperature. The relationship between the thermal expansion coefficient and temperature is similar to that between heat capacity at constant pressure and temperature. Therefore, the relationship between thermal expansion and heat capacity at constant pressure can be expressed as follows:

$$\alpha = \frac{\gamma X c_P}{V}, \tag{1.46}$$

where γ is a parameter, $\gamma \approx 2$; X the coefficient of compressibility; c_P the heat capacity at constant pressure; and V the molar volume. The change in the ratio X/V is very small. Equation (1.46) indicates that the thermal expansion coefficient and heat capacity at constant volume are approximately proportional.

Tab. 1.10 lists optical, mechanical, and thermal properties of common infrared optical materials.

References

[1] Adams I, AuCoing TR, Wolff A. Luminescence in the system Al_2O_3-AlN. Journal of the Electrochemical Society, 1962, 109(11): 1050–1054.

[2] Agrawal DC, Raj R. Measurement of the ultimate shear strength of a metal-ceramic interface. Acta Metallurgica, 1989, 37(4): 1265–1270.

[3] Bandyopadhyay S, Rixeeker G, Aldinger F, et al. Effect of reaction parameters on γ-AlON formation from Al_2O_3 and AlN. Journal of the American Ceramic Society, 2002, 85(4): 1010–1012.

[4] Cai X, Wang JF, Zhou PN, et al. Ultramicrohardness measurement of silicon nitride films. Key Engineering Materials, 1994, (89–91): 542–552.

[5] Chalker PR, Bull SJ, Rickerby DS. A review of the methods for the evaluation of coating-substrate adhesion. Materials Science and Engineering: A, 1991, 140(7): 583–592.

[6] Cheng J, Agrawal D, Roy R. Microwave synthesis of aluminium oxynitride (AlON). Journal of Material Science Letters, 1999, 18(24): 1989–1990.

[7] Corbin ND. Aluminium oxynitride spinel: A review. Journal of European Ceramic Society, 1989, 5(3): 143–154.

[8] Gentilman RL, Dibenedetoo BA, Tustison RW, et al. Chemically Vapour Deposited Coatings. Westerville, OH, The American Ceramic Society Inc., 1981.

[9] Gentilman RL, Maguire EA, Dolher LE. US patent 4520116. 1985.

[10] Gentilman R, McGuire PT, Fiore D. Large-area sapphire windows. Proc SPIE, 2003, 5078: 54–60.

[11] Goela JS, Taylor RT. Monolithic material fabrication by chemical vapour deposition. Journal of Materials Science, 1998, 23(12):4331-4339.

[12] Harris DC. Durable 3-5 μm transmitting infrared window materials. Infrared Physics & Technology, 1998, 39(4): 185–201.

[13] Hartnett TM. Optical properties of AlON. Infrared Physics & Technology, 1998, 39(4): 203–211.

[14] Hilton AR. Chalcogenide Glasses for Infrared Optics. New York, McGraw-Hill Companies, Inc., 2010.

[15] Jayatilaka AS. Fracture Engineering Materials. London, Applied Science Publishers, 1973.

[16] Khattak CP, Schmid F. Growth of the world's largest sapphire crystals. Crystal Growth, 2001, 225: 572–579.

[17] Krell A, Baur G, Dähne C. Transparent sintered sub-μm Al_2O_3 with IR transmissivity equal to sapphire. Window and Dome Technologies VIII. Proc. SPIE. 2003, 5078: 100–207.

[18] Krell A. Transparent Sintered Sub-μm Al_2O_3 for Armour and IR Applications, EUCLID CEPA3 Workshop on Nanomaterials (Session: Novel Smart Materials for Active Structural Applications), Lisbon, Portugal, 2004.
[19] Krell A, Blank P, Ma HW, et al. Processing of high-density submicrometer Al_2O_3 for new applications. Journal of the American Ceramic Society, 2003, 86(4): 546–553.
[20] Krell A, Klimke J, Hutzler T. Advanced spinel and sub-μm Al_2O_3 for transparent armour applications. Journal of the European Ceramic Society, 2009, 29(2): 275–281.
[21] Kumar RS, Hareesh US, Ramavath P, et al. Hydrolysis control of alumina and AlN mixture for aqeous colloidal processing of aluminium oxynitride. Ceramics International, 2011, 37(7): 2583–2590.
[22] Kumar RS, Rajeswari K, Praveen B, et al. Processing of aluminium oxynitride through aqueous colloidal forming techniques. Journal of the American Ceramic Society, 2010, 93(2): 429–435.
[23] Long G, Foster LM. Journal of the American Ceramic Society, 1961, 44(11):255-258.
[24] Medvedovski E. Ballistic performance of armour ceramics: Influence of design and structure. Part 2. Ceramics International, 2010, 36(7): 2117–2127.
[25] Mencl J, Swain MV. Errors associated with depth-sensing microindentation test. Journal of Materials Research, 1995, 10(6): 1491–1501.
[26] Richard P, Thomas J, Landolt D, et al. Combination of scratch-test and acoustic microscopy imaging for the study of coating adhesion. Surface & Coatings Technology, 1997, 91(1–2): 83–90.
[27] Robert M, Swinehart H, et al. Hot-pressable magnesium fluoride powder: US patent 404412.0823, 1977.
[28] Sánchez JM, et al. Cross-sectional nanoindentation: A new technique for thin film interfacial adhesion characterization. Aca Materialia, 1999, 47(17): 4405–4413.
[29] Savage JA. Infrared Optical Materials and their Antireflection Coatings. Adam Hilger, Bristol, UK, 1985: 95–111, 121–126.
[30] Seinen PA. High intensity discharge lamp with ceramic envelopes: A key technology for the lighting future. Proc. 7th Int. Symp. Sci. Tech. Light Sources, The Illuminating Engineering Society of Japan, Kyoto, Japan, 1995.
[31] Shchurov AF, Gavrishchuk EM, Ikonnikov VB, et al., Effect of hot isostatic pressing on the elastic and optical properties of polycrystalline CVD ZnS. Inorganic Materials, 2004, 40(4): 336–339.
[32] Stern KH. Metallurgical and Ceramic Protective Coating. London, Chapman & Hall, 1996.
[33] Straßburger E. Ballistic testing of transparent armour ceramics. Journal of the European Ceramic Society, 2009, 29(2): 267–273.
[34] Super Strong Transparent Alumina Ceramics for Energy-Efficient Lighting (STARELIGHT), http://cordis.europa.eu.
[35] Tatarchenko VA. Sapphire crystal growth and applications. Optical and Optoelectronic Materials, 2005, 99: 338.
[36] Wang X, Li W, Seetharaman S. Thermodynamic study and synthesis of γ-aluminium oxide. Scandinavian Journal of Metallurgy, 2002, 31(1): 1–6.
[37] Yamaguchi G, Yanagida H. Study on the reductive spinel – A new spinel formula AlN – Al_2O_3 instead of the previous one Al_3O_4. Bulletin of the Chemical Society of Japan, 1959, 32(11): 1264–1265.
[38] Yashina EV, Gavrishchuk EM, Ikonnikov VB. Mechanisms of polycrystalline CVD ZnS densification during hot isostatic pressing. Inorganic Materials, 2004, 40(9): 901–904.
[39] Yu HZ. Infrared Optical Materials. Beijing, National Defence Industry Press, 2007.
[40] Wei GC, Hecker A, Goodman DA. Translucent polycrystalline alumina with improved resistance to sodium attack. Journal of the American Ceramic Society, 2001, 84(12): 2853–2862.

[41] Willingham CB, Pappis J. US patent, 5126081, 1992.
[42] Wu ZF, Liu ML, Zhang SJ, et al. Infrared and Low-Light Technology. Beijing, National Defence Industry Press, 1998.
[43] Zhang J, Ardell AJ. Measurements of the fracture toughness of CVD – grown ZnS using a miniaturized disk-bend test. Journal of Materials Research, 1991, 6(9): 1950–1957.
[44] Zhenyi F, Yichao C, Yongliang H, et al. CVD growth of bulk polycrystalline ZnS and its optical properties. Journal of Crystal Growth, 2002, 237–279: 1707–1710.

2 Service environment of infrared antireflective and protective coatings

2.1 Overview

An infrared (IR) window is composed of an infrared antireflective/protective film and substrate material. The film is a vital component in the structural and functional integration of high-speed missiles, and it transmits infrared signals of the target, forms the aerodynamic geometry, and protects the imaging system. It is extensively used in numerous applications. A structural schematic of an IR window is shown in Fig. 2.1.

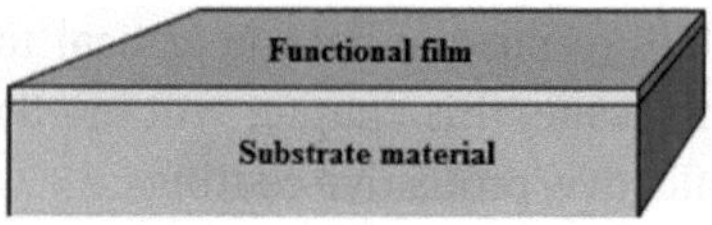

Fig. 2.1: Schematic of the components of an infrared window.

In terms of exterior geometry, infrared windows can be divided into two categories: flat windows and spherical windows, as shown in Fig. 2.2.

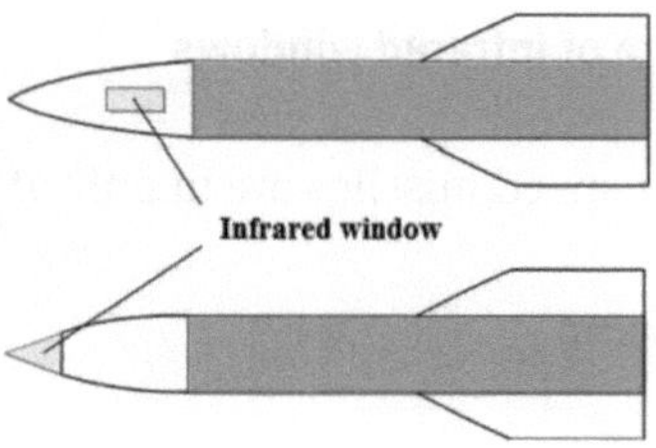

Fig. 2.2: Schematic representation of the positions of infrared windows.

Missiles are subject to the combined effects of aerodynamic heating and force when flying in the atmosphere at a supersonic speed. Therefore, while guaranteeing optical performance, infrared window materials must also satisfy certain strength requirements. After years of research and exploration, zinc sulphide and sapphire are currently the most widely used substrate material for infrared windows, whereas common functional film materials include diamond-like carbon (DLC), germanium (Ge), germanium carbide (Ge_xC_{1-x}), zirconium nitride (ZrN), yttrium fluoride (YF_3), and boron phosphide (BP).

During the high-speed flight of missiles, owing to the action of aerodynamic heat/strength, the infrared window is in a complex, combined thermal/mechanical state. The microscopic structures and thermodynamic properties of the materials will change, possibly leading to failures such as window rupture, coating separation, radiation disturbance saturation, and optical distortion, as shown in Fig. 2.3. More-

https://doi.org/10.1515/9783110489514-005

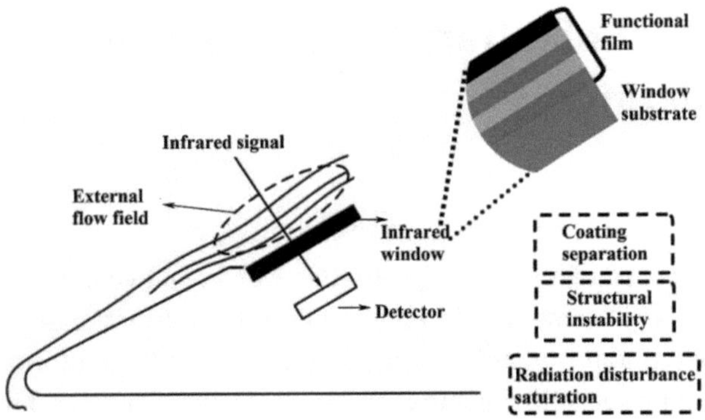

Fig. 2.3: Schematic of infrared window and its antireflective/ protective film under aerodynamic heat/strength.

over, the on-board infrared components might experience damages due to sand and rain erosion. Therefore, in-depth research on the failure mechanisms is critical to the design of infrared windows. In the context of this issue, this chapter will briefly introduce the service environment of infrared antireflective/protective coatings.

2.2 Aerodynamic heat/strength environment and aerodynamic heat/strength failures

2.2.1 Basic forms of aerodynamic heat/strength failure of infrared windows

In their service environment, infrared windows of high-speed missiles are mainly affected by two factors:
(1) mechanical stress induced by aerodynamic forces, and
(2) thermal stress induced by aerodynamic heating.

In the context of the special application of the missile-use infrared window, its failure modes can be divided into two categories: structural failure, including rupture and coating separation, and functional failure, including self-radiation and optical distortion.

2.2.1.1 Rupture

Under the practical service environment, window materials are subject to aerodynamic forces of huge magnitudes. Moreover, the dramatic temperature rise driven by aerodynamic heating can lead to a great temperature gradient inside the material, causing thermal shock. Consequently, an enormous tensile stress would occur on the inner surface of the window and a huge compressive stress would occur on the outer surface, jointly leading to the rupture of the infrared window.

2.2.1.2 Coating separation

Functional films often bear the huge internal stress formed during the deposition process. Under the joint action of a severely nonuniform temperature field and stress field, variation would occur to the functional film itself, as well as the interfaces between the functional film and the substrate and between films, causing a decrease of the binding force and, thereby, eventual coating separation.

2.2.1.3 Radiation disturbance saturation

Infrared window materials also generate spontaneous radiation at high temperatures, leading to interference and decrease in the signal-to-noise ratio (SNR). Consequently, the photoelectric detector may saturate and not be able to receive signals from the target. In other words, it would be impossible to image or detect the target.

2.2.1.4 Optical distortion

In the service environment, a sharp temperature increase and aerodynamic stress could cause the geometrical and optical parameters of infrared window materials to change in a nonuniform manner, resulting in the deviation of transmitted light from the focal point, imaging indistinction and shifting, and even the total failure of target imaging and detection.

Owing to the complexity of material structures and heat/stress states during flight, researchers have encountered numerous challenges when investigating material failure mechanisms. In terms of structural failures, because of the high temperature gradient, local nonlinear characteristics of materials become quite noticeable, and material features including thermal conductivity, coefficient of thermal expansion, and elastic properties would show a remarkable correlation with temperature. Meanwhile, the delaminating extension and buckling of the functional film under the action of thermal shock would cause the film to separate from the infrared window. Moreover, the microstructure of the window influences the failure. For functional failures, self-radiation and optical distortion have a close relation with the internal structures and defect features of materials. Failure mechanisms include intrinsic excitation, multiphonon excitation, free-carrier excitation, and impurity-level excitation. These mechanisms can also couple. Moreover, high temperature and stress could lead to the microstructure variation of materials, further complicating the excitation mechanisms.

2.2.2 The development of infrared window materials

Infrared guidance is a critical approach in precision guidance. The essence of this technology is to realize precision guidance by detecting the radiant energy emitted by a tracked infrared target with the use of an infrared detector. It has been extensively used in national defence science and technology for the infrared guidance systems of various missiles, such as antitank missiles, air-to-ground missiles, and air-to-air missiles. The infrared imaging system mainly consists of an infrared window, an infrared sensing/imaging system, and a signal processing system. Before reaching the infrared sensing/imaging system to form images, the infrared radiation generated by the target needs to be transmitted through the atmosphere and the infrared window. Located between the infrared sensing/imaging system and the external target environment, the infrared window protects the whole infrared imaging system. Therefore, a high transmittance of infrared windows at the appropriate infrared wavelengths is essential to ensure that a sufficient infrared radiation signal can be captured. Hence, the infrared window is a vital part of the infrared imaging system.

Based on the transmission spectrum of electromagnetic radiation in the atmosphere, atmosphere windows can be divided into three bands: near infrared (NIR, 0.76–1.1 μm), mid infrared (MIR, 3–5 μm), and far infrared (FIR, 8–12 μm). All matter at a temperature higher than the thermodynamic zero radiates a certain amount of infrared energy. The radiant energy is distributed based on the frequency, and the waveband of radiation is related to the object temperature. It can be known from Wien's displacement law that, when the temperature of an object is greater than 300 °C, the maximum wavelength of its radiation spectrum mainly lies in the band of 3–5 μm; when the temperature of the object is less than 100 °C, the maximum wavelength of its radiation spectrum is principally in the band of 8–12 μm. Tab. 2.1 shows the band of infrared radiation of several representative targets. It can be seen that weapons in their working state, such as aircraft, missiles, and tanks, have a typical temperature of hundreds of degrees and a band with a maximum infrared radiation wavelength of 3–5 μm; ground targets at the room temperature, including human bodies, buildings, and highways, generally have a temperature lower than 100 °C, leading to a maximum infrared radiation wavelength of 8–12 μm. Therefore, research on infrared detectors is mainly conducted for these three bands. Early investigations of infrared detectors were mostly focused on the near-infrared band. With the rapid development of infrared detection technology, infrared detector studies have already extended to the mid- and far-infrared bands.

Tab. 2.1: Infrared radiation wavebands of typical targets.

Target	Temperature (°C)	Radiation waveband (μm)
Missile body surface and engine plume	600–700	3–5
Aircraft engine nozzle	500–600	3–5
Engine exhaust	400	3–5
Tank exhaust	200–400	3–5
High-rise building	30–90	8–12
Human body	37	5–20

The term "infrared optical materials" mainly refers to materials that are transparent to infrared radiation at a wavelength of 3–5 μm and 8–12 μm. Their mechanical properties include the fracture strength and figure of merit of thermal shock resistance, whereas their optical properties include transmittance, refractive index, dispersion, and emissivity.

2.2.2.1 Fracture strength

During high-speed flight, the infrared window is principally affected by two types of forces: mechanical stress generated under the action of aerodynamic force and thermal stress generated under the action of aerodynamic heat. If these stresses exceed the ultimate stress of optical window materials, i.e., the fracture strength, then mechanical failure of the window will occur. Infrared optical materials are classified as ceramic brittle materials. Thus, it is assumed that they would show elastic deformation under the action of external tensile forces, with the stress-strain relationship following the Hooke's law. Meanwhile, as the elastic strain energy accumulates inside the material, fracture failure will occur when the stress exceeds the corresponding ultimate value. From the perspective of energy, the elastic strain energy stored inside the material is released and transformed into the surface energy of the two new faces formed after fracture.

2.2.2.2 Figure of merit of thermal shock resistance

When an aircraft flies at a supersonic speed, a hot retardation layer, namely the hot-air boundary layer, exists around the surface of the window, separating the window from the high-speed airflow. The heat flux is transmitted to the window through the retardation layer and consequently heats the window surface. In the case where the edge of the infrared window is fixed, a compressive stress will be formed on the heated outer surface, while a tensile stress will be formed on the cold inner surface. The thermal shock resistance factor is usually used to evaluate the ability of an infrared window to withstand thermal shock. A high figure of merit of thermal shock resistance indicates a strong ability of the material to resist thermal shock. The figure of merit of thermal shock resistance can be written as

$$\mathrm{FOM} = \begin{cases} \dfrac{\sigma_f(1-\mu_{\mathrm{po}})}{\alpha E_Y}, & Bi \geq 1 \\ \dfrac{\sigma_f(1-\mu_{\mathrm{po}})\lambda}{\alpha E_Y}, & Bi < 1, \end{cases} \tag{2.1}$$

where σ_f denotes the fracture strength, λ the thermal conductivity, α the coefficient of thermal expansion, μ_{po} the Poisson ratio, E_Y the elastic modulus, Bi the Biot number $Bi = \frac{h\delta}{\lambda}$, h the coefficient of heat transfer, and δ the thickness.

It can be seen from equation (2.1) that thermal shock resistance of the material is associated with the fracture strength σ_f, thermal conductivity λ, heat expansion coefficient α, and elastic modulus E_Y. The larger the values of σ_f and λ, the smaller the values of α and E_Y, and the stronger the thermal shock resistance.

2.2.2.3 Transmittance

Infrared optical materials demand a high transmittance within the working wavebands as well as a wide infrared spectrum range to satisfy the working-wavelength requirement of infrared detection devices. The range of transmitting wavelengths is dependent on the intrinsic structure and properties of infrared materials. The short-wavelength limit is determined by the energy gap of the energy band structure of the material, whereas the long-wavelength limit is determined by the crystal structure and thermal lattice vibration. These are intrinsic properties of a material that cannot be changed by varying the material preparation process.

2.2.2.4 Refractive index and dispersion

The optical constants of a solid are related to the frequency of the electromagnetic wave, indicating that the transmittance speed of the electromagnetic wave (light) in a certain medium is associated with the frequency. This relationship is named dispersion. The design of an imaging system following the principles of geometrical optics is mainly based on the refractive index of the optical material. The refractive index and dispersion of infrared window materials are demanded to be sufficiently low in order to reduce the reflective loss. When materials work at high temperatures, the relationship of refractive index varying with temperature also needs to meet certain requirements.

2.2.2.5 Emissivity

All matter generates thermal radiation when heated. When the temperature of a window increases because of aerodynamic heating, the detector will receive noise from the thermal radiation produced by the window. Once the temperature increases to a certain level, the noise will submerge the target signal or will at least be comparable to the target signal, preventing the infrared detector from imaging or resulting in indistinct images. Therefore, the thermal emissivity of the infrared window must be as low as possible such that the effect of noise generated under high-temperature working conditions on the SNR of the whole infrared imaging system is acceptable. The emissivity of the material is mainly determined by the material type, surface temperature, and surface morphology.

According to the transmission waveband, infrared optical materials can be roughly divided into two categories: mid-wavelength materials (0.9–5 μm) and long-

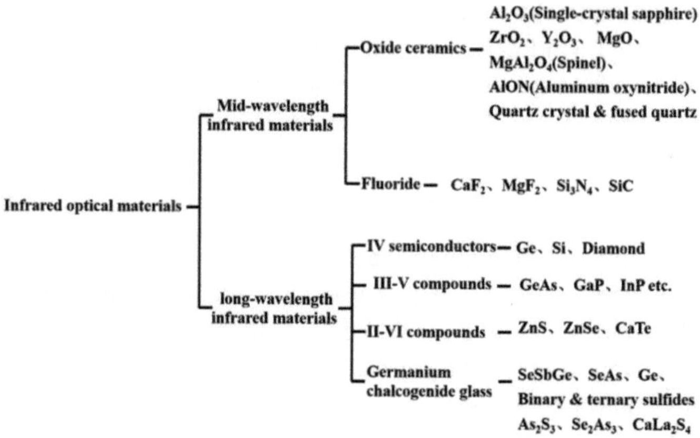

Fig. 2.4: Classification of infrared optical materials.

Tab. 2.2: Properties of infrared optical materials.

Material	Fracture strength (MPa)	Hardness (GPa)	Transmittance (%)	Refractive index (4 μm)	FOM of thermal shock resistance (10^{-3} W/m)
MgF_2	150	57.5	91 (4 μm)	1.349	0.9
Sapphire	700	150–220	88 (4 μm)	1.677	4.3
Y_2O_3	150	72	82 (4 μm)	1.859	1.3
Spinel	100–200	140	88 (4 μm)	1.635	1.9
ALON	30	195	85 (4 μm)	1.702	1.5
CaF_2	37	17	95 (8 μm)	1.410	0.28
ZnSe	52	10.5	71 (8 μm)	2.433	1.08
ZnS	97	23	72 (8 μm)	2.252	2.60
Ge	90	78	47 (8 μm)	4.025	6.08
GaAs	130	75	55 (8 μm)	3.304	8.02

wavelength materials (8–12 μm). Fig. 2.4 shows a classification diagram of infrared optical materials, and Tab. 2.2 lists the properties of representative infrared optical materials.

2.2.3 Design method for infrared window functional films

In the early days, the design and fabrication of infrared windows mostly depended on tests and empirical equations. Since the end of the 20th century, the development of theories on the strength and optical properties of infrared windows has been significantly promoted with the progress of missile technology. Today, the use of theoretical analysis to design infrared windows has gradually become the primary approach.

2.2.3.1 Heat/stress of single-layer film window

In the case of a single-layer film window, window failures often occur owing to the occurrence of an excessive tensile stress that is greater than the fracture strength. For the window material under thermal shock, the instantaneous peak of tensile stress that could be reached is

$$\sigma_P = \frac{1}{TSF}\frac{\alpha E}{1-\nu}\Delta T_P, \tag{2.2}$$

where ΔT_P is the maximum temperature difference between the two sides; TSF is the thermal stress factor, which is related to the geometrical parameters and lateral temperature distribution of the window; α and E are the coefficient of thermal expansion and elastic modulus, respectively; and ν is the Poisson's ratio.

The parameter ΔT_P depends on time and the type of heat flux, and it is associated with the Biot number, Bi, which is defined as

$$Bi = \frac{hL}{\kappa}, \tag{2.3}$$

where h is the coefficient of heat transfer, L the thickness of the window, and κ the heat conductivity.

According to the lumped parameter approximation, ΔT_P can be written in a simpler form:

$$\Delta T_P = \begin{cases} Q/h, & Bi > 1, \\ Q/(L/\kappa), & Bi < 1. \end{cases} \tag{2.4}$$

If $(\sigma_P)_{\lim}$ is the maximum allowable stress, then by considering the assurance factor n_s, one can obtain

$$(\sigma_P)_{\lim} = \frac{\sigma_f}{n_s}. \tag{2.5}$$

Combining the equations above, we arrive at the formula of the maximum allowable thermal flux:

$$Q_{\lim} = \begin{cases} \dfrac{TSF}{n_s}\dfrac{\sigma_f(1-\nu)}{\alpha E}h, & Bi > 1, \\ \dfrac{TSF}{n_s}\dfrac{\sigma_f(1-\nu)}{\alpha E}\dfrac{\kappa}{L}, & Bi < 1. \end{cases} \tag{2.6}$$

Equation (2.6) is the calculation formula of the maximum allowable thermal flux of thermal shock, where TSF/n_s is a parameter associated with the geometrical structure of the material. By separating this parameter, the thermal shock sensitivity with practical guiding significance can be achieved:

$$FOM = \begin{cases} \dfrac{\sigma_f(1-\nu)}{\alpha E}h, & Bi > 1, \\ \dfrac{\sigma_f(1-\nu)}{\alpha E}\dfrac{\kappa}{L}, & Bi < 1. \end{cases} \tag{2.7}$$

Based on previous studies, the situation of $B_i < 1$ can be simplified into

$$(\mathrm{FOM})_{Bi<1} = \begin{cases} \dfrac{\sigma_\mathrm{f}^{3/2}(1-\nu)\kappa}{\alpha E}, & \text{if window is flat,} \\ \dfrac{\sigma_\mathrm{f}^{5/3}(1-\nu)\kappa}{\alpha E}, & \text{if window is spherical.} \end{cases} \tag{2.8}$$

Above are the conditions of stress and thermal flux that need to be satisfied for single-layer film windows under thermal shock. Based on the results above, the design formulas of other parameters required in the window design process can be derived. For example, in the research of Klein et al., the computational formula of thickness optimization of the missile dome was derived based on the formula of the maximum allowable thermal shock, and the dome was assessed using the Biot number. Their work provided a significant reference to practical applications. By using integral transformation, M. Q. Liu et al. theoretically solved the heat conduction equation and thermal-elasticity equation of optical thin-film specimens under the action of pulse or square-wave modulated excitation light; thus, they obtained the temperature fields and deformation fields of the specimens in the transient and steady states, respectively.

In practical applications, a similar derivation can be carried out using various models or empirical data to obtain the calculation methods of strength-related parameters under different conditions, as summarized in Tab. 2.3.

2.2.3.2 Optical theory of single-layer film window

Window materials need to have a certain set of optical properties for the realization of corresponding functions. Theoretical studies on the optical properties of window materials had been essentially completed in the 20th century, and they have played an important role in practical applications.

For the propagation of light through the window, Andres et al. conducted detailed diagrammatizing analyses on the characteristics of transmission, reflection, and energy conservation, as well as the effects of window presence/absence on light propagation. However, apart from the optical properties, i.e. transmission, reflection, and refraction, there also exist some microwave properties that have significant influences under certain operating conditions, such as the formula of thin-film microwave shielding effect derived by Klein.

There are several major parameters for the evaluation of optical properties of windows, as shown below.

Signal-to-noise (SNR) ratio:

$$\mathrm{SNR} = \int_0^\infty \frac{H_\lambda \,\mathrm{d}\lambda}{\mathrm{NEI}(\lambda)}, \tag{2.9}$$

$$\mathrm{NEI}(\lambda) = \frac{\mathrm{NEP}(\lambda, f_0, 1)\sqrt{(\Delta f)_\mathrm{n}}}{A_\mathrm{c}\eta_0\eta_\mathrm{e}}, \tag{2.10}$$

Tab. 2.3: Summary of computational formulas of strength-related parameters.

Approach	Results	Notes
Modified Weibull model	$P(\sigma) = 1 - \exp\left\{-\left(\frac{S}{s}\right)\left[\Gamma\left(1+\frac{1}{m}\right)\right]^m\left(\frac{\sigma}{\sigma_c}\right)^m\right\}$, where $P(\sigma)$ is the cumulative failure probability of the material, m the Weibull coefficient, S/s the ratio of effective area to the total area, $\Gamma(z)$ the Gamma factional function, σ the peak stress, and σ_c the effective stress, which is dependent on stress distribution.	The action of stress is considered in the computational formula of statistical probability for window failures.
Spatial axisymmetric finite-element model	$\{\sigma\} = [D]_e\{\varepsilon\}$, $[D]_e = \frac{E}{(1+\mu)(1-2\mu)}\begin{bmatrix} 1-\mu & \mu & \mu & 0 \\ \mu & 1-\mu & \mu & 0 \\ \mu & \mu & 1-\mu & 0 \\ 0 & 0 & 0 & (1-2\mu)/2 \end{bmatrix}$, where $\{\sigma\}$ and $\{\varepsilon\}$ are the compression and tension, respectively; $[D]_e$ the elasticity matrix; E the elastic modulus; and μ the Poisson's ratio.	Based on deformation, displacement, and load of the material at the moment of $t + \Delta t$, the stress can be solved through iteration with the two equations.
Semiempirical formula	$h_{st} \approx 1.2c_p\sqrt{\frac{\rho_\infty \mu_\infty a_\infty}{R}}Ma_\infty\left[1+\left(\frac{\gamma-1}{2}\right)Ma_\infty^2\right]^{0.1}$, where h_{st} is the heat-transfer coefficient; c_p the specific heat at constant pressure; ρ_∞, μ_∞, and a_∞ the height-related density, viscosity, and local velocity of sound, respectively; γ the specific heat ratio, and Ma_∞ the Mach number.	Using the empirical formula along with the classic formula of thermal shock, the curve of the maximum allowable Mach number against the flying height of the missile can be obtained.

Tab. 2.3: (continued)

Approach	Results	Notes
Ruark–Young estimation method	$\sigma_{\max} = \frac{3F}{2\pi t^2}\left[(1+v)\ln\left(\frac{r_1}{r_0}\right) + \frac{1-v}{2}\left(1 - \frac{r_0^2}{r_1^2}\right)\right],$ where F is the acting force, t the thickness of sample, v the Poisson's ratio, r_1 the radius of the supporting ring, and r_0 the radius of the loading ring.	CVC-diamond is taken to calculate the maximum allowable stress.
Thin-film approximation model	$\sigma_s(z) = \frac{6}{t_s^2}\left(\frac{t_s}{3} - z\right)\sum_{l=1}^{n}\sigma_l t_l, \quad \sigma_s\|_{\text{inter}} = -\frac{4}{t_s}\sum_{l=1}^{n}\sigma_l t_l,$ where $\sigma_s(z)$ is the z base stress, z the distance between the current position and the ground, t_s the thickness of the base, t_l the thickness of the film layer, σ_l the macro stress, and $\sigma_s\|_{\text{inter}}$ the peak stress at the interface.	Round window, without edge effects. Computational equations of the variation and peak value of the normal stress along the thickness direction are obtained.
	$\sigma_{s,\rho}(r)\|_{\text{inter}} \simeq \sigma_s\|_{\text{inter}}[1 - \chi(r)],$ $\sigma_{s,\theta}(r)\|_{\text{inter}} \simeq \sigma_s\|_{\text{inter}}[1 - v_s\chi(r)], \quad \chi(r) = \frac{I_0(kr)}{I_0(kr_0)},$ where $\sigma_{s,\theta}$ and $(r)\|_{\text{inter}}$ are the peak stresses of the radial and azimuthal interfaces; $\sigma_s\|_{\text{inter}}$ the peak stress of the interface; k and v_s the elastic modulus and Poisson's ratio of the base material, respectively; r_0 the radius of the edge; and $I_0(z)$ the modified Bessel function of the first kind.	Round window, with no edge effects. Computational equation of the stress in the whole region is achieved.
Generalized Gaussian approximation model	$\sigma_\theta(1, 1/2, t) = \sigma_\theta^{(p)}(1, 1/2) + \overline{\sigma_\theta}(1, t) + \Delta\sigma(1, \pm 1/2, t),$ where σ_θ is the stress of azimuthal angle change, t the time, $\sigma_\theta^{(p)}$ the principle stress induced by compression that changes with the azimuthal angle, $\overline{\sigma_\theta}$ the principle stress induced by incident light that changes with the azimuthal angle, and $\Delta\sigma$ the compression increment.	By considering the effects of compression- and incident-light-induced stresses on the total stress, the computational equation of the low-pressure boundary is obtained.

where SNR is the signal-to-noise ratio, H_λ the spectral peak of spontaneous radiation, λ the wavelength, NEI(λ) the radiation strength corresponding to the wavelength λ of monochromatic light, NEP(λ, f_0, 1) the noise-equivalent energy value received by the detector, f_0 the sampling frequency, $(\Delta f)_n$ the bandwidth of electronic noise, A_c the area of collecting region, η_0 the effectiveness coefficient of light, and η_e the coefficient of electrical response.

The photon noise coefficient is

$$\sqrt{\frac{\sum_i \bar{\phi}_i}{\bar{\phi}_0}} = \frac{1}{\sqrt{2}}\left[(1-\bar{E}_W)\frac{Q(T_B)}{Q(T_O)} + \bar{E}_W\frac{Q(T_B)}{Q(T_O)} + \frac{Q(T_S)}{Q(T_O)}\right]^{1/2}, \tag{2.11}$$

where $\sqrt{\sum_i \bar{\phi}_i/\bar{\phi}_0}$ denotes the photon noise coefficient, $\bar{E}_W$ the emissivity of radiation,

$$Q(T) = \int_{\lambda_2}^{\lambda_1} Q_\lambda(T)\,d\lambda,$$

and $Q_\lambda(T)$ the black-body background radiation.

For infrared windows, functional indexes also include the figure of merit of optical distortion:

$$X = \frac{dn}{dT} + (1+\upsilon)\alpha(n-1) + 0.5n^3\alpha\langle\pi\rangle, \tag{2.12}$$

where n is the refractive index, T the Kelvin temperature, υ the Poisson's ratio, $\langle\pi\rangle$ the optical coefficient of average stress, and α the coefficient of thermal expansion.

In terms of the optical property theory of materials in the working environment, Yaping Zhang et al. analyzed the relationships among the material refractive index, temperature field, and stress field in the aerodynamic heating environment, where the aircraft flew within the atmosphere. Subsequently, the relational expression of the effects of temperature and stress on the refractive index of materials was obtained, providing certain guiding significance for applications.

2.2.3.3 Stress theory of multilayer film windows

In order to better adapt to the working environment in terms of function or strength, most windows have more than one functional-film layer. Therefore, the theory of multilayer-film windows has naturally emerged.

The theory of multilayer film windows is established by considering the mechanical and thermodynamic interactions between layers based on the theory of single-layer film windows. The calculation equation of compressive stress is

$$\sigma_i(z) = E_i'\left[-\ln(l_i) + \frac{\sum_{j=1}^{n} E_i' t_j \ln(l_i)}{\sum_{j=1}^{n} E_i' t_j} + (z_N - z)K\right], \tag{2.13}$$

where $\sigma_i(z)$ denotes the compressive stress changing with thickness, E_i' the biaxial modulus, t_j the thickness of the film layer, l_i the characteristic length of the film layer that is in the completely loose state, z_N the neutral plane, and K the curvature.

Klein discussed the computational equations of the interfacial stress between layers of a single-layer film and multilayer film. From the analysis, it was concluded that for the interfaces of multilayer films, the normal stress reflects the linear sum of stresses of all the layers. Therefore, in multilayer-film calculations, the effects of all the related layers must be considered.

In design processes, researchers adopted various simplified methods to carry out the calculation and successfully gained solutions while ensuring a certain level of precision. For example, using different approximation approaches to derive the strength-related computational formulas of multilayer films is a common way.

2.2.4 Ground failure tests of infrared window

2.2.4.1 Structural failure tests

When a missile flies in air at a supersonic/hypersonic speed, the Mach number is generally 1–6. Air in front of the missile is intensively compressed. Severe friction between the missile body and the surrounding air causes most of the kinetic energy of air to be converted to thermal energy, making the air temperature increase drastically. The huge temperature gradient on the surface of the missile causes the heat to be rapidly transmitted to the missile, which promotes temperature increase at the missile surface. Moreover, the sharp temperature increase can instantly lead to a huge temperature rise inside the infrared window, resulting in thermal shock.

Thermal shock has attracted much attention as a major factor causing structural failures. Researchers around the world have conducted testing simulations under different conditions. By summarizing pertinent literature, it is known that tests for thermal shock are mostly implemented in wind tunnels or directly on practical flying missiles to acquire the window temperature, gas pressure, and thermal flux for further investigating the change of material performance.

The American aerospace manufacturer McDonnell Douglas (Boeing) carried out a survivability investigation on the infrared window of a high-speed missile flying within the atmosphere. Window-failure issues in the aerodynamic heating environment were explored from two perspectives, i.e. theoretical prediction and wind tunnel test. The window-failure issue under window cooling in the two crucial periods, i.e. the moment of dome separation and after dome separation, was studied, and numerous statistical trends were observed. In recent years, in order to build a theoretical model of infrared window failures, the US implemented a large number of wind tunnel experiments on the aerodynamic thermal failure of infrared windows. As reported, tests of aerodynamic thermal failures and cooling issues of infrared windows have been conducted in multiple types of wind tunnels, including the 60-MW arc-heated wind tunnel at NASA Ames, the VKFB and VKFC supersonic wind tunnels of AEDC, Wind Tunnel 9 of NSWC, and the large-energy national shock tunnels (LENS), acquiring a great abundance of test data. However, data for real flight have barely been published.

Russell et al. conducted tests on of sapphire, ALON, dynasil, and grafoil under three different working conditions including baseline flight, practical testing flight, and T-range flight. Curves of recovery temperature, edge pressure, and cold-wall heat flux of window materials over time under the three working conditions were achieved, and response speeds of temperature in different conditions were compared. Hingst et al. carried out tests on the temperature variation of an infrared window at Mach 2.8–3.5 at a height of 1000 m above the sea level by using sapphire, Al_2O_3, and zinc sulphide.

In terms of strain analysis, Klein et al. gathered and analyzed the bending-strength data of infrared transmitting materials from four mechanical testing devices during a certain period, discovering that fracture always occurred in a certain ring located in the middle section of stretching surfaces, which subsequently propagated outwards. Based on the assumption that fracture is derived from randomly distributed surface detects, Weber's two-parameter model was established to describe the test failure probability. Infrared materials, such as sapphire, diamond, and zinc sulphide, were investigated, and all the tested materials except for oxyfluoride glass were shown to be well predicted by the model. Yubin Li and Cunbing Huang tested the mechanical properties of materials prepared under certain environments, including fracture toughness, fracture strength, relative density, microhardness, and compressive strength.

2.2.4.2 Functional failure tests

The primary function of the missile infrared window is to ensure that the infrared signal within a certain wavelength range can pass through the window with minimum interference so that detectors can identify the signal and satisfy precision guidance. Therefore, the change of optical properties of the window is a dominant factor causing functional failures. Researchers worldwide have implemented numerous tests on the optical properties of windows, as summarized in Tab. 2.4.

With the existing technologies, there are three main approaches for measuring the optical properties of materials, i.e., MRF, MR JET, and SSI. For further investigation of the functional characteristics of materials, the development and applications of new techniques are necessary.

2.2.4.3 Comprehensive tests

Comprehensive tests are the ones conducted on window materials in a practical working environment or under conditions close to the practical conditions. This type of

Tab. 2.4: Summary of tests on the optical properties of window materials.

Material	Example	Results	Pertinent literature	Notes
Sapphire	Wavelength change Temperature change	Refractive index Estimated thermo-optic coefficient		Base property test
Sapphire	Sand and rain erosion ARM disposal & nondisposal	NIR transmission Laser-induced damage threshold (LiDT) test		Environment durability test
Sapphire	Interference conditions	Refractive index	The Johns Hopkins University Applied Physics Laboratory	Search for temperature-affected optical properties
Zinc sulphide	Wavelength change Temperature change	Refractive index Elasto-optic coefficient		
Zinc sulphide	296 K, 400 K, 506 K, 600 K, and 700 K	Refractive index Thermo-optic coefficient Relative stress – optical coefficient		Search for temperature-affected optical properties
Diamond – zinc sulphide	Fourier transformation infrared spectroscopy	Transmission & reflection		
AR-coated zinc sulphide base	NIR test, MIR test	Transmission		Transmission test after rain erosion
Zinc sulphide & zinc selenide	Forward looking infrared (FLIR) window	Correlation between reflection and transmission		Investigation of the relationship between optical coefficients and damage
Diamond	Sand and rain erosion ARM disposal & nondisposal	Longwave infrared (LWIR) transmission LiDT test		Environment durability test

Tab. 2.4: (continued)

Material	Example	Results	Pertinent literature	Notes
Diamond	Wavelength change, temperature change	Transmission & absorption coefficient		
Monocrystal diamond	UV-visible spectroscopy	Absorption coefficient & temperature		Absorption coefficient test under different conditions
CVD diamond	Wavelength change, various CVD diamonds	Transmission lost	Data from Texas Instruments & Norton™	Development of optical diamond and dome
Polycrystalline neodymium	Fourier transformation infrared spectroscopy	Transmission		YAG test on various degrees of aluminized coating
Carbon-coated germanium	Infrared search & track (IRST) window	Correlation between reflection and transmission		Investigation of the relationship between optical coefficients and damage
ALON	Sand and rain erosion ARM disposal & nondisposal	LiDT test LiDT test		Environment durability test
Spinel	Wavelength change	Refractive index, transmission		Spinel is fabricated by Materials & Electrochemical Research (MER) Corporation
Yttrium (III) oxide	Wavelength change	Refractive index		Yttrium (III) oxide is manufactured with different methods
Tantalum pentoxide	X-ray diffraction pattern			
Infrared fluorite crystal 80, infrared fluorite crystal 85	Wavelength change, 7–15 μm	Transmission & reflection, radiation curve, emissivity		
Irtran-1	Thickness of 1.0 mm, 3.8 mm, and 7.6 mm at 300–700 °C	Emissivity		Test on new material Irtran

tests can provide direct reference for the practical working performance of the material; therefore, such tests are popular among researchers in this field.

To determine the causes of window failure, Osborne et al. analyzed the samples of 68 damaged infrared windows, classified the depredations of the window surfaces, and measured the surface roughness of the window materials. Their conclusions were as follows: the degree of window damage was associated with the pressure in the assemblage, whether gaskets were assembled, and the window structure. The critical factor causing window damage was stress; more specifically, the principle cause of window damage was the residual stress occurring during installation and the gradually increased stress occurring during the working process. Hui Gong et al. studied the surface-damage characteristics of several infrared window materials using a continuous CO_2 laser, and they indicated that the failure mechanism could be described as the thermal shock stress damage caused by impurity/defect absorption. A damage model was also built for the analysis of the failure mode of thermal shock stress.

To study the damage of window materials in the practical working environment, Hobbs investigated the specific damage phenomena of sapphire, ALON, and diamond in water and sand, observed the surface abrasion using an electron microscope, and detected functional measurement by adopting optical devices. Joseph et al. implemented the rain erosion test on the functional film of the base. Considering the rain in the practical working environment, Klocek proposed a window structure with gallium phosphide as the major material, which could well resist rain erosion in real applications. Xiao Han, Jiaqi Zhu, and coworkers discussed the erosion/abrasion behaviors of high-speed raindrops on the infrared transmitting window/dome, stress-wave propagation patterns of a special surface load, and categories and mechanical mechanism of damage cracks. Influencing factors, evaluation approach, and indexes of rain erosion damage were introduced, and a direction of current research work for impact wear behaviors of high-speed raindrop on the infrared transmitting window was indicated. Zhigao Ju et al. introduced the phenomenon, mechanism, and antioxidant measures of CVD diamond oxidization at high temperatures under the action of aerodynamic heating for a missile in flight at a high speed.

Huang made appropriate improvements to test methods by analysing the existing measurement approaches, suggested a method without compensation to measure windows of special shapes, and designed a test system that could well satisfy the practical measurement requirement.

2.2.5 Numerical simulation of failure of infrared window

In order to verify the failure theory, a numerical simulation is indispensable. However, owing to the complexity of the working environment and thermal properties of window materials, the pertinent literature is not abundant. Moreover, most of the existing data are obtained from simulations in separate environments or window simulations instead of simulations as a whole.

2.2.5.1 Numerical simulation of working environment

To investigate the thermodynamic working environment of general missiles, Russel et al. conducted simulations on the recovery temperature, pressure, cold-wall heat flux, and response temperature under the baseline flight, testing flight, and T-Range flight of a missile by using the aerodynamic heating analysis code ATAC3D. Consequently, their variation trends were obtained over time to support the subsequent simulations. Stefan adopted the commercial software FLUENT 5TM of FLUENT Inc. to temporally and spatially solve the 2nd-order Navier–Stokes equations of finite volume, and used the Reynolds-averaged Navier–Stokes (RANS) method to solve the turbulence problem, when simulating parameters of airflow around the window. Based on the flow around the window, John et al. carried out grid generation using Genie++ and then conducted computational fluid dynamics (CFD) calculations using the general aerodynamic simulation program (GASP). Wan employed the equivalent cone method to simulate the aerodynamic/thermal working conditions around the missile. Jialing Le et al. studied the viscous effects of flow around a body at a hypersonic speed by using the 2D boundary layer approximation method and engineering approach, obtaining the frictional resistance and heat-transfer coefficient of the hypersonic aircraft surface. Gang Xia et al. successfully simulated the thermodynamic environment by using the finite volume approach based on Navier–Stokes equations for the flow field and combining the AUSM+ scheme and the temporal explicit iterative Runge–Kutta method. Chunsheng Huang et al. solved the supersonic flow field by adopting the high-precision finite-volume upwind TVD scheme, achieving particular parameters of the working environment. Qingmei Ren et al. employed the engineering algorithms to simulate the aerodynamic heating. In engineering algorithms, the aerodynamic heat flow exerted on typical aircraft parts is mainly calculated using the empirical computational formula of aerodynamic heating. These algorithms are simple and timesaving, suitable for the calculation of aerodynamic heating of a simple-shaped aircraft surface, but dependent on the support of an aerodynamic test database.

For other environments, Adler et al. defined and simulated the raindrop model of windows during working and built a model for the condensation of vapor. Both a hemispheric window and flat window were simulated, and the action of damage by droplets was analyzed.

2.2.5.2 Numerical simulation of window thermodynamic response

Researchers have used various approaches in simulations of infrared windows. Russell et al. simulated particular behaviors of the window by using the finite-element analysis software ANSYS using thermodynamic parameters obtained by ATAC3D. They also compared the simulation results with test results. Stefan obtained the CFD parameters by using FLUENT 5TM and achieved a comparison diagram of simulation and test temperature distribution through ANSYS. John et al. calculated the system-average turbulence effects as well as the nonuniformity and abnormality of mean flow by feeding the CFD parameters into the OPD code to acquire the pressure distribution, turbulence properties, and aerodynamic & optical properties. Gang Xia et al. coupled the flow field and structural region through the interface heat flow and temperature boundary condition, and implemented a simulation of the structural part of heat transfer by using the Galerkin finite-element method based on the 2D heat-conduction equation. Avsar et al. directly simulated the window failures by using finite-element software. They also obtained modified numerical results of the raindrop effect on the window by using the SAAB aerosystems rain erosion test device and determined the maximum allowable moving speed of window film layers before failure occurred. Dapei Tang et al. constructed a 3D axisymmetric computational model using finite-element calculation software to simulate the residual stress at the interface between the base and coating of the window with molybdenum as the base material and diamond as the coating material. Based on the distribution patterns of the residual stress component in each direction, the effects of the residual stress on the film failure were analyzed. For the reusable thermal protection structure of spacecraft, Deyi Yang et al. particularly investigated the thermal response characteristics of the leading-edge structure through numerical simulations. The influence of factors such as the structural scheme and material physical performance on the structure temperature was analyzed, and the results could serve as reference for further structure design.

In conventional flow field/heat/structure calculations, the flow field is usually calculated before the thermal response and thermal structure, which obviously decomposes the naturally uniform successive process and leads to inaccurate computational results. Tang Huang et al. suggested a numerical simulation approach based on the coupled flow field, heat, and structure, and gained superior results.

Thermodynamic results achieved from the simulations above roughly match the corresponding test results. Therefore, the approach of conducting simulations to obtain a reference framework before optimizing the window parameters has become the major window design method.

2.3 Analysis of typical aerodynamic heat/strength environment

When a supersonic aircraft flies in the atmosphere, intensive friction is generated between the incoming flow and aircraft surface. In the boundary layer, the lost kinetic energy is transformed into thermal energy. Owing to the severe compression and rapid temperature increase of air around the aircraft, heat is transferred from air to the aircraft surface, and intensive aerodynamic heating is produced. Therefore, in the research of material failure under the action of aerodynamic heating, the aerodynamic heating distribution of the external flow field of the supersonic aircraft should be precisely calculated first. This section will introduce the basic equations and turbulence governing equations of the external flow field of supersonic aircraft, analyse the external flow field of aircraft with a blunt cone and pointed cone under different flow conditions, and predict the temperature distribution, pressure distribution, surface heat flux, and convective heat-transfer coefficient.

2.3.1 Calculation model of supersonic aircraft outer flow field

Generally, in flow calculations, both the conservation of mass and conservation of momentum must be ensured. When heat transfer or compressible flow is involved, the conservation of energy must also be ensured. When turbulence is involved in the flow, the turbulence transport equation must be solved as well. Without considering chemical reactions or mixing effects, based on the compressibility and viscous properties of supersonic flow, in this section, we will solve the three basic governing equations in the Cartesian coordinate system – conservation of mass, conservation of momentum, and conservation of energy – and solve the turbulence equation by using the Reynolds averaging approach.

2.3.1.1 Basic governing equations

Conservation of mass

The equation of conservation of mass, also known as the continuity equation, expresses that the mass increment per unit volume per unit time is equal to the net mass flowing into this volume during the same period of time. The vector form of the equation is

$$\frac{\partial \rho}{\partial t} + \nabla \cdot (\rho \mathbf{v}) = 0, \tag{2.14}$$

where ρ is the density, t the time, and v the velocity vector.

For a 2D model, the conservation of mass can be expressed as

$$\frac{\partial \rho}{\partial t} + \frac{\partial}{\partial x}(\rho v_x) + \frac{\partial}{\partial y}(\rho v_y) = 0. \tag{2.15}$$

Conservation of momentum

The equation of conservation of momentum expresses that the rate of change of momentum with respect to time of the fluid per unit volume is equal to the sum of all the forces exerted on the unit volume. Neglecting body forces, its vector form is expressed as

$$\frac{\partial}{\partial t}(\rho \mathbf{v}) + \nabla \cdot (\rho \mathbf{v} \mathbf{v}) = -\nabla p + \nabla \cdot (\boldsymbol{\tau}), \tag{2.16}$$

where p denotes the pressure and $\boldsymbol{\tau}$ the stress tensor. $\boldsymbol{\tau}$ is expressed as

$$\boldsymbol{\tau} = \mu\left[(\nabla \mathbf{v} + \nabla \mathbf{v}^T) - \frac{2}{3}\nabla \cdot \mathbf{v}\boldsymbol{I}_\mu\right], \tag{2.17}$$

where μ denotes the viscosity coefficient and $\boldsymbol{I}_\mu$ the unit tensor.

For a 2D model, the axial and radial equations for conservation of momentum are

$$\frac{\partial}{\partial t}(\rho v_x) + \frac{\partial}{\partial x}(\rho v_x v_x) + \frac{\partial}{\partial y}(\rho v_y v_x) = -\frac{\partial p}{\partial x} + \frac{\partial \tau_{xx}}{\partial x} + \frac{\partial \tau_{yx}}{\partial y}, \tag{2.18}$$

$$\frac{\partial}{\partial t}(\rho v_y) + \frac{\partial}{\partial x}(\rho v_x v_y) + \frac{\partial}{\partial y}(\rho v_y v_y) = -\frac{\partial p}{\partial y} + \frac{\partial \tau_{xy}}{\partial x} + \frac{\partial \tau_{yy}}{\partial y}, \tag{2.19}$$

where

$$\begin{cases} \tau_{xx} = \mu\left(2\dfrac{\partial v_x}{\partial x} - \dfrac{2}{3}\left(\dfrac{\partial v_x}{\partial x} + \dfrac{\partial v_y}{\partial y}\right)\right), \\ \tau_{yy} = \mu\left(2\dfrac{\partial v_x}{\partial y} - \dfrac{2}{3}\left(\dfrac{\partial v_x}{\partial x} + \dfrac{\partial v_y}{\partial y}\right)\right), \\ \tau_{xy} = \tau_{yx} = \mu\left(\dfrac{\partial v_y}{\partial x} + \dfrac{\partial v_x}{\partial y}\right). \end{cases} \tag{2.20}$$

Conservation of energy

The equation of conservation of energy expresses that the rate of increase of energy per unit volume is equal to the sum of the net quantity of heat entering the unit volume and the work done by body forces on the volume. Neglecting the action of external forces, its vector form is

$$\frac{\partial(\rho e)}{\partial t} + \nabla \cdot (\rho \mathbf{v} e) + \nabla \cdot (p\mathbf{v}) = \nabla \cdot (\mathbf{v} \cdot \boldsymbol{\tau}) + \nabla \cdot (\lambda \nabla T) + S_e, \tag{2.21}$$

where e denotes the total energy per unit mass, T the temperature, λ the thermal conductivity of the fluid, and S_e the source term, e.g., thermal radiation.

For a 2D model, the energy equation can be expressed as

$$\begin{aligned} &\frac{\partial(\rho e)}{\partial t} + \frac{\partial(\rho e v_x)}{\partial x} + \frac{\partial(\rho e v_y)}{\partial y} + \frac{\partial(p v_x)}{\partial x} + \frac{\partial(p v_y)}{\partial y} \\ &= \frac{\partial}{\partial x}(v_x \tau_{xx} + v_y \tau_{xy}) + \frac{\partial}{\partial y}(v_x \tau_{yx} + v_y \tau_{yy}) + \frac{\partial}{\partial x}\left(\lambda\frac{\partial T}{\partial x}\right) + \frac{\partial}{\partial y}\left(\lambda\frac{\partial T}{\partial y}\right) + S_e. \end{aligned} \tag{2.22}$$

Equation of state

For a perfect gas, the equation of state is

$$p = \rho R_g T, \tag{2.23}$$

where R_g denotes a particular gas coefficient, $R_g = c_p - c_v$; $c_p/c_v = \gamma$; c_p is the specific heat at constant pressure, c_v is the specific heat at constant volume, and γ is the specific heat ratio. Therefore,

$$e = c_v T + \frac{1}{2}(v_x^2 + v_y^2). \tag{2.24}$$

2.3.1.2 Turbulence equations

There are three types of numerical simulations for turbulence: direct numerical simulation (DNS), large eddy simulation (LES), and the Reynolds-averaged Navier–Stokes (RANS) method. Among these, RANS is thus far the most extensively used method. In this book, the shear stress transport (SST) κ-ω model is adopted as the RANS model. It was developed by Menter on the basis of the κ-ω model. The standard κ-ω model is used in the near-wall boundary layer, and the standard κ-ε model is used in the shear flow region far from the wall. The model can accurately simulate both the low Reynolds number (*Re*) flow within the boundary layer as well as the full turbulent flow beyond the boundary layer. It is capable of not only dealing with a large range of inflow velocities, but also strongly simulating the flow separation induced by an adverse pressure gradient. The cross-diffusion term is integrated into this model, making it suitable for applications both near and far from the wall. The lateral dissipation derivative term is added, and the effects of turbulent shear stress is considered in the definition of eddy viscosity, which enables this model to be used in a variety of cases including the calculation of supersonic shock waves.

Similar to the standard κ-ω model, the turbulent kinetic energy κ and the dissipation rate ω in the SST κ-ω model can be expressed as

$$\frac{\partial}{\partial t}(\rho k) + \frac{\partial}{\partial x_i}(\rho k v_i) = \frac{\partial}{\partial x_i}\left(\Gamma_k \frac{\partial k}{\partial x_i}\right) + \tilde{G}_k - Y_k + S_k, \tag{2.25}$$

$$\frac{\partial}{\partial t}(\rho \omega) + \frac{\partial}{\partial x_i}(\rho \omega v_i) = \frac{\partial}{\partial x_i}\left(\Gamma_\omega \frac{\partial \omega}{\partial x_i}\right) + G_\omega - Y_\omega + D_\omega + S_\omega, \tag{2.26}$$

where Γ_κ and Γ_ω are the effective diffusivities of κ and ω, respectively, $\tilde{G}_\kappa$ is the turbulent kinetic energy caused by the laminar velocity gradient, G_ω is generated by ω, Y_κ and Y_ω denote the dissipation of κ and ω, respectively, D_ω represents the cross diffusion term, and S_κ and S_ω are source terms defined by the user. These variables can be written as

$$\Gamma_k = \mu + \frac{\mu_t}{\sigma_{t,k}}, \tag{2.27}$$

$$\Gamma_\omega = \mu + \frac{\mu_t}{\sigma_{t,\omega}}, \tag{2.28}$$

$$\tilde{G}_k = \min(G_k, 10\rho\beta^* k\omega), \tag{2.29}$$

$$G_\omega = \frac{\alpha}{v_t}\tilde{G}_k, \tag{2.30}$$

$$Y_k = \rho\beta^* k\omega, \tag{2.31}$$

$$Y_\omega = \rho\beta\omega^2, \tag{2.32}$$

$$D_\omega = 2(1 - F_1)\rho\sigma_{\omega,2}\frac{1}{\omega}\left(\frac{\partial k}{\partial x}\frac{\partial \omega}{\partial y} + \frac{\partial k}{\partial y}\frac{\partial \omega}{\partial x}\right), \tag{2.33}$$

where $\sigma_{t,\kappa}$ and $\sigma_{t,\omega}$ are the turbulent Prandtl numbers (Pr) of κ and ω, respectively, and μ_t is the turbulent viscosity, which can be calculated from the following formulas:

$$\mu_t = \frac{\rho k}{\omega} \frac{1}{\max\left[\frac{1}{\alpha^*}, \frac{S_\Omega F_2}{\alpha_1 \omega}\right]}, \tag{2.34}$$

$$\sigma_{t,k} = \frac{1}{F_1/\sigma_{k,1} + (1 - F_1)/\sigma_{k,2}}, \tag{2.35}$$

$$\sigma_{t,\omega} = \frac{1}{F_1/\sigma_{\omega,1} + (1 - F_1)/\sigma_{\omega,2}}, \tag{2.36}$$

where S_Ω represents the strain rate magnitude, α^* is the turbulent viscosity correction factor for low-Re flow, and F_1 and F_2 are the blending functions. They can be obtained from

$$\alpha^* = \frac{0.024 + Re_t/6}{1 + Re_t/6}, \tag{2.37}$$

$$Re_t = \frac{\rho k}{\mu \omega}, \tag{2.38}$$

$$F_1 = \tanh \phi_1^4, \tag{2.39}$$

$$\phi_1 = \min\left[\max\left(\frac{\sqrt{k}}{0.09\omega y}, \frac{500\mu}{\rho y^2 \omega}\right), \frac{4\rho k}{\sigma_{\omega,2} D_\omega^+ y^2}\right], \tag{2.40}$$

$$D_\omega^+ = \max\left[2\rho \frac{1}{\sigma_{\omega,2}} \frac{1}{\omega} \frac{\partial k}{\partial x_j} \frac{\partial \omega}{\partial x_j}, 10^{-10}\right], \tag{2.41}$$

$$F_2 = \tanh \phi_2^2, \tag{2.42}$$

$$\phi_2 = \max\left(2\frac{\sqrt{k}}{0.09\omega y}, \frac{500\mu}{\rho y^2 \omega}\right), \tag{2.43}$$

where y is the distance to the next surface.

The constants in the model are as follows:

$$\sigma_{k,1} = 1.176, \quad \sigma_{k,2} = 1.0, \quad \sigma_{\omega,1} = 2.0, \quad \sigma_{\omega,2} = 1.168, \quad \alpha_1 = 0.31.$$

2.3.1.3 Discrete scheme

Up to this point, the conservation of mass (2.14), conservation of momentum (2.16), conservation of energy (2.22), turbulence equation (2.25), and equation (2.26) can be expressed together in one equation:

$$\frac{\partial \boldsymbol{Q}_N}{\partial t} + \frac{\partial \boldsymbol{E}_N}{\partial x} + \frac{\partial \boldsymbol{F}_N}{\partial y} = \frac{\partial \boldsymbol{E}_N^v}{\partial x} + \frac{\partial \boldsymbol{F}_N^v}{\partial y} + \boldsymbol{S}, \tag{2.44}$$

where $\boldsymbol{Q}_N$ is a conservative variable; $\boldsymbol{E}_N$ and $\boldsymbol{F}_N$ denote the inviscid fluxes in the x and y directions, respectively, $\boldsymbol{E}_N^v$ and $\boldsymbol{F}_N^v$ denote the viscous fluxes in the x and y directions, respectively, and $\boldsymbol{S}$ is the source term:

$$\boldsymbol{Q} = \begin{bmatrix} \rho \\ \rho \upsilon_x \\ \rho \upsilon_y \\ \rho e \\ \rho k \\ \rho \omega \end{bmatrix}, \quad \boldsymbol{E}_{\mathrm{N}} = \begin{bmatrix} \rho \upsilon_x \\ \rho \upsilon_x \upsilon_x + p \\ \rho \upsilon_x \upsilon_y \\ \rho e \upsilon_x + p \upsilon_x \\ \rho k \upsilon_x \\ \rho \omega \upsilon_x \end{bmatrix}, \quad \boldsymbol{F}_{\mathrm{N}} = \begin{bmatrix} \rho \upsilon_y \\ \rho \upsilon_y \upsilon_x \\ \rho \upsilon_y \upsilon_y + p \\ \rho e \upsilon_y + p \upsilon_y \\ \rho k \upsilon_y \\ \rho \omega \upsilon_y \end{bmatrix},$$

$$\boldsymbol{E}_{\mathrm{N}}^{v} = \begin{bmatrix} 0 \\ \tau_{xx} \\ \tau_{xy} \\ \upsilon_x \tau_{xx} + \upsilon_y \tau_{xy} + \lambda \frac{\partial T}{\partial x} \\ \Gamma_k \frac{\partial k}{\partial x} \\ \Gamma_\omega \frac{\partial \omega}{\partial x} \end{bmatrix}, \quad \boldsymbol{F}_{\mathrm{N}}^{v} = \begin{bmatrix} 0 \\ \tau_{yx} \\ \tau_{yy} \\ \upsilon_x \tau_{yx} + \upsilon_y \tau_{yy} + \lambda \frac{\partial T}{\partial y} \\ \Gamma_k \frac{\partial k}{\partial y} \\ \Gamma_\omega \frac{\partial \omega}{\partial y} \end{bmatrix},$$

$$\boldsymbol{S} = \begin{bmatrix} 0 \\ 0 \\ 0 \\ S_e \\ \tilde{G}_k - Y_k + S_k \\ G_\omega - Y_\omega + D_\omega + S_\omega \end{bmatrix}.$$

Then, the finite-volume method is used to construct the discrete equation. First, the computational domain must be divided into a finite number of control volumes with a finite size. Then, the governing differential equation inside the control volume V can be integrated to obtain

$$\frac{\partial}{\partial t} \int_V \boldsymbol{Q}_{\mathrm{N}} \, \mathrm{d}V + \oint \boldsymbol{Q}_{\mathrm{N}} \cdot \boldsymbol{n} \, \mathrm{d}\boldsymbol{A} = \oint \boldsymbol{P}_{\mathrm{N}}^{v} \cdot \boldsymbol{n} \, \mathrm{d}\boldsymbol{A} + \int_V \boldsymbol{S} \, \mathrm{d}V, \tag{2.45}$$

where $\boldsymbol{P}_{\mathrm{N}} = \boldsymbol{E}_{\mathrm{N}} \boldsymbol{i} + \boldsymbol{F}_{\mathrm{N}} \boldsymbol{j}$ is the inviscid flux, $\boldsymbol{P}_{\mathrm{N}}^{v} = \boldsymbol{E}_{\mathrm{N}}^{v} \boldsymbol{i} + \boldsymbol{F}_{\mathrm{N}}^{v} \boldsymbol{j}$ is the viscous flux, $\boldsymbol{n}$ is the outer-pointing normal vector of the boundary, and $\boldsymbol{A}$ is the area vector of the surface.

Discrete scheme

Flux-difference splitting (FDS), proposed by Roe, is a type of Godunov method and has been proven to show excellent performance in the simulation of a stationary shock wave and contact discontinuity with the accurate capturing of the shock wave, high resolution of viscosity, and superior computational results for thermal flux. This book adopts the Roe scheme to conduct discretization.

By using (i, j) to represent the central point of the computing space unit $V_{i,j}$, the semidiscrete scheme of the unit is

$$\frac{\partial \boldsymbol{Q}_{\mathrm{N},i,j}}{\Delta t} + \frac{\boldsymbol{E}_{\mathrm{N},i+1/2,j} - \boldsymbol{E}_{\mathrm{N},i-1/2,j}}{\Delta x} + \frac{\boldsymbol{F}_{\mathrm{N},i+1/2,j} - \boldsymbol{F}_{\mathrm{N},i-1/2,j}}{\Delta y} = \frac{\boldsymbol{E}^{v}_{\mathrm{N},i+1/2,j} - \boldsymbol{E}^{v}_{\mathrm{N},i-1/2,j}}{\Delta x} + \frac{\boldsymbol{F}^{v}_{\mathrm{N},i+1/2,j} - \boldsymbol{F}^{v}_{\mathrm{N},i-1/2,j}}{\Delta y} + \boldsymbol{S}_{i,j}. \quad (2.46)$$

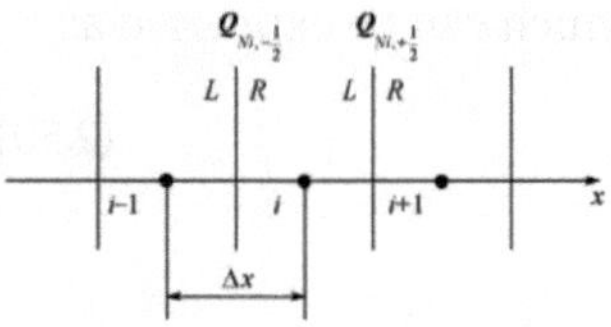

Fig. 2.5: Schematic of difference grids in the x direction.

Take the inviscid flux in the x direction as an example, as shown in Fig. 2.5. The Roe-FDS scheme can be expressed as

$$\boldsymbol{E}_{i+1/2} = \frac{1}{2}\left[\boldsymbol{E}(\boldsymbol{Q}^{\mathrm{L}}_{\mathrm{N}}) + \boldsymbol{E}(\boldsymbol{Q}^{\mathrm{R}}_{\mathrm{N}}) - |\tilde{\boldsymbol{A}}|(\boldsymbol{Q}^{\mathrm{R}}_{\mathrm{N}} - \boldsymbol{Q}^{\mathrm{L}}_{\mathrm{N}})\right]_{i+1/2}, \quad (2.47)$$

where superscripts L and R denote variables on the left and right sides of the unit boundary, respectively. Therefore, the net inviscid flux of the unit $V_{i,j}$ in the x direction is

$$(\delta_x \boldsymbol{E})_{i,j} = \frac{1}{2}\left[\boldsymbol{E}(\boldsymbol{Q}^{\mathrm{L}}_{\mathrm{N}}) + \boldsymbol{E}(\boldsymbol{Q}^{\mathrm{R}}_{\mathrm{N}}) - |\tilde{\boldsymbol{A}}|(\boldsymbol{Q}^{\mathrm{R}}_{\mathrm{N}} - \boldsymbol{Q}^{\mathrm{L}}_{\mathrm{N}})\right]_{i+1/2} - \frac{1}{2}\left[\boldsymbol{E}(\boldsymbol{Q}^{\mathrm{L}}_{\mathrm{N}}) + \boldsymbol{E}(\boldsymbol{Q}^{\mathrm{R}}_{\mathrm{N}}) - |\tilde{\boldsymbol{A}}|(\boldsymbol{Q}^{\mathrm{R}}_{\mathrm{N}} - \boldsymbol{Q}^{\mathrm{L}}_{\mathrm{N}})\right]_{i-1/2}, \quad (2.48)$$

where $|\tilde{\boldsymbol{A}}|$ is the Jacobian matrix of flux calculated using the Roe average. The detailed derivation is presented in the research work of Roe and other researchers.

Flux limiter

Variables on the left and right sides of the boundary, $\boldsymbol{Q}^{\mathrm{L}}_{\mathrm{N}}$ and $\boldsymbol{Q}^{\mathrm{R}}_{\mathrm{N}}$, are used in the scheme mentioned above. If $\boldsymbol{Q}^{\mathrm{L}}_{\mathrm{N}}$ and $\boldsymbol{Q}^{\mathrm{R}}_{\mathrm{N}}$ take the values of the control volume on both sides of the boundary, it will be the 1st-order upwind scheme, where numerical oscillation will not occur but the solution accuracy will be low, i.e. the scheme will not be able to capture the shock wave well. Therefore, the 2nd-order scheme must be employed to construct high-precision $\boldsymbol{Q}^{\mathrm{L}}_{\mathrm{N}}$ and $\boldsymbol{Q}^{\mathrm{R}}_{\mathrm{N}}$. However, the 2nd-order upwind scheme will generate numerical oscillation near the shock wave. In order to restrain this numerical oscillation in the solving process, a flux limiter is introduced.

In the 2nd-order upwind scheme, variables on the left and right sides of the boundary, $\boldsymbol{Q}^{\mathrm{L}}_{\mathrm{N},i+1/2}$ and $\boldsymbol{Q}^{\mathrm{R}}_{\mathrm{N},i+1/2}$, are written as

$$\begin{cases} \boldsymbol{Q}^{\mathrm{L}}_{\mathrm{N},i+1/2} = \boldsymbol{Q}_{\mathrm{N},i} + \left\{ \dfrac{\phi_{\mathrm{lim}}}{4} [(1+\phi_{\mathrm{lim}})\Delta_- + (1-\phi_{\mathrm{lim}})\Delta_+] \right\}_i , \\ \boldsymbol{Q}^{\mathrm{R}}_{\mathrm{N},i+1/2} = \boldsymbol{Q}_{\mathrm{N},i+1} + \left\{ \dfrac{\phi_{\mathrm{lim}}}{4} [(1+\phi_{\mathrm{lim}})\Delta_- + (1-\phi_{\mathrm{lim}})\Delta_+] \right\}_{i+1} . \end{cases} \tag{2.49}$$

Here, the increments are defined as $(\Delta_-)_i = \boldsymbol{Q}_{\mathrm{N},i} - \boldsymbol{Q}_{\mathrm{N},i-1}$ and $(\Delta_+)_i = \boldsymbol{Q}_{\mathrm{N},i+1} - \boldsymbol{Q}_{\mathrm{N},i}$, and ϕ_{lim} is the flux limiter. We use the minmod limiter, which can be expressed as

$$\phi_{\mathrm{lim}}(x) = \max\{0, \min(x, 1)\}. \tag{2.50}$$

Discrete scheme of viscous flux

The treatment of viscous flux is similar to the method we used for inviscid flux, with the major difference of the gradient term existing in $\boldsymbol{P}^{v}_{\mathrm{N}}$. The gradient value of the unit area f of the given control volume is discretized using Green–Gauss theorem:

$$\nabla \boldsymbol{Q}_{\mathrm{N}} = \frac{1}{V} \sum_f \overline{\boldsymbol{Q}_{\mathrm{N},f}} \boldsymbol{A}_f , \tag{2.51}$$

$$\overline{\boldsymbol{Q}_{\mathrm{N},f}} = \frac{1}{N_f} \sum_n^{N_f} \overline{\boldsymbol{Q}_{\mathrm{N},n}} , \tag{2.52}$$

where V is the volume, $\boldsymbol{A}_f$ is the area of face element f, and $\overline{\boldsymbol{Q}_{\mathrm{N},n}}$ is the nodal value.

Subsequent to obtaining the gradient of variables, the viscous flux can be solved using an approach similar to the one used for inviscid flux.

The governing equation is discretized using the 2nd-order upwind scheme based on the Roe method, and the minmod limiter is introduced. The algebraic equation derived from the discretization of the governing equation can be solved using the implicit Gauss-Seidel method. With the measures above, numerical dissipation in the calculation of supersonic flow can be effectively restrained to improve the accuracy of shock-wave simulation.

2.3.1.4 Boundary conditions

Wall boundary condition

For a supersonic aircraft with viscous flow around, the no-slip condition is adopted on the wall, arriving at

$$v_x = 0, \quad v_y = 0. \tag{2.53}$$

Another wall boundary condition is that the normal pressure gradient is zero:

$$\frac{\partial p}{\partial \boldsymbol{n}} = 0. \tag{2.54}$$

Far-field boundary condition

For supersonic flow, the far field boundary condition is used on both the inlet and outlet of inflow. Adopting the far-field boundary condition makes it possible to set the free boundary condition at infinity, which requires a sufficiently large distance between the boundary of the computational domain and the aircraft wall. In the simulation shown here, three airflow parameters, namely Mach number, aerostatic pressure, and temperature, are set.

Symmetric boundary condition

The computational model here has an axisymmetric structure. A 2D axisymmetric model is adopted for the calculation with the symmetric boundary condition set on the axis of symmetry. The speed perpendicular to the boundary is regarded as zero on the symmetric boundary, and other physical quantities have equal values inside and outside the boundary.

2.3.2 Verification of supersonic aircraft outer flow-field model

In order to study the deviation of test results achieved in various wind tunnels, the European Space Agency (ESA) used the Electre blunt cone as the standard model to carry out a flight test in 1971, obtaining highly significant data for aerothermodynamics. Data of heat flux in the re-entry stage was measured using thermal couples assembled on the model vehicle at a flight altitude of approximately 20–60 km and a flight speed of 3800–4200 m/s.

In this section, the 2D blunt cone model is selected for the calculation. The SST κ-ε turbulence model is adopted for the simulation conducted using the commercial software FLUENT® 14.0. The results will be compared with the data in the reference to evaluate the accuracy of this aerodynamic heating model.

The Electre blunt cone has a spherical-head radius of 0.175 m, a half-cone angle of 4.5°, and a length of 2 m. Its geometrical dimensions are shown in Fig. 2.6, and the mesh generation is illustrated in Fig. 2.7.

Test data at the time instant of 293 s are selected; the free inflow conditions are $\upsilon_\infty = 4230.0\,\text{m/s}$, $p_\infty = 53.0\,\text{Pa}$, and $T_\infty = 265.0\,\text{K}$ with a zero angle of attack, and an isothermal wall is adopted with a temperature $T_\omega = 343.0\,\text{K}$.

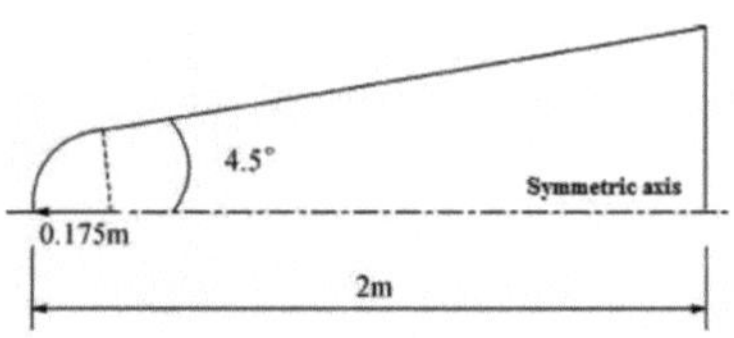

Fig. 2.6: Geometrical dimensions of the Electre blunt cone.

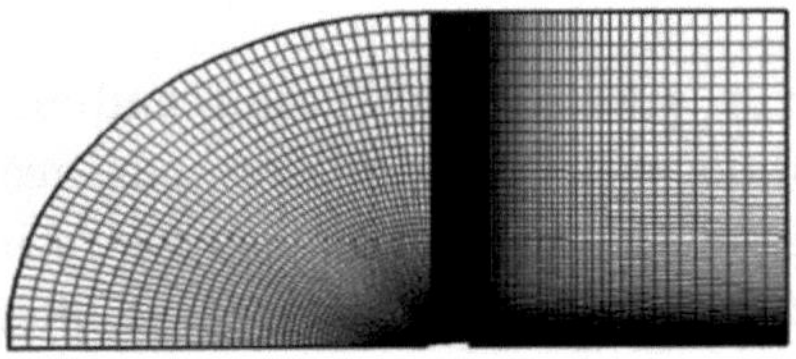

Fig. 2.7: Computational grids for the calculation of the outer flow field.

Three different grid densities are employed to verify the grid independence. Their $y+$ values are 5.16, 1.65, and 0.451. With different values of $y+$, a comparison between the heat-flow density on the wall of the blunt cone head obtained from numerical simulation and the measured results in the test of the reference are shown in Fig. 2.8; the temperatures on the line of stagnation points and the results of the reference are illustrated in Fig. 2.9. It can be seen that when $y+ = 0.451$, the computational results are in good agreement with the test results, verifying the accuracy of the aerodynamic heating model. The temperature distribution of the outer flow field is shown in Fig. 2.10, from which it can be observed that a bow shock wave is formed in front of the blunt cone.

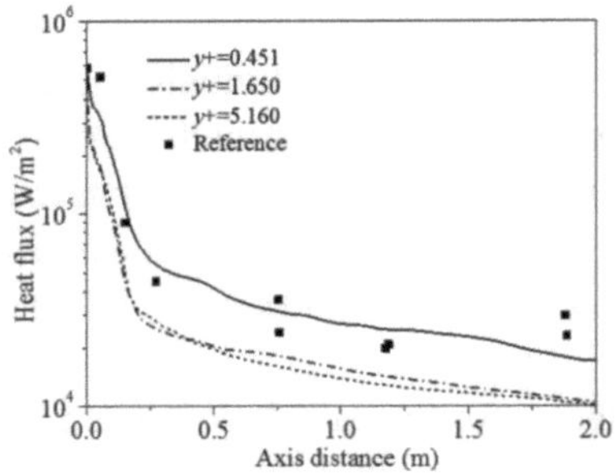

Fig. 2.8: Comparison of computational results of heat-flow density on the blunt cone head.

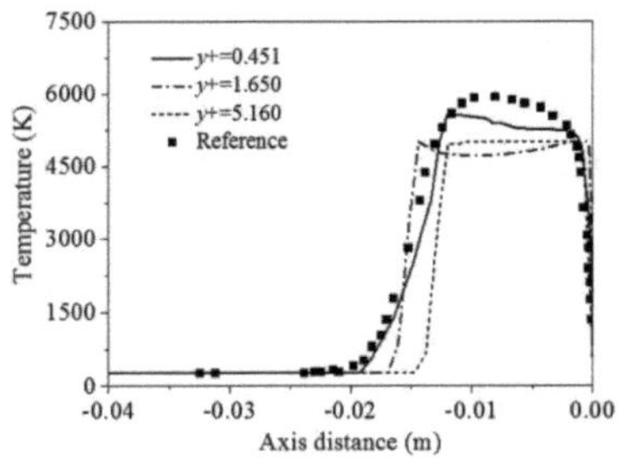

Fig. 2.9: Comparison of computational results of temperatures on the line of stagnation points.

Fig. 2.10: Aerodynamic heating distribution of the blunt cone in the flow field.

2.3.3 Calculation of blunted cone outer flow field

A typical supersonic aircraft geometry is considered for the calculation of its outer flow field. It is simplified as a 2D model in calculation owing to the axisymmetric structure, as shown in Fig. 2.11. The overall length of the aircraft is 0.65 m; the head employs the structure of a blunt cone with a half-cone angle of 5°; the radius of the optical dome is 0.06 m.

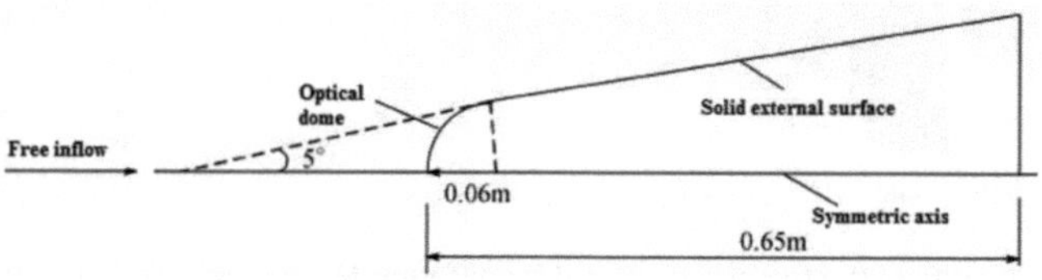

Fig. 2.11: Representative geometry of a typical supersonic aircraft.

In the flow-field calculation, standard atmosphere parameters at a flight height of 15 km are adopted. Parameters of the free inflow include $p_\infty = 12112.0$ Pa and $T_\infty = 216.8$ K. The outer flow-field distributions at Mach 3, 4, 5, and 6 are calculated using the commercial software FLUENT® 14.0.

Based on the computational model of the heating environment of the infrared window, the temperature distribution and pressure distribution of the outer flow field of the blunt cone are calculated, as shown in Fig. 2.12 and 2.13. It can be seen that when the aircraft flies at a supersonic velocity, a detached bow shock appears at the leading edge owing to the strong retardation effect of the blunt head exerted on the airflow. When the airflow traverses the shock wave, a dramatic variation of airflow parameters occurs before and after the shock. The pressure and temperature rapidly increase as the velocity of airflow behind the shock suddenly decreases. If the front end of the detached shock is perpendicular to the airflow direction, it is named a normal shock wave; if the outer surface of the detached shock forms an oblique angle with the airflow direction, it is named an oblique shock wave. Air is most intensively compressed when the airflow passes through the normal shock. Owing to the action of viscous resistance, almost all of the kinetic energy of incident airflow is transformed into thermal energy, making the temperature of the leading edge of the blunt cone the highest. The strength of the oblique shock wave is much weaker than that of the normal shock wave. Therefore, on the wall far away from the stagnation point of the blunt head, the temperature gradually declines, and the pressure also gradually decreases to the static pressure. Meanwhile, it can be seen from the comparison of temperature and pressure distributions for different Mach numbers that both the temperature and pressure of the flow field increase with increasing Mach number.

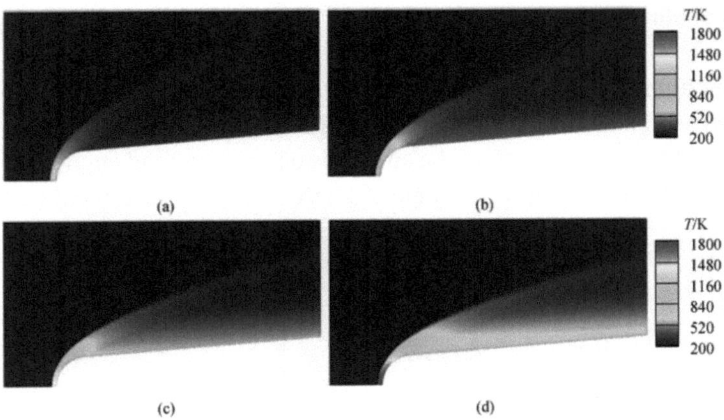

Fig. 2.12: Numerical simulation results of the temperature field for different Mach numbers: (a) $Ma = 3$, (b) $Ma = 4$, (c) $Ma = 5$, (d) $Ma = 6$.

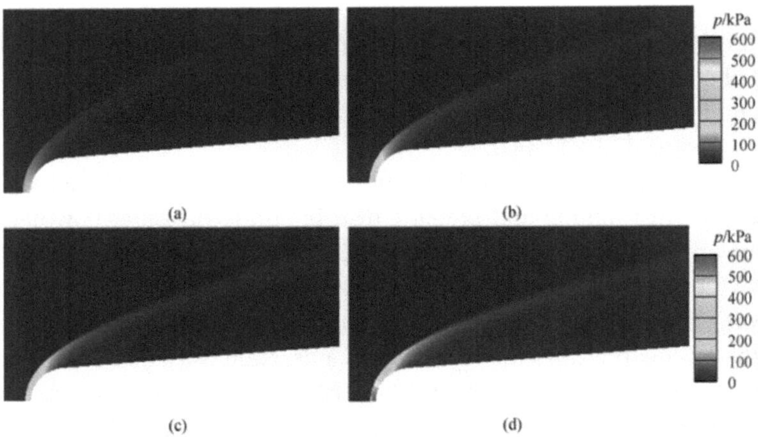

Fig. 2.13: Numerical simulation results of the pressure field for different Mach numbers: (a) $Ma = 3$, (b) $Ma = 4$, (c) $Ma = 5$, (d) $Ma = 6$.

For different Mach numbers, the ratio of the temperature at the $x = 0.325$ m section to the temperature of the far field, T/T_∞, is shown in Fig. 2.14; the ratio of the pressure of this section to the pressure of the far field, p/p_∞, is shown in Fig. 2.15. It can be observed that both the temperature and pressure change dramatically at the near wall, which is exactly the shock layer. When the incident airflow traverses the shock wave, a steady parameter change of the airflow occurs before and after the shock wave. Airflow parameters before the shock remain approximately constant, whereas the airflow speed after the shock rapidly decreases, leading to the sudden increase of both temperature and pressure. The comparison of the temperature ratio and pressure ratio for different Mach numbers shows that, with the increase of Mach number, the thinner

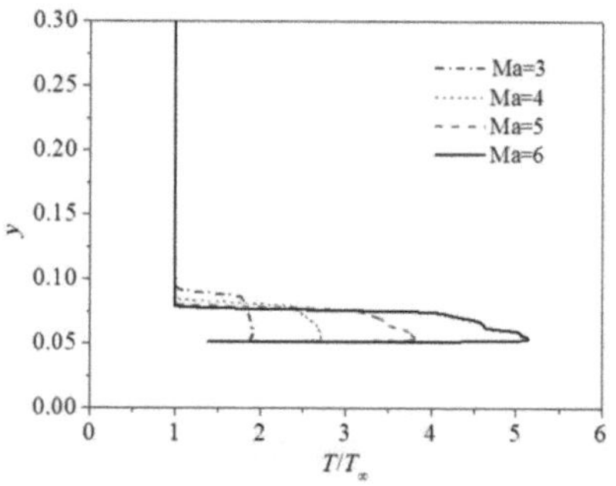

Fig. 2.14: Temperature distribution of the $x = 0.325$ m profile for different Mach numbers.

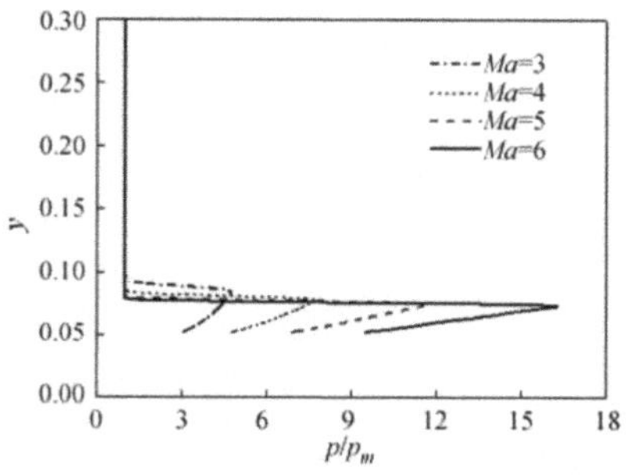

Fig. 2.15: Pressure distribution of the $x = 0.325$ m profile for different Mach numbers.

the shock wave, the more severe the parameter variation before and after the shock layer. This is because for the normal shock wave, when the specific heat ratio γ of the perfect gas is considered a constant, the strength of the shock wave can be expressed by the ratio of overall pressure before and after the shock,

$$\sigma_p = \frac{2\gamma Ma^2 - (\gamma - 1)}{\gamma + 1}.$$

Therefore, the ratios of airflow parameters before and after the shock are only associated with the Mach number of the inflow in front of the shock and the specific heat ratio of the air, γ. The higher the Mach number of the inflow, the greater is the strength of the shock wave, and the larger the difference of airflow parameters before and after the shock. The physical properties of oblique shock are entirely the same as those of normal shock, except for a weaker intensity.

The distributions of heat-flux density and convective heat-transfer coefficient along the symmetric axis for different Mach numbers are shown in Fig. 2.16 and 2.17, respectively. It can be readily observed that, under the same Mach number, the heat-flux density and convective heat-transfer coefficient peak at the stagnation point and dramatically drop after the stagnation point; these two parameters become nearly constant 0.06 m and 0.16 m downstream of the cone head, respectively. The underlying reason for this distribution can be explained as follows. The inflow air is strongly compressed at the head of the cone. The resultant detached shock wave leads to a great decrease in velocity. Owing to the viscous retardation, the kinetic energy of the

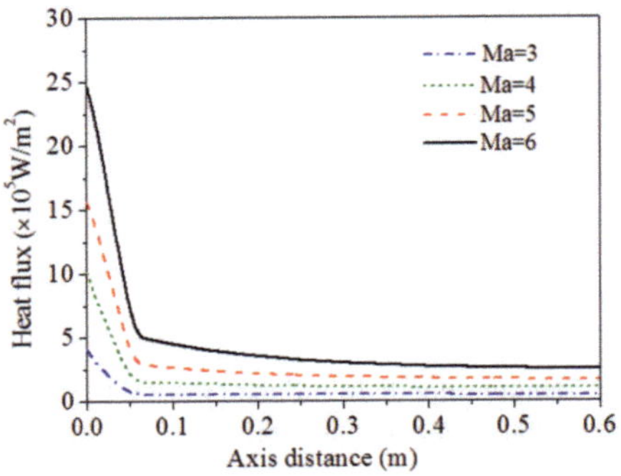

Fig. 2.16: Distribution of heat-flux density along the wall for different Mach numbers.

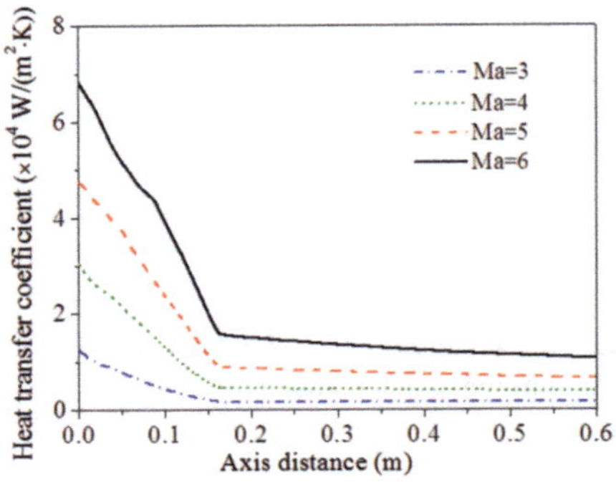

Fig. 2.17: Distribution of convective heat-transfer coefficient along the wall for different Mach numbers.

inflow will be completely converted to the internal energy, which accounts for the rather large values of heat-flux density and convective heat-transfer coefficient at the cone head. However, over the cone body, only a small part of the kinetic energy will be converted to internal energy. Hence, the heat-flux density and convective heat-transfer coefficient are significantly decreased at this part compared to the stagnation point. Moreover, in contrast to the heat-flux density, the variation of the convective heat-transfer coefficient shows a certain amount of delay, mainly because transfer of heat takes time. A comparison among different Mach numbers shows that a larger *Ma* corresponds to larger values of heat-flux density and convective heat-transfer coefficient, which leads to greater heat transfer to the aircraft. The optical dome, located at the blunt cone head of the aircraft, is strongly affected by aerodynamic heating and is susceptible to damage. Convective heat-transfer coefficients at the optical dome for different Mach numbers are listed in Tab. 2.5.

2.3.4 Calculation of the pointed cone outer flow field

In this section, another typical supersonic aircraft geometry is adopted to calculate the outer flow field. Considering the axisymmetric geometry of the aircraft, it is simplified into a 2D model in the calculation, as shown in Fig. 2.18. With a total length of 3 m, the

Tab. 2.5: Convective heat-transfer coefficient at the stagnation point.

Mach number	3	4	5	6
Convective heat-transfer coefficient (W/(m²·K))	12596.3946	30744.16211	47834.8633	68451.1484
Stagnation-point temperature (K)	605.2946	906.4111	1296.5503	1774.2997

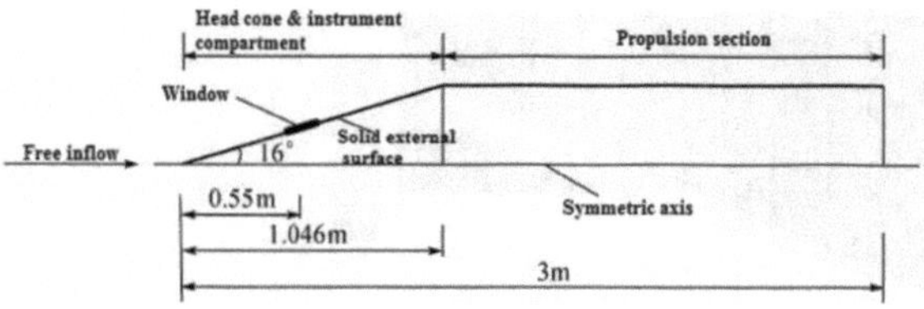

Fig. 2.18: Geometry of a typical supersonic aircraft.

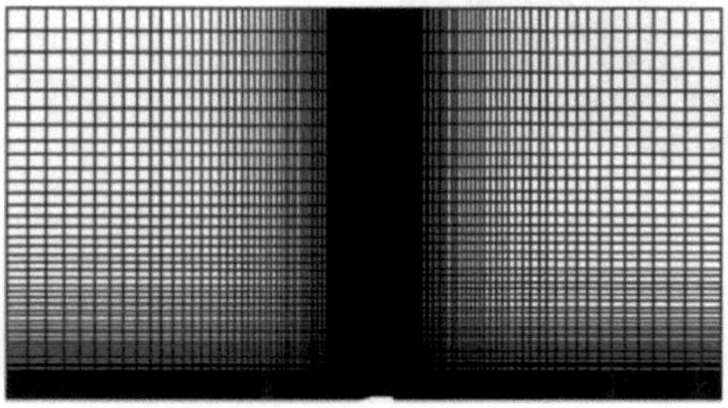

Fig. 2.19: Meshing of the computational domain of the flow field.

aircraft mainly contains two parts: the cone/instrument compartment (1.046 m) and the propulsion section (1.954 m). The head cone employs the structure of a pointed cone with a half-cone angle of 16°. The distance between the infrared window and the head-cone tip is 0.55 m. Fig. 2.19 shows the meshing.

The standard atmosphere parameters at a flight height of 15 km are adopted in the flow-field calculation. The parameters of the free inflow are $p_\infty = 12112.0$ Pa and $T_\infty = 216.8$ K. The commercial software FLUENT® 14.0 is used to calculate the outer flow-field distributions under Mach numbers of 3, 4, 5, and 6, respectively.

The temperature distribution and pressure distribution of the outer flow field of the pointed cone model are shown in Fig. 2.20 and 2.21, respectively. It can be observed that when the aircraft flies at a supersonic speed, an attached oblique shock is generated at the leading edge of the pointed cone. When the airflow traverses the shock wave, the speed suddenly decreases, leading to an abrupt increase of both the temperature and pressure. A comparison of the temperature and pressure distributions for different Mach numbers shows that both the temperature and pressure within the shock layer increase with the increase of the Mach number, while the thickness of the layer decreases. According to the oblique shock theory in aerodynamics, under a fixed airflow deflection angle (equal to the half-cone angle), the density increment after the shock rapidly increases with the increase of the inflow Mach number. Based on the conservation of mass, it can be demonstrated that a larger density after the shock requires a smaller area, which means that a smaller distance between the shock wave and the object surface will make the shock wave closer to the object surface. On approaching the object surface, the shock drives the compressed airflow to move downstream from the leading edge, causing the increase of pressure and temperature in this domain.

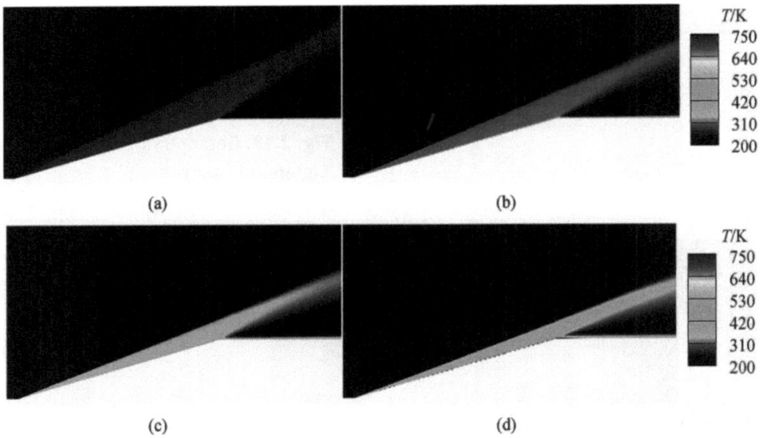

Fig. 2.20: Simulation results of the temperature field for different Mach numbers: (a) *Ma* = 3, (b) *Ma* = 4, (c) *Ma* = 5, (d) *Ma* = 6.

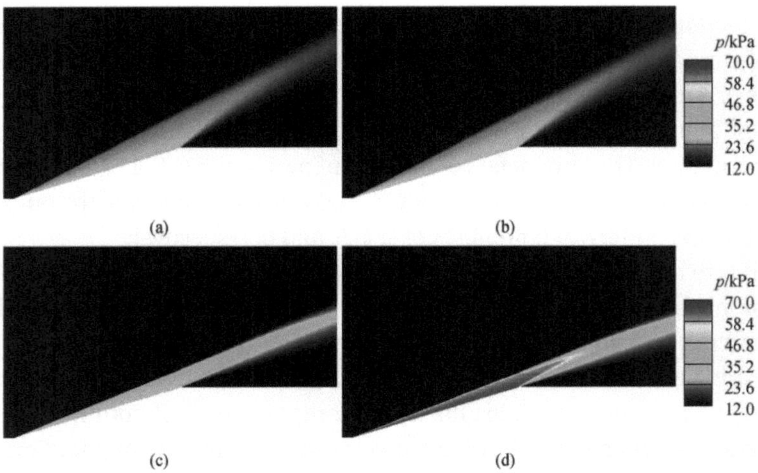

Fig. 2.21: Simulation results of the pressure field for different Mach numbers: (a) *Ma* = 3, (b) *Ma* = 4, (c) *Ma* = 5, (d) *Ma* = 6.

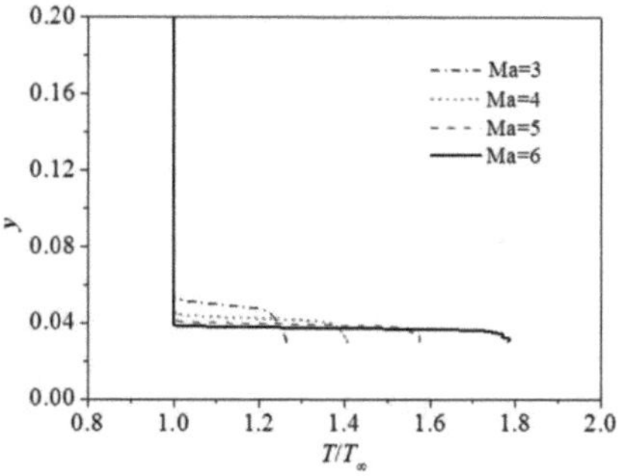

Fig. 2.22: Temperature distribution at the profile of $x = 0.1$ m for different Mach numbers.

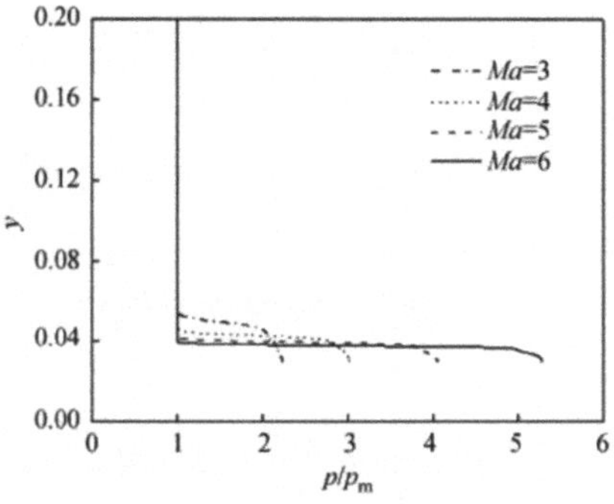

Fig. 2.23: Pressure distribution at the profile of $x = 0.1$ m for different Mach numbers.

At Mach 3, 4, 5, and 6, the ratio between the flow-field temperature at the profile of $x = 0.1$ m and the far-field temperature, T/T_∞ is shown in Fig. 2.22. The corresponding pressure ratio, p/p_∞, is shown in Fig. 2.23. It can be seen that both the temperature and pressure remarkably increase near the wall. In other words, the flow-field parameters experience drastic variations within the shock layer. The variation trends of temperature and pressure are similar. Moreover, a comparison of the results of different Mach numbers demonstrates that the shock-layer thickness decreases with increasing Mach number. The shock-wave angle (angle between the gas flow before the shock and the oblique shock), β_j, is only dependent on the Mach number before shock, Ma, and the gas-flow deflection angle, δ. From

$$\tan_{\delta_j} = \frac{Ma^2 \sin^2 \beta_j - 1}{\left[1 + Ma^2\left(\frac{\gamma+1}{2} - \sin^2 \beta_j\right)\right] \tan \beta_j},$$

at a given half-cone angle, an increase of the Mach number leads to a decrease of the shock-wave angle. Consequently, the shock layer will become rather thin, and the flow parameters inside and outside the shock layer will exhibit more intense variations.

The distributions of heat-flux density and convective heat-transfer coefficient along the symmetric axis under Mach numbers 3, 4, 5, and 6 are shown in Fig. 2.24 and 2.25, respectively. As shown in these two figures, under the same Mach number, the heat-flux density and convective heat-transfer coefficient peak at the tip of the pointed cone and dramatically decrease downstream of the tip. These two parameters

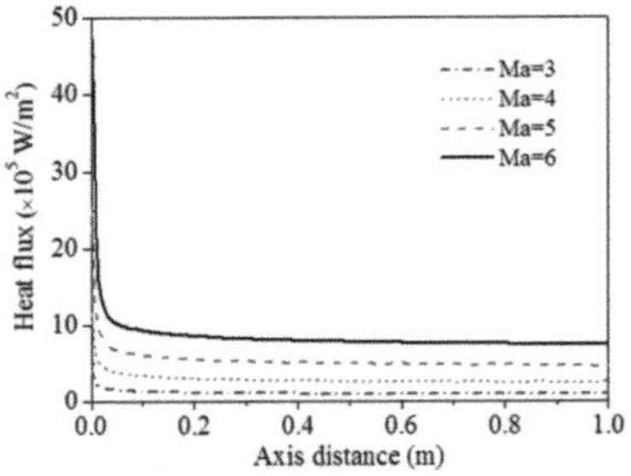

Fig. 2.24: Distribution of heat-flux density along the symmetric axis for different Mach numbers.

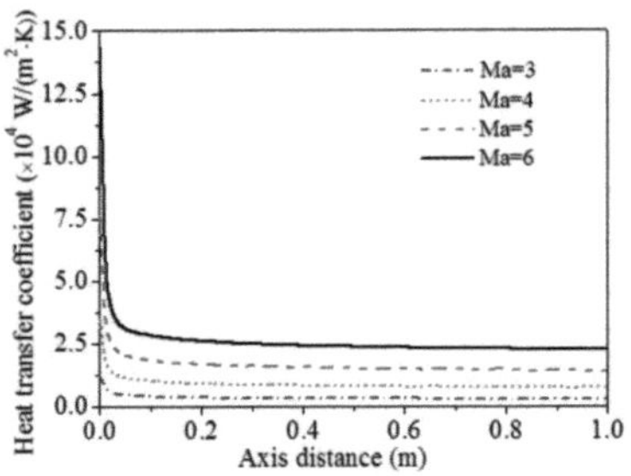

Fig. 2.25: Distribution of convective heat-transfer coefficient along the symmetric axis for different Mach numbers.

reach an almost constant state 0.2 m downstream of the cone tip. The underlying reason for this distribution can be explained as follows. The inflow air is strongly compressed at the cone tip. The generated shock leads to rapid pressure increase and velocity reduction. The kinetic energy of the inflow will be almost completely converted to internal energy, causing a dramatic temperature increase near the wall, which accounts for the rather large values of heat-flux density and convective heat-transfer coefficient at the cone tip. However, over the cone body, as the oblique shock extends, its intensity weakens. Thus, only a small part of kinetic energy will be converted to internal energy. Therefore, the heat-flux density and convective heat-transfer coefficient at this part are significantly less than those at the cone tip. By comparing the results for different Mach numbers, it can be observed that a larger *Ma* corresponds to larger values of heat-flux density and convective heat-transfer coefficient. The infrared window should be configured at the position 0.55 m downstream of the cone tip, as the heat-flux density and convective heat-transfer coefficient reach their minimum values and are fairly steady at this position. Such a configuration can minimize the possibility of damage to the infrared window. Convective heat-transfer coefficients at the optical dome for different Mach numbers are listed in Tab. 2.6.

Tab. 2.6: Convective heat-transfer coefficient at the position of the infrared window.

Mach number	**3**	**4**	**5**	**6**
Convective heat-transfer coefficient (W/(m^2·K))	3084.57	7940.36	14848.22	23440.18
Environment temperature (K)	361.78	458.74	605.65	815.18

In summary, this section introduced the basic governing equations and the turbulence governing equations of the 2-equation SST κ-ω model for the outer flow field of a supersonic aircraft. The blunt cone model was used to verify the validity of the computational model of the outer flow field of the supersonic aircraft. Computational results were close to the test results. Flow-field distributions of the blunt cone and pointed cone at a flight height of 15 km and Mach numbers of 3, 4, 5, and 6 were calculated to arrive at the following conclusions:

(1) The 2-equation SST κ-ω turbulence model can accurately simulate the flowing state of the boundary layer of the supersonic aircraft and achieve a high precision of calculation of heat-flux density and convective heat-transfer coefficient.
(2) As the computational results show, the temperature and heat flux of the aircraft head reach their peak values, dramatically decrease downstream of the head, and then remain steady afterwards. The position 0.55 m downstream of the pointed cone tip has the minimum heat-flux density and convective heat-transfer coefficient that are in the steady state. The infrared window can be protected from damage at this position.
(3) The computational results of the convective heat-transfer coefficient at the window positions of the blunt cone head and the pointed cone for different Mach numbers can be used as the input condition for calculating the structural failure of windows under the action of aerodynamic heating.

2.4 Rain erosion

Under practical working conditions, front-mounted protective components of aircraft detectors and guidance systems, including the dome and side window of an infrared-guided missile, window of an airborne forward-facing infrared pod, and infrared-detecting dome of the aircraft nose, are required for the high transmittance of the signal in the working waveband and superior comprehensive mechanical performance in terms of strength, hardness, and heat conductivity. With the increase of the aircraft velocity, high-speed impact from solid particles present in the harsh natural environment, such as sand and ice particles, as well as liquid particles, especially raindrops and floating droplets, can cause cracks, erosion, and even severe damage and deformation on the external surface of infrared windows/domes. Consequently, the infrared transmission and material strength are reduced, which will eventually affect the detecting and guiding performance. Based on the discussion above, it can be seen that the requirement of modern military infrared technology is the investigation of the rain erosion damage of the protective coating of the infrared window from the perspective of tribology in order to predict the ability to adapt to the high-speed flying environment.

2.4.1 Rain erosion and influential factors

The rain erosion of materials refers to the damage caused by high-speed raindrops through the hydrodynamic force acting on the solid material surface. It belongs to the phenomenon of erosive wear. Early research on this topic mainly focused on the abrasion of turbine blades of steam engines caused by liquid droplets. Since the 1940s and 1950s, the rain erosion of infrared windows has gradually attracted the attention of researchers worldwide. Field et al. studied the rain erosion of materials and their damage mechanisms by using the water-jet impact test, and they analyzed and described the stress-wave pattern of jet-flow impact by using high-speed schlieren photography. Adler et al. investigated the formation mechanism of rain erosion cracks on ZnS and optical glass through the whirling arm test, which provided a theoretical reference for the study of rain erosion of the infrared window/dome.

2.4.1.1 Rain erosion damage behavior

The impact process of high-speed water drops on the surface of rigid materials primarily involves two stages: liquid compression and jetting. In the liquid compression stage, since the expansion rate of the contact surface between the raindrop/jet-flow and material surface is less than the propagation rate of the shock wave inside the liquid, the shock wave is impeded by the contact surface and then forms the envelope surface. Liquid above the envelope surface is compressed to generate the "water hammer" pressure. Then, the shock wave escapes the boundary restraint and releases the pressure, leading to the sudden jetting of compressed liquid and the extension of the contact surface towards the free surface with the velocity of sound in water, as shown in Fig. 2.26.

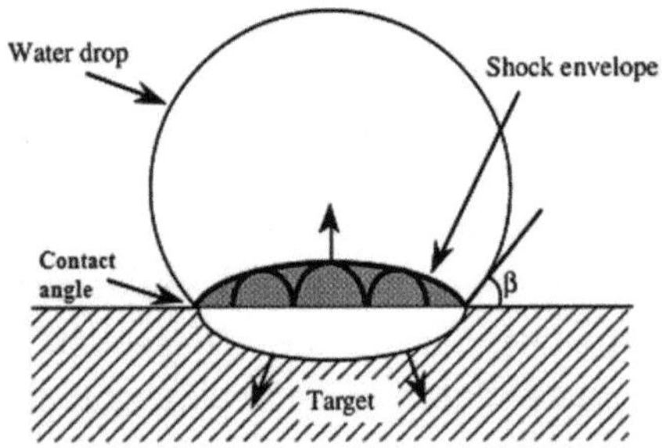

Fig. 2.26: Shock envelope of a water drop in the initial impact stage.

Haller et al. calculated and simulated the flow behavior of small-sized (200 μm) water drops impacting at a supersonic speed (500 m/s) by using the Euler's formula, and they validated the compressible flow model in the early stage of impact. Results obtained from the edge lateral jetting are in agreement with the computational prediction results.

When the impact speed is greater than the damage threshold, the water-hammer force will exceed the elastic limit of the material, resulting in the deformation and cracking of the surface. The water-hammer force perpendicular to the incident central axis is

$$P = \rho C v, \tag{2.55}$$

where ρ represents the density of water, C the propagation rate of the shock wave inside liquid, and v the impact velocity.

The instantaneous pressure at the contact interface is approximately $3\rho C v$. However, when the high pressure release is completed and the incompressible stage begins, the pressure at the central axis reduces to the Bernoulli stagnation pressure, $\rho v^2/2$. Di Zhang et al. simulated the action process of the water-hammer force by using a nonlinear wave model, and their results matched the analytical solution well. During the release process of the water-hammer pressure, three stress waves will be generated: a Rayleigh wave, longitudinal expansion wave, and transverse shear wave. During spreading, the expansion wave is manifested as compressive stress, causing no damage, whereas the Rayleigh wave and shear wave produce tensile stress, which leads to damage cracks. Particularly, the Rayleigh wave accounts for the majority of the energy (around 67 %) and barely decays during propagation. It interacts with the defects on the surface of a brittle material to create rings of cracks. Under single impact, the microcracks generated around the central axis are not continuous owing to the low duration of the Rayleigh wave. In contrast, under multiple impacts, the Rayleigh wave impulse can cause the incremental expansion of the cracks, forming a ring pattern of cracks. For relatively thin samples, the expansion wave with a fairly high speed is rebounded from the sample surface to the impact surface to generate tensile stress. In the same time, a stress-reinforced area is developed owing to the overlapping of the expansion wave and the Rayleigh wave. Thus, the damage ring is enlarged and exceeds the loading area. The diameter of the damage ring can be written as

$$y_1 = \frac{2x}{[(C_1/C_R)^2 - 1]^{1/2}}, \tag{2.56}$$

where x denotes the sample thickness, and C_1 and C_R represent the propagation speeds of the expansion wave and the Rayleigh wave, respectively.

Within the crack ring, no damage occurs owing to the presence of compressive stress. Wang et al. predicted the diameter of the undamaged central area within the damage ring and obtained theoretical results that were in agreement with the experimental data. Fig. 2.27 shows the circular pattern of cracks on the ZnS surface under jet impact. Bourne et al. determined that the propagation of stress waves could also cause subsurface crack rings. Adler et al. studied the impact damage of ZnS by a high-speed water drop and detected two types of subsurface cracks (Fig. 2.28). The type-I crack, caused by a shear wave, was located around the undamaged area of the impact centre ($d = 2a_c$) and had a small angle with respect to the impact axis. The type-II crack was associated with the formation and propagation of the Rayleigh wave. With

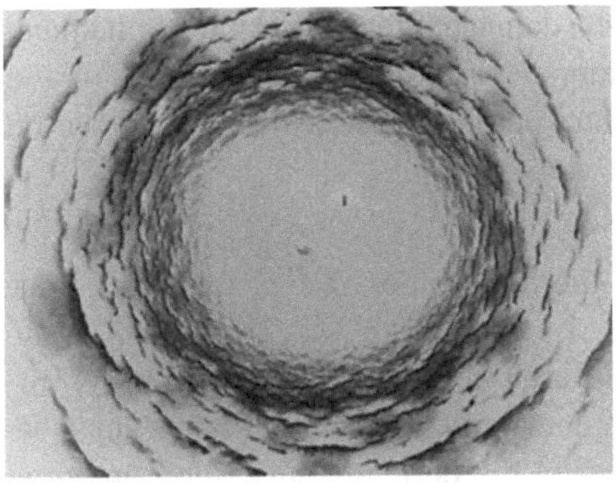

Fig. 2.27: Damage rings on the ZnS surface under multiple jet impacts at a speed of 300 m/s.

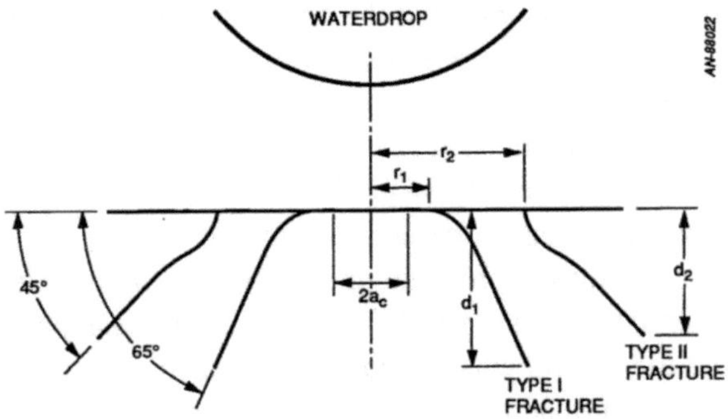

Fig. 2.28: Subsurface crack model under the vertical impact of water drops.

a fairly large radius, this type of crack was located near the contact interface and had a relatively large angle with respect to the central axis as well as a reversed curvature. However, the formation mechanisms of subsurface cracks have not been reported in the literature thus far. In a practical flight environment, the subsurface cracks tend to extend in an overlapping manner under large-scale raindrop impact, leading to the loss of optical transmission of the infrared window/dome.

Seward et al. discovered the damage mode of central cracks when studying a jet impacting brittle materials. The central cracks occur because during the successive impact process, the release of the stress inside the material causes axial tension. These cracks, formed before the ring cracks, can be connected into a web pattern and can affect the formation of ring cracks and their damage thresholds. However, this is only considered as a secondary damage pattern. Furthermore, if the impact site is sufficiently close to the sample edge, the Rayleigh surface wave and volume wave will be reflected on the edge of the sample to form a stress-reinforced area, leading to boundary-expanding cracks.

In summary, the rain erosion of materials is a special type of abrasion, which is associated with distinct damage mechanisms. Research in this area should start from liquid flow, stress-wave propagation, and crack formation mechanism in order to simulate the process of raindrop impact and damage for further prediction.

2.4.1.2 Factors influencing rain erosion

The impact velocity, incident angle, and geometrical dimension of raindrops or jet flow all influence the damage degree of materials. A higher impact velocity can promote a faster occurrence of ring cracks with a larger ring diameter. Under a low impact velocity, the damage degree increases with the decrease of the incident angle θ (angle between the impact direction and the surface normal), whereas the impact of a supersonic, obliquely incident water drop or jet flow forms the cutting flow that eventually leads to more severe damage. Small raindrops are scattered into vapor by the airflow around the front of the infrared window, causing no abrasion. However, the curvature radius of deformed raindrops under the aerodynamic action is greater than their initial diameter and is more capable of destruction.

The number and distribution of defects and the grain-boundary strength of the material surface affect the geometry and expansion rate of ring cracks. For example, because cracks are liable to extend along the cleavage plane, single-crystal Ge is more susceptible to damage cracks than monocrystal ZnS. Moreover, the dimensions of the sample also affect the damage threshold. For instance, a small diamond sample is more susceptible to damage with a lower damage threshold, inside which the stress wave exhibits a faster propagation rate and slower energy decay. In the high-speed flight process, the friction between the infrared window and air usually leads to a high temperature. As discussed by Deom et al., thermal shock generated by the temperature difference between raindrops and the window can aggravate the damage. Moreover, the surface microcracks generated by the washout from tiny sand dust are the origin of ring cracks, which will accelerate the crack extension and lead to a reduction of infrared transmission.

In conclusion, the impact damage to the infrared window materials caused by high-speed raindrops is rather complicated and affected by multiple factors. Various impact conditions and material surface structures can be analyzed using computer simulations. Adler simulated high-speed raindrops impacting an infrared window with a protective coating by using a 3D dynamic finite-element method. He analyzed the stress distribution and dynamic response of the material and predicted the deformation of raindrops during the impact process. His work has a guiding significance for the subsequent research on rain-erosion-resistant protective coatings for infrared windows.

2.4.2 Test method and facilities of rain erosion

Section 2.2.4.3 roughly introduced tests on rain erosion. Overall, the main existing methods for testing rain erosion include the rocket sled, whirling arm, water jet impact, ballistic hydrometeor, and nylon bead impact methods.

2.4.2.1 Rocket sled

The rocket sled test is a method in which a rocket propulsion unit is used to propel a sled loaded with the sample to traverse an artificial water-spraying zone along a ground rail at a high speed, and thereby simulate the process of a subsonic or supersonic aircraft traversing the rain field. Fig. 2.29 shows the rocket sled test equipment in the US Holloman Air Force Base. The rocket sled test can accurately simulate the practical flight environment. However, owing to the large equipment size and the difficulty in implementing the technique, this type of test is only suitable for testing and evaluating final products.

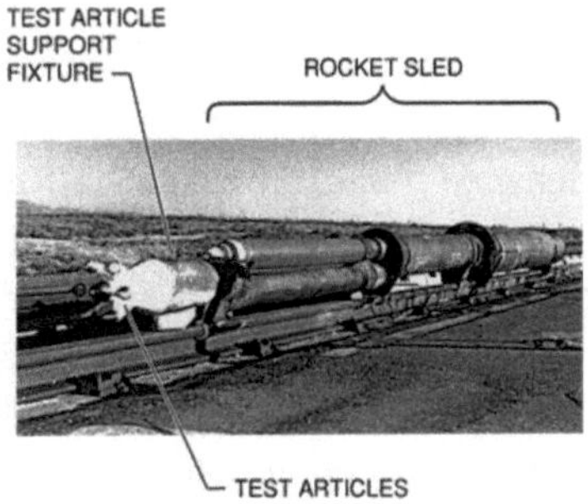

Fig. 2.29: Rail of the rocket sled.

2.4.2.2 Whirling-arm rain erosion

A whirling arm facility, equipped with simple structures and allowing for easy operations, provides an indoor simulation environment for research on material rain erosion. Fig. 2.30 shows a structural schematic of the whirling arm rain erosion test facility run by the University of Dayton Research Institute at the US Air Force Research Laboratory (AFRL). Specimens are fixed on both sides of the whirling arm in the closed chamber, where the velocity can reach up to 290 m/s (0.87 *Ma*). Above the whirling arm, some spray heads arranged in circles are used to simulate the rainfall environment. However, owing to factors including the disturbance airflow generated by the high-speed rotation, the design of the inside wall, head distribution, the spec-

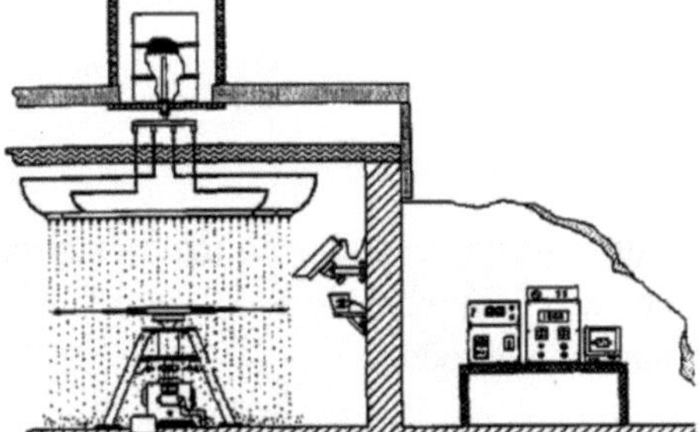

Fig. 2.30: Whirling-arm rain erosion test facility.

imen table of the closed chamber, and the length of the whirling arm, experimental results from different facilities are not in good agreement with each other. Therefore, the whirling arm is only appropriate for the performance evaluation of rain erosion under subsonic or transonic conditions.

2.4.2.3 Water jet impact

At the initial stage of impact, a water jet can generate a damage pressure similar to that of high-speed raindrops. Moreover, the jet flow has a relatively strong controllability, making it capable of simulating raindrop impact at different speeds with different dimensions. Cavendish Laboratory simulated the raindrop impact on infrared transmitting materials by using a water jet. They investigated the jet impact behavior and the distribution, propagation, and damage mechanisms of impact stress by using a high-speed camera to obtain rain-erosion data for multiple types of materials. There are two types of water-jet impact tests: single-impact jet and multiple-impact jet tests.

The single-impact jet facility drives an external metal disk to impact water in a nozzle by compressing the air, generating a water jet with a speed of 90–1000 m/s. The multiple-impact jet facility expels the water in the nozzle through the reciprocating motion of the nylon piston, forming two water-jet impacts with a speed of 30–600 m/s every minute. Fig. 2.31 shows the photograph of a 600 m/s jet impact captured by a high-speed camera. In the research of Seward et al., the test approach of water-jet impact was particularly investigated. Jet impact is able to effectively simulate the raindrop impact during supersonic flight. With reliable experimental data, this technique has been proven to be a practically simple method for research on the rain-erosion damage of infrared window/dome materials at the laboratory scale. However, it should be noted that a water jet and raindrop are not completely identical. Consequently, the calculation of the equivalent diameters under different impact velocities is needed.

Fig. 2.31: Single impact jet (nozzle diameter = 0.8 mm).

2.4.2.4 Hydrometeor impact

Neither the whirling-arm rain erosion nor water-jet impact test can generate spherical raindrops. However, the ballistic hydrometeor impact apparatus developed by GRC International is capable of simulating the impact of supersonic spherical or deformed water drops on the material; the working principle of this apparatus is based on the collision between the accelerated specimen and an individual water drop produced through a hypodermic needle in a pressurized helium atmosphere. This apparatus is able to create stable test conditions that can be easily characterized. However, due to the structural complexity and high operational costs of the equipment, it is not suitable for collecting data on a large scale.

2.4.2.5 Nylon bead impact

In 1987, Adler et al. developed a supersonic multiparticle impact apparatus that made use of nylon beads to simulate raindrop impact. The analysis of the impact stress waves and observation of the damage morphology demonstrated that, within a certain velocity range, the impact damage of nylon beads to some materials (especially ceramics) was highly similar to that caused by raindrops. In this study, nylon beads were also utilized to simulate the impact of supersonic raindrops on ZnS, where a vivid circular pattern of cracks was obtained. However, it was also found that material damage was related to multiple factors including the impact velocity of nylon beads and shape of the specimen in a complicated manner.

Based on the above discussion, we conclude that the whirling-arm and jet-impact tests can be considered as superior approaches for research on material erosion due to high-speed raindrops. However, the precision of such experimental equipment must be improved, and the impact loading conditions must be better stabilized.

2.4.3 Evaluation index of rain erosion resistance

The damage threshold velocity determined from the water-jet impact test is a major measure of the rain-erosion resistance of brittle materials. Damage threshold velocity, including mechanical threshold, single impact threshold, and absolute threshold, refers to the minimum impact velocity at which the microcracks on the material surface can grow into a ring pattern of cracks. A higher damage threshold velocity corresponds to better rain-erosion resistance of the material. Evans et al. obtained the following theoretical expression for the absolute threshold of brittle materials by using both analytical and numerical methods:

$$v_{\mathrm{ADT}}^3 = 1.4\frac{K_{\mathrm{Ic}}^2 c_{\mathrm{R}}}{(r\rho^2 c^2)_\omega}, \tag{2.57}$$

where K_{Ic} is the material fracture toughness, c_R the propagation speed of the Rayleigh surface wave, c the propagation speed of the compressive stress wave in water, r the raindrop radius, and ρ the water density.

Moreover, durability is an important evaluation indicator of infrared windows/domes. It refers to the performance failure limit of a material that is continuously exposed in a high-speed rain field, which can be measured based on the damage accumulation rate and the infrared transmission loss rate.

2.4.4 Protection from rain erosion

Rain erosion has become a constraint for the military application of the infrared window/dome. Thus far, research has confirmed that a film with a certain thickness, high transmission, and high modulus that is coated using physical or chemical methods can cover the surface defects of the substrate material. Thus, the Rayleigh surface wave and stress at the crack tip can be reduced, and the damage threshold and rain-erosion resistance can be improved. Based on the phenomenological model of the coating-protected substrate, Goldman et al. described the damage threshold ratio between coated and uncoated substrates as

$$\frac{v_{\text{dam}}^{\text{coa}}}{v_{\text{dam}}^{\text{unc}}} = \frac{1.122\sqrt{2}}{F_B} + \frac{2}{3}\left(\frac{M_c}{M_s}\right)\left(\frac{C_w^2}{C_R v_{\text{dam}}^{\text{unc}}}\right)\frac{t}{r_w}, \tag{2.58}$$

where M_c and M_s are the elastic moduli of the coating and substrate, respectively, C_w the speed of sound in water; C_R the propagation speed of the Rayleigh surface wave, F_B the structural factor associated with coating thickness, r_w the radius of the water drop, and t the coating thickness. Equation (2.58) suggests that a certain thickness is required for the coating to be properly protective.

Today, rain erosion resistant films of infrared windows have become a hot research topic. Commonly used films include CVD diamond, diamond-like carbon, germanium carbide, and boron phosphide. With high infrared transmission, low internal stress, and good attachability to the window base, germanium carbide (Ge_xC_{1-x}), boron phosphide (BP), and gallium phosphide (GaP) films can effectively enhance the rain-erosion resistance of infrared windows. Diamond-like carbon (DLC) films have superior infrared optical properties, mechanical performance, and abrasion resistance, in addition to a mature manufacturing technique. Thus, it can be used to fabricate a composite coating with germanium-carbide and boron-phosphide films to improve the infrared transmission performance and rain-erosion resistance of windows/domes.

The amorphous diamond (α-D) film prepared using the filtered cathodic vacuum arc technique has high modulus, hardness, and infrared transmission comparable to diamond, and it can be regarded as a high-performance antireflective and protective coating for infrared window/dome.

In summary, the erosion wear of infrared windows caused by high-speed raindrops has drawn extensive attention in the field of tribology. However, experimental research is still challenging owing to the inevitably strict precision and stability requirements for the experimental apparatus. Thus far, research in this area has mainly focused on the formation and propagation of the impact stress waves, formation mechanisms of damage cracks, and evaluation of rain-erosion resistance. In recent years, the development of computer simulation technology has provided an effective tool for research on high-speed raindrop-impact damage theory. Modern military technology urgently demands high-performance, rain-erosion-resistant infrared windows/domes. The research on novel infrared materials and protective coatings, therefore, has extensive application prospects in the future. Research in the following aspects should be strengthened in the future:

(1) development of high-precision rain-erosion test apparatus for accurately simulating the impact of raindrops under more practical conditions;
(2) generation process of high pressure inside the raindrop due to high-speed impact and the damaging effects of lateral-jetting-induced shear stress on the infrared window materials;
(3) influence of raindrop-impact conditions and material-surface structure on the rain-erosion behavior for constructing surface and subsurface damage crack growth models;
(4) rain-erosion resistance evaluation and coating-protection mechanism of infrared window coatings.

2.5 Sand erosion

In the normal working process, in addition to the erosion of raindrops, an aircraft also faces the high-speed impact of solid particles, such as sand and dust particles, on the window. Consequently, the external surface of the infrared antireflective/protective coating could be damaged and its optical transmittance could decrease, which would eventually affect the missile detection and guidance performance. Research on the sand-erosion damage of infrared antireflective/protective coatings can help improve their sand-erosion resistance. This section briefly analyses the sand-erosion resistance of infrared antireflective/protective coatings.

2.5.1 Sand erosion damage behavior

When a protective coating is impacted by high-speed sand and dust particles at the surface, plastic penetration will occur on its external surface, forming an approximately semispherical plastic zone surrounded by elastic zones. Residual tensile stress exists on the boundary of the plastic zone. The outward expansion of the residual stress field will lead to two forms of microcracks, i.e. radial cracks and lateral cracks. Lateral cracks, parallel to the coating surface, has a greater effect on the external surface damage of the protective coating and the reduction of transmission. The sizes of lateral cracks are dependent on the residual tensile stress field. They interact with the original microcracks on the material surface. If the velocity of the incident sand particle is sufficiently high, the volumes enclosed by lateral cracks will break and shed, leaving small craters and scratches on the outer surface of the protective film. Thus, damage is generated on the surface of the optical dome.

Since the development of the first erosion theory, namely the microcutting theory, a variety of erosion models have been proposed by researchers attempting to elucidate or predict the material erosion behavior including sand erosion. However, thus far, no model has been reported that can reveal the underlying mechanisms in a comprehensive manner.

2.5.2 Factors influencing sand erosion

Due to the complexity of the sand erosion process, the sand-erosion resistance of a material can be affected by different external conditions and the intrinsic properties of the material itself. Hence, current research on the factors influencing sand erosion mostly focuses on two dimensions: external environment and material properties.

2.5.2.1 Impact angle

The impact angle is the angle between the target material surface and the trajectory of the incident particle, also known as the angle of incidence or the angle of attack. The sand-erosion resistance of the material is closely related to the impact angle.

2.5.2.2 Impact velocity

Hesheng Shao et al. suggested that an impact velocity threshold exists for material erosion. That is, when the impact velocity is less than this threshold, only elastic deformation will occur, without erosion damage. Specifically for sand erosion, there also exists a critical value for the impact velocity. Similarly, sand erosion does not occur when the impact velocity is less than this value. Other studies have shown that the maximum impact angle of the material is not related to the particle impact velocity, indicating that the particle impact velocity does not significantly influence the sand-erosion mechanism.

2.5.2.3 Environment temperature

The variation of sand-erosion resistance of materials with temperature is rather complicated. Roughly speaking, three categories can be observed from experiments:
(1) erosion rate decreases with the increase of temperature and rises again after reaching a minimum;
(2) erosion rate is approximately constant when the temperature is below a threshold but rapidly increases with the increase of temperature beyond the threshold; and
(3) erosion rate rises monotonically with increasing temperature.

2.5.2.4 Sand shape

Under the same conditions, angular grains can lead to more serious erosion than mellow spherical ones. It is generally believed that ploughing deformation dominates for mellow spherical grains, whereas cutting deformation dominates for angular grains.

2.5.2.5 Sand granularity

The effects of sand granularity on erosion rate can be categorized into two types: the particle-size effect and the erosion property variation of brittle materials. It has been demonstrated that the erosion rate of a material increases with increasing particle size, but it exhibits very slow or nearly no variation when the granularity increases to a critical value. This phenomenon is called the size effect. A reasonable explanation for this effect is provided here. When the particle size is less than a limit, the majority of the sand particles will be deflected around the optical dome, instead of impacting. Consequently, there is only a very small impact effect on the dome, and sand-particle breaking will not occur. The erosion rate increases with the increase of particle size. When the sand-particle size reaches a critical value, the effects of secondary erosion caused by sand-particle breaking also reach saturation, resulting in a constant erosion rate.

2.5.2.6 Sand hardness

The hardness ratio between the sand particle and the material surface has a remarkable influence on the sand-erosion behavior of materials. As pointed out by Tabor, when this ratio is greater than 1.2, the erosion rate of a plastic material is large and approaches saturation; when this ratio is less than 1.2, the erosion rate decreases with the decrease of this ratio.

2.5.3 Introduction of sand erosion test

Thus far, there are two types of sand erosion test facilities to conduct quantitative analyses of sand-erosion damage. One is the Defence Nuclear Agency (DNA) sand erosion testing machine, which is schematically shown in Fig. 2.32. When using the DNA test facility to test the protective coating, 17 specimens with a dimension of 50 mm × 50 mm are tested each time under the same conditions. Then, the average values of the experimental results are adopted as the final measurement data for further analysis. The other type of sand erosion test facility makes use of commercial sand-blasting machines.

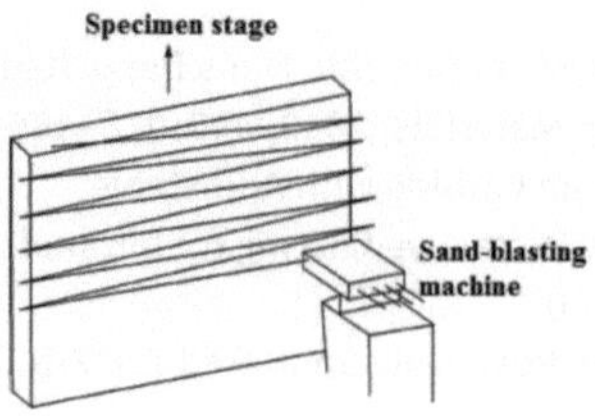

Fig. 2.32: Schematic of the DNA sand erosion test machine.

Owing to the complex nature of the sand-erosion behavior, a consensus has not yet been reached regarding the calculation method for assessing the sand-erosion resistance of materials. Therefore, the evaluation of sand-erosion resistance of a material should not be based on a single measurement. Instead, it should be determined by combining the practical working environment. In recent years, many researchers have observed phenomena different from those of conventional materials when conducting studies on protective films offering erosion resistance for materials, which are related to the sand-erosion research of infrared antireflective/protective coatings. However, the detailed mechanisms and influencing factors need further exploration.

References

[1] Russell GW, Cayson SC, Jones MM, et al. Army Hypersonic Compact Kinetic-Energy Missile Laser Indow Design. Aerosense 2003. International Society for Optics and Photonics, 2003, 5078: 28–39.

[2] Qi YW. Research on Aerodynamic Force/Heat of Missile Wing. Nanjing, Nanjing University of Science & Technology, 2008.

[3] Lyu LL. Engineering Algorithm Study of Hypersonic Aerodynamic Heating. Xi'an, Northwestern Polytechnical University, 2005.

[4] Harris DC. Development of Hot-Pressed and Chemical-Vapor-Deposited Zinc Sulfide and Zinc Selenide in the United States for Optical Windows. Proc. SPIE. 2007, 6545: 654502–654502-27.

[5] Mccloy J. Semi-empirical Scaterring Model for Chemical Vapor Deposited Zinc Sulfide. SPIE Defense, Security and Sensing. International Society for Optics and Photonics, 2009, 7302: 73020W–73020W-8.

[6] Harris DC, Anagnostakis E A. Materials for Infrared Windows and Domes. Optics & Photonics News, 2001, 12: 54–62.

[7] Yan MG. China Aviation Materials Manual. Beijing, Standard Press of China, 2002.

[8] Harris DC, Cambrea L. Assessment of Low-expansion Tungstates for Thermal-shock-resistance Infrared Windows. SPIE Defense, Security and Sensing. International Society for Optics and Photonics, 2013, 8708: 870809–870809-14.

[9] Chen YF, Xu JQ, Su JH, et al. The Optical Properties of DLC Films Processed by Laser Radiation. Advanced Materials Research, 2011, 337: 120–124.

[10] Sullivan RM. A Historical View of Germanium as an Infrared Window Material. SPIE Defense, Security and Sensing. International Society for Optics and Photonics, 2009, 7302: 73020L–73020L-11.

[11] Shafi HZ, Mahmood A, Ali Z, et al. Optical and Structural Analysis of Gec Thin Films Deposited by Reactive Pulsed Laser Ablation Technique. Key Engineering Materials, 2010, 442: 178–186.

[12] Han J, Jiang C, Zhu J. Non-Hydrogenated Amorphous Germanium Carbide with Adjustable Microstructure and Properties: a Potential Anti-Reflection and Protective Coating for Infrared Windows. Surface and Interface Analysis, 2013, 45(3): 685–690.

[13] Edlou SM, Simons JC, et al. Optical and Electrical Properties of Reactively Sputtered Tin, ZrN, and HfN Thin Films. SPIE's 1994, 2262: 96–106.

[14] Pearce SJ, Charlton MDB, et al. Structural and Optical Properties of Different Dielectric Thin Films for Planar Waveguiding Applications. SPIE OPTO. International Society for Optics and Photonics, 2011, 7934: 79341N–79341N-8.

[15] Harris DC. Frontiers in Infrared Window and Dome Materials. SPIE's 1995 International Symposium on Optical Science, Engineering, and Instrumentation. International Society for Optics and Photonics, 1995, 2552: 325–335.

[16] Han J, Zhu J, Niu L, et al. Superexcellent Infrared Protective Coatings: Amorphous Diamond Films Deposited by Filtered Cathodic Vacuum Arc Technology. Surface and Coatings Technology, 2007, 201(9): 5323–5325.

[17] Klocek P, Hoggins JT, Wilso M. Broadband IR Transparent Rain Erosion Protective Coating for IR Windows. San Diego '92. International Society for Optics and Photonics, 1992, 1760: 210–223.

[18] Joseph S, Marcovintch O, Yadin Y, et al. High Durability Antireflection Coatings for Silicon and Multispectral ZnS. Defense and Security Symposium. International Society for Optics and Photonics, 2007, 6545: 65450T–65450T-10.

[19] Zhang GF. New IR Antireflective and Protective Coatings. Infrared technology, 1995, 17(5): 23–28.

[20] Liu Y, He Y, Zhu J, et al. Numerical Simulation of Aerodynamic Heating and Stresses of Chemical Vapor Deposition ZnS for Hypersonic vehicles. Journal of Zhejiang University Science A, 2014, 15(3): 185–196.

[21] Yu HZ. Infrared Optical Material. Beijing, National Defense Industry Press, 2007.

[22] Klein CA. Infrared Missile Domes: Heat Flux and Thermal Shock. SPIE's 1993, 1997: 150–169.

[23] Xuan Z, Goo NS. Analysis of Agas Cooling System for an Infrared Sensor Windows. Thermal and Thermomechanical Phenomena in Electronic Systems (Itherm), 2012 13th IEEE Intersociety Conference on, IEEE, 2012: 1420–1424.

[24] Wu Z, Cheng H, Zhang W, et al. Effects of Thermal Environment on Dynamic Properties of Aerospace Vehicle Panel Structures. Acta Aeronautica et Astronautica Sinica, 2013, 34(2): 334–342.

[25] Klein CA. Flexural Strength of Infrared-transmitting Window Materials: Bimodal Weibull Statistical Analysis. Optical Engineering, 2011, 50(2): 023402–023402-10.

[26] Zhong X, Zhao H, Zhou X, et al. Thermal Shock Behavior of Toughened Gadolinium Zirconate/YSZ Double-ceramic-layered Thermal Barrier Coating. Journal of Alloys and Coupounds, 2014, 593: 50–55.

[27] Shi MR, Cao YF, Zhao JZ. Influence of Heat Treatment on the Metal Films of the Infrared Window. Laser and Infrared, 2011, 41(6): 657–659.

[28] Zhang YF, Dai JM, Zhang L, et al. Spectral Emissivity and Transmissivity Measurement for Zinc Sulphide Infrared Window Based on Integrating-sphere Reflectometry. Optical Engineering, 2013, 52(8): 087107–087107-10.

[29] Turley WD, Holtkamp DB, Veeser LR, et al. Infrared Emissivity of Tin Upon Release of a 25 GPa Shock into a Lithium Fluoride Window. Journal of Applied Physics, 2011, 110(10): 103510–103510-9.

[30] Guo L, Noda N. Investigation methods for Thermal Shock Crack Problems of Functionally Graded Materials – Part I: Analytical Method. Journal of Thermal Stresses, 2014, 37(3): 292–324.

[31] Zhang Y, Guo L, Noda N. Investigation for Thermal Shock Crack Problems of Functionally Graded Materials – Part II: Combined Analytical-numerical Method. Journal of Thermal Stresses, 2014, 37(3): 325–339.

[32] Na X, Lin C, Xiu Q H, et al. Nonlinearity of FY-2D Infrared Detector in Thermal Window Channels and Its Correction Method. Journal of Infrared and Millimeter Waves, 2013, 32(4): 337–343.

[33] Chen S, Ma H, Yi X, et al. Smart VO_2 Thin Film for Protection of Sensitive Infrared Detectors from Strong Laser Radiation. Sensors and Actuators A: Physical, 2004, 115(1): 28–31.

[34] Sonawane HR, Mahulikar SP. Tactical Air Warfare: Generic Model for Aircraft Susceptibility to Infrared guided Missile. Aerospace Science and Technology, 2011, 15(4): 249–260.

[35] Feng LP. Design, Preparation and Properties of Middle-Wave Infrared Antireflective and Protective Films on Sapphire Substrates. Xi'an, Northwestern Polytechnical University, 2006.

[36] Wu ZX. Investigation on Design and Preparation of Y_2O_3/SiO_2 Antireflective Film Deposited on Sapphire for Infrared Window Application. Harbin, Harbin Institute of Technology, 2009.

[37] Chen ZJ. Controllable Preparation and Performance of Infrared Transmittance Materials. Beijing, Beijing University of Chemical Technology, 2012.

[38] Klein CA. Diamond Windows and Domes: Flexural Strength and Thermal Shock. Diamond and Related Materials, 2002, 11(2): 218–227.

[39] Kaiarova M, Tatarko P, Burik P, et al. Thermal Shock Resistance of Si_3N_4 – SiC Ceramics with Rare-earch Oxide Sintering Additives. Journal of the European Ceramic Society, 2014, 34(14): 3301–3308.

[40] Harris D C. Durable 3–5 μm Transmitting Infrared Window Materials. Infrared Physics & Technology, 1998, 39(4): 878–884.

[41] Li YL, Li JM, Su XP, et al. Research Trends and Current Status in Infrared Window and Dome Materials. Journal of Synthetic Crystals, 2007, 36(4): 878–884.

[42] Cross EF. Window-heating Effects on Airborne Infrared System Calibration. San Diego's 92, International Society for Optics and Photonics, 1993, 1762: 576–583.

[43] Klein CA. Infrared Missile Domes: Is There a Figure of Merit for Thermal Shock. SPIE proceedings. 1992, 1739: 23–53.

[44] Klein CA. Thermal Shock Resistance of Infrared Transmitting Windows and Domes. Optical Engineering, 1998, 37(10): 2826–2836.

[45] Klein CA, Gentilman RL. Thermal Shock Resistance of Convectively Heated infrared Windows and Domes. Aerosense'97. International Society for Optics and Photonics, 1997, 3060: 115–129.
[46] Liu MQ, Li BC. Analysis of Temperature and Deformation Fields in an Optical Coating Sample. Acta Physica Sinica, 2008, 57(6): 3404–3409.
[47] Andres ER. Infrared Windows in Industrial Applications I. Proceedings of SPIE 2005, 5782, 438–466.
[48] Klein CA. Simple Formulas for Estimating the Microwave Shielding Effectiveness of EC – Coated Optical Windows. 1989 Orlando Symposium. International Society for Optics and Photonics, 1989, 1112: 234–243.
[49] Klein CA. How Infrared Missile Windows Degrade the Noise – Equivalent Irradiance of Infrared Seeker Systems. SPIE's 1994 International Symposium on Optics, Imaging and Instrumentation. International Society for Optics and Photonics, 1994, 2286: 458–470.
[50] Zhang YP, Fan ZG, Liu JQ. Study on Aero-optical Effect in the Terminal Infrared Guidance. Laser and Infrared, 2006, 36(6): 487–490.
[51] Klein CA. Multilayered Optical Windows: Strains, Stresses and Curvature. Aerosense'99. International Society for Optics and Photonics, 1999, 3705: 250–265.
[52] Townsend PH, Barnett DM, Brunner TA. Elastic Relationships in Layered Composite Media with Approximation for the Case of Thin Films on a Thick Substrate. Journal of Applied Physics, 1987, 62(11): 4438–4444.
[53] Suhir E. An Approximate Analysis of Stresses in Multilayered Elastic Thin Films. Journal of Applied Mechanics, 1988, 55(1): 143–148.
[54] Klein CA, Miller RP. Strains and Stresses in Multilayered Elastic Structures: The Case of Chemically Vapor-deposited ZnS/ZnSe Laminates. Journal of Applied Physics, 2000, 87(5): 2265–2272.
[55] Chen Y, Lu J, Ni X. Analysis of Thermal Stress Damage in Single-crystal Silicon Induced by 1064 nm Long Pulse Laser. Photonics Asia 2007. International Society for Optics and Photonics, 2007, 6835: 68351X–68351X-7.
[56] Klein C A. Normal and Interfacial Stresses in Thin-film Coated Optics: the Case of Diamond-coated Zinc Sulphide Windows. Optical Engineering, 2001, 40(6): 1115–1124.
[57] Klein CA. Toward an Athermal HEL Optical Window: is Oxyfluoride Glass the Way to Go. Boulder Damage Symposium XXXVII: Annual Symposium on Optical Materials for High Power Lasers. International Society for Optics and Photonics, 2005, 5991: 59910S–59910S-22.
[58] Hingst U, Koerber S. IR Window Design for Hypersonic Missile Seekers: Thermal Shock and Cooling Systems. Aerospace/Defense Sensing, Simulation and Controls. International Society for Optics and Photonics, 2001, 4369: 662–672.
[59] Klein CA, Miller RP, Gentilman RL. Characteristic Strength and Weibull Modulus of Selected Infrared-transmitting Materials. Optical Engineering, 2002, 41(12): 3151–3160.
[60] Li YB. Study on Hot-press Technology and Properties of Poly-crystalline ZnS as Infrared Optical Materials. Wuhan: Wuhan University of Technology, 2003.
[61] Huang CB. A Study on Preparation and Properties of $MgO \cdot n\ Al_2O_3$ Transparent Ceramics of Different Stoichiometries Condensed State Physics. PhD Thesis, Chengdu, Sichuang University, 2005.
[62] Hallock B, Shorey A. Technologies for Precision Manufacture of Current and Future Windows and Domes. SPIE Defense, Security and Sensing. International Society for Optics and Photonics, 2009, 7302: 73020V–73020V-11.
[63] Osborne N R, et al. LANTIRN Infrared Window Failure Analysis. SPIE's 1994 International Symposium on Optics, Imaging and Instrumentation. International Society for Optics and Photonics, 1994, 2286: 444–455.

[64] Gong H, Li CF. CW CO_2 Laser Induced Damage to Infrared Window Materials. Acta Optica Sinica, 1996, 16(3): 332–335.

[65] Klocek P, Tabored P. Infrared Transmissive Protective Window: US patent 6, 160, 661. 2000-12-12.

[66] Han X, Zhu JQ, Han JC, et al. Progress of Rain Erosion for IR Windows and Domes. Tribology, 2006, 26(5): 498–504.

[67] Qu ZG, Liu ZT, Li YP. High Temperature Oxidation and Protection of CVD Diamond. Materials Review, 2005, 19(4): 64–66.

[68] Yang D, Thomas ME, Kaplan SG. Measurement of the Infrared Refractive Index of Sapphire as a Function of Temperature. Aerospace/Defense Sensing, Simulation and Controls. International Society for Optics and Photonics, 2001, 4375: 53–63.

[69] Hobbs DC. Study of the Environment and Optical Durability of AR Micro-structures in Sapphire, ALON, and Diamon. SPIE Defense, Security and Sensing. International Society for Optics and Photonics, 2009, 7302: 73020J–73020J-12.

[70] Blodgett DW, Yang D, Walts SC, et al. Measurement of the Temperature-dependent Refractive Index and Relative Photoelastic Constant of Zinc Sulphide. Aerospace/Defense Sensing, Simulation and Controls. International Society for Optics and Photonics, 2001, 4375: 224–234.

[71] Fujii A, Wada H, Shibata K, et al. Diamond-ZnS Composite Infrared Window. Aerospace/Defense Sensing, Simulation and Controls. International Society for Optics and Photonics, 2001, 4375: 206–217.

[72] Bernt ML, Stover JC. Infrared Window Damage Measured by Reflective Scatter. San Diego '91, San Diego, CA. International Society for Optics and Photonics, 2011, 9016: 80160H–80160H-7.

[73] Thomas ME. Multiphoton Difference Band Absorption in Diamond. SPIE Defense, Security and Sensing. International Society for Optics and Photonics, 1991, 1530: 42–49.

[74] Friel I, Geoghegan SL, Twitchen DJ, et al. Development of High Quality Single Crystal Diamond for Novel Laser Applications. Security, Defense. International Society for Optics and Photonics, 2010, 7838: 783819–783819-8.

[75] Springer RM, Thomas ME, Hahn DV, et al. Optical Properties of Poly-crystalline Nd: YAG for Different Levels of Nd Doping. SPIE Defense, Security and Sensing. International Society for Optics and Photonics, 2009, 7302: 730212–730212-8.

[76] Sepulveda JL, Loutfy RO, Chang S, et al. High-performance Spinel Ceramics for IR Windows and Domes. SPIE Defense, Security and Sensing. International Society for Optics and Photonics, 2011, 8016: 801604–801604-12.

[77] Hatch SE. Emittance Measurements on Infrared Windows Exhibiting Wave Length Dependent Diffuse Transmittance. Applied Optics, 1962, 1(5): 595–601.

[78] Huang C. The New Null Testing Method for the Special Optical Window. International Symposium on Photoelectronic Detection and Imaging 2009. International Society for Optics and Photonics, 2009: 73831H–73831H-6.

[79] Stefan RK. Coupled Aerokinetic Heating of Missile Structures at High Velocities. RTO AVT Symposium on Missile Aerodynamics, 1998, 38-38-10.

[80] John EP, Charles TW, George WS. Side Mounted IR Window Aero-optic and Aerothermal Analysis. Proceedings of SPIE, 1999, 3705, 266–275.

[81] Wan ZM. The Calculation Models of Cooling IR Window and Window Background Radiation. Proceedings of SPIE, 1998, 3375, 195–202.

[82] Jia-ling L, Ganimedov VL, Muchnaja MI, et al. The Calculations of Aerodynamic Heating and Viscous Friction Forces on the Surface of Hypersonic Flightvehicle. Experiments and Measurements in Fluid mechanics, 2002, 16(L): 8–20.

[83] Xia G, Liu JX, Cheng WK, et al. Numerical Simulation of Coupled Aeroheating and Solid Heat Penetration for a Hypersonic Blunt Body. Journal of National University of Defense Technology, 2003, 25(1): 35–39.
[84] Huang CS, Wu J, Fan XQ. Coupled Numerical Simulation of Supersonic Flow Field and Heated Structure of Flying Vehicles. Mechanics in Engineering, 2004, 26(2): 24–26.
[85] Ren QM, Yang ZB, Cheng Z, et al. Aeroheating/Thermal response coupled analysis and experimental Research. Spacecraft Environment Engineering, 2007, 2: 20–32.
[86] Adler WF, Mihora DJ. Infrared-transmitting Window Survivability in Hydrometeor Environments. Proceedings of SPIE, 1992, 1760: 291–302.
[87] Salman HA, Sinan AE, et al. Development and Verification of High Durability Infrared Transparent Coating for Airborne Usage. Proceedings of SPIE, 2010, 7786: 77860K–77860K-10.
[88] Tang D P, Gao Q, Li Y H. 3-D Simulation of Thermal Residual Stresses and Failure Analysis of Thick Diamond Films. Laser and Infrared, 2008, 38(1): 81–83.
[89] Yang DY, Gui WY. Numerical Simulation on Thermal Response Characteristic in the Thermal Protection Layer of the Leading Edge. Mass and Heat Transfer Conference, China Society of Engineering Thermophysics, 1999, VI–58–63.
[90] Huang T, Mao GL, Jiang GQ, et al. Two Dimensional Coupled Flow-Thermal Structural Numerical Simulation. Acta Aerodynamica Sinica, 2000, 18(1): 115–119.
[91] Shahmardan MM, Norouzi M, Kayhani MH, et al. An Exact Analytical Solution for Convective Heat Transfer in Rectangular Ducts. Journal of Zhejiang University Science A, 2013, 13(10): 768–781.
[92] Tang K, Yu J, Jin T, et al. Influence of Compression-expansion Effect on Oscillating-flow Heat Transfer in a Finned Heat Exchanger. Journal of Zhejiang University Science A, 2013, 14(6): 427–434.
[93] Menter FR. Two-equation Eddy-viscosity Turbulence Models for Engineering Applications. AIAA Journal, 1994, 32(8): 1598–1605.
[94] Lee SH. Effects of Condition Number on Preconditioning for Low Mach Number Flows. Journal of Computational Physics, 2012, 231(10): 4001–4014.
[95] Roe PL. Approximate Riemann Solvers, Parameter Vectors, and Difference Schemes. Journal of Computational Physics. 1981, 43: 357–372.
[96] Ouahsine A, Smaoui H. Flux-limiter Schemes for Oceanic Tracers: Application to the English Channel Tidal Model. Computer methods in Applied mechanics and Engineering, 1999, 179(3–4): 307–325.
[97] Muylaert J, Walpot L, Hauser J, et al. Standard Model Testing in the European High Enthalpy Facility F4 and Extrapolation to Flight. AIAA 17th Aerospaceground Testing Conference. Nashville: AIAA Paper 92-3905, 1992, 1–16.
[98] Lu ZL. Aerodynamics. Beijing, Beihang University Press, 2008.
[99] Adler WF. Development of Design Data for Rain Impact Damage in Infrared-transmitting Windows and Radomes. Optical Engineering, 1987, 26(2): 262143–262143.
[100] Field JE. Stress Waves, Deformation and Fracture Caused by Liquid Impact. Philosophical Transactions for the Royal Society of London. Series A, Mathematical and Physical Sciences, 1966: 86–93.
[101] Dear JP, Field JE. High-speed Photography of Surface Geometry Effects in Liquid/Solid Impact. Journal of Applied Physics, 1988, 63(4): 1015–1021.
[102] Field JE, Dear JP, Ogren JE. The Effects of Target Compliance on Liquid Drop Impact. Journal of Applied Physics, 1989, 40(13): 5113–5122.
[103] Hand RJ, Field JE. Liquid Impact on Toughened Glasses. Engineering Fracture Mechanics, 1990, 37(2): 293–311.
[104] Adler WF, Hooker SV. Water Drop Impact Damage in Zinc Sulfide. Wear, 1978, 48(1): 103–119.

[105] Adler WF, Hooker SV. Rain Erosion Mechanisms in Brittle Materials. Wear, 1978, 50(1): 11–38.

[106] Heymann FJ. High-speed Impact Between in Brittle Materials. Journal of Applied Physics, 1969, 40(13): 5113–5122.

[107] Lasser M. Thirty Years of Liquid Impact Research: A tutorial Review. Wear, 1995, 186: 28–34.

[108] Bounre NK. On Impacting Liquid Jets and Drops Onto Polymethyl Methacrylate Targets. Proceedings of the Royal Society A: Mathematical, Physical and Engineering Science, 2005, 461(2056): 1129–1145.

[109] Haller KK, Ventikos Y, Poulikakos D, et al. Computational Study of High-speed Liquid Droplet Impact. Journal of Applied Physics, 2002, 92(5): 2821–2828.

[110] Seward CR, Coad EJ, Pickles CSJ, et al. Rain Erosion Resistance of Diamond and Other Window Materials. SPIE's 1994 International Symposium on Optics, Imaging and Instrumentation. International Society for Optics and Photonics, 1994. 285–300.

[111] Field JE. ELSI Conference: Invited Lecture. Liquid Impact: Theory, Experiment, Applications. Wear, 1999, 233: 1–12.

[112] Shi HH, Takayama K, Nagayasu N. The Measurement of Impact Pressure and Solid Surface Response in Liquid-solid Impact up to Hypersonic Range. Wear, 1995, 186: 352–359.

[113] Zhang D, Zhou QL. Study on Nonlinear Wave Model for Liquid Droplet-Solid Impact. Xi'an Jiaotong University Xuebao, 2002, 36(11): 1138–1141.

[114] Jackson MJ, Field JE. Liquid Impact Erosion of Single-crystal Magnesium Oxide. Wear, 1999, 233: 39–50.

[115] Miller GF, Pursey H. On the Partition of Energy Between Elastic Waves in a Semi-infinite Solid. Proceedings of the Royal Society of London. Series A. mathematical and Physical Sciences, 1955, 233(1192): 55–69.

[116] Seward CR, Pickles CSJ, Marrah R, et al. Rain Erosion Data on Window and Dome Materials. San Diego'92. International Society for Optics and Photonics, 1992, 280–290.

[117] Seward CR, Coad EJ, pickles CSJ, et al. The Liquid Impact Resistance of a Range of IR-transparent Materials. Wear, 1995, 186: 375–383.

[118] Goldman LM, Gremin PE, Varitimos TE, et al. Damage Resulting From Single and Multiphe-waterdrop Impacts on Coated and Uncoated LWIR Substrates. San Diego '92. International Society for Optics and Photonics, 1992, 224–242.

[119] Shi HH, Field JE. Stress Wave Propagation in Solid Materials under High Speed Liquid Impact. Science China Series G, 2004, 34(5): 577–590.

[120] Debotton G. The interaction of A Coated Target and An Impinging Waterdrop. Wear, 1998, 219(1): 60–72.

[121] Wang LL, Field JE, Sun Q, et al. Surface Damage of Polymethylmethacrylate Plates by Ice and Nylon Ball Impacts. Journal of Applied Physics, 1995, 78(3): 1643–1649.

[122] Bourne NK, Obara T, Field JE. High-speed Photography and Stressgauge Studies of Jet Impact Upon Surfaces. Philosophical Transaction of the Royal Society of London Series A Mathematical, Physical and Engineering Sciences, 1997, 335(1724): 607–623.

[123] Adler WF, Flavin JW, Richards JP, et al. Multiple Simulated Waterdrop Impact Damage in Zinc Sulfide at Supersonic Velocities. San Diego's 92. International Society for Optics and Photonics, 1992, 303–315.

[124] Hand RJ, Field JE, Townsend D. The Use of Liquid Jets to Simulate Angled Drop Impact. Journal of Applied Physics, 1991, 70(11): 7111–7118.

[125] Adler WF. Particulate Impact Damage Predictions. Wear, 1995, 186: 35–44.

[126] Coad EJ, Pickles CSJ, et al. Aerospace Erosion of Diamond and Diamond Coatings. Diamond and Related Materials, 1996, 5(6): 640–643.

[127] Deom AA, Luc A, Amara S, et al. Towards More Realistic Erosion Simulation Tests for High Velocity EM and IR Windows. Wear, 1999, 23: 13–24.

[128] Jilbertg H, Field JE. Synergistic Effects of Rain and Sand Erosion. Wear, 2000, 243(1): 6–17.
[129] Adler WF. Waterdrop Impact Modeling. Wear, 1995, 186: 341–351.
[130] Adler WF. Rain Impact Retrospective and Vision for the Future. Wear, 1999, 233: 25–38.
[131] Westmark C, Lawless GW. A Discussion of Rain Erosion Testing at the United States Air Force Rain Erosion Test Facility. Wear, 1995, 186: 384–387.
[132] Adler WF, Boland PL. Multiparticle Supersonic Impact Test Program. San Dieg-DL Tentative. International Society for Optics and Photonics, 1990: 268–279.
[133] Evans AG, Ito YM, Rosonblatt M. Impact Damage Thresholds in Brittle Materials Impact by Water Drops. Journal of Applied Physics, 1980: 51(5): 2473–2482.
[134] Hasan W, Propst SH. Durability Testing of Hard Carbon Coatings Forge and ZnS Substrates. SPIE's 1994 International Symposium on Optics, Imaging, and Instrumentation. International Society for Optics and Photonics, 1994: 354–363.
[135] Wang ZJ, Chen XM, Huang Q. Rain Erosion Test on Aircraft Optics Window's and Dome's Material. Proceedings of the 6th China National Conference on Experimental Fluid Mechanics, 2004.
[136] Goldman LM, Tustison RW. High Durability Infrared Transparent Coatings. Proc. SPIE. 1994, 2286(3): 16–324.
[137] Vilcarromero J, Marques FC, Freire FL. Optoelectronic and Structural Properties of a-$Ge_{1-x}C_x$:H prepared by RF Reactive Cosputtering. Journal of Applied Physics, 1998, 84(1): 174–180.
[138] Song JQ, Liu ZT, Geng DS, et al. Design and Preparation of Ge_xC_{1-x} Inhomogeneous Antireflective and Protective Films. Laser and Infrared, 2001, 31(4): 253–255.
[139] Hu CQ, Zheng WT, et al. Chemical Bonding of a-Ge_xC_{1-x}:H Films Grown by RF Reactive Sputtering. Vacuum, 2004, 77(1): 63–68.
[140] Gibson DR, Waddell EM, Lewis KL. Advances in Ultradurable Phosphide-based Broadband Antireflection Coatings for Sand and Rain Erosion Protection of Infrared Windows and Domes. SPIE's 1994, International Symposium on Optics, Imaging, and Instrumentation. International Society for Optics and Photonics, 1994: 335–346.
[141] Robertson J. Diamond-like Amorphous Carbon. Materials Science and Engineering: R: Reports, 2002, 37(4): 129–281.
[142] Huang LY, Xu KW, Lyu J, et al. Experimental Study of Tribological and Scratch Properties of a-C:H Films. Tribology, 2001, 21(4): 129–139.
[143] Xu S, Tay BK, Tan HS, et al. Properties of Carbon Ion Deposited Tetrahedral Amorphous Carbon Films as A Function of Ion Energy. Journal of Applied Physics, 1996, 79(9): 7234–7240.
[144] Yu X, Zhang X, Wang CB, et al. Preparation and Study of the Tribological Properties of Tetrahedral Amorphous Carbon Film. Tribology, 2005, 25(2): 135–139.
[145] Gou YN, Sun H, Huang N, et al. Fabrication and Tribological Characterization of Multilayer C/C Films Prepared by Filtered Cathodic Vacuum Arc Technology. Tribology, 2006, 26(2): 12–124.
[146] Machowski JM, Cimma B. Optical and mechanical Behaviour of Gec and BP Antireflection Coatings Under Rain Erosion Tests. In: Window and Dome Technologies and Materials IV, Klocek Peds, SPIE. San Diego, CA, USA, 1994, 2286.
[147] Han JC, Zhu JQ. Amorphous diamond films deposited by filtered cathodic vacuum arc technology as infrared protective coatings. The Chinese Journal of Nonferrous Metals, 2005, 15(11): 1795–1799.
[148] Zhu, Han JC, Meng SH, et al. Correlations Between Substrate Bias, Microstructure and Surface Morphology of Tetrahedral Amorphous Carbon Films. Vacuum, 2004, 72(3): 285–290.
[149] Zhu JQ, Wang JH, Meng HS, et al. The microstructure and properties of tetrahedral amorphous carbon films deposited by filtered arc with accelerating at different energetic grades. Acta Physica Sinica, 2004, 53(4): 1150–1156.

[150] Dong G, Zhang JY. Developments of Research on the Solid Particle Erosion of Materials. Journal of Materials Science and Engineering, 2003, 21(2): 307–312.
[151] Shao HS, Qu JS. Friction and Wear. Beijing, Coal Industry Press, 1992.
[152] Ma Y, Ren J, Li YD, et al. Development of Research on Erosion of Materials. Journal of Lanzhou University of Technology, 2005, 31(1): 21–25.
[153] Levy AV, Yan J, Patterson J. Elevated Temperature Erosion of Steels. Wear, 1986, 108(1): 43–60.
[154] Gat N, Tabakoff W. Some Effects of Temperature on the Erosion of Metals. Wear, 1978, 50(1): 85–94.
[155] Liu YJ, Cheng KQ. Analysis Foundation of Wear failure. Beijing, China Machine Press, 1991.
[156] He JA, Wang YW. Material Abrasion and Wear Resistant Materials. Shenyang, Northeastern University Press, 2002.
[157] Dong G. Study on the Erosion Wear Behaviors and Mechanisms of Several Materials. PhD Thesis, Hangzhou, Zhejiang University of Technology, 2004.
[158] Wellman RG, Nicholls JR. Some Observations on Erosion Mechanisms of FB PVD TBCS. Wear, 2000, 242(1): 89–96.
[159] Trezona RI, Hutchings IM. Resistance of Paint Coatings to Multiple Solid Particle Impact: Effect of Coating Thickness and Substrate Material. Progress in Organic Coatings, 2001, 41(1): 85–92.
[160] Murthy JKN, Rao DS, Venkataraman B. Effect of Grinding on the Erosion Behaviour of A WC-Co-Cr Coating Deposited by HVOF and Detonation Gun Spray Processes. Wear, 2001, 249(7): 592–600.

3 Optical design of infrared antireflective and protective coating system

3.1 Overview

This chapter introduces the principles and methods for the optical design of infrared antireflective and protective coatings. We begin with the most basic monolayered dielectric film, and then analyse periodic multilayered dielectric films. It is hoped that the reader can gain basic knowledge on common optical thin films with the introduction of the principles and design methods for multilayered dielectric films.

The structure of monolayered dielectric films is the foundation of coating design and constitutes a typical example. In general, a three-layered medium can be formed by coating a glass substrate with a MgF_2 thin-film layer of thickness less than one wavelength, with the top of the coating exposed to air. This forms a simple monolayered dielectric coating structure. Light entering the monolayered dielectric film causes a distribution of electromagnetic field in the coating layer. Based on Maxwell's theory of electromagnetism, the field values of the three-layered medium can be analyzed, and mathematical methods can be applied to calculate the transmittance, reflectance and other parameters of the entire coating system. The use of Maxwell's theory of electromagnetism for the analysis of field values usually involves much work; hence, the results of the theory will be applied directly to discuss the electric-field and magnetic-field distributions in different regions of the dielectric film, as well as their boundary conditions, thereby leading us to the required coating design results.

In addition, within the actual application environment of infrared antireflective coatings, infrared detectors usually need to monitor the target signals of infrared transmission from different directions. As the oblique incidence of light waves causes a sharp decrease in the intensity of transmitted light, design methods for conventional antireflective coatings based on normal incidence are not applicable. Therefore, in order to ensure that infrared windows have good transmittance under oblique incidence, this chapter also introduces an infrared antireflective coating design for oblique incidence, following which an outline of the theoretical basis and methods for designing and developing oblique-incidence infrared antireflective coating is presented.

3.2 Basic theories of coating design

3.2.1 Mathematical method

Fig. 3.1 is a schematic diagram of a monolayered dielectric film. Let the incident monochromatic plane wave be linearly polarized light, which can be decomposed

https://doi.org/10.1515/9783110489514-006

into transverse-electric (TE) waves, with the electric vector perpendicular to the incident surface, and transverse-magnetic (TM) waves, with the electric vector parallel to the incident surface. As the boundary conditions for the vertical and parallel components of E are mutually independent at the medium interface, these two waves are mutually independent as well. Only TE waves are discussed in Fig. 3.1. Similar methods can be applied to TM waves in order to obtain the corresponding results.

In Fig. 3.1 (a), $E_{r\mathrm{I}}$ and $E_{r\mathrm{II}}$ represent the sums of all possible wave-vector components transmitted in the indicated directions at their respective points in the medium. In Fig. 3.1 (b), when points A_{I} and A_{II} are compared, the phase shift between the two points of a plane wave entering interface II in medium II is given by

$$b = \frac{2\pi}{\lambda_0} n_2 h \cos\theta_2. \tag{3.1}$$

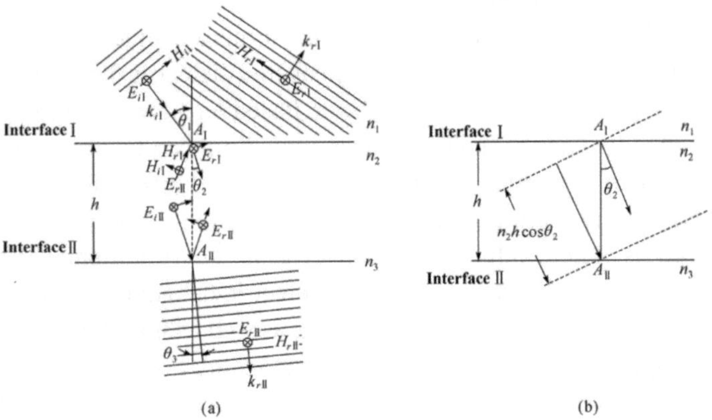

Fig. 3.1: Schematic diagram of monolayered dielectric film: (a) electromagnetic field at the boundary; (b) the distance between two equiphase surfaces passing through points A_{I} and A_{II} is $n_2 h \cos\theta_2$.

According to the theory of electromagnetism, the tangential component of the electric field $\boldsymbol{E}$ and magnetic field $\boldsymbol{H}$ at the interface between two media should be continuous. In specific terms, at interface I,

$$E_{\mathrm{I}} = E_{i\mathrm{I}} + E_{r\mathrm{I}} = E_{t\mathrm{I}} + E'_{r\mathrm{II}} \tag{3.2}$$

and

$$H_{\mathrm{I}} = H_{i\mathrm{I}} \cos\theta_1 - H_{r\mathrm{I}} \cos\theta_1 = H_{t\mathrm{I}} \cos\theta_2 - H'_{r\mathrm{II}} \cos\theta_2. \tag{3.3}$$

In nonmagnetic media, the relationship between $\boldsymbol{E}$ and $\boldsymbol{H}$ is given by

$$\boldsymbol{H} = \sqrt{\frac{\varepsilon_0}{\mu_0}} n\boldsymbol{s} \times \boldsymbol{E}, \tag{3.4}$$

where $\boldsymbol{s}$ is the unit vector in the direction of wave vector $\boldsymbol{k}$ and n is the refractive index of the dielectric layer involved.

H_{I} can be expressed as

$$H_{\mathrm{I}} = \sqrt{\frac{\varepsilon_0}{\mu_0}}(E_{i\mathrm{I}} - E_{r\mathrm{I}})n_1 \cos\theta_1 = \sqrt{\frac{\varepsilon_0}{\mu_0}}(E_{t\mathrm{I}} - E'_{r\mathrm{II}})n_2 \cos\theta_2. \tag{3.5}$$

By using the same method, the electric-field and magnetic-field components at interface II are obtained as

$$E_{\mathrm{II}} = E_{i\mathrm{II}} + E_{r\mathrm{II}} = E_{t\mathrm{II}} \tag{3.6}$$

and

$$H_{\mathrm{II}} = \sqrt{\frac{\varepsilon_0}{\mu_0}}(E_{i\mathrm{II}} - E_{r\mathrm{II}})n_2 \cos\theta_2 = \sqrt{\frac{\varepsilon_0}{\mu_0}}E_{t\mathrm{II}}n_3 \cos\theta_3, \tag{3.7}$$

where n_3 is the refractive index of the substrate.

When the electric vectors of points A_{I} and A_{II} on both interfaces are compared and the phase shift in equation (3.1) is considered, then

$$E_{i\mathrm{II}} = E_{t\mathrm{I}}\, \mathrm{e}^{\mathrm{j}\beta}, \tag{3.8}$$

$$E_{r\mathrm{II}} = E'_{r\mathrm{II}}\, \mathrm{e}^{-\mathrm{j}\beta}. \tag{3.9}$$

By using equations (3.8) and (3.9), equations (3.6) and (3.7) can be expressed as

$$E_{\mathrm{II}} = E_{t\mathrm{I}}\mathrm{e}^{+\mathrm{j}\beta} + E'_{r\mathrm{II}}\mathrm{e}^{-\mathrm{j}\beta} \tag{3.10}$$

and

$$H_{\mathrm{II}} = (E_{t\mathrm{I}}\mathrm{e}^{+\mathrm{j}\beta} - E'_{r\mathrm{II}}\mathrm{e}^{-\mathrm{j}\beta})\sqrt{\frac{\varepsilon_0}{\mu_0}}n_2 \cos\theta_2. \tag{3.11}$$

$E_{t\mathrm{I}}$ and $E'_{r\mathrm{II}}$ can be solved using equations (3.10) and (3.11), and the results can be substituted into equations (3.2) and (3.5) to obtain the electric-field and magnetic-field components at interface I as follows:

$$E_{\mathrm{I}} = E_{\mathrm{II}} \cos\beta - H_{\mathrm{II}}\frac{\mathrm{j}\sin\beta}{P_2}, \tag{3.12}$$

$$H_{\mathrm{I}} = -E_{\mathrm{II}}P_2(\mathrm{j}\sin\beta) + H_{\mathrm{II}} \cos\beta, \tag{3.13}$$

$$P_2 = \sqrt{\frac{\varepsilon_0}{\mu_0}}n_2 \cos\theta_2. \tag{3.14}$$

Equations (3.12) and (3.13) express the linear relationship between the field values at the two interfaces. These relations can also be expressed in the matrix form as follows:

$$\begin{bmatrix} E_{\mathrm{I}} \\ H_{\mathrm{I}} \end{bmatrix} = \begin{bmatrix} \cos\beta & -\mathrm{j}\frac{\sin\beta}{P_2} \\ -\mathrm{j}P_2 \sin\beta & \cos\beta \end{bmatrix} \begin{bmatrix} E_{\mathrm{II}} \\ H_{\mathrm{II}} \end{bmatrix} \tag{3.15}$$

or

$$\begin{bmatrix} E_{\mathrm{I}} \\ H_{\mathrm{I}} \end{bmatrix} = \boldsymbol{M}_{\mathrm{I}} \begin{bmatrix} E_{\mathrm{II}} \\ H_{\mathrm{II}} \end{bmatrix}, \tag{3.16}$$

where $\boldsymbol{M}_{\mathrm{I}}$ is a characteristic matrix, which can be used to establish the relation between the field components at the two interfaces.

Based on the incidence of the TE wave, the characteristic matrix $\boldsymbol{M}_{\mathrm{I}}$ can be expressed as

$$\boldsymbol{M}_{\mathrm{I}} = \begin{bmatrix} \cos\beta & -\mathrm{j}\frac{\sin\beta}{P_2} \\ -\mathrm{j}P_2 \sin\beta & \cos\beta \end{bmatrix}. \tag{3.17}$$

By substituting P_2 in equation (3.17) with $q_2 = \sqrt{\varepsilon_0/\mu_0}\, n_2/\cos\theta_2$, the same calculation methods can be employed to obtain the characteristic matrix under TM wave incidence.

The above discussion based on monolayered films can easily be generalized to multilayered films. For example, if the monolayer film shown in Fig. 3.1 (a) is coated with a new film layer, it will form three interfaces. The relationship between the field quantities of interfaces II and III (in sequence from top to bottom) can be established using the characteristic matrix $\boldsymbol{M}_{\mathrm{II}}$:

$$\begin{bmatrix} E_{\mathrm{II}} \\ H_{\mathrm{II}} \end{bmatrix} = \boldsymbol{M}_{\mathrm{II}} \begin{bmatrix} E_{\mathrm{III}} \\ H_{\mathrm{III}} \end{bmatrix}. \tag{3.18}$$

Multiplying both sides of equation (3.18) by $\boldsymbol{M}_{\mathrm{I}}$ gives

$$\begin{bmatrix} E_{\mathrm{I}} \\ H_{\mathrm{I}} \end{bmatrix} = \boldsymbol{M}_{\mathrm{I}}\boldsymbol{M}_{\mathrm{II}} \begin{bmatrix} E_{\mathrm{III}} \\ H_{\mathrm{III}} \end{bmatrix}. \tag{3.19}$$

Under general conditions, if p denotes the number of layers in the coating system, then each layer will have its specific n and h. Thus, the field values at the first and last interfaces are linked by the following equation:

$$\begin{bmatrix} E_{\mathrm{I}} \\ H_{\mathrm{I}} \end{bmatrix} = \boldsymbol{M}_{\mathrm{I}}\boldsymbol{M}_{\mathrm{II}} \dots \boldsymbol{M}_{P} \begin{bmatrix} E_{(p+1)} \\ H_{(p+1)} \end{bmatrix}. \tag{3.20}$$

The characteristic matrix of the entire coating system is the product of the 2 × 2 matrix of each layer in a given sequence, as shown below:

$$\boldsymbol{M} = \boldsymbol{M}_{\mathrm{I}}\boldsymbol{M}_{\mathrm{II}} \dots \boldsymbol{M}_{P} = \begin{bmatrix} m_{11} & m_{12} \\ m_{21} & m_{22} \end{bmatrix}. \tag{3.21}$$

At this point, equation (3.20) can be simplified to

$$\begin{bmatrix} E_{\mathrm{I}} \\ H_{\mathrm{I}} \end{bmatrix} = \boldsymbol{M} \begin{bmatrix} E_{(p+1)} \\ H_{(p+1)} \end{bmatrix}. \tag{3.22}$$

3.2.2 Monolayered uniform dielectric film

3.2.2.1 Reflectance and transmittance at equivalent interfaces in monolayered uniform dielectric film

Assume that the monolayered dielectric film under discussion is uniform and nonmagnetic. When a monochromatic plane wave is incident on the film surface, as shown in Fig. 3.2, then in terms of the TE wave, equations (3.16) and (3.17) can be used to obtain

$$\begin{bmatrix} E_{\mathrm{I}} \\ H_{\mathrm{I}} \end{bmatrix} = \begin{bmatrix} m'_{11} & m'_{12} \\ m'_{21} & m'_{22} \end{bmatrix} \begin{bmatrix} E_{\mathrm{II}} \\ H_{\mathrm{II}} \end{bmatrix}, \tag{3.23}$$

where

$$m'_{11} = m'_{22} = \cos\beta, \quad m'_{12} = -\mathrm{j}\frac{\sin\beta}{P_2}, \quad m'_{21} = -\mathrm{j}P_2\sin\beta,$$

$$P_2 = \sqrt{\frac{\varepsilon_0}{\mu_0}}n_2\cos\theta_2.$$

Let

$$P_i = \sqrt{\frac{\varepsilon_0}{\mu_0}}n_i\cos\theta_i \quad (i = 1, 2, 3 \text{ which is the sequence of the film layers}).$$

By using equations (3.2) and (3.5)–(3.7), the following results can be obtained:

$$E_{\mathrm{I}} = E_{i\mathrm{I}} + E_{r\mathrm{I}}.$$

At interface I:

$$H_{\mathrm{I}} = \sqrt{\frac{\varepsilon_0}{\mu_0}}(E_{i\mathrm{I}} - E_{r\mathrm{I}})n_1\cos\theta_1 = P_1(E_{i\mathrm{I}} - E_{r\mathrm{I}}),$$

$$E_{\mathrm{II}} = E_{t\mathrm{II}}.$$

At interface II:

$$H_{\mathrm{II}} = \sqrt{\frac{\varepsilon_0}{\mu_0}}n_3E_{t\mathrm{II}}\cos\theta_3 = P_3E_{t\mathrm{II}}.$$

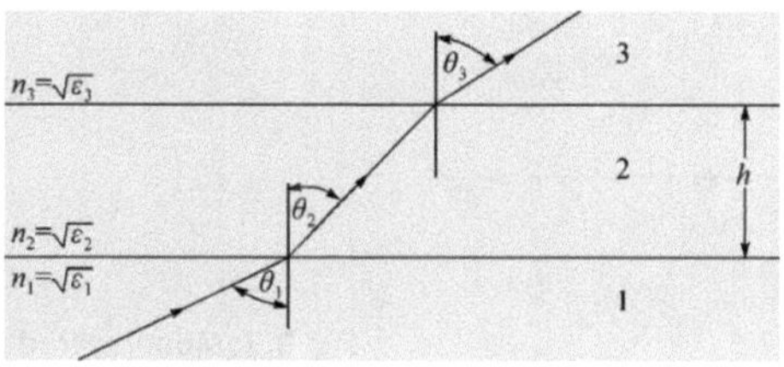

Fig. 3.2: Schematic of the transmission path of an electromagnetic wave passing through a uniform dielectric film.

Substituting the results above into equation (3.23) gives

$$\begin{bmatrix} E_{i\mathrm{I}} + E_{r\mathrm{I}} \\ P_1(E_{i\mathrm{I}} - E_{r\mathrm{I}}) \end{bmatrix} = \begin{bmatrix} m'_{11} & m'_{12} \\ m'_{21} & m'_{22} \end{bmatrix} \begin{bmatrix} E_{t\mathrm{II}} \\ P_3 E_{t\mathrm{II}} \end{bmatrix}. \tag{3.24}$$

By defining the amplitude reflectance $r = E_{r\mathrm{I}}/E_{i\mathrm{I}}$ and the amplitude transmittance $t = E_{t\mathrm{II}}/E_{i\mathrm{I}}$ of the monolayer film, the following can be obtained from equation (3.24):

$$1 + r = m'_{11}t + m'_{12}P_3 t, \tag{3.25a}$$

$$(1 - r)P_1 = m'_{21}t + m'_{22}P_3 t. \tag{3.25b}$$

The amplitude reflectance r and transmittance t of the equivalent layers can thus be obtained by solving the above simultaneous equations as follows:

$$r = \frac{(m'_{11} + m'_{12}P_3)P_1 - (m'_{21} + m'_{22}P_3)}{(m'_{11} + m'_{12}P_3)P_1 + (m'_{21} + m'_{22}P_3)}, \tag{3.26}$$

$$t = \frac{2P_1}{(m'_{11} + m'_{12}P_3)P_1 + (m'_{21} + m'_{22}P_3)}. \tag{3.27}$$

The values of r and t in equations (3.26) and (3.27), respectively, are based on the entire monolayered coating system; Fig. 3.3 shows this equivalence relation. The reflection and transmission of a monolayer film described by r and t behave as if they occur on a single equivalent interface, and their values are generally complex.

Let

$$r = |r|\mathrm{e}^{\mathrm{j}\delta_r}, \quad t = |t|\mathrm{e}^{\mathrm{j}\delta_t}.$$

where δ_r is the reflection phase shift and δ_t is the transmission phase shift, which characterize the reflection and transmission occurring on the equivalent interface.

In terms of TM waves, the r and t of TM waves incident on a monolayered equivalent interface can be obtained by replacing P_i in equations (3.26) and (3.27) with $q_i = \sqrt{\varepsilon_0/\mu_0}\, n_i/\cos\theta_i$ $(i = 1, 2, 3)$.

The results obtained from equations (3.26) and (3.27) can also be applied in multilayered systems by replacing m_{11}', m_{12}', m_{21}', and m_{22}' in these two equations with the matrix elements m_{11}, m_{12}, m_{21}, and m_{22} in equation (3.21). At this point, r and t

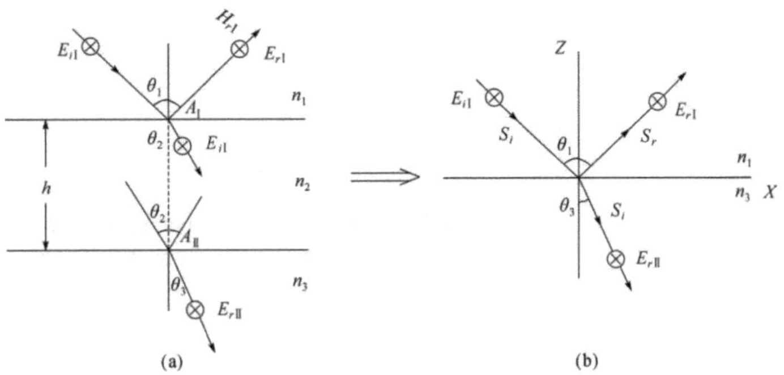

Fig. 3.3: Monolayered dielectric film and equivalent interface.

denote the amplitude reflectance and transmittance of the equivalent interface in a multilayered film. The values of the matrix elements can be computed by calculating the effects of all film layers using equation (3.21).

As for TE waves, Fresnel's equations can be applied to obtain the amplitude reflectance r_1 and transmittance t_1 at Interface I, given respectively by

$$r_1 = \frac{n_1 \cos\theta_1 - n_2 \cos\theta_2}{n_1 \cos\theta_1 + n_2 \cos\theta_2} = \frac{P_1 - P_2}{P_1 + P_2}, \tag{3.28a}$$

$$t_1 = \frac{2n_1 \cos\theta_1}{n_1 \cos\theta_1 + n_2 \cos\theta_2} = \frac{2P_1}{P_1 + P_2}. \tag{3.28b}$$

The equations for the amplitude reflectance r_2 and transmittance t_2 at Interface II are similar to equations (3.28a) and (3.28b). The amplitude reflectance r and transmittance t of a monolayered dielectric film can also be expressed using r_1, t_1 and r_2, t_2 at these two interfaces and phase shift $\beta = (2\pi/\lambda_0)n_2 h \cos\theta_2$; the results are as follows:

$$r = \frac{r_1 + r_2 e^{2j\beta}}{1 + r_1 r_2 e^{2j\beta}}. \tag{3.29a}$$

Both TE and TM modes can satisfy

$$t = \frac{t_1 t_2 e^{j\beta}}{1 + r_1 r_2 e^{2j\beta}}. \tag{3.29b}$$

3.2.2.2 Intensity reflectance and transmittance of equivalent interface in monolayered uniform dielectric film

A Cartesian coordinate system was established on the equivalent interface, as shown in Fig. 3.3 (b). In the figure, $\boldsymbol{s}_i$, $\boldsymbol{s}_r$, and $\boldsymbol{s}_t$ represent the unit vectors of the incident wave, reflected wave, and transmitted wave at the equivalent interface. In an isotropic medium, their directions are consistent with those of the energy flux density vectors $\boldsymbol{S}_i$, $\boldsymbol{S}_r$, and $\boldsymbol{S}_t$ of the corresponding light waves.

The time-averaged energy flux density vectors of the incident wave, reflected wave, and transmitted wave at the equivalent surface are given by

$$\langle \boldsymbol{S}_i \rangle = \frac{1}{2}\sqrt{\frac{\varepsilon_0}{\mu_0}} n_1 |E_{i\text{I}}|^2 \boldsymbol{s}_i,$$

$$\langle \boldsymbol{S}_r \rangle = \frac{1}{2}\sqrt{\frac{\varepsilon_0}{\mu_0}} n_1 |E_{r\text{I}}|^2 \boldsymbol{s}_r,$$

$$|\boldsymbol{S}_t| = \frac{1}{2}\sqrt{\frac{\varepsilon_0}{\mu_0}} n_3 |E_{t\text{II}}|^2 \boldsymbol{s}_t.$$

Using $\boldsymbol{s}_i = \boldsymbol{i}\sin\theta_1 - \boldsymbol{k}\cos\theta_1$, $\boldsymbol{s}_r = \boldsymbol{i}\sin\theta_1 + \boldsymbol{k}\cos\theta_1$, and $\boldsymbol{s}_t = \boldsymbol{i}\sin\theta_3 - \boldsymbol{k}\cos\theta_3$, the time-averaged z-components of the energy flux density vectors for the three waves above are given, respectively, by

$$\langle \boldsymbol{S}_i \rangle_z = \frac{1}{2}\sqrt{\frac{\varepsilon_0}{\mu_0}} n_1 |E_{i\mathrm{I}}|^2 \cos\theta_1 = \frac{1}{2}P_1|E_{i\mathrm{I}}|^2,$$

$$\langle \boldsymbol{S}_r \rangle_z = \frac{1}{2}\sqrt{\frac{\varepsilon_0}{\mu_0}} n_1 |E_{r\mathrm{I}}|^2 \cos\theta_1 = \frac{1}{2}P_1|E_{r\mathrm{I}}|^2,$$

$$\langle \boldsymbol{S}_t \rangle_z = \frac{1}{2}\sqrt{\frac{\varepsilon_0}{\mu_0}} n_3 |E_{t\mathrm{II}}|^2 \cos\theta_3 = \frac{1}{2}P_3|E_{t\mathrm{II}}|^2.$$

The intensity reflectance R and transmittance T of the monolayered uniform dielectric film are given, respectively, by

$$R = \frac{\langle \boldsymbol{S}_r \rangle_z}{\langle \boldsymbol{S}_i \rangle_z} = \frac{|E_{r\mathrm{I}}|^2}{|E_{i\mathrm{I}}|^2} = |r|^2, \tag{3.30a}$$

$$T = \frac{\langle \boldsymbol{S}_t \rangle_z}{\langle \boldsymbol{S}_i \rangle_z} = \frac{P_3|E_{t\mathrm{II}}|^2}{P_1|E_{i\mathrm{I}}|^2} = \frac{P_3}{P_1}|t|^2. \tag{3.30b}$$

The calculation method for the intensity reflectance and transmittance of TM waves is similar. The corresponding R and T can be obtained by replacing P_i in equations (3.28a), (3.28b) and (3.29a), (3.29b) with $q_i = \sqrt{\varepsilon_0/\mu_0}\, n_i/\cos\theta_i$ $(i = 1, 3)$.

By substituting the r and t expressed in equations (3.29a) and (3.29b) into equations (3.30a) and (3.30b), we can obtain the reflectance and transmittance of a monolayered dielectric film:

$$R = |r|^2 = \frac{r_1^2 + r_2^2 + 2r_1r_2\cos 2\beta}{1 + r_1^2r_2^2 + 2r_1r_2\cos 2\beta}, \tag{3.31a}$$

$$T = \frac{P_3}{P_1}|t|^2 = \frac{n_3\cos\theta_3}{n_1\cos\theta_1}\frac{t_1^2t_2^2}{1 + r_1^2r_2^2 + 2r_1r_2\cos 2\beta}. \tag{3.31b}$$

Through equations (3.31a) and (3.31b), it can be seen that $R + T = 1$. The reflection phase shift δ_r and transmission phase shift δ_t are the arguments of complex numbers t and r, which can be calculated using equations (3.29a) and (3.29b).

3.2.2.3 Reflection and transmission characteristics of monolayered uniform dielectric films

Monolayered dielectric films are the most basic example of optical films. Hence, further discussion on their reflection and transmission characteristics is necessary. With further reference to Fig. 3.3 (a), the refractive indices of the topmost layer, coating layer, and substrate are n_1, n_2, and n_3, respectively, and the material is assumed to be nonmagnetic ($\mu_r = 1$). Given that the incident light is a TE wave, the matrix elements in the characteristic matrix of a monolayered film calculated from equations (3.1) to (3.17) are as follows:

$$m'_{11} = m'_{22} = \cos\beta, \quad m'_{12} = -\mathrm{j}\frac{\sin\beta}{P_2}, \quad m'_{21} = -\mathrm{j}P_2\sin\beta,$$

where

$$\beta = \frac{2\pi}{\lambda_0} n_2 h \cos\theta_2, \quad \text{and} \quad P_i = \sqrt{\frac{\varepsilon_0}{\mu_0}} n_i \cos\theta_i \quad (i = 1, 2, 3).$$

By substituting the above parameters into equations (3.26) and (3.27), the amplitude reflectance and transmittance of a monolayered film can be obtained, respectively, as follows:

$$t = \frac{2P_1}{(P_1 + P_3)\cos\beta - \mathrm{j}\left(\frac{P_1 P_3}{P_2} + P_2\right)\sin\beta}. \tag{3.32}$$

If β is replaced with $\beta + \pi$ in equation (3.32), i.e., if h is replaced by $h + \Delta h$, then the values of r and t will remain unchanged, and

$$\Delta h = \frac{\lambda_0}{2n_2 \cos\theta_2}.$$

Therefore, when the difference in thickness of the dielectric film is an integral multiple of $\lambda_0/2n_2 \cos\theta_2$, the reflectance and transmittance will remain invariant.

Let us now consider a special situation in which light is normally incident onto the surface of the monolayered dielectric film, i.e., $\theta_1 = \theta_2 = \theta_3 = 0$. Thus, $\beta = (2\pi/\lambda_0)n_2 h$, and $P_i = \sqrt{\varepsilon_0/\mu_0}\, n_i$ $(i = 1, 2, 3)$. If $q_i = \sqrt{\varepsilon_0/\mu_0}\, n_i$, we can see that TE and TM waves are indistinguishable at this point. If the parameter above is substituted into equation (3.32) and $R = |r|^2$, the following can be obtained:

$$R = \frac{(n_1 - n_3)^2 \cos^2\beta + \left(\frac{n_1 n_3}{n_2} - n_2\right)^2 \sin^2\beta}{(n_1 + n_3)^2 \cos^2\beta + \left(\frac{n_1 n_3}{n_2} + n_2\right)^2 \sin^2\beta}. \tag{3.33}$$

The above is the expression for the intensity reflectance of a monolayered uniform dielectric film under normal incidence of light.

Fig. 3.4 shows the trend of variation in the thin-film reflectance R with the optical thickness of the coating layer $n_2 h$, which was plotted according to equation (3.33). The medium at the incidence side is air, with $n_1 = 1$, and the substrate material is glass, with $n_3 = 1.5$. The characteristics of normally incident light of wavelength λ_0 entering a monolayered uniform dielectric film can be obtained through equation (3.13) combined with Fig. 3.4. The conclusions are as follows.

(1) The reflectance R of the thin film is a function of $\sin^2\beta$ and $\cos^2\beta$, with a period of π. Let $\beta = (2\pi/\lambda_0)n_2 h$. When the change in the geometric thickness of the coating layer is

$$\Delta h = \frac{\lambda_0}{2n_2}, \tag{3.34}$$

then the reflectance R of the thin film will remain unchanged. In more general terms, if the thickness of the coating layer h is an integral multiple of the phase shift $\lambda_0/2n_2$, then the reflectance of the thin film, R, will remain unchanged. Thus, it is not necessary for the coating layer to be very thick when coating a monolayered reflective or antireflective film. Fig. 3.4 also shows that the thin-film reflectance R is the same for every $\lambda_0/2$ increase in $n_2 h$.

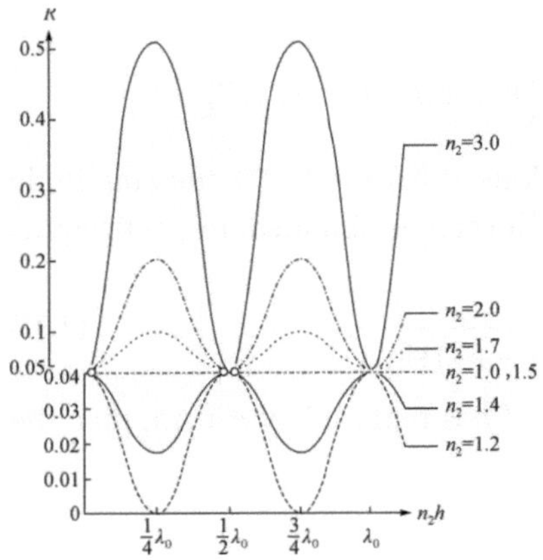

Fig. 3.4: Variations in reflectance of monolayered dielectric film with optical thickness.

(2) When $\beta = m\pi$, or in other words, when the optical thickness of the coating layer is

$$n_2 h = m\frac{\lambda_0}{2} \quad (m = 1, 2, 3, \ldots), \tag{3.35}$$

based on equation (3.33), with $\sin^2 \beta = 0$, the reflectance R of the thin film will be

$$R = \left(\frac{n_1 - n_3}{n_1 + n_3}\right)^2. \tag{3.36}$$

Equation (3.36) indicates that when light passes "directly" from a medium with a refractive index of n_1 to a substrate with a refractive index of n_3, if the optical thickness of the thin film, $n_2 h$, is an even multiple of $\lambda_0/4$ (i.e., integral multiple of $\lambda_0/2$), then the thin-film layer will have no effect on light reflection, as though the light has passed directly from a medium with a refractive index of n_1 to a substrate with a refractive index of n_3.

(3) When $\beta = (2m + 1)\pi/2$, or in other words, when the optical thickness of the coating layer is

$$n_2 h = (2m + 1)\frac{\lambda_0}{4} \quad (m = 0, 1, 2, \ldots)$$

with $\cos^2 \beta = 0$, then the reflectance of the thin film will be

$$R = \left(\frac{n_1 n_3 - n_2^2}{n_1 n_3 + n_2^2}\right)^2 = \left(\frac{n_1 - \frac{n_2^2}{n_3^2} n_3}{n_1 + \frac{n_2^2}{n_3^2} n_3}\right)^2. \tag{3.37}$$

With reference to Fig. 3.4, it can be seen that when the optical thickness of the thin-film, $n_2 h$, is an odd multiple of $\lambda_0/4$ (generally $n_2 h = \lambda_0/4$), then the reflectance R of the thin-film layer will have maxima or minima. However, whether the value is a maximum or a minimum will depend on whether the refractive index of the coating layer, n_2, is greater or smaller than that of the substrate, n_3.

(a) For a $\lambda_0/4$ coating layer with higher refractive index, i.e., $n_2 > n_3$, the maximum of the thin-film reflectance R should be obtained. For example, if the coating material is ZnS, $n_2 = 2.34$, and the refractive index of the glass substrate is $n_3 = 1.5$; thus, $n_2 > n_3$. This $\lambda_0/4$ coating layer will increase the reflectance of the substrate surface and thus is known as a monolayer high-reflective coating.

(b) For a $\lambda_0/4$ coating layer with lower refractive index, i.e., $n_2 < n_3$, the minimum of the thin-film reflectance R should be obtained. For example, if the coating material is MgF_2, $n_2 = 1.38$, which is lower than the refractive index of the glass substrate, $n_3 = 1.5$. This $\lambda_0/4$ coating layer will decrease the reflectance of the substrate surface and thus is known as a monolayer antireflective coating. According to equation (3.37), let $R = 0$, which gives

$$n_2 = \sqrt{n_1 n_3}. \tag{3.38}$$

If a material with such a refractive index is selected, it should, in theory, enable the complete transmission of light without reflection. However, in reality, it is very difficult to find coating materials that fully comply with the refractive index described in equation (3.38) for a given substrate and topmost layer.

3.2.3 Multilayered uniform dielectric film

For a periodic medium with a period of h, its dielectric constant ε and magnetic permeability μ are functions of z, and they should fulfil the following periodicity conditions:

$$\varepsilon(z + nh) = \varepsilon(z), \quad \mu(z + nh) = \mu(z).$$

If there are N periods, then n will be a whole number within $1 \leq n \leq N$.

$\boldsymbol{M}(h)$ denotes the characteristic matrix of one period:

$$\boldsymbol{M}(h) = \begin{bmatrix} m_{11} & m_{12} \\ m_{21} & m_{22} \end{bmatrix}. \tag{3.39}$$

The overall characteristic matrix of a multilayered film with N number of periods will be the product of N number of $\boldsymbol{M}(h)$ matrices, which can be expressed as follows:

$$\boldsymbol{M}(Nh) = \boldsymbol{M}(h)\boldsymbol{M}(h)\ldots\boldsymbol{M}(h) = [\boldsymbol{M}(h)]^N. \tag{3.40}$$

According to matrix theory, we know that

$$[\boldsymbol{M}(h)]^N = \begin{bmatrix} m_{11}u_{N-1}(a) - u_{N-2}(a) & m_{12}u_{N-1}(a) \\ m_{21}u_{N-1}(a) & m_{22}u_{N-1}(a) - u_{N-2}(a) \end{bmatrix} \tag{3.41}$$

$$a = \tfrac{1}{2}(m_{11} + m_{22}), \tag{3.42}$$

where u_N is a Chebyshev polynomial of the second kind, given by

$$u_N(x) = \frac{\sin[(N+1)\arccos x]}{\sqrt{1-x^2}}. \tag{3.43}$$

If $x = \cos\theta$, this polynomial can also be expressed as

$$u_N(x) = u_N(\cos\theta) = \frac{\sin[(N+1)\theta]}{\sin\theta}.$$

The first six polynomials are explicitly expressed as follows:

$$u_0(x) = 1, \quad u_1(x) = 2x, \quad u_2(x) = 4x^2 - 1,$$
$$u_3(x) = 8x^3 - 4x, \quad u_4(x) = 16x^4 - 12x^2 + 1, \quad u_5(x) = 32x^5 - 32x^3 + 6x.$$

The following recursive formula was applied:

$$u_i(x) = 2xu_{i-1}(x) - u_{i-2}(x).$$

Specifically, when $|x| > 1$, θ in $x = \cos\theta$ is an imaginary number. Let $\theta = \arccos x = \mathrm{j}\chi$, and for any real number χ,

$$u_N(\cos\theta) = \frac{\sin[\mathrm{j}(N+1)\chi]}{\sin(\mathrm{j}\chi)} = \frac{\mathrm{j}sh[(N+1)\chi]}{\mathrm{j}sh\chi} = \frac{e^{(N+1)\chi} - e^{-(N+1)\chi}}{e^{\chi} - e^{-\chi}}.$$

Fig. 3.5 is a schematic of a periodic multilayered film. The entire coating system is situated between two uniform media with refractive indices of n_0 and n_G. This coating system uses a double-layered film as its basic periodic unit. Each period is composed of media with two different refractive indices. For example, a low-refractive-index coating with refractive index n_2 and thickness h_2 can be selected with a high-refractive-index coating with refractive index n_3 and thickness h_3. Assume that all media are nonmagnetic ($\mu_r = 1$).

As shown in Fig. 3.5, assuming that the light wave, which passes from a medium with refractive index n_0 into a multilayered film, is a TE wave, its parameters will be

$$\begin{cases} \beta_2 = \dfrac{2\pi}{\lambda_0} n_2 h_2 \cos\theta_2, \quad \beta_3 = \dfrac{2\pi}{\lambda_0} n_3 h_3 \cos\theta_3, \\ P_2 = \sqrt{\dfrac{\varepsilon_0}{\mu_0}} n_2 \cos\theta_2, \quad P_3 = \sqrt{\dfrac{\varepsilon_0}{\mu_0}} n_3 \cos\theta_3, \quad h = h_2 + h_3. \end{cases} \tag{3.44}$$

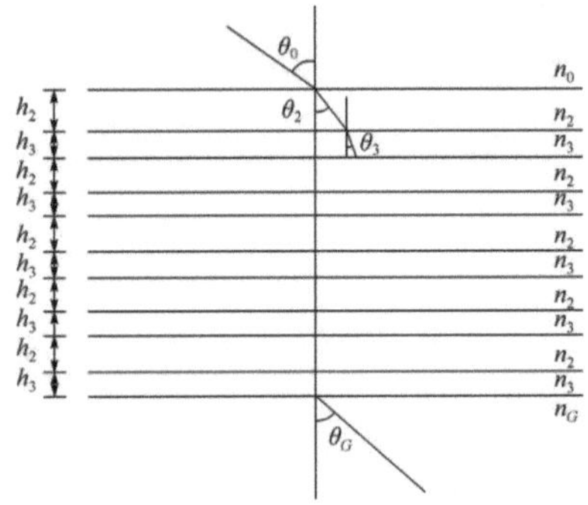

Fig. 3.5: Schematic diagram of the structure of a periodic multilayered film.

The characteristic matrix of one period is as follows:

$$
\begin{aligned}
\boldsymbol{M}_2(h) &= \begin{bmatrix} \cos\beta_2 & -\mathrm{j}\frac{\sin\beta_2}{P_2} \\ -\mathrm{j}P_2\sin\beta_2 & \cos\beta_2 \end{bmatrix} \begin{bmatrix} \cos\beta_3 & -\mathrm{j}\frac{\sin\beta_3}{P_3} \\ -\mathrm{j}P_3\sin\beta_3 & \cos\beta_3 \end{bmatrix} \\
&= \begin{bmatrix} \cos\beta_2\cos\beta_3 - \frac{P_3}{P_2}\sin\beta_2\sin\beta_3 & \frac{-\mathrm{j}}{P_3}\cos\beta_2\sin\beta_3 - \frac{\mathrm{j}}{P_2}\sin\beta_2\cos\beta_3 \\ -\mathrm{j}P_2\sin\beta_2\cos\beta_3 - \mathrm{j}P_3\cos\beta_2\sin\beta_3 & \cos\beta_2\cos\beta_3 - \frac{P_2}{P_3}\sin\beta_2\sin\beta_3 \end{bmatrix}.
\end{aligned}
\tag{3.45}
$$

Therefore, we can see from equation (3.40) that $2N$ layers of media exist between the substrate and coating layers of a multilayer film with N number of periods. Thus, its characteristic matrix $\boldsymbol{M}_{2N}(Nh)$ is given by the following equation:

$$
\boldsymbol{M}_{2N}(Nh) = [\boldsymbol{M}(h)]^N = \begin{bmatrix} \mu_{11} & \mu_{12} \\ \mu_{21} & \mu_{22} \end{bmatrix}. \tag{3.46}
$$

By using equations (3.41) and (3.45), the parameters in equation (3.46) can be obtained as follows:

$$
\begin{aligned}
\mu_{11} &= \left[\cos\beta_2\cos\beta_3 - \frac{P_3}{P_2}\sin\beta_2\sin\beta_3\right]u_{N-1}(a) - u_{N-2}(a), \\
\mu_{12} &= -\mathrm{j}\left[\frac{1}{P_3}\cos\beta_2\sin\beta_3 + \frac{1}{P_2}\sin\beta_2\cos\beta_3\right]u_{N-1}(a), \\
\mu_{21} &= -\mathrm{j}\left[P_2\sin\beta_2\cos\beta_3 + P_3\cos\beta_2\sin\beta_3\right]u_{N-1}(a), \\
\mu_{22} &= \left[\cos\beta_2\cos\beta_3 - \frac{P_2}{P_3}\sin\beta_2\sin\beta_3\right]u_{N-1}(a) - u_{N-2}(a),
\end{aligned}
\tag{3.47}
$$

$$
a = \frac{m_{11} + m_{22}}{2} = \cos\beta_2\cos\beta_3 - \frac{1}{2}\left(\frac{P_2}{P_3} + \frac{P_3}{P_2}\right)\sin\beta_2\sin\beta_3. \tag{3.48}
$$

At this point, the periodic multilayered film can be converted using a characteristic matrix into a monolayered film for equivalence treatment. The refractive indices of the topmost layer and substrate will still be n_0 and n_G. With reference to equation (3.26), the matrix element m'_{ij} can be replaced by μ_{ij} in the equation, and the amplitude reflectance r and transmittance t of the periodic multilayered film can be obtained as follows:

$$
r = \frac{(\mu_{11} + \mu_{12}P_G)P_0 - (\mu_{21} + \mu_{22}P_G)}{(\mu_{11} + \mu_{12}P_G)P_0 + (\mu_{21} + \mu_{22}P_G)}, \tag{3.49}
$$

$$
t = \frac{2P_0}{(\mu_{11} + \mu_{12}P_G)P_0 + (\mu_{21} + \mu_{22}P_G)}, \tag{3.50}
$$

where $P_0 = \sqrt{\varepsilon_0/\mu_0}\, n_0 \cos\theta_0$ and $P_G = \sqrt{\varepsilon_0/\mu_0}\, n_G \cos\theta_G$.

Based on equations (3.49) and (3.50), the energy reflectance and transmittance of the periodic multilayered film can also be determined:

$$
R = |r|^2, \quad T = \frac{P_G}{P_0}|t|^2. \tag{3.51}
$$

3.3 Oblique incidence theory and coating design

Imagine a light wave passing through an interface at a given angle θ_0. As can be seen from Fig. 3.6 and 3.7, the electric field E of an S-polarized wave is parallel to the interface. Based on the continuity on both sides of the interface, we know that

$$E_i + E_r = E_t, \tag{3.52}$$

$$H_i \cos\theta_0 + H_r \cos\theta_0 = H_t \cos\theta_1, \tag{3.53}$$

where E_i, E_r, and E_t are the electric-field components of the incident wave, reflected wave, and transmitted wave, respectively; H_i, H_r, and H_t are the magnetic-field components of the incident wave, reflected wave, and transmitted wave, respectively; and θ_0 and θ_1 are the angles of incidence and refraction, respectively.

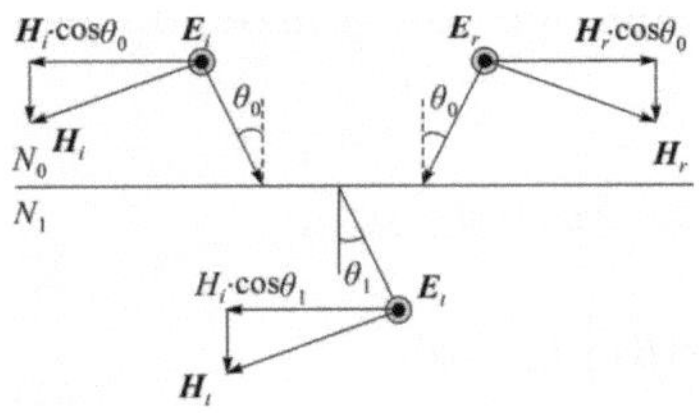

Fig. 3.6: Schematic diagram of the electric- and magnetic-field distribution of S-polarized waves incident on both sides of a single interface.

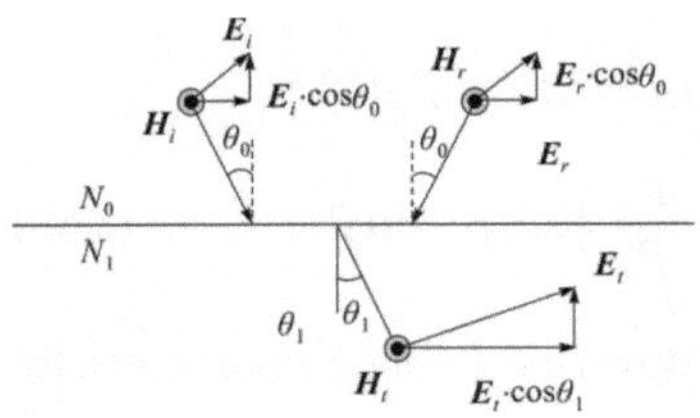

Fig. 3.7: Schematic diagram of the electric- and magnetic-field distribution of P-polarized waves incident on both sides of a single interface.

The relationships between optical admittances Y_0 and Y_1 with the electric- and magnetic-field intensity are as follows:

$$\begin{cases} \boldsymbol{H}_i = Y_0(\boldsymbol{k}_0 \times \boldsymbol{E}_i) \\ \boldsymbol{H}_r = Y_0(-\boldsymbol{k}_0 \times \boldsymbol{E}_r) \\ \boldsymbol{H}_t = Y_1(\boldsymbol{k}_0 \times \boldsymbol{E}_t), \end{cases} \tag{3.54}$$

where Y_0 and Y_1 are the light admittances of the incident medium and transmission medium, respectively, and $\boldsymbol{k}_0$ is the unit vector along the direction of light transmission.

Based on equation (3.54), the relationship between the admittances on both sides of the interface with the angle of incidence and the components of electric-field intensity can be obtained:

$$Y_0 E_i \cos\theta_0 - Y_0 E_r \cos\theta_0 = Y_1 E_t \cos\theta_1. \tag{3.55}$$

Based on equations (3.52)–(3.54), the amplitude transmission and reflection coefficients of S-polarized waves can be obtained:

$$r_s = \frac{E_r}{E_i} = \frac{Y_0 \cos\theta_0 - Y_1 \cos\theta_1}{Y_0 \cos\theta_0 + Y_1 \cos\theta_1}, \tag{3.56}$$

$$t_s = \frac{E_t}{E_i} = \frac{2Y_0 \cos\theta_0}{Y_0 \cos\theta_0 + Y_1 \cos\theta_1}. \tag{3.57}$$

Similarly, for P-polarized waves, H will be parallel to the interface. Based on the continuity of the magnetic field on both sides of the interface, the following can be obtained:

$$r_p = \frac{E_r}{E_i} = \frac{Y_0 \cos\theta_1 - Y_1 \cos\theta_0}{Y_0 \cos\theta_1 + Y_1 \cos\theta_0}, \tag{3.58}$$

$$t_p = \frac{E_t}{E_i} = \frac{2Y_0 \cos\theta_0}{Y_0 \cos\theta_1 + Y_1 \cos\theta_0}. \tag{3.59}$$

It can be seen that when electric and magnetic waves are obliquely incident on an interface, S-polarized and P-polarized waves will have different effective optical admittances:

$$Y_s = \frac{H}{E} = Y \cos\theta = y_0 N \cos\theta, \tag{3.60}$$

$$Y_p = \frac{H}{E} = \frac{Y}{\cos\theta} = \frac{y_0 N}{\cos\theta}, \tag{3.61}$$

where Y is the optical admittance of the medium, $y_0 = 1/377$ is the optical admittance of free space, θ is the angle of incidence, and N is the refractive index of the medium.

Correspondingly, we can deduce that the reflection and transmission of S-polarized and P-polarized waves at the interface will be different:

$$r_s = \frac{Y_0 \cos\theta_0 - Y_1 \cos\theta_1}{Y_0 \cos\theta_0 + Y_1 \cos\theta_1} = \frac{N_0 \cos\theta_0 - N_1 \cos\theta_1}{N_0 \cos\theta_0 + N_1 \cos\theta_1}, \tag{3.62}$$

$$t_s = \frac{2Y_0 \cos\theta_0}{Y_0 \cos\theta_0 + Y_1 \cos\theta_1} = \frac{2N_0 \cos\theta_0}{N_0 \cos\theta_0 + N_1 \cos\theta_1}, \tag{3.63}$$

$$r_p = \frac{Y_0 \cos\theta_1 - Y_1 \cos\theta_0}{Y_0 \cos\theta_1 + Y_1 \cos\theta_0} = \frac{N_0 \cos\theta_1 - N_1 \cos\theta_0}{N_0 \cos\theta_1 + N_1 \cos\theta_0}, \tag{3.64}$$

$$t_p = \frac{2Y_0 \cos\theta_0}{Y_0 \cos\theta_1 + Y_1 \cos\theta_0} = \frac{2N_0 \cos\theta_0}{N_0 \cos\theta_1 + N_1 \cos\theta_0}. \tag{3.65}$$

It can be seen from equations (3.62) to (3.65) that oblique incidence leads to differences in effective admittance between S-polarized and P-polarized light within the same medium. Thus, S-polarized and P-polarized light will show different reflective and refractive effects on the same interface. This can be seen in Fig. 3.8, which shows the variations in the transmittances of S- and P-polarized light with the angle of incidence. With regards to the design of antireflective coatings, during the oblique incidence of light, the transmittance of S-polarized light is drastically reduced with increase in polarization due to the polarization splitting of S- and P-polarized light.

Consequently, the transmittance of the antireflective coating is significantly reduced compared to that for normal incidence, thereby deteriorating the performance of the optical system, especially in infrared window components with high transmittance requirements. Therefore, methods to reduce or eliminate the degree of polarization are necessary to enhance the transmittance of the S-polarization component for achieving high-intensity infrared transmission.

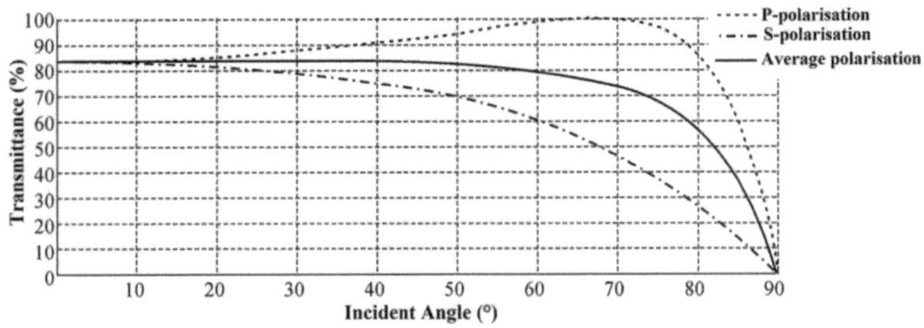

Fig. 3.8: Variations in the transmittance of S- and P-polarized light with angle of incidence.

In order to simplify our calculations, all thin films considered in the coating system mentioned in this book are nonabsorbing dielectric films ($k \approx 0$), unless specified otherwise. Hence, the admittances of thin films are all real numbers. When a wave is obliquely incident on a monolayered antireflective coating, the phase shifts of the S- and P-polarized components for the coating layer are both $\delta_1 = 2\pi/\lambda\, n_1 d_1 \cos\theta_1$. However, the corrected admittances, η_s and η_p, (slightly different from the aforementioned admittance Y) for the same coating layer are different. Furthermore, since transmittance is also a function of the refractive index n, the deviation component Δn of the coating system for transmittance T and reflectance R is given by

$$\Delta n = \frac{\eta_P}{\eta_S} = \frac{n/\cos\theta}{n\cos\theta} = \frac{1}{\cos^2\theta} = \frac{1}{\left(1 - \frac{n_0^2 \sin^2\theta_0}{n^2}\right)}, \tag{3.66}$$

where Δn is the change in refractive index, η_P and η_S are the (corrected) admittances of P- and S-polarized light waves, respectively, n_0 and n are the refractive indices of the incident medium and coating layer, respectively, and θ is the angle of incidence.

Evidently, for a monolayered film, Δn will always be greater than 1, wherein $\eta_P > \eta_S$. Thus, it is not possible to achieve depolarization in a monolayered film. Fig. 3.9 shows the polarization state of a monolayered film. In other words, a multilayered coating system is required for depolarization. Polarization is inevitable in monolayered antireflective coatings and will lead to decreased transmittance. Fig. 3.10 shows the variations in transmittance with wavelength at different angles of incidence for an antireflective coating. For a known n and n_0, Δn will increase with increasing angle of incidence. In addition, for a known refractive index n_0 and angle of incidence θ_0

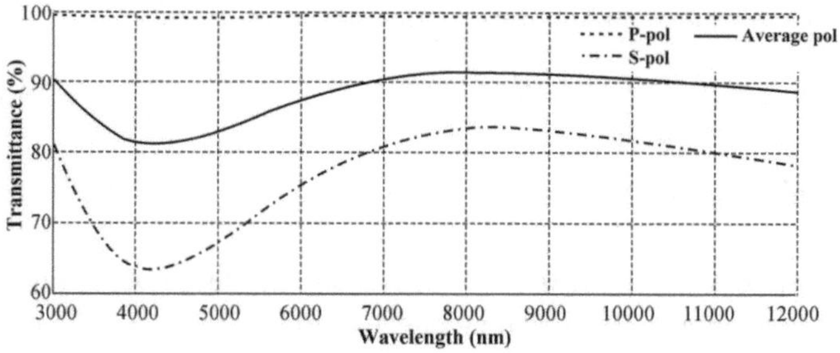

Fig. 3.9: Polarization state of monolayered film.

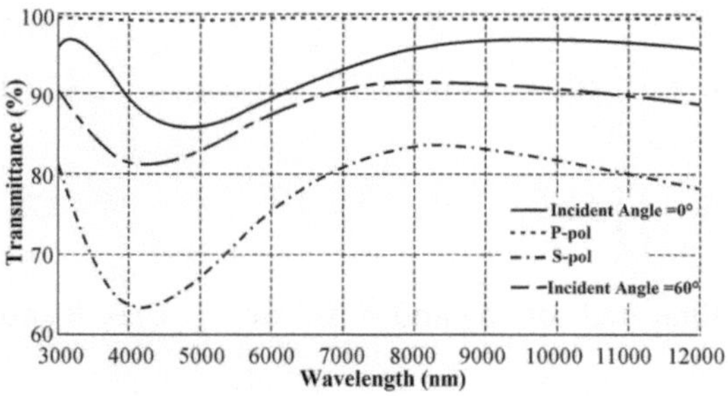

Fig. 3.10: Variation of transmittance at different angles of incidence in anti-reflective coating.

of the incident medium, a larger refractive index of the thin-film material will result in a smaller Δn. Therefore, as coating materials, multilayered materials that are compatible with the substrate and have relatively high equivalent admittances should be selected.

Similarly, for multilayered films under oblique incidence, the admittance values for S- and P-polarized light are different for each thin-film layer; the admittance values at the j-th layer are, respectively

$$\eta_{Pj} = \frac{n_j}{\cos\theta_j}, \tag{3.67}$$

$$\eta_{Sj} = n_j \cos\theta_j, \tag{3.68}$$

where n_j is the refractive index of the j-th layer, η_{Sj} is the corrected admittance of S-polarization at the j-th layer, η_{Pj} is the corrected admittance of P-polarization at the j-th layer, and θ_j is the angle of incidence at the j-th layer.

θ_j can be obtained using Snell's law:

$$n_0 \sin\theta_0 = n_S \sin\theta_S = n_j \sin\theta_j, \tag{3.69}$$

where n_0 and n_S are the refractive indices of the incident medium and substrate, respectively, and θ_0, θ_S, and θ_j are the angles of incidence and refraction.

For both P- and S-polarization, the effective optical thickness ($n_j d_j \cos\theta_j$) and effective phase shift ($\delta_j = 2\pi n_j d_j \cos\theta_j/\lambda$) of the coating layer are identical. The optical matrix at the j-th layer in a multilayered film is expressed as follows:

$$\boldsymbol{M} = \begin{pmatrix} \cos\theta_j & (\mathrm{i}/\eta_j)\sin\theta_j \\ \mathrm{i}\eta_j\sin\theta_j & \cos\theta_j \end{pmatrix}, \tag{3.70}$$

where θ_j is the angle of incidence at the j-th layer and η_j is the optical admittance at the j-th layer.

As can be seen, discrepancy exists between the characteristic matrix of P- and S-polarization in terms of the difference between their effective admittances, $\eta_{j\mathrm{p}}$ and $\eta_{j\mathrm{s}}$.

The characteristic matrices of the entire coating layer, A_P and A_S, are the products of m characteristic matrices $M_{\mathrm{P}j}$ and $M_{\mathrm{S}j}$:

$$A_\mathrm{S} = \prod_{j=1}^{m} M_{\mathrm{S}j}, \tag{3.71}$$

$$A_\mathrm{P} = \prod_{j=1}^{m} M_{\mathrm{P}j}. \tag{3.72}$$

The combined admittance of the multilayered coating and substrate $Y = C/B$. B and C are given by

$$\begin{bmatrix} B \\ C \end{bmatrix} = A \begin{bmatrix} 1 \\ \eta_\mathrm{S} \end{bmatrix}, \tag{3.73}$$

where η_S is the admittance of the substrate.

The amplitude reflection coefficient is

$$r = \frac{\eta_0 - Y}{\eta_0 + Y} = \frac{1 - Y/\eta_0}{1 + Y/\eta_0}. \tag{3.74}$$

The energy reflectance is

$$R = rr^* = \left|\frac{1 - Y/\eta_0}{1 + Y/\eta_0}\right|^2. \tag{3.75}$$

Substituting the effective admittance of P- and S-polarization, R_P and R_S, into the equation above will allow us to calculate their reflectance and transmittance. Under large angles, the difference between R_P and R_S will also be greater. Thus, such deviations need to be minimized for both P- and S-polarized light to achieve high antireflective effects for a given wavelength.

Currently, precise calculations for the optical characteristics of optical multilayered films are mostly performed using characteristic matrices. Even the reflection and transmission characteristics of coating systems having up to a hundred layers within a given wavelength range can be computed very rapidly. However, the opposite problem, i.e. designing a coating combination with specific characteristics and its parameters, is extremely difficult. A few effective analytical design techniques have been developed based on the model above for certain elements of the coating system, including the symmetrical coating, periodic structure, and quarter-wave stack. A common method used to obtain large-angle oblique-incidence antireflective coatings that allow oblique incidence while also maximizing transmittance, is to apply analytical design techniques to obtain a preliminary structure. Optical thin-film theory is applied to construct an appropriate performance evaluation function. Then, the preliminary film structure is tested and corrected using numerical optimization in order to minimize the evaluation function, thereby achieving the required characteristics.

Therefore, the design framework in this chapter will be as follows: design theory → preliminary coating design → coating optimization → satisfy requirements.

The section above discussed oblique incidence in monolayered films, which revealed that monolayered films are not able to achieve depolarization. Therefore, the next section considers a double-layered periodic film structure with alternating high and low refractive indices.

The characteristic matrix of a basic period for a coating system with a structure of $n_0|\alpha\mathrm{H}\beta\mathrm{L}|^s n_g$ or $n_0|\beta\mathrm{L}\alpha\mathrm{H}|^s n_g$ is given by

$$\begin{aligned} M &= \begin{pmatrix} M_{11} & M_{12} \\ M_{21} & M_{22} \end{pmatrix} \\ &= \begin{bmatrix} \cos(\frac{\pi}{2}g\alpha) & \frac{i}{\eta_H}\sin(\frac{\pi}{2}g\alpha) \\ i\eta_H\sin(\frac{\pi}{2}g\alpha) & \cos(\frac{\pi}{2}g\alpha) \end{bmatrix} \begin{pmatrix} \cos(\frac{\pi}{2}g\beta) & \frac{i}{\eta_L}\sin(\frac{\pi}{2}g\beta) \\ i\eta_L\sin(\frac{\pi}{2}g\beta) & \cos(\frac{\pi}{2}g\beta) \end{pmatrix}, \end{aligned} \tag{3.76}$$

where α and β are the effective optical thicknesses with $\lambda_0/4$ as the unit.

Let

$$g = \frac{\lambda_0}{\lambda}, \quad \delta = \frac{2\pi}{\lambda}\frac{\lambda_0}{4} = \frac{\pi}{2}\frac{\lambda_0}{\lambda}.$$

Thus,

$$\begin{aligned} \zeta &= \frac{1}{2}(M_{11} + M_{22}) \\ &= \cos\Big(\frac{\pi}{2}g\alpha\Big)\cos\Big(\frac{\pi}{2}g\beta\Big) - \frac{1}{2}\Big(\frac{\eta_H}{\eta_L} + \frac{\eta_L}{\eta_H}\Big)\sin\Big(\frac{\pi}{2}g\alpha\Big)\sin\Big(\frac{\pi}{2}g\beta\Big). \end{aligned} \tag{3.77}$$

The equation below can be used to measure the polarization deviation. At the cut-off of the transmission wave range, $|M_{11}, M_{22}| \to 1$. Substituting $\zeta = -1$ into equation (3.77), we obtain

$$\frac{\rho + \frac{1}{\rho} - 2}{\rho + \frac{1}{\rho} + 2} = \frac{\cos^2\left[\frac{\pi}{2}g\left(\frac{\alpha+\beta}{2}\right)\right]}{\cos^2\left[\frac{\pi}{2}g\left(\frac{\alpha-\beta}{2}\right)\right]} \tag{3.78}$$

$$\rho = \frac{\eta_H}{\eta_L}.$$

For a given design and angle of incidence, the difference between the cut-off wavelengths of S- and P-polarization can be used to measure the polarization effects. The positions ρ_S and ρ_P can be calculated based on equation (3.77):

$$\rho_P = \left(\frac{\eta_H}{\eta_L}\right)_P = \frac{n_H\cos\theta_L}{n_L\cos\theta_H}, \tag{3.79}$$

$$\rho_S = \left(\frac{\eta_H}{\eta_L}\right)_S = \frac{n_H\cos\theta_H}{n_L\cos\theta_L}, \tag{3.80}$$

where n_H is the refractive index of the high-refractive-index material, n_L is the refractive index of the low-refractive-index material, η_H is the admittance of the high-refractive-index material, and η_L is the admittance of the low-refractive-index material.

It is only when $n_H = n_L$ that $\rho_S = \rho_P$ will be established and δ_g will be equal to 0. The results obtained via proof by contradiction indicate that a multilayered film with a two-layer period cannot eliminate the polarization effects completely and is only able to achieve partial reduction. This further verifies the fact that designing a depolarising thin film should start with at least three layers.

The issue of depolarization is an extremely complex problem in optical thin-film design, which has attracted the attention of many researchers. A series of studies have been conducted worldwide, and they have produced numerous valuable results. Costich, Mahlein, Thelen, Gilo, and others have developed and applied several methods to design nonpolarising thin films, including the equivalent refractive index, a thin film with a basic period of two-layers, and a $\lambda/4$ wave stack of three materials with different refractive indices. In domestic research, Prof. Lin Yong-chang, and Prof. Tang Jin-fa have conducted preliminary investigations.

The following section will discuss the design of a large-angle oblique-incidence antireflective coating with a ZnS substrate based on the key theories.

3.3.1 Depolarization design of incident medium, coating and substrate combinations

Based on the equivalence theory of optical thin films, any multilayered dielectric film can be equivalent to a two-layer film. Only multilayered dielectric films with a symmetrical structure can be equivalent to a monolayered film of equivalent refractive index and thickness with chromatic dispersion. For a symmetrical coating system $n_0|L_2HL_2|n_g$ or $n_0|H_2LH_2|n_g$,

$$\frac{E}{n_H} = \left\{\frac{\left[\sin\delta\cos\delta + \frac{1}{2}(n_H/n_L + n_L/n_H)\sin\delta\cos\delta - \frac{1}{2}(n_H/n_L - n_L/n_H)\sin\delta\right]}{\left[\sin\delta\cos\delta + \frac{1}{2}(n_H/n_L + n_L/n_H)\sin\delta\cos\delta + \frac{1}{2}(n_H/n_L - n_L/n_H)\sin\delta\right]}\right\}^{1/2}, \tag{3.81}$$

$$\frac{E}{n_L} = \left\{\frac{\left[\sin\delta\cos\delta + \frac{1}{2}(n_L/n_H + n_H/n_L)\sin\delta\cos\delta - \frac{1}{2}(n_L/n_H - n_H/n_L)\sin\delta\right]}{\left[\sin\delta\cos\delta + \frac{1}{2}(n_L/n_H + n_H/n_L)\sin\delta\cos\delta + \frac{1}{2}(n_L/n_H - n_H/n_L)\sin\delta\right]}\right\}^{1/2}, \tag{3.82}$$

where E is the equivalent refractive index and δ is the equivalent phase shift.

When $\lambda = \lambda_0/2$, the equivalent refractive index obtained after substitution of this λ into equation (3.81) is as follows:

$$E = n_H \sqrt{\frac{n_H}{n_L}} \tag{3.83}$$

or

$$E = n_L \sqrt{\frac{n_L}{n_H}}. \tag{3.84}$$

For a coating system with the $n_0|H_2LH_2|n_g$ combination, the polarization-splitting component at the position when wavelength $\lambda = \lambda_0/2$ is given by

$$\Delta E = \frac{E_P}{E_S} = \Delta n_H \sqrt{\Delta n_H / \Delta n_L}. \tag{3.85}$$

In order to produce polarization splitting for this particular coating layer at this wavelength position, i.e., $\Delta E = 1$, it can be deduced that

$$1 - \frac{n_0^2 \sin^2 \theta_0}{n_L^2} = \left(1 - \frac{n_0^2 \sin^2 \theta_0}{n_H^2}\right)^3. \tag{3.86}$$

When the incident medium is air, i.e., when $n_0 = 1$, and the angle of incidence $\theta_0 = 60°$, then the relationship between the high and low refractive indices is given by

$$n_H \approx \sqrt{3} n_L. \tag{3.87}$$

A symmetrical coating system with matched film materials that satisfy the conditions in equation (3.87) will not have polarization splitting at the $\lambda = \lambda_0/2$ wavelength position. However, the polarization effect is not completely eliminated in the entire thin-film system. The substrate and incident medium will still exhibit polarization splitting, both of which will still require depolarization in sequence. A coating system with a depolarization combination designed using this method does not take transmittance into account or might be limited by the need for a number of materials, implying that this method cannot be widely applied. Furthermore, this design involves performing depolarization thrice, which will increase the calculation complexity. Therefore, the quarter-wave stack method is often employed during the actual design process.

3.3.2 Quarter-wave stack depolarization design

In order to compensate for the shortcomings of the theoretical design methods above, the quarter-wave stack depolarization design was gradually developed, which considers the polarization characteristics of the coating layer, substrate, and incident medium in an integrated manner.

The characteristic matrix of a quarter-wave stack with an odd number of layers $(2k+1)$ is given by

$$\begin{aligned}\begin{pmatrix}B\\C\end{pmatrix} &= \begin{pmatrix}0 & \frac{i}{\eta_1}\\ i\eta_1 & 0\end{pmatrix}\begin{pmatrix}0 & \frac{i}{\eta_2}\\ i\eta_2 & 0\end{pmatrix}\begin{pmatrix}0 & \frac{i}{\eta_3}\\ i\eta_3 & 0\end{pmatrix}\\ &\quad\cdots\begin{pmatrix}0 & \frac{i}{\eta_{2k-1}}\\ i\eta_{2k-1} & 0\end{pmatrix}\begin{pmatrix}0 & \frac{i}{\eta_{2k}}\\ i\eta_{2k} & 0\end{pmatrix}\begin{pmatrix}0 & \frac{i}{\eta_{2k+1}}\\ i\eta_{2k+1} & 0\end{pmatrix}\begin{pmatrix}1\\ \eta_g\end{pmatrix}\\ &= \begin{pmatrix}\frac{-\eta_2}{\eta_1} & 0\\ 0 & \frac{-\eta_1}{\eta_2}\end{pmatrix}\begin{pmatrix}\frac{-\eta_4}{\eta_3} & 0\\ 0 & \frac{-\eta_3}{\eta_4}\end{pmatrix}\cdots\begin{pmatrix}\frac{-\eta_{2k}}{\eta_{2k-1}} & 0\\ 0 & \frac{-\eta_{2k-1}}{\eta_{2k}}\end{pmatrix}\begin{pmatrix}i\frac{\eta_g}{\eta_{2k+1}}\\ i\eta_{2k+1}\end{pmatrix}\\ &= \begin{pmatrix}i\frac{\eta_g\eta_2\eta_4\ldots\eta_{2k}}{\eta_1\eta_3\ldots\eta_{2k+1}}\\ i\frac{\eta_1\eta_3\ldots\eta_{2k+1}}{\eta_2\eta_4\ldots\eta_{2k}}\end{pmatrix}\end{aligned} \tag{3.88}$$

$$Y = \frac{C}{B} = \frac{\eta_1^2\eta_3^2\cdots\eta_{2k+1}^2}{\eta_g\eta_2^2\eta_4^2\cdots\eta_{2k}^2} \tag{3.89}$$

$$R = \left(\frac{\eta_0\eta_g\eta_2^2\eta_4^2\cdots\eta_{2k}^2 - \eta_1^2\eta_3^2\cdots\eta_{2k+1}^2}{\eta_0\eta_g\eta_2^2\eta_4^2\cdots\eta_{2k}^2 + \eta_1^2\eta_3^2\cdots\eta_{2k+1}^2}\right)^2 \tag{3.90}$$

$$\begin{aligned}T = 1 - R &= \frac{4\eta_0\eta_g\eta_1^2\eta_2^2\eta_3^2\eta_4^2\cdots\eta_{2k+1}^2}{(\eta_0\eta_g\eta_2^2\eta_4^2\cdots\eta_{2k}^2 + \eta_1^2\eta_3^2\cdots\eta_{2k+1}^2)^2}\\ &= \frac{4}{2 + \frac{\eta_0\eta_g\eta_2^2\eta_4^2\cdots\eta_{2k}^2}{\eta_1^2\eta_3^2\cdots\eta_{2k+1}^2} + \frac{\eta_1^2\eta_3^2\cdots\eta_{2k+1}^2}{\eta_0\eta_g\eta_2^2\eta_4^2\cdots\eta_{2k}^2}}.\end{aligned} \tag{3.91}$$

Let

$$x = \sqrt{\eta_0\eta_g}\frac{\eta_2\eta_4\cdots\eta_{2k}}{\eta_1\eta_3\cdots\eta_{2k+1}}. \tag{3.92}$$

Thus, we can infer that

$$\begin{cases} T = \dfrac{4}{2 + x^2 + x^{-2}}\\ R = \left(\dfrac{1 - x^2}{1 + x^2}\right)^2. \end{cases} \tag{3.93}$$

By the same principle, for an even number of layers $(2k)$, the following can be derived:

$$R = \left(\frac{\eta_0\eta_2^2\eta_4^2\cdots\eta_{2k}^2 - \eta_1^2\eta_3^2\cdots\eta_{2k-1}^2\eta_g}{\eta_0\eta_2^2\eta_4^2\cdots\eta_{2k}^2 + \eta_1^2\eta_3^2\cdots\eta_{2k-1}^2\eta_g}\right)^2 \tag{3.94}$$

$$\begin{aligned}T = 1 - R &= \frac{4\eta_0\eta_g\eta_1^2\eta_2^2\eta_3^2\eta_4^2\cdots\eta_{2k}^2}{(\eta_0\eta_2^2\eta_4^2\cdots\eta_{2k}^2 + \eta_1^2\eta_3^2\cdots\eta_{2k-1}^2\eta_g)^2}\\ &= \frac{4}{2 + \frac{\eta_0\eta_2^2\eta_4^2\cdots\eta_{2k}^2}{\eta_1^2\eta_3^2\cdots\eta_{2k-1}^2\eta_g} + \frac{\eta_1^2\eta_3^2\cdots\eta_{2k-1}^2\eta_g}{\eta_0\eta_2^2\eta_4^2\cdots\eta_{2k}^2}}.\end{aligned} \tag{3.95}$$

Let

$$x = \sqrt{\frac{\eta_0}{\eta_g}} \frac{\eta_2\eta_4 \dots \eta_{2k}}{\eta_1\eta_3 \dots \eta_{2k-1}}. \tag{3.96}$$

Thus,

$$\begin{cases} T = \dfrac{4}{2 + x^2 + x^{-2}} \\ R = \left(\dfrac{1 - x^2}{1 + x^2}\right)^2. \end{cases} \tag{3.97}$$

It can be seen that, regardless of whether the quarter-wave stack has an odd or even number of layers (including the substrate and incident medium), its reflectance and transmittance have the same expression. Only the implication of x is different within these expressions.

When the x term for P and S are equal, it will enable the entire combination to achieve depolarization; i.e., when $x_{(P)} = x_{(S)}$, then $T_{(P)} = T_{(S)}$ and $R_{(P)} = R_{(S)}$.

To achieve $x_{(P)} = x_{(S)}$ in an odd-numbered layer,

$$\sqrt{\eta_0^{(P)}\eta_g^{(P)}}\eta_2^{(P)}\eta_4^{(P)} \dots \frac{\eta_{2k}^{(P)}}{\eta_1^{(P)}\eta_3^{(P)}} \dots \eta_{2k+1}^{(P)} = \sqrt{\eta_0^{(S)}\eta_g^{(S)}}\eta_2^{(S)}\eta_4^{(S)} \dots \frac{\eta_{2k}^{(S)}}{\eta_1^{(S)}\eta_3^{(S)}} \dots \eta_{2k+1}^{(S)}. \tag{3.98}$$

Thus, the polarization splitting between each medium should satisfy the following relationship:

$$\sqrt{\Delta n_0 \Delta n_g}\, \Delta n_2 \Delta n_4 \dots \frac{\Delta n_{2k}}{\Delta n_1 \Delta n_3} \dots \Delta n_{2k+1} = 1. \tag{3.99}$$

Similarly, for an even-numbered layer,

$$\sqrt{\Delta n_0 / \Delta n_g}\, \Delta n_2 \Delta n_4 \dots \frac{\Delta n_{2k}}{\Delta n_1 \Delta n_3} \dots \Delta n_{2k-1} = 1. \tag{3.100}$$

Equations (3.99) and (3.100) are necessary and sufficient conditions for a quarter-wave stack to achieve nonpolarising effects at the central wavelength.

For the depolarization design for oblique incidence onto a ZnS substrate, the quarter-wave stack method can be used to design a depolarising and highly anti-reflective coating when the angle of incidence $\theta = 60°$. A schematic of this design is shown in Fig. 3.11.

A quarter-wave stack was selected for the construction of this coating system to ensure that the coating showed depolarization while maintaining a relatively high transmittance at the central wavelength $\lambda_0 = 9.6\,\mu m$. A necessary and sufficient condition for depolarization at λ_0 is that the polarization splitting of the refractive indices between the incident medium (air), ZnS substrate, and each coating layer should satisfy the relation expressed in equation (3.98). For our design of a three-layered coating,

$$\frac{\sqrt{\Delta n_0 \Delta n_g}\, \Delta n_2}{\Delta n_1 \Delta n_3} = 1. \tag{3.101}$$

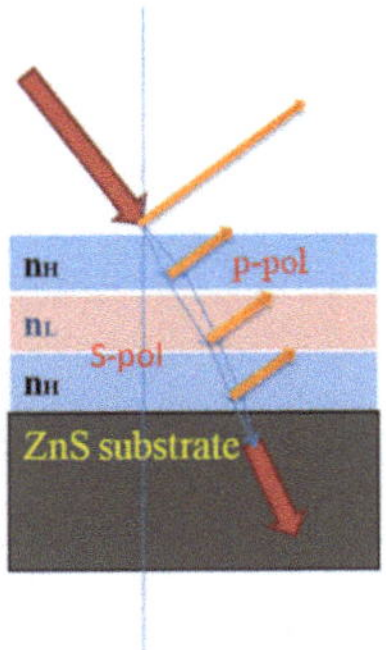

Fig. 3.11: Quarter-wave stack depolarizing coating system.

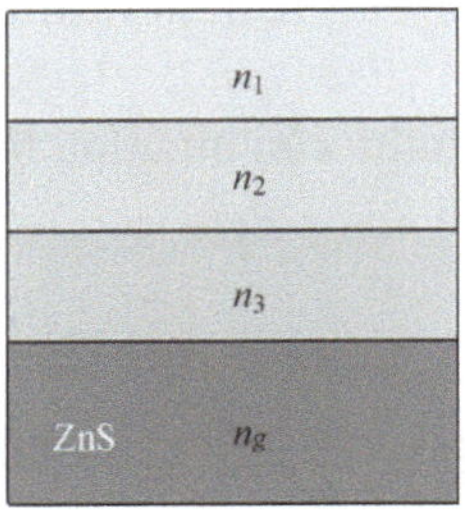

Fig. 3.12: Design of a three-layered depolarizing coating system.

Based on the central wavelength and the types of usable materials in practical application, three types of materials were selected to construct these three coating systems, as shown in Fig. 3.12. Firstly, low- and high-refractive-index materials were chosen to design the coating system, which was based on currently available infrared-transparent materials. The refractive index of the low-refractive-index material was approximately 1.4, and the material had to be near-, mid-, and far-infrared transparent. The refractive index of the high-refractive-index material was approximately 1.8, and the material had to have a broad waveband and high transmittance. Among them, low-refractive-index materials are mostly fluoride-based systems. Yttrium fluoride (YF_3) has a relatively low refractive index and absorption coefficient, with a relatively high transmittance from visible light to the far-infrared range. Furthermore, compared to other fluorides, YF_3 has a very high level of hardness. Therefore, with considerations for aerothermodynamics, we selected YF_3 as the low-refractive-index material. There are several choices for the high-refractive index material, including nitrides, oxides, and carbides. Owing to considerations for good compatibility within the coating system, yttrium oxide (Y_2O_3, ≈ 1.8) was selected as the high-refractive-index material.

In Fig. 3.12, $n_0 = 1$ (air), $n_1 = 1.4$ (YF_3), $n_2 = 1.8$ (Y_2O_3), $n_3 = ?$ (to be determined), $n_g = 2.3$ (ZnS substrate), and $\theta_0 = 60°$.

Equation (3.101) can then be applied to determine which film material n_3 should be in order to form a depolarizing coating system.

With

$$\Delta n = \frac{1}{\left(1 - \frac{n_0^2 \sin^2 \theta_0}{n^2}\right)},$$

we know that

$$\Delta n_0 = \frac{1}{\left(1 - \frac{n_0^2 \sin^2 \theta_0}{n^2}\right)} = 4$$

$$n_0 \sin \theta_0 = 1 \cdot \sin 60° = 0.866$$

$$\Delta n_1 = \frac{1}{\left(1 - \frac{n_0^2 \sin^2 \theta_0}{n_1^2}\right)} = 1.62.$$

Based on equation (3.79), the following can be obtained:

$$n_0 \sin \theta_0 = n_1 \sin \theta_1 = 0.866; \quad \cos \theta_1 = 0.7856.$$

Similarly, the following can also be obtained:

$$\Delta n_2 = \frac{1}{\left(1 - \frac{n_0^2 \sin^2 \theta_0}{n_2^2}\right)} = 1.301; \quad \Delta n_g = \frac{1}{\left(1 - \frac{n_0^2 \sin^2 \theta_0}{n_g^2}\right)} = 1.165,$$

$$\cos \theta_2 = 0.8766; \quad \cos \theta_g = 0.926.$$

By substituting all the data above into equation (3.50), we know that

$$\Delta n_3 = 1.734.$$

Further calculations will reveal that $n_3 = 1.33$. Based on the usable materials in practical application, the material we have selected that is closest to our calculation results is YF_3 ($n = 1.4$).

The following can be obtained from equation (3.92):

$$x = \frac{\sqrt{\eta_0 \eta_g \eta_2}}{\eta_1 \eta_3} = 1.466.$$

Substitution into equation (3.32) will give

$$\begin{cases} T = \dfrac{4}{2 + x^2 + x^{-2}} = 86.68\,\% \\ R = \left(\dfrac{1 - x^2}{1 + x^2}\right)^2 = 13.32\,\%. \end{cases}$$

The above calculation is based on a nonabsorbing medium. It is already known that the optical admittances of S-polarization and P-polarization, η_S and η_P, during oblique incidence are different. Hence, the following can be obtained through equations (3.77) and (3.78):

$$x^{\mathrm{S}} = \frac{\sqrt{\eta_0^{\mathrm{S}} \eta_{\mathrm{g}}^{\mathrm{S}} \eta_2^{\mathrm{S}}}}{\eta_1^{\mathrm{S}} \eta_3^{\mathrm{S}}} = 1.4666,$$

$$x^{\mathrm{P}} = \frac{\sqrt{\eta_0^{\mathrm{P}} \eta_{\mathrm{g}}^{\mathrm{P}} \eta_2^{\mathrm{P}}}}{\eta_1^{\mathrm{P}} \eta_3^{\mathrm{P}}} = 1.4653.$$

Substitution into equation (3.92) will give

$$\begin{cases} T^{\mathrm{S}} = \dfrac{4}{2 + x^2 + x^{-2}} = 86.663\,\%, \\ T^{\mathrm{P}} = \dfrac{4}{2 + x^2 + x^{-2}} = 86.7\,\%. \end{cases}$$

It can be seen from the design results above that this calculation not only satisfies the requirements for depolarization at a given wave range, but also provides relatively high transmittance. These results were simulated using Macleod software, and further optimization was performed (Fig. 3.13).

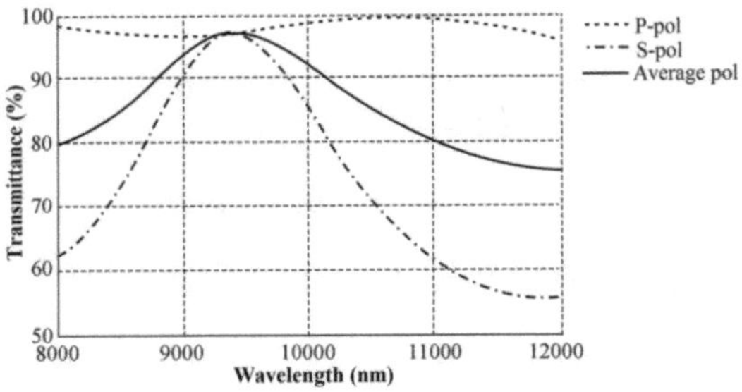

Fig. 3.13: Design results of a three-layered depolarizing coating system.

3.4 Discussion on the parametric variations of coating systems

During the actual fabrication of thin films, certain discrepancies might exist between the performance of the fabricated thin film and our thin-film design owing to the influence of different factors. Therefore, it is necessary to discuss the influence of thin-film parameters on the transmittance and polarization states of the depolarization coating design.

3.4.1 Influence of incident angle

As can be seen from the curves showing the variation of thin-film polarization with the angle of incidence, at 0°–50°, the polarization splitting of the thin film was not sensitive to changes in the angle of incidence. However, when the angle of incidence was 50°–80°, a very large polarization splitting was produced. Therefore, we will explore the changes in polarization state and transmittance within this range of incident angles. As shown in Fig. 3.14, the level of the average transmittance and the transmission wavelength range varied with changes in the angle of incidence. Transmittance decreased gradually with increasing angle of incidence. When the angle of incidence was 80°, the maximum transmittance was only 70 %. As the angle of incidence increased, the reflection of light waves also increased. In addition, the effective optical path length of light waves in the thin-film increased, which caused a substantial decrease in transmitted energy. Compared to normal incidence, polarization had a smaller effect on transmittance at this point. Furthermore, there were changes in the maximum transmission wavelength range. As the angle of incidence increased, the central wavelength shifted toward shorter wavelengths because the changes in the angle of incidence led to variations in the polarization splitting component, which affected the effective thickness of the thin film, thereby causing a shift in the central wavelength.

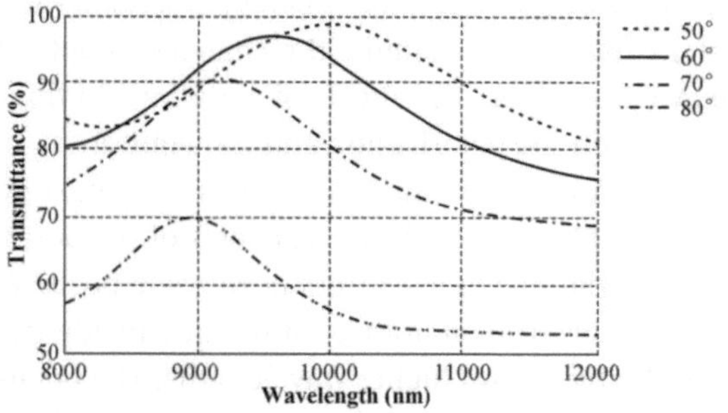

Fig. 3.14: Variations in transmittance under different angles of incidence.

It can be seen from Fig. 3.15 and 3.16 that the variations in S- and P-polarized light with changing angles of incidence are mutually consistent, as both showed a gradual decrease. Fig. 3.14–3.16 indicate that for a depolarization coating system designed for 60° oblique incidence, its polarization characteristics are not sensitive to changes in the angle of incidence. Hence, the designed coating system has depolarization characteristics within a relatively large range.

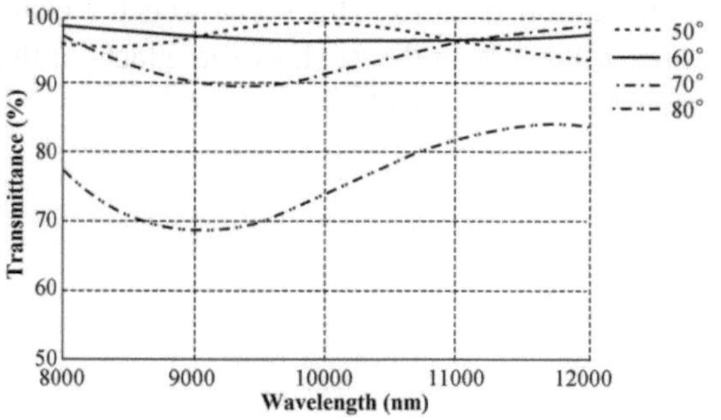

Fig. 3.15: Variations in transmittance of P-polarized light under different angles of incidence.

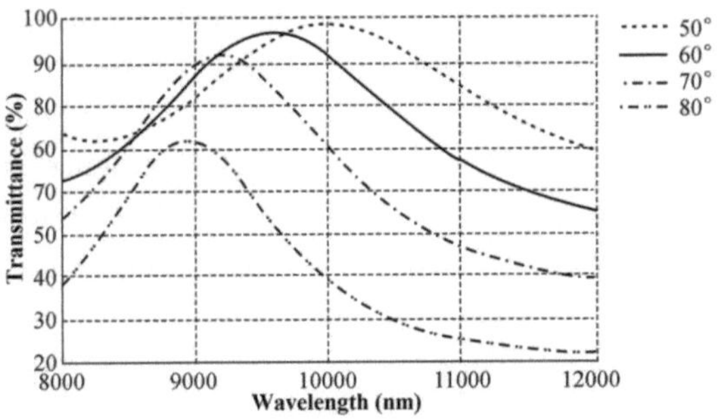

Fig. 3.16: Variations in transmittance of S-polarized light under different angles of incidence.

3.4.2 Influence of thin-film optical thickness

Optical thickness is a crucial parameter in thin-film optical design. It is the product of the refractive index and physical thickness of the thin film, which characterizes the combined effects of the two parameters. In order to examine the changes in optical thickness for different coating layers, we investigate the effects of variations in thin-film optical thickness at each coating layer on the polarization and transmittance.

Fig. 3.17–3.22 indicate that the transmittance and polarization states of the coating system varied with optical thickness. Fig. 3.17 and 3.20 show the effects of the outermost film layer. As the optical-thickness deviation changed from negative to positive, no significant changes occurred in the average transmittance for oblique incidence. Only the central wavelength shifted gradually toward higher wavelengths. The polarization state showed greater variations with changes in the optical thickness of the outer coating. At negative deviation, polarization will produce splitting. Increases toward the negative direction will lead to greater deviation. For the optical thickness at equilibrium, polarization splitting will disappear. As optical thickness increases, the changes in polarization splitting are not significant. Fig. 3.18 and 3.21 show the effects of the intermediate film layer. The optical thickness of the intermediate layer does not have a substantial impact on transmittance, and its effects are consistent with that of the outermost layer. However, unlike the outermost layer, polarization splitting occurs in the same direction as that of the increase in optical thickness of the intermediate layer: greater increases imply larger polarization splitting. This scenario is the exact opposite of the case of the outermost layer. Fig. 3.19 and 3.22 show the effects of the innermost layer, which are broadly consistent with those of the outermost layer.

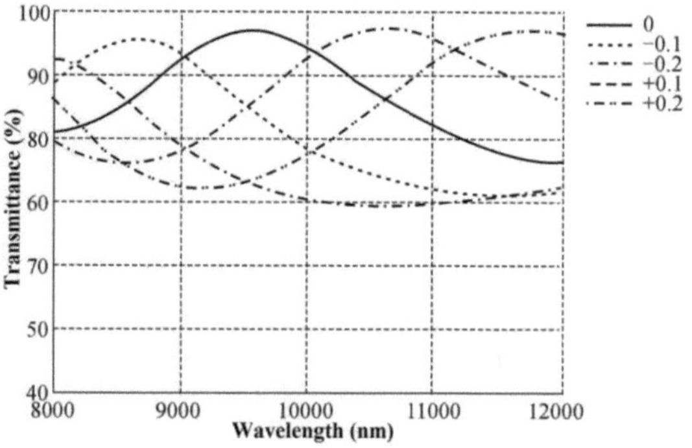

Fig. 3.17: Effects of changes in the optical thickness of the outermost layer on transmittance.

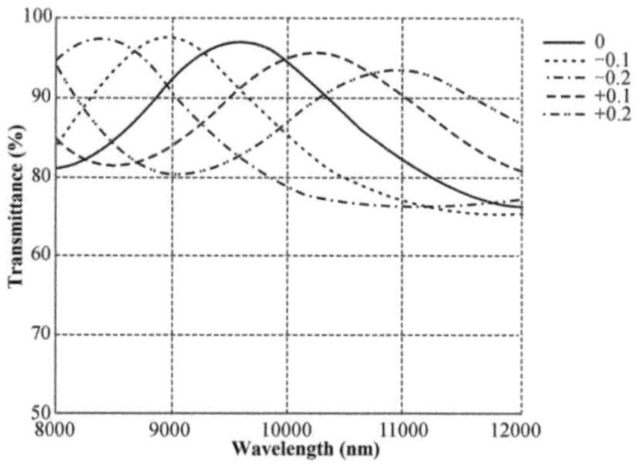

Fig. 3.18: Effects of variations in optical thickness of the intermediate layer on transmittance.

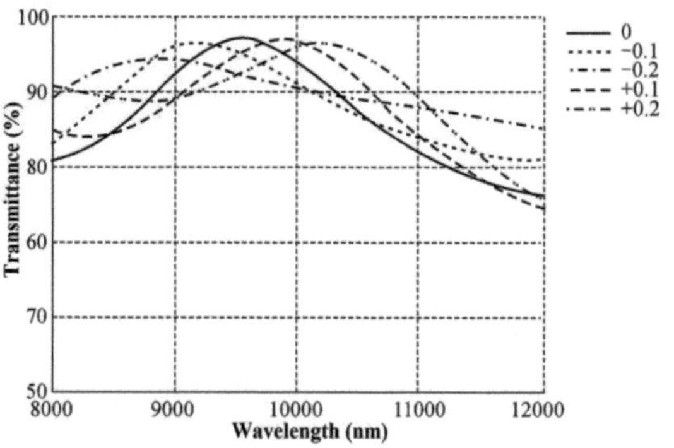

Fig. 3.19: Effects of variations in optical thickness of the innermost layer on transmittance.

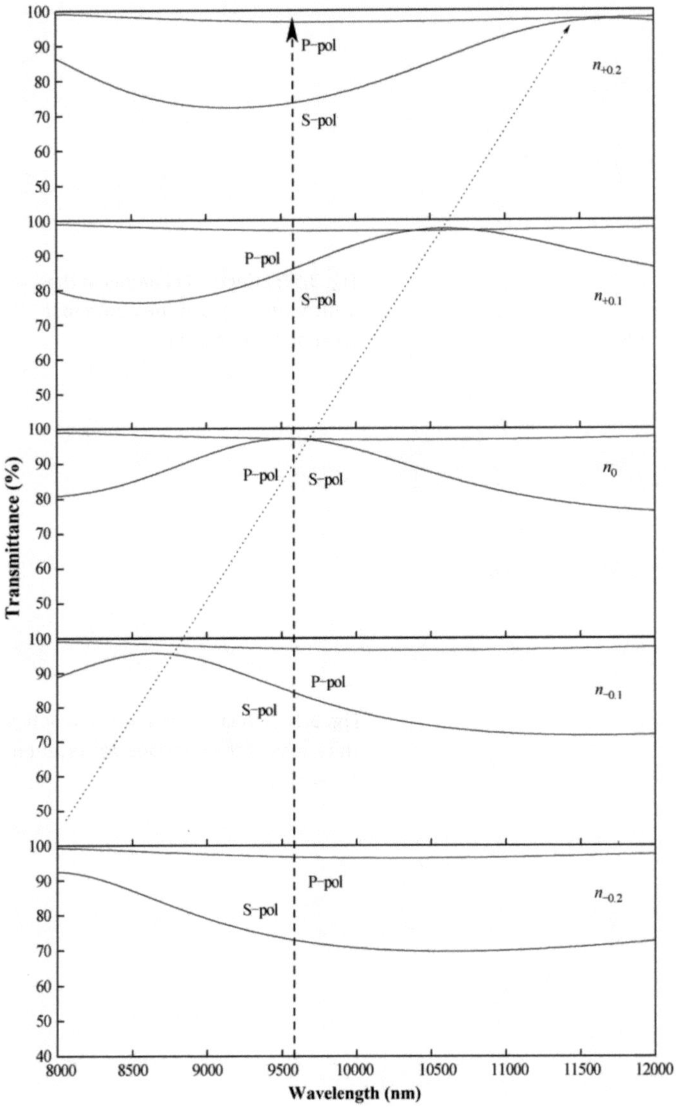

Fig. 3.20: Effects of variations in optical thickness of the outermost layer on polarization splitting.

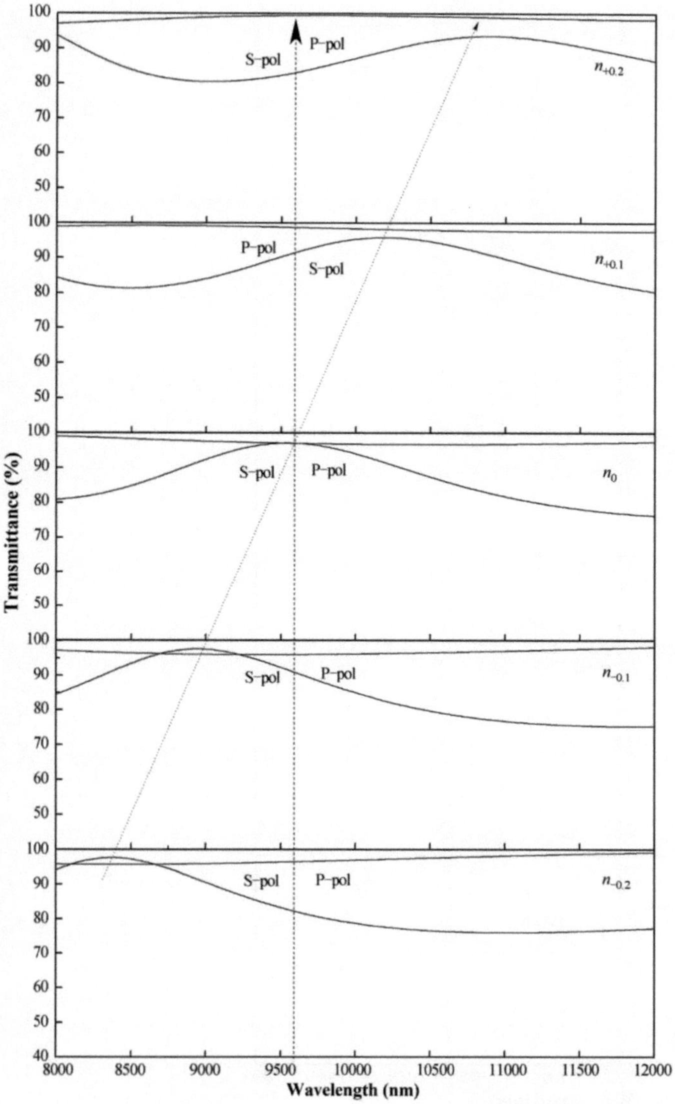

Fig. 3.21: Effects of variations in optical thickness of the intermediate layer on polarization splitting.

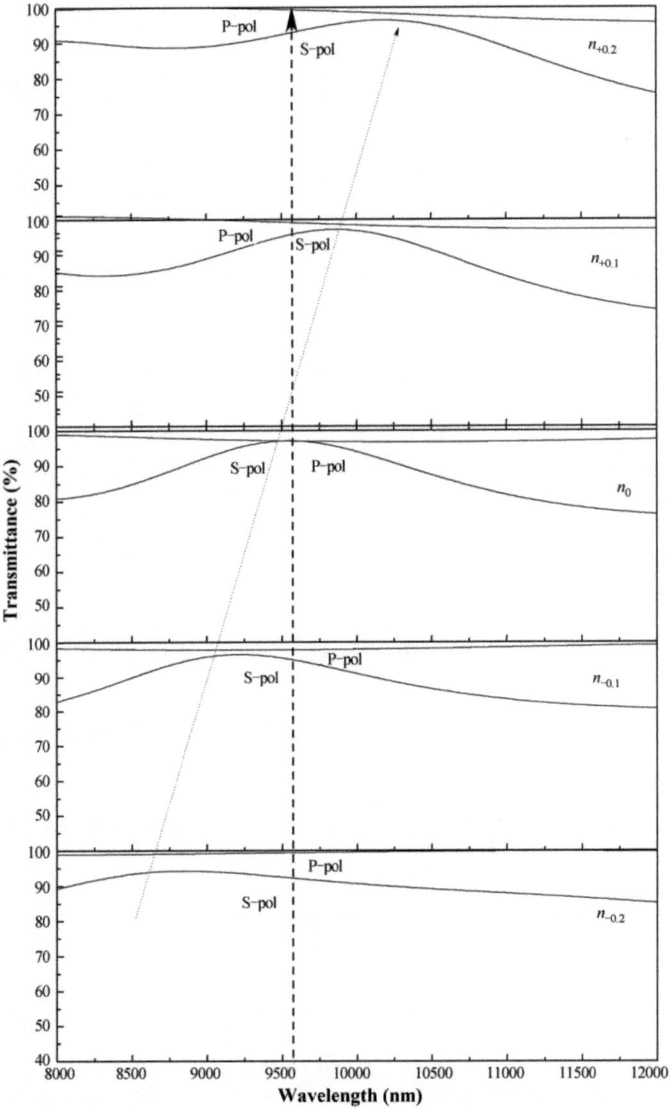

Fig. 3.22: Effects of variations in optical thickness of the innermost layer on polarization splitting.

3.4.3 Influence of thin-film refractive index

During the process of thin-film fabrication, the optical performance of thin films produced under different processing conditions could exhibit substantial differences. One of these factors is its refractive index. Different compositions and structures will lead to different refractive indices. Hence, the investigation of the changes in refractive index within a given range plays a crucial role in thin-film design.

It can be seen from Fig. 3.23–3.25 that the transmittance and polarization states of the coating system showed different degrees of changes with variations in refractive index. When the refractive indices of the first and third coating layer increased or decreased, the transmittance was not significantly affected. Only the central wavelength shifted to the left or right. However, when the refractive index of the intermediate layer increased by 0.2, not only did the central wavelength of the transmittance shift to the right, the transmittance was also reduced by a large extent. This indicates that the intermediate layer had a greater impact on the transmittance of the coating system than the inner and outer layers. Similarly, the polarization state was also most sensitive to the intermediate layer.

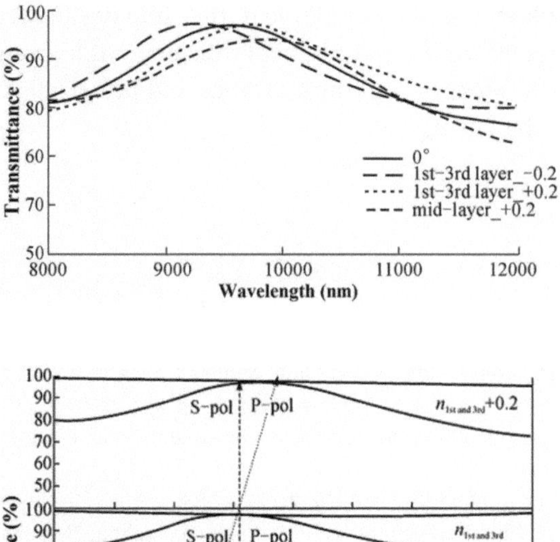

Fig. 3.23: Effects of variations in refractive index on transmittance.

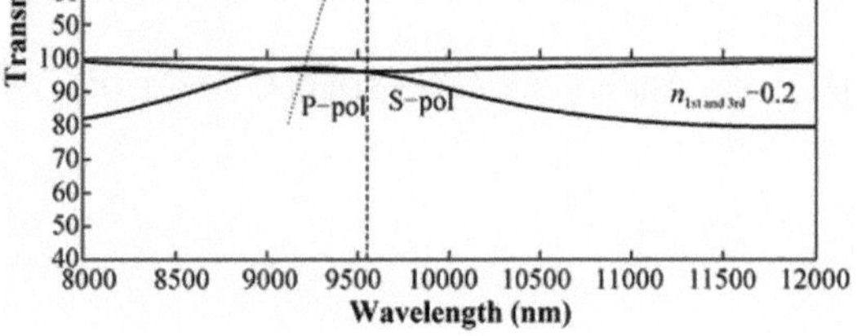

Fig. 3.24: Effects of variations in the refractive index of the first and third layers on polarization splitting.

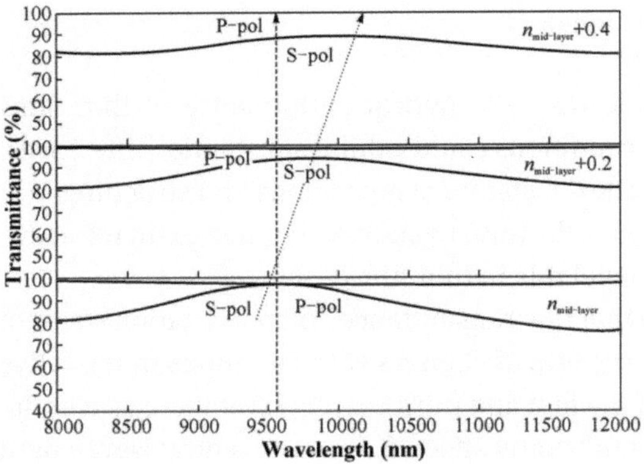

Fig. 3.25: Effects of variations in refractive index of the intermediate layer on polarization splitting.

3.4.4 Influence of thin-film physical thickness

The physical thickness of the thin film directly determines its optical thickness, thereby influencing its function and structure. Owing to the influence of various factors during the fabrication process, the differences in thickness might lead to different characteristics. Therefore, the investigation of the effects of thickness on optical performance within a specific range plays a fundamental role in thin-film design. As optical thickness is a function of refractive index and physical thickness, at a given refractive index, the impact of physical thickness on transmittance and polarization state is consistent with that of optical thickness.

References

[1] Costich V. Reduction of Polarization Effects in Interference Coatings. Applied Optics, 1970, 9(4): 866–870
[2] Gilo G. Design of a Non-polarizing Beam Splitter Inside a Glass Cube. Applied Optics, 1992, 31 (25): 5345–5349.
[3] Lu JJ, Liu WG. Optical Thin-film Technology. Xi'an, Northwestern Polytechnical University Press, 2005.
[4] Mahlein HF. Non-polarizing Beam Splitters. Applied Optics, 1974, 21: 577–583.
[5] Tang JF, Gu PF, Liu X, et al. Modern Optical Thin-film Technology. PhD thesis, Hangzhou, Zhejiang University Press, 2006.
[6] Tang WZ. Fabrication Theory and Application of Thin-film Materials. 2nd edn. Beijing, Metallurgical Industry Press, 2005.
[7] Thelen A. Nonpolarizing Interference Films Inside a Glass Cube. Applied Optics, 1976, 15(12): 2983–2985.

4 Preparation methods of infrared antireflective and protective coatings

4.1 Overview

According to the different application requirements in practice, a variety of thin-film preparation methods have been developed. The methods can be divided into two categories, dry-type and wet-type methods, according to the phase of the film-forming medium (gaseous phase or liquid phase). Based on whether chemical reactions occur on the film-forming interface, the dry film preparation methods can be further classified as physical vapor deposition (PVD) and chemical vapor deposition (CVD) methods. In terms of the generation path of deposited particles, preparation can be carried out using multiple methods, including vacuum evaporation, sputter deposition, and ion plating. Wet film preparation methods include electroplating, chemical plating, anodic oxidation, and the sol-gel method. Typical applications of different thin-film preparation methods have already been discussed in detail in the previous studies. In this chapter, common preparation approaches of infrared antireflective/protective coating materials, in particular, will be introduced. The preparation approaches of thin-film materials such as amorphous diamond, germanium carbide, and boron phosphide will be briefly discussed. Moreover, we will illustrate how to achieve uniform deposition for infrared window/dome coatings.

4.2 Magnetron sputtering

Magnetron sputtering (MS) has been extensively and successfully applied in many areas as a highly efficient thin-film deposition method. Especially in the field of microelectronics, optical thin films and coatings, and material surface processing, this technology is commonly adopted to deposit thin films and prepare coatings. Grove first described the physical phenomenon of sputtering in 1852. In the 1940s, sputtering started to be applied and developed as a film deposition technique. It was with the rapid rise of the semiconductor industry in the 1960s that this technique obtained its truly popularized and extensive use, as a method for depositing the metal electrode layer of transistors in integrated circuits. After the invention and development of magnetron sputtering and its application for manufacturing the reflecting layers of compact discs (CDs) in the 1980s, the application fields of this technology were dramatically expanded, and it gradually became a common method to fabricate a wide range of products. Over the past decade, a series of new sputtering techniques have been developed.

https://doi.org/10.1515/9783110489514-007

The magnetron sputtering system was developed based on the two-pole sputtering system by solving the problems suffered by the original system, including low plasma ionization, notable substrate heating, and low depositing speed. The design of the magnetron device forces the high-voltage ionized electrons to move around the magnetic lines of force, along the spiral trajectory, under the action of the Lorentz force (Fig. 4.1).

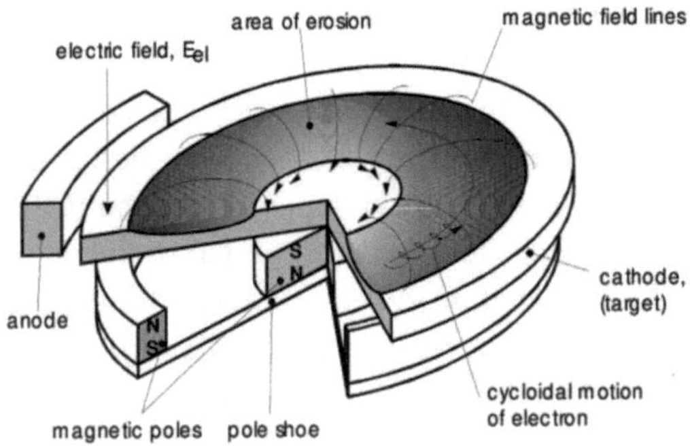

Fig. 4.1: Magnetic control device.

The motion of these electrons causes them to collide continuously with the Ar atoms, ionizing a great amount of Ar^+, with a rapidly increasing ionization rate. Therefore, the plasma density in this area is fairly high. Subsequent to multiple collisions, the energy of the electrons gradually decreases, making the electrons escape the constraint of magnetic lines of force and finally approach the substrate plate, internal wall of the vacuum chamber, and the target anode. On the other hand, the Ar^+ ions impact the target material and release energy under the action of a high-voltage electric field, causing the atoms on the target material surface to absorb the kinetic energy of Ar^+ ions and detach from the original lattice's restraint. The neutral target atoms escape from the surface of the target material and move towards the substrate, thereby depositing and forming the thin film (Fig. 4.2). The particle energy of sputter deposition is usually in the range of 1–10 eV, and the theoretical density can reach 98 %. Therefore, the thermal effect on the substrate is relatively weak, and the depositing speed is dramatically increased.

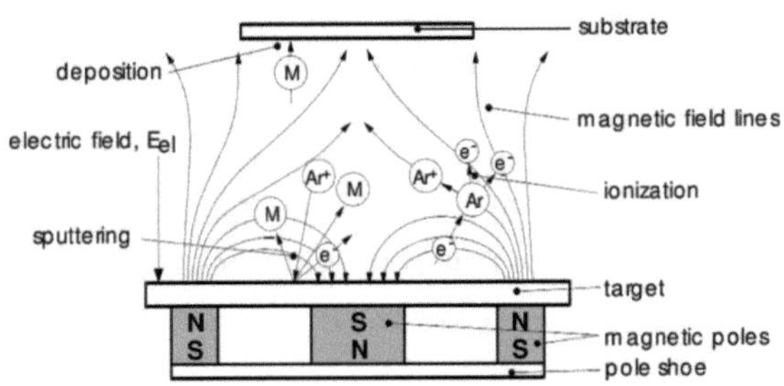

Fig. 4.2: Schematic of the working principle of magnetron sputtering.

Based on the application requirements of magnetron sputtering, a variety of cathode magnetrons with diverse structures and variable magnetic fields have been developed to enhance the coating quality and the target-use efficiency. The widespread applications of magnetron sputtering can be attributed to the properties of this technology that are different from those of other deposition methods. The properties can be summarized as follows.

(1) Materials to be fabricated into target materials can also be used as the coating materials, including various metals, semiconductors, ferromagnetic materials, insulating oxides, ceramics, and polymers, making this technology especially suitable for the deposition of materials with high melting point and low vapor pressure.
(2) Under appropriate conditions, the cosputtering of a multielement target can deposit mixture and compound coatings with desired compositions.
(3) Adding oxygen, nitrogen, or some other active gases into the discharge atmosphere of sputtering can generate compound coatings consisting of the target material and gas molecules.
(4) A stable depositing speed can be essentially obtained by controlling the pressure of the vacuum chamber and sputtering power, and high-precision coating with uniform thickness and superior repeatability can be achieved through the accurate monitoring of the sputter deposition time.
(5) Sputtering particles are almost free from the effect of gravity, and the positions of the target and substrate can be configured without restriction.
(6) The adhesive strength of the substrate and coating is more than ten times that of general vapor-deposition coatings; the high energy of sputtered particles leads to continuous surface diffusion, forming a hard and dense coating on the film surface. In the meantime, the high-energy particles enable the substrate to gain a crystallized film at a relatively low temperature.
(7) Because of the high nucleation density of coating in the initial forming stage, an extremely thin, continuous film with a thickness less than 10 nm can be created.

Multielement alloy and compound coatings can be deposited through direct sputtering on the target manufactured using the desired alloy and compound materials. Moreover, the reactive sputtering and co-sputtering techniques are frequently adopted to deposit coatings of compound, mixture, and alloy with multiple elements. Compared to those sputtering methods with a compound target, reactive sputtering and co-sputtering have the ability to adjust the composition of coating materials by regulating the sputtering parameters and, thereby, to deposit coatings including stoichiometric and nonstoichiometric materials with various required constituents.

Reactive sputtering adds a proportion of reactive gas into the sputtering atmosphere originally filled with inert gas. In this case, oxygen and nitrogen are commonly used as the reactive gas. With the existence of a reactive gas, when the target is sputtered, a compound will be generated from the reaction between the target material and reactive gas, which will subsequently be deposited on the substrate. When an inert gas is used in compound-target sputtering, owing to the chemical instability, the generated coating often lacks one or more components of the target. At this time, the supply of a reactive gas can compensate for the missing components; this technique is regarded as reactive sputtering. The DC reactive magnetron sputtering process of dielectric films often exhibits a high degree of instability because, with the increase of reactive gas flow rate, operation will deviate from the metal sputtering mode. The interaction of the target and reactive gas will cover the layer of insulating medium in the nonetching area of the target surface, leading to an abrupt decrease of deposition rate and a great accumulation of electric charge on the dielectric film. Discharging will occur as the electric charge excessively aggregates, and therefore, the target surface will be contaminated; this phenomenon is called cathode poisoning. The shielded anode near the cathodic magnetron may also be covered by the dielectric film, resulting in the disappearance of the anode.

Fig. 4.3 shows the hysteresis loop of the cathodic target voltage against the flow rate of reactive gas. Generally, the high-speed reactive deposition process occurs in the transition mode. For operation in this zone, a rapid reactive gas control system is demanded. Moreover, a pulsed power supply or medium-frequency AC power supply needs to be used to overcome the electric-charge accumulation and discharging of the dielectric film during the sputtering process.

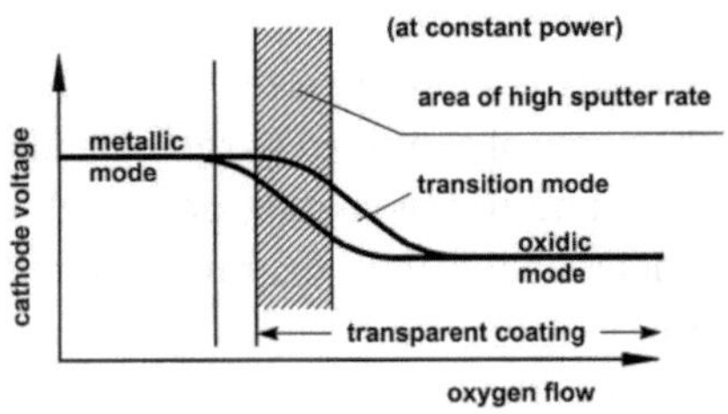

Fig. 4.3: Hysteresis loop of the cathodic target voltage against the flow rate of reactive gas.

4.3 Thermal evaporation deposition

Thermal evaporation deposition (TED) refers to the approach described below (see Fig. 4.4). The source material situated on the evaporation source in the vacuum chamber is heated to evaporate and release vapor atoms and molecules from the surface. When the mean free path of vapor molecules is higher than the linear dimension of the vacuum chamber, the vapor molecules can be hardly impacted or impeded by foreign molecules or atoms. Instead, they are directly deposited on the substrate and condense into a solid coating owing to the low temperature of the substrate.

In order to enhance the adhesive force between vapor molecules and the substrate, appropriate heating for the substrate is necessary. The smooth implementation of evaporation deposition needs to satisfy the following two conditions.

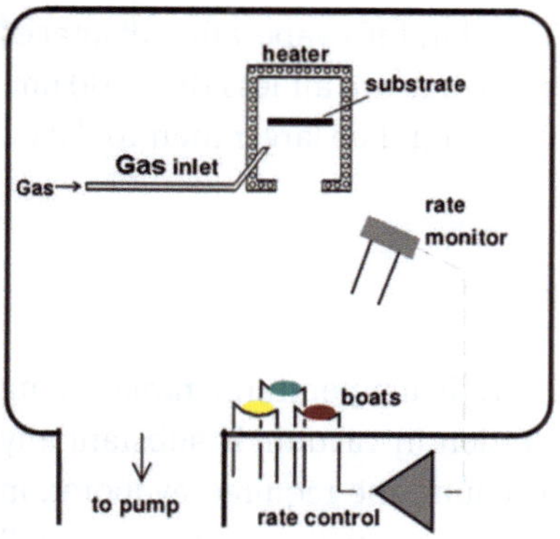

Fig. 4.4: Structural schematic of the evaporation deposition chamber.

4.3.1 Vacuum condition of the evaporation process

When the mean free path of vapor molecules in the vacuum chamber exceeds the distance between the evaporation source and the substrate, a sufficient degree of vacuum can be obtained. Therefore, it is necessary to create a high-vacuum environment in the chamber so as to increase the mean free path of residual gas, thereby reducing the probability of collisions between vapor molecules and residual gas molecules. Assume that L is the evaporation distance, which is considered as a known actual distance; l the mean free path of gas molecules; N_0 the number of vapor molecules evaporated from the evaporation source; and N_1 the number of vapor molecules that undergo collisions within the distance L. Then,

$$\frac{N_1}{N_0} = 1 - \exp\left(\frac{-L}{l}\right). \tag{4.1}$$

Fig. 4.5 shows the relationship between the probability of collision during the deposition process, N_1/N_0 (%), and the nominal distance of gas molecules, L/l.

It can be seen from the figure that when $L = l$, 63 % of the vapor molecules will experience collisions. Therefore, collisions between the vapor molecules in the travelling process and the residual gas molecules can be avoided only when the mean free

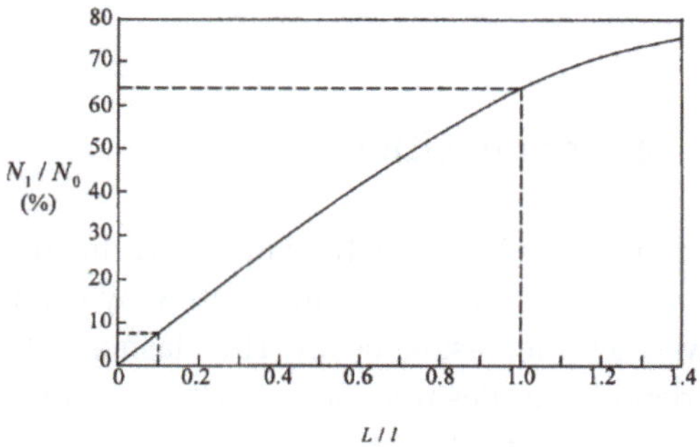

Fig. 4.5: Relationship between the molecule collision probability and the nominal distance.

path is far greater than the evaporation distance. Thus far, the evaporation distances of commonly used vacuum evaporation deposition devices are all less than 500 nm. Hence, the vacuum degree of the deposition chamber should be larger than 10^{-2} Pa.

4.3.2 Evaporation condition in the deposition process

The source material will vaporize when heated to a certain temperature, transforming from the solid phase to the gaseous phase. Evaporation in vacuum is substantially easier than that under atmospheric pressure; in vacuum, the required evaporation temperature of an object will be dramatically decreased, the evaporation process will be significantly shortened, and the evaporation rate will be considerably increased. The number of molecules evaporated from unit area of the source material in unit time in the form of vapor can be expressed as

$$N = 2.64 \times 1024p\left(\frac{1}{T\mu}\right)^{1/2} \alpha_1, \tag{4.2}$$

where p denotes the saturated vapor pressure (Pa) of the source material at temperature T, T is the absolute temperature (K), μ is the relative molecular weight of the source material, and α_1 is the coefficient used to characterize the surface cleanness of the source material during evaporation, which is named the evaporation coefficient.

Therefore, in order to accelerate the deposition forming process of the coating and increase the evaporation rate of the source material, the following measures can be adopted: the maintenance of a sufficiently low pressure within the vacuum chamber, accurate control of the evaporation temperature, and removal of the impurities in the source material. Meanwhile, under a certain degree of vacuum in the chamber, the appropriate increase of the evaporation rate of the source material can help reduce the contamination of the source material by the residual gas.

At present, as a pollution-free, "green" coating technology, vacuum coating is gradually showing its superiority in terms of energy saving, cost reduction, and the additional function expansion of products. It plays an increasingly important role in our daily life as well as in scientific & technological development, with growing popularity.

4.4 Plasma-enhanced chemical vapor deposition

Plasma-enhanced chemical vapor deposition (PECVD) is a method involving the ionization of the atoms that contain the components required for the coating by using microwaves or radio-frequency (RF) waves to form plasma locally. The plasma, with strong chemical activity, reacts with and deposits the desired coating on the substrate. Fig. 4.6 shows a structural schematic of parallel-plate PECVD.

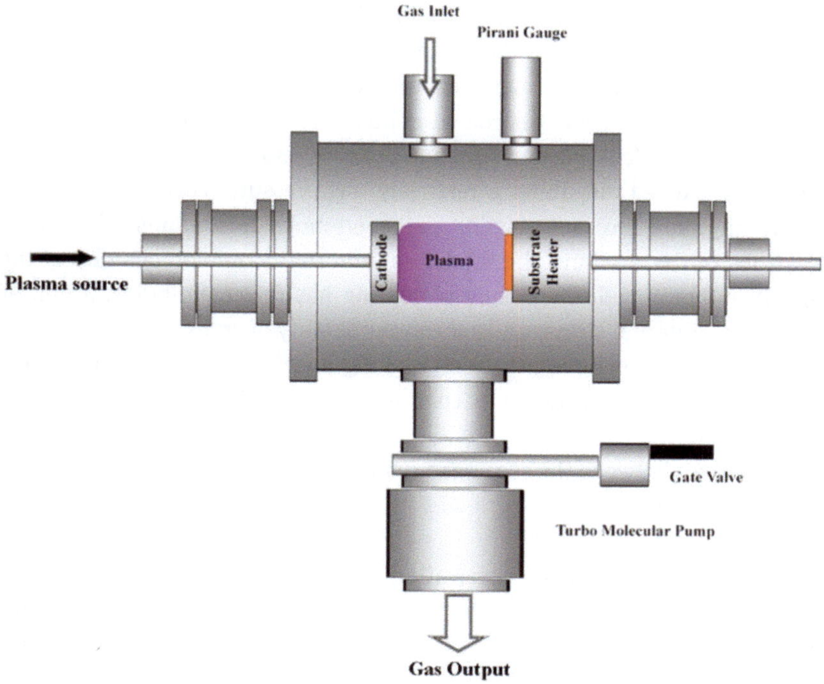

Fig. 4.6: Structural schematic of PECVD.

PECVD is a chemical vapor deposition process with an enhancing effect of glow discharge. During the PECVD process, in addition to the thermochemical reactions, extremely complex plasma-chemical reactions are involved. The types of plasma used to enhance CVD include radio-frequency plasma, direct-current plasma, pulsed plasma, microwave plasma, and electron cyclotron resonance plasma, which are obtained from the photolysis and discharge processes by using radio frequency, high DC voltage, pulse, microwave, and electron cyclotron resonance, respectively, to stimulate the rarefied gas. Generally, the discharge gas pressure is of the order of several Pascals. The plasma has the following functions in chemical vapor deposition:

(1) stimulating gas molecules in the reactant into reactive ions, thereby decreasing the temperature required for the reaction;
(2) accelerating the diffusion of reactant on the surface (surface mobility), thereby increasing the film-forming speed;
(3) cleaning the surfaces of the substrate and coating, i.e., removing the particles with weak attachment formed by sputtering, thereby enhancing the adhesive force between the formed coating and the substrate;
(4) forming the coating with a uniform thickness through collisions and scatterings between atoms, molecules, ions, and electrons in the reactant.

As a result, compared with conventional CVD, PECVD has the following advantages.

(1) Coatings can be formed under a low temperature (most common temperature is 300–350 °C) with a smaller thermal effect on the substrate. Problems induced by high-temperature filming, such as coarse-grain layers and the formation of a brittle phase between the coating and substrate, can therefore be avoided.

(2) PECVD can be carried out under a relatively low pressure. Owing to the collision, scattering, and ionization actions between molecules, atoms, plasmoids, and electrons in the reactant, the thickness uniformity and composition of the coating can be improved. The resultant coating has fewer pin-holes, a denser structure, lower internal stress, and a lower probability of crack occurrence.

(3) The CVD application scope is extended, making it possible to fabricate various metal coatings, amorphous inorganic coatings, and organic polymer coatings on different substrates.

(4) The adhesive force between the coating and the substrate is greater than that produced by conventional CVD.

PECVD is highly appropriate for the passivation of semiconductor components and integrated circuits to enhance their reliability. At present, PECVD has become an indispensable technology for scientific research and production in the fields of microelectronics and photoelectronics.

4.5 Filtered cathodic vacuum arc deposition

In the filtered cathodic vacuum arc (FCVA) deposition method, a high-purity, conductive, solid target is used as the cathode to generate arc plasma, and the fragments or liquid droplets in the cathode are filtered using appropriate methods to deposit the coating from pure target ions. The FCVA technology was initially developed by I. I. Aksenov et al. of the Kharkov Institute of Physics and Technology of the former Soviet Union in the late 1970s. It was then further promoted by research groups from the University of Sydney in Australia, University of Cambridge in the UK, Lawrence Berkeley National Laboratory in the US, and Nanyang Technological University in Singapore. Nowadays, FCVA has become a critical coating deposition technology with high efficiency and quality.

Theoretically, all conductive, solid target materials can be used as the cathode. When necessary, the addition of background gas into the base vacuum and the use of a filtered cathodic vacuum arc can generate the corresponding elementary substance or compound coatings. For example, a tetrahedral amorphous carbon (a-D) coating can be fabricated as a microelectronic mechanical-functional coating or hard-drive protective coating; TiN, TiC, AlN, and WC ceramics can be fabricated to enhance wear resistance or decoration; Al_2O_3, TiO_2, Ta_2O_5, and HfO_2 medium coatings can be fabricated as an optical coating; and a monocrystalline metal coating can be fabricated as a magnetic coating. FCVA coatings have a series of advantages, such as accurate controllability of structural performance, good repeatability of processing, and strong adhesive force. Tab. 4.1 compares FCVA with sputter deposition, which is representative of the class of physical vapor deposition, and with plasma-enhanced chemical deposition, which is representative of the class of chemical vapor deposition. It can be observed that ionic-state FCVA particles have a series of advantages including the adjustability of the peak energy of incident particles, superior adhesion between the coating and substrate, feasibility of deposition at ambient temperature, high growth speed, good uniformity and repeatability, and excellent coating quality.

Tab. 4.1: Comparison of properties of filtered cathodic vacuum arc deposition technology with those of other techniques.

	Coating type		
Index	**PVD**	**CVD, PECVD**	**FCVA**
Deposited particle type	Atom	Free radical	Ion
Incident particle energy	≈ 0.1 eV	≈ 1 eV	20–500 eV
Adhesion	Moderate	Good	Excellent
Deposition pressure	1×10^{-3} Torr	1×10^{-3} Torr	1×10^{-6} Torr
Coating density	50–80 %	80–100 %	≈ 100 %
Temperature	≈ 120 °C	> 120 °C	< 80 °C
Coating size	Large	Small	Medium
Coating speed	Fast	Moderate	Fast
Source material	Solid	Gas	Solid
Uniformity	10 %	5 %	4 %
Repeatability	Moderate	Excellent	Good
Coating quality	Moderate	Excellent	Good
Fabrication cost	Moderate	Extremely high	Relatively low

The ionization rate of regular sputter-deposition plasma is relatively low, often lower than 10 %, and the energy distribution of energetic particles is rather wide. In contrast, the ionization rate of arc plasma is extremely high, up to 100 %, and the half-peak width of energy distribution of energetic particles under a drift potential is fairly narrow. Therefore, the incident energy of deposited particles can be accurately controlled by exerting an electrostatic field (substrate bias voltage), and the coating–substrate adhesion and coating-layer performance can be regulated, as shown in Fig. 4.7.

Two key techniques for FCVA deposition coating need to be developed, namely the arc-spot control technique and the plasma-beam magnetic filtration technique. Fig. 4.8 (a) is the state diagram of arc-spot-emitting plasma and macroscopic particles. The arc spot is micrometre-sized and tends to move in all directions. Once it is detached from the cathodic target surface, break arcs will be generated. Therefore, the motion of the arc spot must be controlled. Arc-spot detachment and arc discontinuity

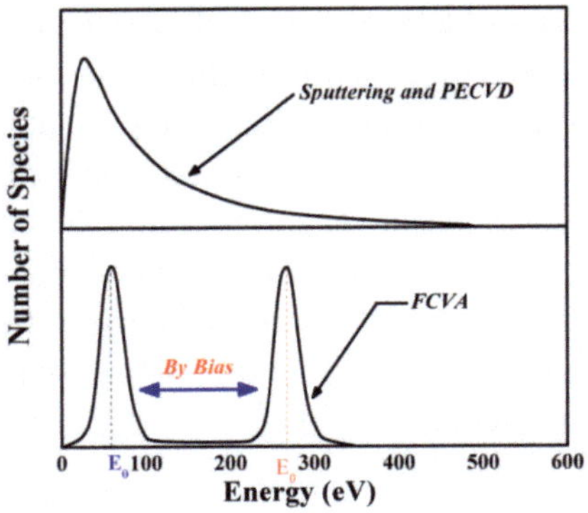

Fig. 4.7: Energy distributions of energetic particles in FVCA/sputtering/PECVD.

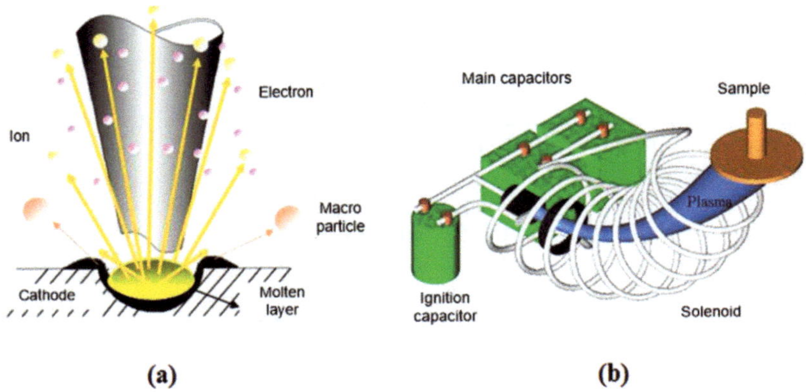

Fig. 4.8: Key techniques for amorphous diamond coating using FVCA: (a) arc-spot control technique; (b) plasma-beam magnetic filtration technique.

can be avoided by using electromagnetic confinement, in which a red copper coil is configured outside the anode and on the cathodic target surface, as well as a power source, to force the arc spot to burn inside the cathodic target surface. However, after a long working period, the cathodic target surface will possibly be burnt and eroded into a concave shape. Hence, the scheduled maintenance and cutting of the cathode outer edge are required to keep the target surface smooth.

Substances emitted from the cathodic spot contain not only ions of target materials but also a great number of uncharged cathodic fragments and droplets, which will be deposited on the coating surface as macroparticles and severely degrade the coating quality, if not filtered. In order to remove the macroparticles, filters with multiple patterns have been designed, such as straight duct filters, angled filters, and bended tube filters. Fig. 4.8 (b) shows a schematic diagram of the most conventional filter, i.e. the quarter torus filter. Arc plasma is eventually focused and deposited at the desired location through the guidance of the magnetic field in the spiral pipe, whereas the uncharged macro-particles are trapped within the filter pipe. Previous tests have indicated that ordinary filter approaches cannot achieve ideal effects in practice. Therefore, we adopted the off-plane-double-bend filter pipe and obtained a high-quality coating with a pure surface.

4.6 Large-area filtered cathodic vacuum arc uniform deposition

When a plasma beam reaches the substrate after filtering and focusing, the spot diameter is only 20–30 mm, which is much less than the diameter of the required uniformly deposited coating area. Therefore, how to obtain a large-area a-D coating with a uniform thickness and homogeneous properties is the key question in practical applications. Based on our application purpose, we demonstrate here the deposition technique of a ⌀300 mm nonuniform a-D coating.

Since the ionization rate of carbon plasma generated by vacuum arc is extremely high, a magnetic field can be used to guide and filter the plasma beam. Similarly, a magnetic field can also be applied to properly deflect and wiggle the plasma beam to control efficiently the accurate drop positions of the pure carbon plasma beam on the substrate and the relative duration on these spots and thereby realize the large-area uniform deposition of the a-D coating. In the practical processing, coating uniformity is affected by multiple factors, including the drop-point density in the deposition area, relative durations of different drop points, and the scanning method. The determination of the appropriate scanning waveform requires abundant experimental data. It has been proven by a large number of tests that homogeneous deposition on the entire area can be achieved by linearly scanning along the radial direction and rotating the substrate relative to the ion beam at a constant angular speed. Fig. 4.9 shows the plasma scanning method with an arc current of 60 A and a substrate rotation speed of 30 r/min. In the diagram, the circles represent the drop points of the plasma beam on the substrate, and the numbers represent the relative duration of the plasma beam. However, in practical situations, the plasma-beam drop points often overlap on the substrate. Therefore, the scanning waveform demands corresponding modification.

A a-D coating is deposited on a 300 mm × 300 mm piece of ordinary glass plated with transparent conductive adhesive using the scanning waveform of the plasma beam shown in Fig. 4.9. The appearance of the deposited coating is illustrated in Fig. 4.10 (a). When the glass is placed at an angle under light, it can be observed

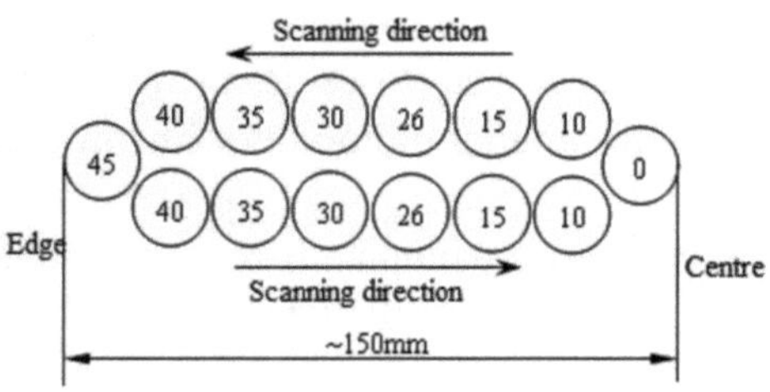

Fig. 4.9: Schematic of control of the scanning waveform of a filtered arc depositing the a-D coating.

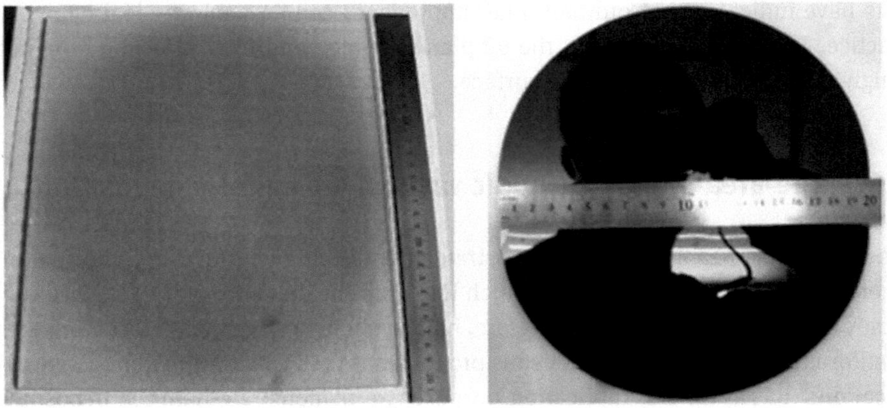

(a) (b)

Fig. 4.10: Specimen of the uniformly deposited a-D coating: (a) on a piece of 300 mm × 300 mm glass; (b) coating specimen on an 8-inch-diameter monocrystalline silicon wafer.

that the coating layer is uniform, indicating that the coating thickness is essentially the same within the entire area. In order to measure the thickness uniformity of the coating quantitatively, the approach shown in Fig. 4.10 (b) is adopted. We measure the thickness values along the radial direction with an interval of 1 cm by using a step profiler and adopt the average value. The results show that the average thickness and the thickness deviation of the coating on a 8-inch-diameter silicon wafer are 69.2 nm and ±3 %, respectively.

4.7 Large-size plane uniform magnetron sputtering

Magnetron sputtering is a simple coating fabrication technique with a fast deposition speed and superior film quality, and it has been widely applied in industrial production. This technique can be used to fabricate multiple coating materials such as conductors, semiconductors, and insulators. However, owing to the nonuniform distribution of the magnetic field, the generated plasma zone is inhomogeneous, resulting in a thicker-center and thinner-edge deposition. This phenomenon has always been the critical problem that restricts the application of this technique. Generally, to increase the area of uniform deposition, a large target is often adopted, which causes the serious wastage of target material and dramatically increases production costs. Hence, the fabrication of large-sized coatings with superior uniformity with low cost is a major topic to be addressed.

For fabricating large-sized homogeneous coatings, we propose here a novel compound motion method of depositing coatings, in which the substrate revolves about its own axis and the magnetron sputtering target moves step by step. Based on this method, a magnetron sputter deposition machine is designed, which consists of a specimen stage that can revolve about its axis and two magnetron sputtering targets with a diameter of 49 mm that can conduct stepping motion. The relational expression between the thickness of the coating fabricated by this magnetron sputter deposition machine and the mechanical control parameters is constructed. The mechanical control parameters include ratios between the duration of each position at which the target stays and the moving step length of the target. Fig. 4.11 shows the schematic of the proposed magnetron sputtering system.

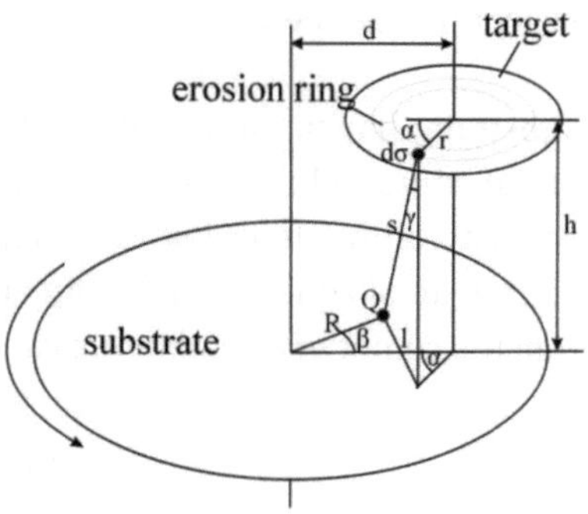

Fig. 4.11: Schematic of the proposed magnetron sputtering system.

In Fig. 4.11, h denotes the vertical distance between the target and the substrate, d the horizontal distance between them, r the distance between a sputtering unit $d\sigma$ of the target and the target center, R the distance between a certain point Q on the substrate and the substrate center, s the distance between a sputtering unit $d\sigma$ on the target and a certain point Q of the substrate, and r the sputtering angle.

According to Fig. 4.11, simple geometrical relations expressed by these parameters can be obtained:

$$\cos\gamma = \frac{h}{s}, \tag{4.3}$$

$$s^2 = h^2 + l^2, \tag{4.4}$$

$$l^2 = (d - R\cos\beta - r\cos\alpha)^2 + (R\sin\beta + r\sin\alpha)^2. \tag{4.5}$$

For most planar magnetron sputtering techniques, incident particles have low energy but high working pressure (several Pascals). Accordingly, we make the following four assumptions:

(1) Considering that the magnetic lines of force on the surface of the magnetron sputtering target are parallel to the target surface and the sputtered particles of ionized gas are accelerated to impact the target surface along the magnetic lines of force, the incident angle of particles towards the sputtering unit $d\sigma$ of the target is zero.

(2) All the particles sputtered from the target are derived from the nonuniform etched ring located within the rage of r_1 to r_2, where r_1 and r_2 represent the inner and outer diameters of the two poles of magnetic cylinder, respectively. The magnetic

field intensity within this area is very strong; thus, most of the secondary electrons are trapped around the ring. As the sputtering gas is ionized around this area, target particles are mainly sputtered from this area.

(3) For the coating sputter deposition method, the angular distribution of sputtered target particles is generally considered to satisfy a simple cosine distribution.

(4) During the process of magnetron sputter deposition, the pressure of the vacuum chamber is often only several Pascals, causing the sputtered particles to be scattered by the gas molecules of the chamber when approaching the substrate. The probability of particles to be deposited on the substrate is inversely proportional to the distance traversed by the particles in the vacuum chamber.

The surface of the mobile target is parallel to the substrate that can rotate about its own axis. Given that m stands for the mass of material sputtered from the zone $d\sigma$ of the etched ring on the target surface and ρ the density of the deposited coating, based on the four assumptions above, the deposition rate $d\upsilon$ of zone $d\sigma$ on the target surface towards the point Q that is at a distance R from the center of the substrate can be expressed as

$$d\upsilon = \frac{m(n+1)\cos^{n+1}\gamma}{2\pi\rho s^2}, \tag{4.6}$$

where n is the angular distribution coefficient.

The angular distribution coefficient is associated with multiple factors, which can be determined by tests if other conditions are known. When the substrate revolves about its own axis, given that the etched ring is located within the range of r_1 to r_2, the total deposition rate of the whole etched ring towards this point is

$$\upsilon = \int_0^{2\pi}\int_0^{2\pi}\int_{r_1}^{r_2} \frac{m(n+1)h^{n+1}r}{2\pi\rho s^{n+3}}\, d\alpha\, d\beta\, dr. \tag{4.7}$$

Fig. 4.12 exhibits the relative motion between the magnetron sputtering target and the substrate. It can be observed that when the magnetron sputtering target intermittently moves along the substrate radius step by step, a series of arrest points P_d will be passed along the radial direction. Given that the duration on each of these points is T_d and that the scanned area on the substrate corresponding to this point is S_d, the thickness distribution of the coating fabricated using the planar magnetron sputter deposition system can be expressed as

$$t = \sum_{d=1}^{n} t_d = \sum_{d=1}^{n} \upsilon_d T_d, \tag{4.8}$$

where υ_d represents the coating deposition rate when the horizontal distance between the target center and the substrate center is d.

In order to obtain the relationship between the coating thickness distribution and control parameters, a series of calculations based on equation (4.9) are performed in

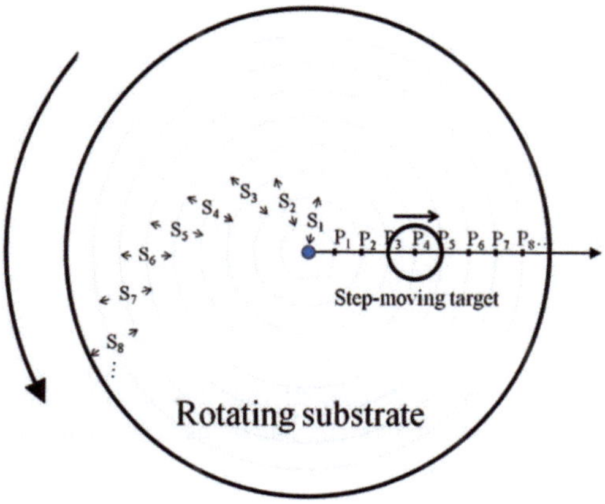

Fig. 4.12: Relative motion between the magnetron sputtering target and the substrate.

which the target–substrate distance and the outer diameter of the etched ring are fixed as 70 mm and 15 mm, respectively. The degree of nonuniformity of the coating is

$$\text{Max} - \text{Min} = \frac{H_{\max} - H_{\min}}{H_{\max} + H_{\min}} \times 100\,\%, \tag{4.9}$$

where $H_{\max}$ is the maximum thickness of the coating and $H_{\min}$ is the minimum thickness.

Fig. 4.13 shows the relative deposition rate distribution of the coating under different horizontal distances between the target center and the substrate center. It can be seen that when the target center is close to the substrate center, the relative deposition rate is a monotonically decreasing function; when the target center is far from the substrate center, the relative deposition rate first increases and then decreases. Moreover, when the distance between the target center and the substrate center is relatively large, variation patterns of the relative deposition rate are similar. As the distance increases, the fluctuation of the relative coating deposition rate gradually decreases, and therefore, the uniformity of the deposited coating is improved.

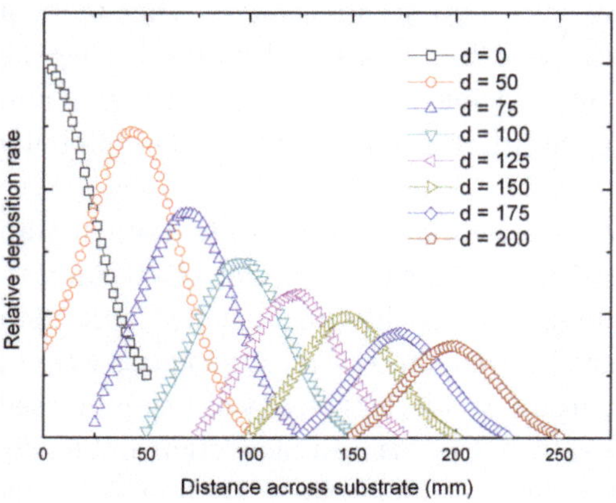

Fig. 4.13: Relative deposition rate distribution of the coating under different horizontal distances between the target center and the substrate center.

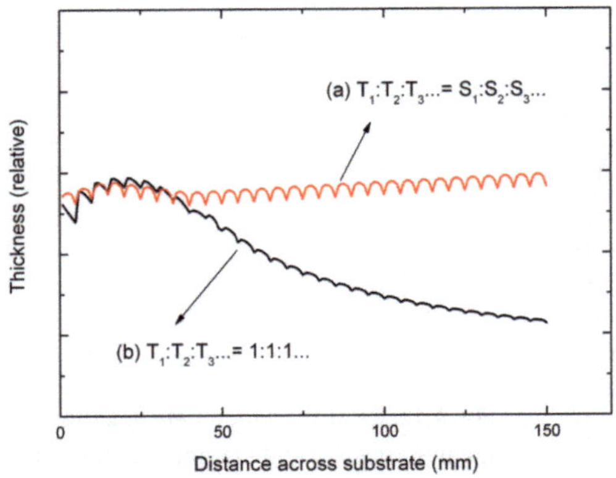

Fig. 4.14: Coating–thickness distribution under different time ratios at a step length of 5 mm.

Fig. 4.14 illustrates the thickness distribution under different time ratios when the step length is fixed as 5 mm. It can be seen that the duration of the target at each position significantly affects the coating uniformity. If the target is controlled to stay at each position for the same amount of time, considering the gradually decreasing coating deposition rate, the coating thickness will gradually decrease along the radial direction from the center towards the edge. In other words, under the same deposition conditions, the number of particles sputtered form the target per unit time is constant. In order to deposit a coating with an identical thickness at each point, the sputtering time received by each position should be directly proportional to the corresponding ring area, i.e.

$$T_1 : T_2 : T_3 \cdots = S_1 : S_2 : S_3 \cdots . \tag{4.10}$$

Fig. 4.15 shows the influence of step length of target motion on the coating uniformity, from which we can see that the coating uniformity is quite sensitive to the step length. Under the same time ratio, the coating has poor uniformity when the step length is set to 25 mm. Because the relative deposition rates along the radial direction are different, when deposition is conducted at any position on the coating, the coating geometry is similar to the lunar crater, which has a raised rim (high) and crater wall (low). Therefore, a thin area will appear between two adjacent concentric craters. When the step length of target motion is set to 5 mm, because of the decrease of distance between the two craters, the former thin area will be filled, giving the coating superior overall uniformity. Therefore, a smaller moving step length will contribute to better uniformity of the deposited coating.

Based on the model described above, a magnetron sputter deposition system with substrate self-rotation and target step-motion is designed to fabricate a ⌀300 mm coating and measure the coating thickness along the radial direction. Detailed fabrication and measurement processes are described below. A piece of ⌀300 mm glass was cut using a glass knife and adopted as the substrate. The glass piece was first processed through ultrasonic cleaning with acetone, then dried using a blow drier, and finally attached with a 5-mm-wide and 150-mm-long tape along the radial direction from

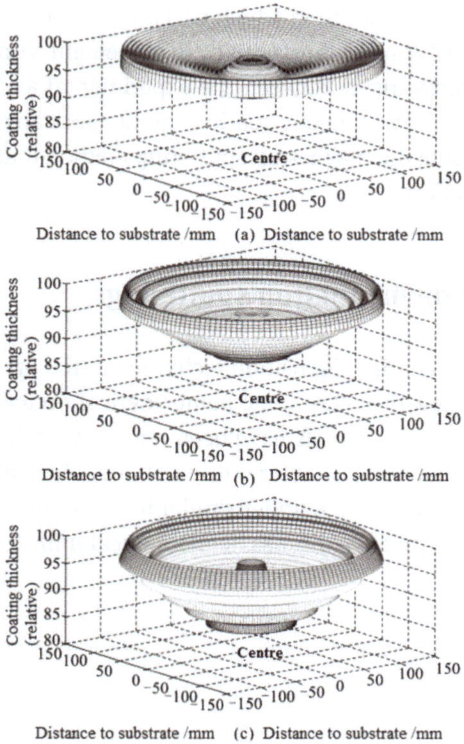

Fig. 4.15: Coating–thickness distribution under different step lengths: (a) step length of 5 mm, (b) step length of 15 mm, (c) step length of 25 mm.

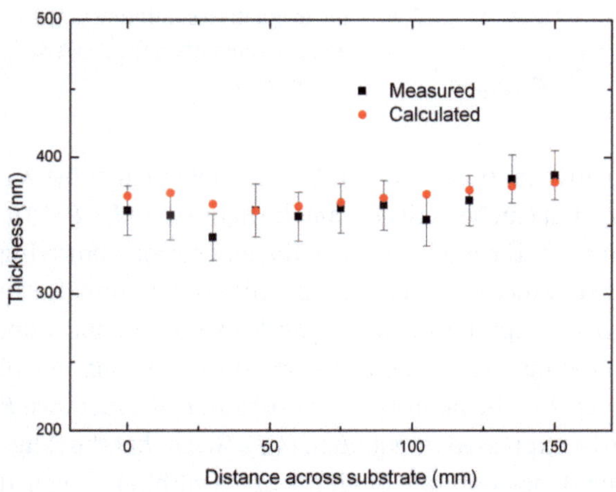

Fig. 4.16: Computational and experimentally measured coating–thickness distributions.

the center to the edge. The deposition was performed at room temperature with a working pressure of 1.0 Pa and a power of 100 W. After the deposition, the tape was removed and the formed step was cleaned. The coating thickness was measured using a Talysurf PGI 1240 profiler. Three measurements were conducted on each test point, the average from which was utilized as the final thickness value of the coating. The coating–thickness distribution is shown in Fig. 4.16, from which we can see that the actual thickness distribution of the fabricated coating is essentially in agreement with the calculated thickness distribution. A comparison between the experimental and computational results indicates that a large homogeneous coating can be fabricated with a relatively small target material by using the magnetron sputtering method that can control the relative substrate-target motion.

4.8 Heavy-calibre spherical uniform magnetron sputtering

We discussed the large-sized plane uniform coating deposition technique in the previous section. In practical applications, there also exist some hemispherical optical components that require coating. However, few methods have been reported for the uniform coating fabrication on these hemispheres. Therefore, research on spherical uniform deposition is of great significance. Here, in a manner similar to the study of the large-sized plane uniform deposition method, we discuss hemispherical uniform deposition in this section.

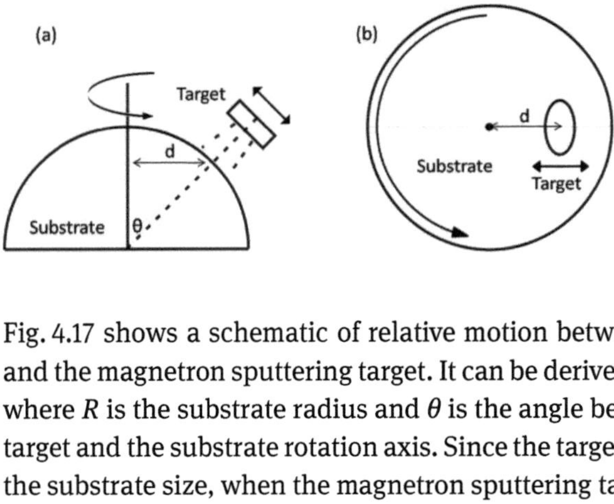

Fig. 4.17: Relative motion between the magnetron sputtering target and the substrate: (a) side view; (b) top view.

Fig. 4.17 shows a schematic of relative motion between the hemispherical substrate and the magnetron sputtering target. It can be derived from the figure that $d = R\sin\theta$, where R is the substrate radius and θ is the angle between the magnetron sputtering target and the substrate rotation axis. Since the target size is rather small compared to the substrate size, when the magnetron sputtering target travels to a certain location, it is approximately equivalent to coating on a plane. Therefore, the deposition rate of the zone $d\sigma$ on the target with respect to the point Q on the substrate at a distance R from the substrate center can also be expressed by equation (4.7). Given that the range of angle β is 0–π, the total coating deposition rate of all the areas within the etched ring towards this point is

$$\upsilon = \int_0^{2\pi}\int_0^{\pi}\int_{r_1}^{r_2} \frac{m(n+1)h^{n+1}r}{2\pi\rho s^{n+3}}\,\mathrm{d}\alpha\,\mathrm{d}\beta\,\mathrm{d}r. \tag{4.11}$$

When the magnetron sputtering target moves in an intermittent step-by-step manner along the longitudinal direction of the external surface of the substrate, a series of stopover points P_θ will be passed on the substrate in the radial direction. Assuming that the duration of each point is T_θ and the scanned area on the substrate corresponding to the point is S_θ, the thickness distribution of the coating deposited by the magnetron sputter deposition system can be expressed by the equation

$$t = \sum_{\theta=1}^{n} t_\theta = \sum_{\theta=1}^{n} v_\theta T_\theta. \tag{4.12}$$

Similar to the case of plane deposition, in order to ensure a uniform thickness for the whole area of the deposited coating, the duration at a certain position should be directly proportional to the corresponding ring area of this position. For example, the angle increment of each step can be set to 5°, and the magnetron sputtering target can be set to move above the substrate and downwards along the longitudinal direction of the substrate. According to this approach, we used a magnetron sputtering target with a diameter of 49 mm to coat a hemispherical substrate with a diameter of 300 mm. Fig. 4.18 displays the transmission curves of the substrate before and after the protective-coating fabrication. It can be observed that the transmission after deposition is slightly greater because the refractive index of the coating is greater than that of the substrate. Meanwhile, three different points on the substrate are selected to measure transmission. It can be seen that the transmission curves of the three points are almost identical, indicative of a successive uniform coating deposition on the entire external surface of the hemispherical substrate. In summary, experimental results show that, by controlling the relative motion between the substrate and the target, the magnetron sputtering method is able to fabricate large-sized hemispherical uniform coatings with fairly small targets.

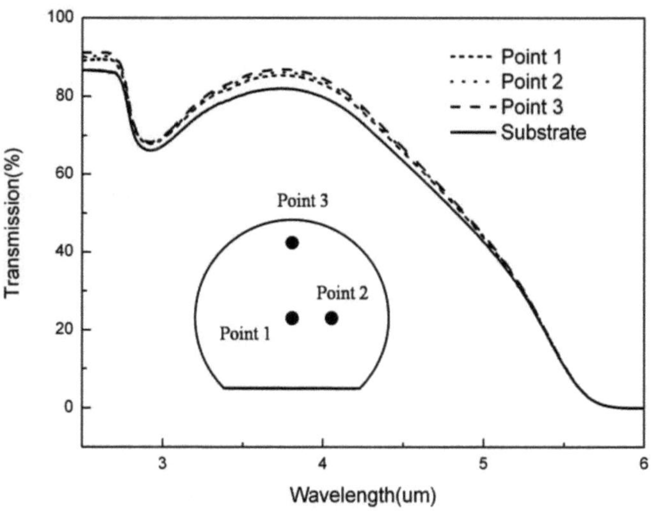

Fig. 4.18: Transmission spectra of different points on the substrate deposited with coating.

References

[1] Aksenov II, Belous VA, Padalka VG, et al. Transport of Plasma Streams in a Curviliner Plasma Optical System. Journal of Plasma Physics, 1978, 4: 425–428.

[2] Aksenov II, Belous VA, Padalka VG, et al. Motion of Cathode Spot of a Vacuum Arc in Inhomogeneous Magnetron Field. Technical Physics Letters, 1978, 3: 525.

[3] Biederman H. R.F. Sputtering of Polymers and Its Potential Application. Vacuum, 2000, 59: 594–599.

[4] Cooke KE, Hampshire J, Southall W, et al. Industrial Application of Pulsed DC Bias Power Supplies in Closed Field Unbalanced Magnetron Sputter Ion Plating. Surface Engineering, 2004, 20(3): 189–195.

[5] Debessaia M, Filipb P, Aouadia SM. Niobium Zirconium Nitride Sputter-Deposited Protective Coatings. Applied Surface Science, 2004, 236: 63–70.

[6] Glocker DA, Romach MM, Lindberg VW. Recent Development in Inverted Cylindrical Magnetron Sputtering. Surface and Coatings Technology, 2001, 146-147: 457–462.

[7] Grove AS, Snow EH, Deal BE, et al. Simple Physical Model for the Space-Charge Capacitance of Metal-Oxide-Semiconductor Structures. Journal of Applied Physics, 1964, 35(8): 2458–2460.

[8] Hakovirta M, Tiainen VM, Pekko P. Techniques for Filteringgraphite Macroparticles in the Cathodic Vacuum Arc Deposition of Tetrahedral Amorphous Carbon Films. Diamond and Related Materials, 1999, 8: 1183–1192.

[9] Heister U, Krempel-Hesse J, Szczyrbowski J, et al. TwinMag II: Improving an Advanced Sputtering Tool. Vacuum, 2000, 59: 424–430.

[10] Kelly PJ, Hisek J, Zhou Y, et al. Advanced Coatings Through Pulsed Magnetron Sputtering. Surface Engineering, 2004, 20(3): 157–162.

[11] Levichkova M, Mankov V, Starbov N, et al. Structure and Properties of Nanosized Electron Beam Deposited Zirconia Films. Surface & Coatings Technology, 2001, 141(1): 70–77.

[12] Li YQ. Vacuum Coating Technology and Equipment. Shenyang: Northeastern University Press, 1989.

[13] Martin PJ, Bendavid A. Review of the Filtered Vacuum Arc Process and Materials Deposition. Thin Solid Films, 2001, 394: 1–14.

[14] Roeck D, Winter GD, Paemel RV, et al. Rotatable Magnetron Sputtering of $YBa_2Cu_3O_{7-x}$ Thin Films on Single Crystal Substrate. Physica C, 2002, 372–376: 1067–1070.

[15] Sanders DM, Anders A. Review of Cathodic Arc Deposition Technology at the Start of the New Millennium. Surface & Coatings Technology, 2000, 133-134: 78–90.

[16] Tang GZ, Ma XX, Sun MR, et al. Mechanical Characterization of Ultra-thin Fluorocarbon Films Deposited by R.F. Magnetron Sputtering. Carbon, 2005, 43: 345–350.

[17] Thian ES, Huang J, Best SM, et al. Magnetron Co-Sputtered Silicon-Containing Hydroxyapatite Thin Films an in Vitro Study. Biomaterials, 2005, 26: 2947–2956.

[18] Yehya M, Kelly PJ. Novel Enhanced Magnetron Sputtering System. Surface Engineering, 2004, 20(3): 177–180.

5 Amorphous diamond films

5.1 Overview

In the early 1970s, S. Aisenberg and R. Chabot from the Space Sciences Division of Whittaker Corporation, USA became the first to employ the plasma ion beam deposition of carbon to form a thin-film material that possessed properties similar to those of diamond crystals, including transparency, insulation, and superhardness, but was also amorphous. The material was referred to as diamond-like carbon (DLC). After 20 years, S. Aisenberg described the motivation behind its nomenclature as follows:

> These carbon films had many of the properties of natural diamond but were predominately amorphous and not crystalline. Professional caution prevented use of the name "diamond" but it was felt that "diamond-like" would identify the properties as being similar to diamond without being presumptuous.

Subsequent research has demonstrated that the significance of this work was epoch-making. As DLC thin films exhibit a series of excellent properties, they have immense application value. In fact, DLC thin films have remained a hot topic of research over the last few decades, and several different processing methods have been employed to fabricate different forms of DLC. DLC thin films can be classified according to the absence or presence of hydrogen into hydrogenated and nonhydrogenated DLC. Compared to hydrogenated DLC thin films, nonhydrogenated DLC has higher performance indicators, better thermal stability, lower fabrication temperature, and other advantages. Hence, the latter has gradually dominated the mainstream research and applications of DLC thin films. A statistical analysis performed by Dr S. S. Eskinsen from Denmark revealed that the number of research articles related to hydrogenated DLC has been declining each year, whereas the number of articles related to nonhydrogenated DLC has been increasing and has surpassed the number of articles related to hydrogenated DLC.

Based on the amount of sp^3 hybridization in the thin films, hydrogenated DLC can be further classified into hydrogenated amorphous carbon (a-C:H) and hydrogenated tetrahedral amorphous carbon (ta-C:H), whereas nonhydrogenated DLC can be classified into amorphous carbon (a-C) and tetrahedral amorphous carbon (ta-C) or amorphous diamond (a-D). Under general conditions, the name "ta-C" has been widely used worldwide; similarly, "amorphous diamond" (an English translation from the Chinese name) has been widely accepted. According to the sp^2-sp^3-H ternary phase diagram shown in Fig. 5.1 and the general consensus in the industry, "a-D" is used to refer to nonhydrogenated DLC with more than 50 % of sp^3 hybridization. Owing to the increased amount of tetracoordinate hybridization, a-D has better photothermal stability and superior mechanical, optical, and electrical properties compared to a-C and

https://doi.org/10.1515/9783110489514-008

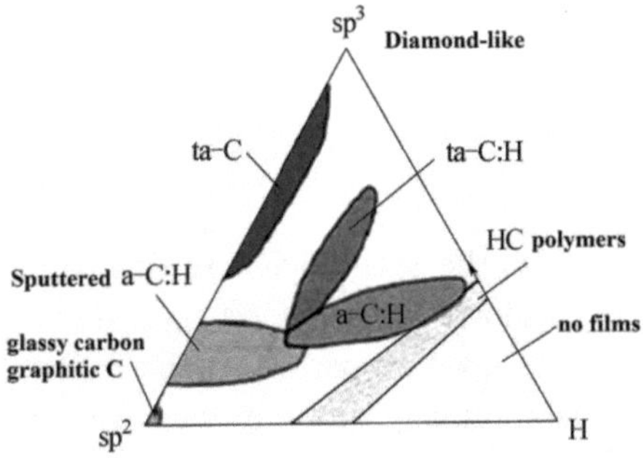

Fig. 5.1: sp^2-sp^3-H ternary phase diagram of diamond-like carbon.

a-C:H. Compared to diamond films, a-D has several advantages, including amendable structure and properties, fabrication at room temperature, smooth surface, and large deposition area. Thus, it is more valuable for practical applications.

5.2 Determination of deposition energy for rich sp^3 hybridization

Generally speaking, the mechanical properties of a-D are predominantly determined by sp^3 hybridization in the thin film, while its photoelectric properties are mainly attributed to sp^2 hybridization. Therefore, this section will begin by investigating the relationship between thin-film mechanical properties, including hardness and elastic modulus, and substrate bias. We then analyse the relationship between thin-film microstructure and substrate bias by using Raman spectroscopy, x-ray photoelectron spectroscopy, and electron energy-loss spectroscopy. The relationship between the actual energy E_i of deposited particles when they arrive at the substrate and the substrate bias V_b is given by

$$E_i = e(V_p - V_b) + E_0, \tag{5.1}$$

where V_p is the plasma potential, e is the electron charge, and E_0 is the initial energy of the plasma.

Without the addition of substrate bias, the substrate placed within the plasma beam has the same electric potential as the plasma beam. A Langmuir probe test will show that the plasma electric potential is approximately −13 V (anodic grounding). Experiments have indicated that the ion energy distribution of carbon plasma at the magnetic filter outlet is a Gaussian curve, and its full width at half maximum (FWHM) is related to the charge number carried by carbon ions. The initial energy of carbon ions is approximately 20–30 eV, which is related to the arc current size and arc spot state. In this case, the deposition energy of carbon ions is primarily controlled through substrate bias, and the relationship between the deposition energy of carbon ions and substrate bias is given by

$$E_i = -eV_b + (7\ldots17)\,\text{eV}. \tag{5.2}$$

For the sake of simplicity, this book will mainly discuss the influence of substrate bias on the a-D thin-film structure and its properties.

5.2.1 Mechanical properties of thin films

Thin-film hardness is an integrated property. Numerically, it is the ratio between the normal force acting on the diamond indenter and the projected area of indentation deformation. However, the property itself does not have a definite physical meaning. As can be seen from Fig. 5.2, with increasing substrate bias, thin-film hardness and elastic modulus initially showed an increase followed by a decrease, reaching their maximum values when the negative bias was 80 V. The variations in deposition energy with hardness and elastic modulus are consistent with other experimental results. Experiments have also shown that the hardness and elastic modulus of a-D thin films are closely related to the amount of tetracoordinate hybridization in the thin films. Within a certain range, the thin-film hardness and elastic modulus showed an almost linear relationship with the amount of sp^3 hybridization.

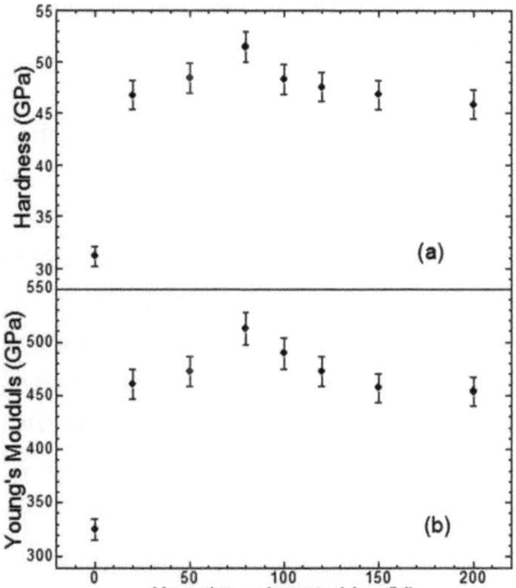

Fig. 5.2: Variation of a-D thin-film hardness and elastic modulus with substrate bias.

It can be seen that deposition energy not only shows the same variations with thin-film hardness and elastic modulus, but also has a definite proportional relationship, i.e. hardness H/elastic modulus $E \approx 0.1$. Compared to certain forms of a-C:H, this ratio is relatively low. The differences in elastic deformation and plastic deformation produced by a diamond indenter on two types of materials should be the main cause of this disparity.

Similar variations can also be observed between the critical scratch load and substrate bias. As shown in Fig. 5.3, the critical scratch load was the highest when the negative substrate bias was 80 V. When the bias was increased or decreased, the critical scratch load decreased. The critical scratch load can be determined by detecting the changes in lateral friction produced when a diamond indenter is scratched across a thin film. Evidently, a thin film with a higher level of hardness will require the application of a higher normal load and will lead to a larger critical load. Therefore, the variations shown in Fig. 5.3 essentially reflect the variations of thin-film hardness.

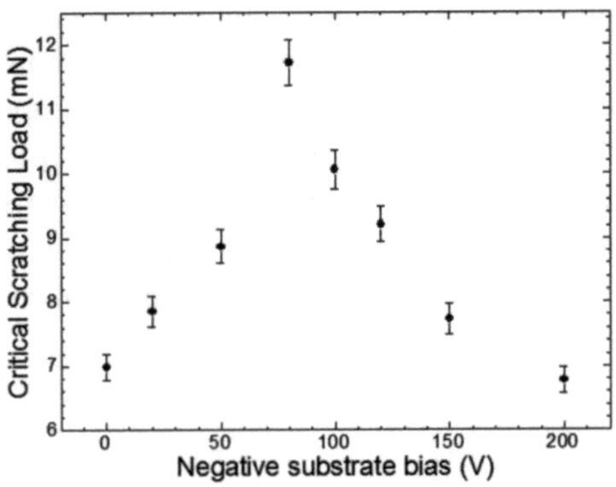

Fig. 5.3: Relationship between a-D thin-film critical scratch load and negative substrate bias.

In summary, thin-film hardness, elastic modulus, and critical scratch load all reached the maximum when the substrate bias is −80 V. The maximum hardness attained is 51.49 GPa, and the minimum hardness without additional bias is 31.17 GPa. The maximum elastic modulus attained is greater than 500 GPa, and the maximum critical scratch load is nearly 12 mN. This indicates that there is good adhesion between the thin film and substrate. The mechanical properties of a-D thin films are determined by the sp^3 hybridization in the thin film, and the variations of mechanical properties with deposition energy reflect the variations in hybridization ratio in the thin film with deposition energy. However, additional research on thin-film microstructure is needed for accurately determining the deposition conditions for rich sp^3 hybridization in thin films.

5.2.2 Raman analysis

As can be seen from Fig. 5.4, all thin films showed Raman spectra with asymmetrical broad peaks centred at approximately 1580 cm^{-1} within the range of 1300–1800 cm^{-1}. The obvious difference from the Raman spectra of other DLC thin films is the absence of D shoulder peaks reflecting the breathing vibration of the ring-like sp^2 field at 1300–1400 cm^{-1}. Furthermore, at 900–1000 cm^{-1}, a square and relatively small second-order peak can be observed, which is caused by a monocrystalline Si substrate. However, the variations of the Si second-order peak with deposition conditions are not significant. Thus, this section will mainly explore the variations of thin-film microstructure with deposition energy based on the asymmetrical broad peak in the range of 1300–1800 cm^{-1}.

The π-bonds in thin films reflect sp^2 hybridization, and its band gap is 2.25 eV, which is roughly equivalent to the photon energy of an Ar^+ ion laser at the wavelength of 514.5 nm (2.36 eV). In contrast, the band gap of σ-bonds is 5.50 eV, which is substantially different from the photon energy of visible light. This implies that the Raman cross section of sp^2 hybridization is 50–230 times higher than that of sp^3 hybridization. Thus, even if an infrared wavelength of 244 nm is employed, it will be difficult to detect the stronger characteristic signals of σ-bonds by using Raman spectroscopy because of the long-range polarization effects of π-bonds. Therefore, the broad peaks shown in Fig. 5.4 essentially reflect the state of sp^2 hybridized bonds in the thin film. They can almost be regarded as G-peaks produced by the stretching vibration of π-bond chains, and are associated with the E_{2g} symmetry of crystalline graphite in optical mode. Thus, they can be used as Raman "fingerprints" to distinguish a-D.

As the D-peak of a-D thin films is relatively weak in visible Raman spectroscopy, it is not possible to identify the D-peak position. However, performing the double-peak fitting of asymmetrical broad peaks seems forced and inaccurate. Therefore, in this book, a monoclinic Lorentzian line shape described by the Breit–Wigner–Fano (BFW) function is applied for the single-peak fitting of the a-D thin-film Raman data.

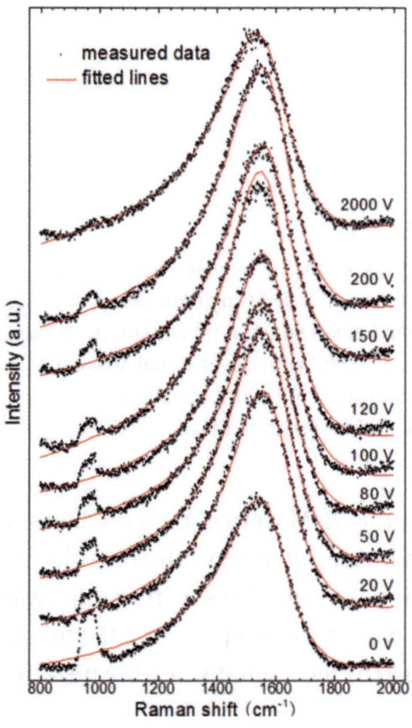

Fig. 5.4: Raman spectra of a-D thin films fabricated under different substrate biases.

Subsequently, the fitting parameters are used to characterize the microstructure of the thin film:

$$I(\omega) = \frac{I_0\left[1 + \frac{2(\omega-\omega_0)}{Q\Gamma}\right]^2}{1 + \left[\frac{2(\omega-\omega_0)}{\Gamma}\right]^2} + (a + b\omega), \tag{5.3}$$

where the photon scattering intensity $I(\omega)$ is a function related to the frequency of Raman shift, I_0 is the maximum peak intensity at Raman frequency ω_0, Γ is the full width at half maximum (FWHM) of the asymmetrical broad peak, $(a + b\omega)$ is the linear background, and Q is the coupling coefficient of the BWF function, the magnitude of which represents the degree of asymmetry of the monoclinic Lorentzian line shape (a symmetrical Lorentzian line shape will be presented as Q^{-1} approaches 0).

The amount of sp^3 bonds in the thin film can be determined using the Q value. A smaller Q ($Q < 0$) indicates a larger proportion of sp^3 bonds. The fitting results are presented as solid lines in Fig. 5.4. The test data from visible Raman spectroscopy coincided well with the fitted curves. In addition, the changes in fitting parameters with substrate bias are plotted in Fig. 5.5, in which the errors originated from the fitting calculation process.

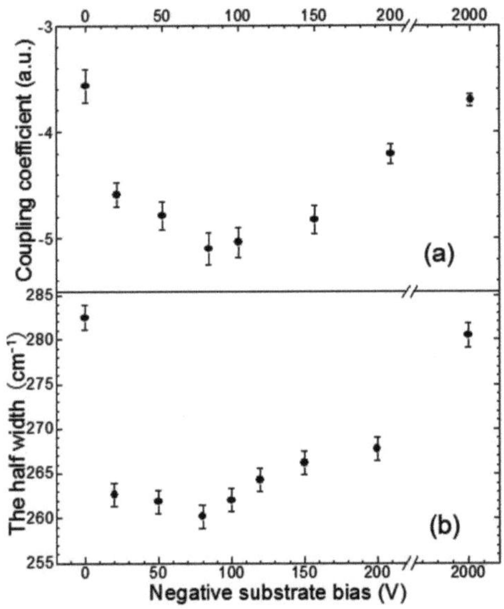

Fig. 5.5: Relationship between negative substrate bias and BWF fitting parameters of a-D thin-film Raman spectra: (a) variations of coupling coefficient with negative substrate bias; (b) variations of full width at half maximum with negative substrate bias.

The minimum values of the coupling coefficient and FWHM were obtained when the negative substrate bias was 80 V. Both parameters increased as the substrate bias increased or decreased. Fig. 5.5 (a) indicates that when the negative substrate bias was 80 V, the visible Raman spectra of the a-D thin film was the most symmetrical, and the amount of sp^3 hybridization was the highest. Fig. 5.5 (b) shows that when the negative substrate bias was 80 V, the visible Raman spectra of a-D thin film had the narrowest FWHM, indicating that the proportion of the D-peak in spectral broadening was the smallest. Furthermore, according to the graphite–microcrystalline graphite–a-C–a-D three-stage model, the amount of sp^2 hybridization in the thin film was also the lowest. Another important factor influencing the FWHM in the visible Raman spectra of a-D thin film is the sp^2 hybrid cluster size. Tamor and Vassell studied over a hundred a-C thin films from five research laboratories, and found that when the FWHM of the G-peak was greater than 50 cm^{-1}, the sp^2 cluster size L_a of the sp^2 field was less than 1 nm. This implies that the cluster size of sp^2 hybrids in the thin film was very small, and its structure should be a tetracoordinate, σ-bonded amorphous network embedded with microscopic sp^2 clusters. Unlike the coupling coefficient and FWHM, the peak position of asymmetrical broad peaks did not show regular variations with substrate bias.

Changes in the Raman line shape profile can also be characterized by the sloping coefficient. Fig. 5.6 reflects the regularity of variations of the sloping coefficient with the negative substrate bias.

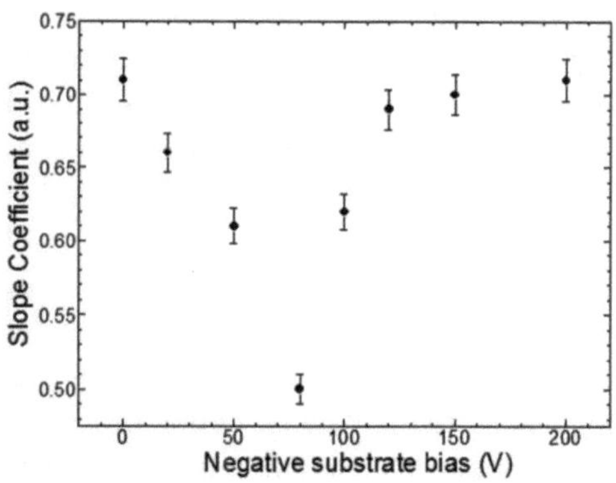

Fig. 5.6: Variations of sloping coefficient of Raman spectra with negative substrate bias.

The sloping coefficient S can be defined using the following equation:

$$S = \frac{I_{1300} - I_{1100}}{I_{1100}}, \tag{5.4}$$

where I_{1300} and I_{1100} are the scattering intensities of Raman shifts at 1300 cm^{-1} and 1100 cm^{-1}, respectively.

As can be seen from Fig. 5.6, the sloping coefficient was the lowest when the negative bias was 80 V, and the sloping coefficient increased as the negative bias increased or decreased. In practice, the sloping coefficient reflects the extent of the bump in the spectral profile at the D-peak position. Despite its lack of precise physical significance, the sloping coefficient is still able to characterize the deposition conditions for rich sp^3 hybridization in a-D thin-films.

In summary, as the deposition energy increases, the visible Raman spectral profile of a-D thin films shows significant changes. Thus, the characterization parameters of spectral line shapes (e.g. coupling coefficient, FWHM, and sloping coefficient) will also exhibit regular patterns of variation. When the negative substrate bias is 80 V, the spectrum will have optimal symmetry, the narrowest FWHM, and minimum sloping coefficient. These parameters will increase as the negative bias increases or decreases. These relationships are able to reflect the variations of thin-film microstructure.

5.2.3 X-ray photoelectron spectrum analysis

The interactions between incident photons and the crystal lattice will excite resonant Raman signals. When the energy of the incident photon is less than 4 eV, the scattering cross section will be dominated by information from sp^2 hybridization. Even if information from the stretching vibrations of sp^3 bonds can be obtained, the weighted quantity of its scattering intensity will be much weaker than that of sp^2 bonds. Therefore, the application of visible Raman spectroscopy to investigate the ratio of hybridization in thin films actually involves the use of information on sp^2 hybridization to reflect the characteristics of sp^3 hybridization. To achieve the accurate detection of a-D thin-film microstructure, other characterization methods are needed to directly obtain information on sp^3 bonds. X-ray photoelectron spectroscopy is a simple and reliable method to obtain direct evidence of different hybridizations in a-C. It is a powerful nondestructive testing method that can overcome the shortcomings of visible Raman spectroscopy in the investigation of structural details in a-C thin films.

Fig. 5.7 shows the C 1s optical emission spectra of a-D thin films fabricated under different substrate bias conditions. The C 1s spectral line shapes did not show significant changes with increasing substrate bias. As the 1s core excitation of carbon atoms with the two types of hybridizations overlapped at the binding energy range of 284–285 eV, peak deconvolution was needed when investigating the microstructure of a-D thin films. From a mathematical standpoint, spectral fitting requires an envelope with the minimum number of components that can maximize the coverage of discrete experimental data points. From a physical standpoint, the C 1s core optical emission spectra of DLC thin films can be decomposed into two components originating from sp^2 and sp^3 hybridizations, respectively. Furthermore, these two peaks will have essentially identical FWHMs and a known energy interval of 0.8–0.9 eV. In order to avoid data distortion, the thin-film surface did not undergo ion etching with inert gases before testing. However, the oxygen adsorbed to the surface of the thin film will affect the profile generation at the high-energy end of C 1s spectral curves. As can be seen from Fig. 5.7, based on the binding energies of standard pyrolytic graphite and diamond crystals, the low-binding-energy peak can be attributed to sp^2 hybridization, while the high-binding-energy peak can be attributed to sp^3 hybridization. Under different deposition conditions, the positions of the two peaks remained unchanged, i.e.

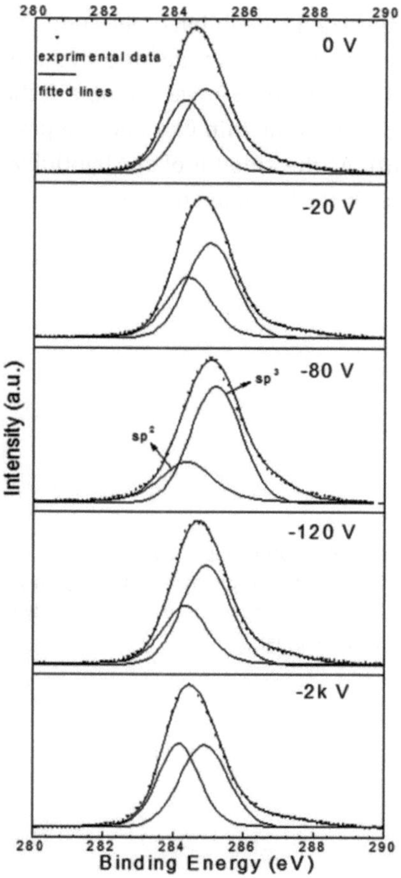

Fig. 5.7: Relationship between x-ray-excited C 1s optical emission spectra and substrate bias (C–O bond characteristics are not included in the figure).

at 284.3 eV and 285.1 eV. As such, the ratio of the two hybridizations in the thin film can be characterized by the integrated areas under the deconvoluted peaks.

As shown in Fig. 5.8, with increasing negative substrate bias, the amount of sp^3 hybridization in the thin film initially increased and then decreased; the amount of sp^3 hybridization was maximum when the negative substrate bias was 80 V. Under conditions of high substrate bias, the amount of sp^3 hybridization declined rapidly. Owing to the constraints of electron escape depth, x-ray photoelectron spectroscopy is a surface-sensitive technique, and its effective detection depth is only 3–5 nm. However, the cross-sectional structure of a-D thin films generally has "sandwich-like" characteristics, and its surface layer usually has a greater amount of sp^2 hybridization compared to the internal layers. The thickness of the surface layer is generally 1–3 nm, and it is related to the deposition energy; a greater substrate bias will imply a thicker surface layer. Thus, owing to the rich tricoordinate cyclic π-bond clusters in the surface, the ratio of sp^3 hybridization will be underestimated when performing x-ray photoelectron spectroscopy on thin films, and it will be drastically reduced at high substrate biases.

The peak position of C 1s optical emission spectra also exhibited variations as shown in Fig. 5.9. That is, the binding energy was maximum when the negative substrate bias was 80 V. By combining the results of Fig. 5.8 and 5.9, we can obtain the relationship between the amount of sp^3 hybridization in the thin film and the peak position of C 1s optical emission spectra (Fig. 5.10). As the amount of sp^3 hybridization decreased, the C 1s spectral peak shifted towards the low band energy.

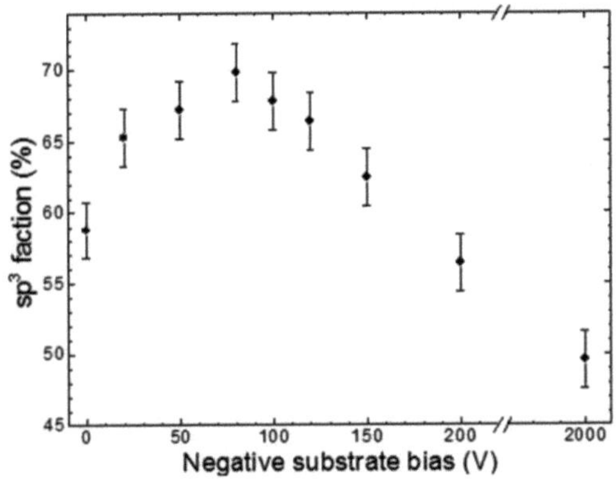

Fig. 5.8: Relationship between the ratio of sp^3 hybridization, detected using x-ray photoelectron spectroscopy, and negative substrate bias.

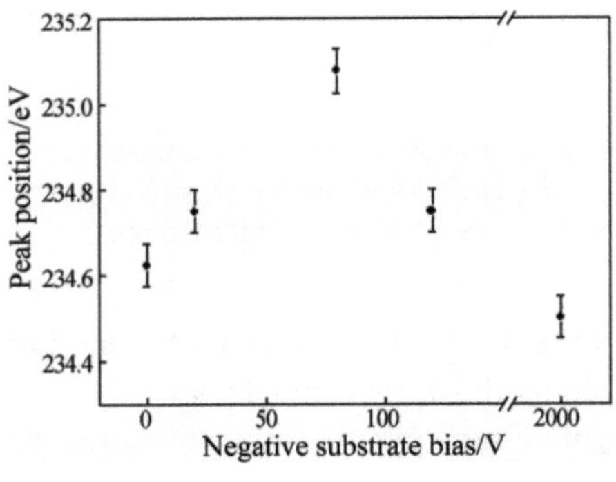

Fig. 5.9: Relationship between C 1s peak position of x-ray photoelectron spectra and negative substrate bias.

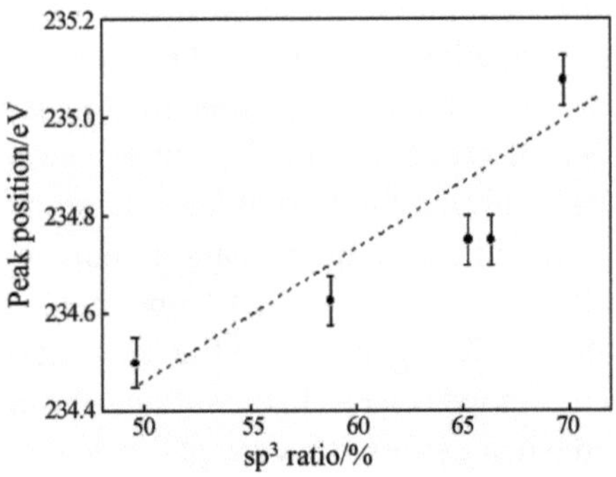

Fig. 5.10: Relationship between the amount of sp^3 hybridization in a-D thin film and C 1s binding energy.

The variation is consistent with the experimental results obtained by S. Turgeon and R. W. Paynter from the analysis of different hybridization ratios in polystyrene-polyethylene copolymers. For a-D thin films, the shift in the C 1s spectral peak is mainly determined by the degree of localization of valence electrons surrounding the excited atoms. Greater delocalization of the valence electrons will lead to a lower binding energy in the excited atoms. Owing to the semimetallic properties of graphite, the increase of sp^2-hybridized atoms in the thin film will cause a stronger delocalization of valence electrons and a corresponding shift in the C 1s spectral peak towards the low-energy end.

As shown in Fig. 5.11, the FWHM of the C 1s spectral peak essentially remained stable for low substrate bias but became significantly narrower under high substrate bias. Under constant test parameters, the FWHM of x-ray photoelectron spectroscopy is mainly determined by the thin-film hybridization composition and crystalline characteristics. When the substrate bias exceeded the deposition conditions for rich sp^3 hybridization and continued to increase, the component of the spectral peak representing tetracoordinate constituents was significantly weakened. More importantly, under high substrate bias, the ordering of the thin-film microstructure was enhanced, causing a significant narrowing of the FWHM of the spectral peak.

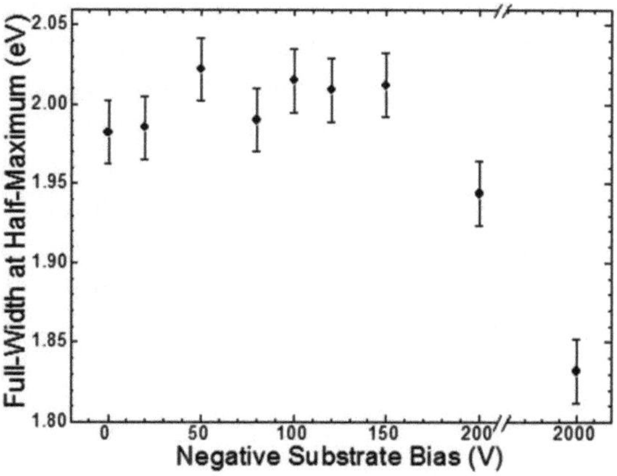

Fig. 5.11: Relationship between FWHM of x-ray photoelectron spectra and negative substrate bias.

In summary, the analysis of x-ray photoelectron spectra verified that deposition conditions for rich sp^3 hybridization occur at a negative substrate bias of approximately 80 V. The peak position of C 1s spectra shifted toward the high-energy end as the amount of sp^3 hybridization in the thin film increased. Furthermore, the FWHM of the C 1s peak essentially remained constant under low substrate bias but showed significant narrowing under high substrate bias.

5.2.4 Electron energy loss spectrum analysis

When applying visible Raman spectroscopy to investigate the hybridization composition of thin films, we cannot break away from the basic assumption of a double-peak model, regardless of whether peak deconvolution is applied to a graphite–microcrystalline graphite–a-C–a-D three-stage model in order to calculate the I_D/I_G ratio; or the sloping coefficient is used to calculate the variations in spectral profile; or a monoclinal Lorentzian curve is used to resolve the spectral parameters. In other words, the visible Raman cross sections of a-D thin films are mainly composed of vibrational information from two types of crystal lattices, stretching vibration from π-bond chains (G peak) and breathing vibrations from π-bond rings (D peak). In reality, Raman characteristics within the range of 1300–1800 cm^{-1} consist of multiple overlapping vibrational forms. For example, semicircular stretching vibrations might occur at 1486 cm^{-1}. When using x-ray photoelectron spectroscopy to study the hybridization ratio in thin films, peak deconvolution is performed to determine the energy interval. The binding energies of pyrolytic graphite and diamond crystals are used to determine the positions of the two deconvoluted peaks. However, the microstructure of a-D thin films is not the simple mixing of two crystal structures, and this analytical method has substantial surface sensitivity. Therefore, the use of Raman spectroscopy and x-ray photoelectron spectroscopy to analyse the hybridization composition of a-D thin films has significant limitations. Other characterization methods are needed for directly and accurately determining the thin-film microstructure. Electron energy loss spectroscopy (EELS) is currently the most commonly applied detection method to measure the amount of sp^3 hybridization in a-C films. Fig. 5.12 shows the influence of substrate bias on the carbon K-edge electron energy loss spectra in a-D thin-films.

Within the energy window of 280–310 eV, the electron energy loss spectrum of carbon included two major characteristics: 1s → π* transition at approximately 284 eV, and 1s → σ* transition at approximately 290 eV. In the absence of substrate bias, the π*

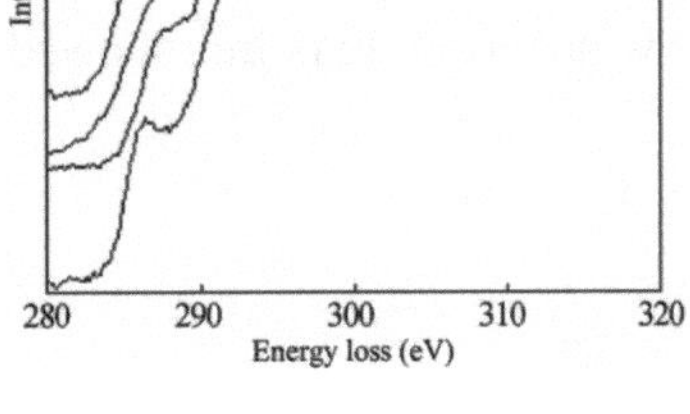

Fig. 5.12: Carbon K-edge electron energy loss spectra of a-D thin films fabricated under different substrate biases.

peak was clearly visible. As substrate bias increased, the π^* peak intensity gradually decreased. When the negative substrate bias was 80 V, the π^* peak was the weakest. As the substrate bias continued to increase, π^* increased gradually once again. This trend of change reflects the variations of sp^2 hybridization with deposition energy.

$$F_{sp^2} = \frac{\left[\dfrac{\text{area}(\pi^*)}{\text{area}(\pi^* + \sigma^*)}\right]_{\text{sample}}}{\left[\dfrac{\text{area}(\pi^*)}{\text{area}(\pi^* + \sigma^*)}\right]_{\text{reference graphite}}}. \tag{5.5}$$

Within a given energy window, by using the integrated area of the π^* K-edge peak of the C 1s electron energy loss spectra with reference to standard graphite, the sp^2 ratio in a-D thin films can be calculated using the relation shown in equation (5.5). The results are plotted in Fig. 5.13. It can be seen that when negative substrate bias was 80 V, sp^3 hybridization in the thin film reached 82 %. Compared to x-ray photoelectron spectroscopy, electron energy loss spectroscopy has a greater detection depth (10–20 mm) and is able to reflect the thin-film microstructure accurately.

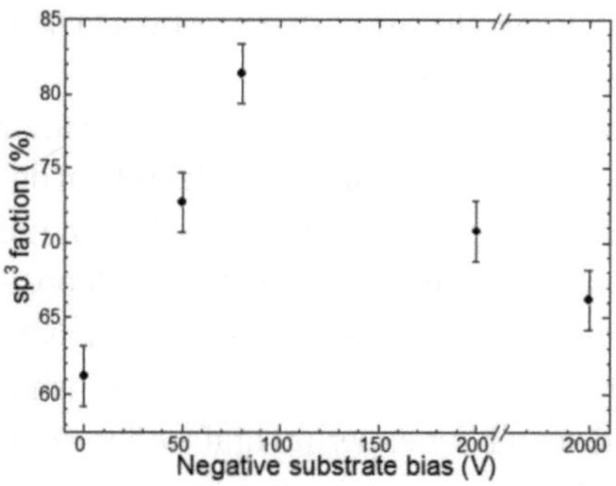

Fig. 5.13: Amount of sp^3 hybridization detected by electron energy loss spectrum in a-D thin films fabricated under different substrate biases.

In summary, from the perspective of analysing the variations of hardness, elastic modulus and other mechanical properties of a-D thin films with deposition energy, visible Raman spectroscopy, x-ray photoelectron spectroscopy, and electron energy loss spectroscopy were applied to investigate the thin-film microstructure and detect its hybridization composition. The investigation confirmed that a negative substrate bias of 80 V is the deposition condition for rich sp^3 hybridization in a-D thin-films fabricated using filtered arc deposition.

5.3 Deposition mechanism of amorphous diamond films

The deposition mechanism of a-D thin films continues to be a topic of great interest. Based on the Berman–Simon phase diagram of carbon, McKenzie et al. suggested that high pressure under room temperature is a necessary condition to produce sp^3

hybridization. Hence, they proposed the basic idea of inducing the formation of a-D thin films through compressive stress. There is a general consensus on the view that compressive stress is able to promote the formation of sp^3 hybridization. However, there are still obvious differences in opinion as to whether high internal compressive stress is a necessary condition for high sp^3 hybridization in the thin film. Recent work has shown that, despite the substantial relaxation of internal compressive stress after low-temperature vacuum annealing, the a-D thin film retained good chemical stability. The use of Monte Carlo simulations to model the internal compressive stress and local rigidity in thin films also confirmed that equilibrated/annealed films that relaxed the external constraints had zero total internal compressive stress while maintaining a high sp^3 hybridization ratio. As a more comprehensive understanding was gained on the impact of compressive stress on the growth process of a-D thin films, researchers began to focus on describing the deposition mechanism on an atomic scale. As the incident ion beam overcomes surface binding energy and undergoes shallow subsurface implantation (or subplantation), the entire process can be divided into three stages: collisional stage, thermalization stage, and thermal relaxation stage. A large number of computer simulations were performed to perfect a theoretical framework for the subplantation deposition mechanism of a-D thin films. However, there is still a lack of rigorous experimental data to support this deposition model.

The filtered arc deposition of a-D thin films essentially involves physical vapor deposition via low-energy ion beams. The deposition energy plays a crucial role in determining the structural composition of the thin films. The research described in the previous section has already clarified the optimal conditions required for the deposition of rich sp^3 hybridization. This section will mainly focus on the deposition mechanism of a-D thin films. We will present a qualitative explanation of the relationship between substrate bias and microstructure, and search for experimental evidence for thin-film growth mechanisms based on density distribution and surface morphology.

5.3.1 Cross-section layered density distribution of thin films

The densification effect of incident ions on the shallow surface layer is a necessary condition for the formation of sp^3 hybridization in a-D thin films. The relation between deposition energy and densification is given by

$$\frac{\Delta\rho}{\rho} = \frac{f\phi}{1 - f\phi + 0.016(E_\mathrm{i}/E_0)^{5/3}}, \tag{5.6}$$

where ϕ is the ratio of ions with energy E_i in the incident particle beam, f is the ratio of energetic ions that overcome the surface binding energy, and E_0 is the diffusion activation energy.

On the other hand, the relation between a-D density and thin-film microstructure can be established based on the density functional theory. We can see that density is the link between thin-film microstructure and deposition energy. Furthermore, because densification occurs at the shallow surface layer, the vertical cross-section density of the thin-film might not be uniform. Here, grazing-incidence x-ray reflectometry (XRR) can be used to detect thin-film density, and the OLYP exchange-correlation energy was applied to establish a thin-film structural model with different densities in order to analyse thin-film growth mechanisms.

Grazing-incidence XRR is a nondestructive detection technique that is widely used in material surface and thin-film analyses. By detecting the reflected x-ray signals on the thin-film surface, we are able to determine several important parameters of monolayered or multilayered thin films, including density, thickness, and surface roughness. XRR is gradually gaining popularity in a-D thin-film research as it is simple, rapid, and nondestructive.

5.3.1.1 Basic theory

The refractive index of materials within the x-ray wavelength range can be expressed as follows:

$$n = 1 - \delta - \mathrm{i}\beta, \tag{5.7}$$

$$\delta = \frac{r_0\lambda^2}{2}\sum_j \frac{\rho_j}{M}(Z_j + f_j'), \tag{5.8}$$

$$\beta = \frac{N_\mathrm{A}}{2\pi} r_0\lambda^2 \sum_j \frac{\rho_j}{M}(f_j''), \tag{5.9}$$

where Z_j is the atomic number, N_A is the Avogadro constant, f_j' is the diffusion correction factor, f_j'' is the absorption correction factor, M is the molar mass, ρ_j is the density of component j, and r_0 is the classical electron radius.

When a beam of grazing-incidence x-rays reaches the interface of two media (with refractive indices of n_{l-1} and n_l), the Fresnel reflectance can be approximated by

$$r_{l-1,l} = \frac{k_{l-1} - k_l}{k_{l-1} + k_l}, \tag{5.10}$$

$$k_l = 2\pi\sqrt{\frac{\sin^2\theta - 2(\delta_i + \mathrm{i}\beta_i)}{\lambda}}. \tag{5.11}$$

This represents the vertical wave vector of x-rays at the l-th layer. Therefore, if the x-rays are completely reflected at the thin-film surface, the critical angle is

$$\theta_\mathrm{c} = \sqrt{2\delta} = \lambda\sqrt{\frac{N_\mathrm{A} r_0}{\pi}\sum_j \frac{\rho_j}{M}(Z_j + f_j')}. \tag{5.12}$$

For a-D thin films, only the carbon element is considered, which has $f_\mathrm{c}' = 10^{-2}$. Hence, the critical angle and thin-film density, respectively, are as follows:

$$\theta_c = \lambda \sqrt{\frac{N_A r_0 \rho (Z_c + f_c')}{\pi M}}, \tag{5.13}$$

$$\rho = \frac{2\pi^2 c^2 \varepsilon_0}{3\lambda^2 N_A e^2} M m \theta_c^2. \tag{5.14}$$

When the angle of incidence exceeds the critical angle θ_c, the reflection curve will show periodic interference peaks. The film thickness d can be calculated using the formula $d = \lambda/(2\Delta\theta_r)$, where $\Delta\theta_r$ is the angular difference between adjacent peaks and troughs.

5.3.1.2 Curve-fitting analysis

The detection of thin-film density using XRR can be divided into three steps: data acquisition, construction of structural model, and curve fitting. Data acquisition involves the measurement of x-ray reflectance under different incident angles using the θ–2θ scanning mode. The construction of the structural model entails observing and measuring the periodic variations in the reflection curves, determining the number of layers in the preliminary model (generally starting with a small number of layers), and then comparing the simulated and experimental curves. If the disparities between the two are excessively large, more layers are added, and the model is reconstructed, until an optimal state of fit is reached.

Fig. 5.14 shows the measured and fitted curves of a-D thin film without substrate bias.

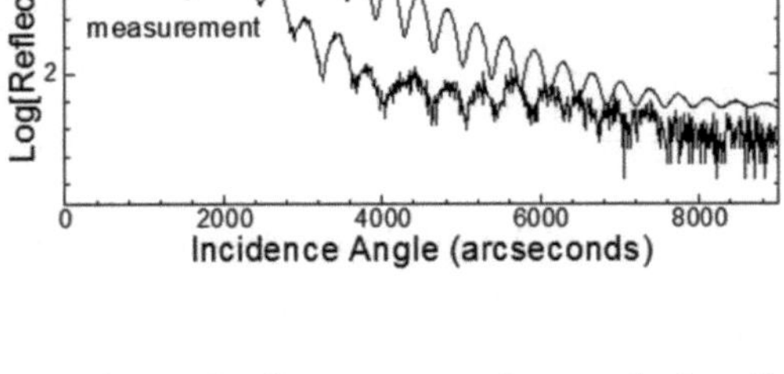

Fig. 5.14: X-ray reflectance curves of a-D thin film fabricated without substrate bias measured experimentally using XRR and calculated through curve fitting.

In order to facilitate comparison, relative displacement along the vertical direction was performed on both curves. As can be seen from Fig. 5.14, within the range of 0–4000″, reflection intensity decreased by five orders of magnitude as the incident angle increased. When the incident angle of light exceeded 4000″, the measured and fitted curves showed deviations due to the influence of background intensity. In the curve, the peak values of reflection intensity are not determined solely by interface roughness; the density difference between the substrate and thin film is also an important factor. Moreover, the fluctuation period is directly associated with the thickness of each layer in the a-D thin film.

The XRR measurement curves of a-D thin films under different substrate biases are shown in Fig. 5.15. According to the physical structure of the thin film and substrate, the cross section along the growth direction of the thin film can be divided into four geometric regions: thin-film surface layer, thin-film body, thin-film and substrate interface, and substrate. The four-layered model was used to simulate the measured curves, and the obtained fitting parameters are listed in Tab. 5.1.

As can be seen from Fig. 5.15, the XRR-measured critical angles of different thin films showed an increase followed by a decrease with increasing substrate bias, reaching the maximum value when the negative substrate bias was 80 V. Furthermore, it

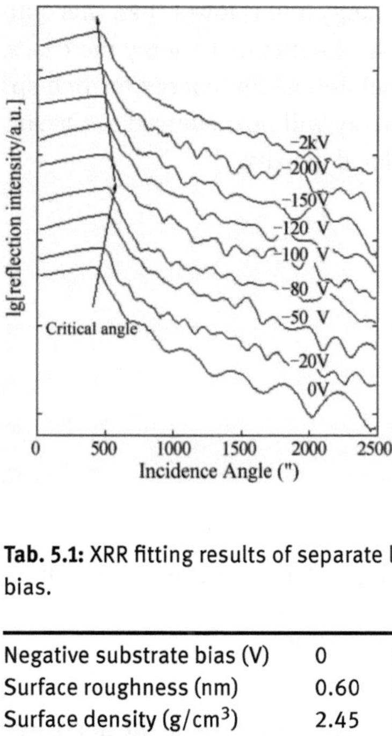

Fig. 5.15: XRR experimental curves of a-D thin films fabricated under different deposition energies.

Tab. 5.1: XRR fitting results of separate layers in a-D thin films fabricated under different substrate bias.

Negative substrate bias (V)	0	20	50	80	100	120	150	200	2000
Surface roughness (nm)	0.60	0.71	0.68	0.40	0.58	0.67	0.60	0.82	1.00
Surface density (g/cm^3)	2.45	1.90	2.50	2.86	2.75	2.46	2.58	2.63	2.61
Surface thickness (nm)	2.00	2.55	6.50	1.90	1.50	1.50	1.50	2.40	1.10
Body density (g/cm^3)	2.55	2.86	2.91	3.26	3.11	3.08	3.18	3.05	2.63
Body thickness (nm)	41.8	68.7	80.2	66.9	65.3	62.3	86.3	64.2	119.2
Interface density (g/cm^3)	1.90	2.52	2.54	2.56	2.58	2.62	2.61	2.54	2.47
Interface thickness (nm)	1.60	3.80	3.50	2.50	2.10	2.48	2.54	3.12	1.20
Si substrate density (g/cm^3)	2.33	2.33	2.33	2.33	2.33	2.33	2.33	2.33	2.33
Si substrate thickness (nm)	0.30	0.30	0.30	0.30	0.30	0.30	0.30	0.30	0.30

can be seen that when the negative substrate bias was 2000 V, the variations in the distances between the peaks and troughs of thin-film reflectance curves became extremely uniform. This might have been related to the formation of a uniform interface between the thin film and substrate promoted by high incident energy.

By using the density of the thin-film body listed in Tab. 5.1 as the density of a-D thin films, the relationship between thin-film density and negative substrate bias can be plotted as shown in Fig. 5.16. With increasing negative substrate bias, thin-film density increased gradually, reached a maximum of 3.26 g/cm^3 when the negative substrate bias was 80 V, and then decreased gradually. As with the variations of the critical angle, the magnitude of increase in thin-film density was greater when the negative substrate bias was 80 V. Thin-film density gradually decreased when the negative substrate bias exceeded 80 V. Based on the subplantation deposition mechanism, the densification effect that promotes the maximum sp^3 hybridization was used to calculate the deposition energy in order to obtain the optimal value. If the energy of the deposited ions is able to overcome surface binding energy and is lower than this optimal energy, then the thin-film density will increase as the incident energy increases. If the incident energy that is higher than the optimal deposition energy for rich sp^3 hybridization continues to increase, the residual energy will be dispersed to the surroundings as phonons, leading to the relaxation of local density.

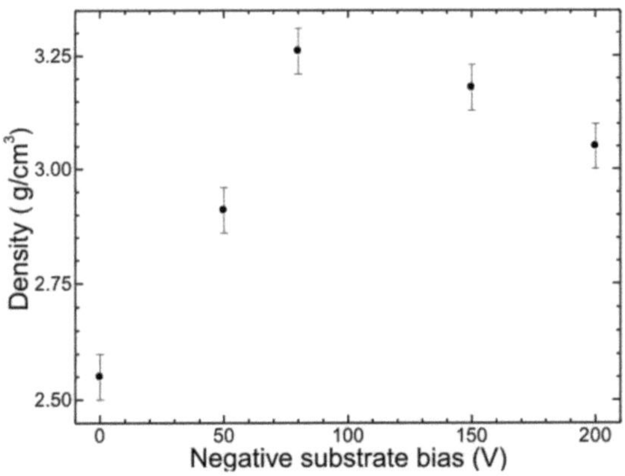

Fig. 5.16: Relationship between a-D thin-film density and negative substrate bias.

Based on the theory of subplantation growth, if the local densification of the deposited ions occurs a few atomic layers below the surface, then the a-D thin film must have a low-density surface layer. As summarized in Tab. 5.1, the surface densities of a-D thin films fabricated under different substrate biases are lower than the densities of the thin-film body. To express the layered density distribution of these film growth cross sections more clearly, the representative data in the table were used to plot Fig. 5.17. We can see from the figure that the surface/film-body/substrate-film interface showed a clear low/high/low layered density distribution. The substrate-film interface has a thickness of approximately 1–3 nm. As the density of monocrystalline Si is lower than

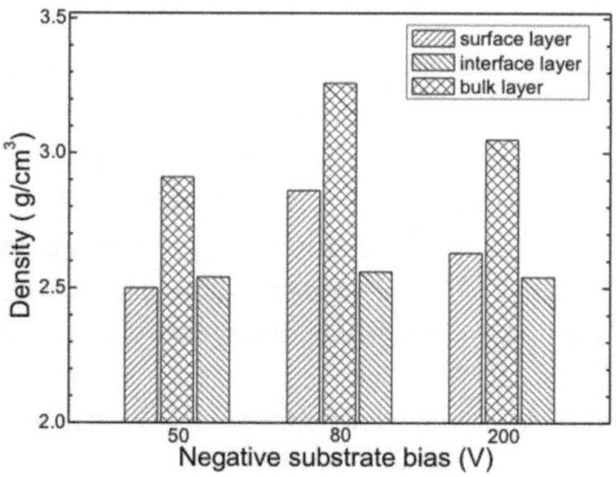

Fig. 5.17: Layered density distribution of a-D thin films fabricated under different substrate biases.

that of diamond crystals, and carbon/silicon interfaces usually possess many defects; hence, the substrate-film interface will have a relatively low density. The fact that the thin-film surface has lower density than the film body is direct evidence for subplantation growth in a-D thin films.

In summary, research based on grazing-incidence XRR indicates that with increasing substrate bias, the density of a-D thin films will present an initial increase followed by decrease. Thin-film density is the highest under conditions for rich sp^3 hybridization. A fitting analysis of reflection curves revealed that the thin-film growth cross section showed a clear low/high/low layered density distribution from the surface to the substrate layer. The low density of the surface layer can be regarded as rigorous evidence for thin-film subplantation deposition.

5.3.2 Structural model of amorphous diamond films

Despite confirming the layered density distribution of a-D thin films along the direction of growth, we are still interested in the microstructural characteristics at the atomic scale reflected in this type of layered density differences. There remain difficulties in conducting experimental research on the structure of a-C at the atomic scale. Nevertheless, molecular dynamics simulation is an effective approach. Different methods can be adopted based on the differences in interatomic interactions, including empirical potential, tight-binding approximation, Monte Carlo simulation, and density functional theory. Among them, density functional theory, which treats the system electron density as a variable, is able to reflect the physical properties of materials more accurately, and has begun to receive increasing attention in the structural research of a-D thin films. In this book, we will apply molecular dynamics based on density functional theory for the structural modelling of a-C with different densities in order to explore the variations in thin-film microstructure reflected by the different densities.

5.3.2.1 Calculation methods

With the use of a rapid liquid-quenching model, the simulation process involved the application of ab initio Car–Parrinello molecular dynamics. The planar wave cut-off energy was 40 Ry, and the time step was 3 a.u. (0.072 fs). The accuracy of calculations in density functional theory is determined by the precision of the exchange-correlation functional. This is particularly so for complex structures such as a-C, in which the choice of exchange-correlation functional will severely influence computational accuracy. To this end, the OLYP exchange-correlation energy was adopted in this book for structural modelling.

The entire simulation process can be divided into four parts: spontaneous melting, liquid equilibration, nonequilibrium cooling, and solid equilibration. Each simulation system consists of 64 carbon atoms in a simple cubic supercell, and the volume remained constant throughout the simulation process. In the initial structure, the carbon atoms are uniformly distributed in the cell, and they are subsequently given a random displacement of 0.2 Å. This type of structure is extremely unstable, which will give rise to spontaneous melting. Kinetics adjustments were not performed during the simulation process. Within 0.05 ps, the temperature rose rapidly from 0 K to 5000 K, and the process of liquid equilibration began. During liquid equilibration, the ion temperature and hypothetical electron kinetic energy could be controlled using a Nosé thermostat. This process was continued for 0.5 ps in order to ensure the adequate diffusion of the liquid. During the process of nonequilibrium cooling, the system temperature decreased exponentially with time from 5000 K to 300 K within 0.5 ps. The application of rapid cooling not only satisfies the computational requirements, but also complies with actual physical processes and enables empirical verification. The control method for ion and electron temperature during the process of solid equilibration is the same is that during liquid equilibration, and the equilibration time is 0.5 ps. By changing the cell size at the initial stage of simulation and repeating the four processes above, the structures of different densities can be obtained.

5.3.2.2 Calculation results

As shown in Fig. 5.18 (a)–(e), computer simulations of the a-C network structure were performed for five different densities (2.0, 2.3, 2.6, 2.9, and 3.2 g/cm^3). The cut-off distance of interatomic bonding defined based on the atomic radius of carbon and radial distribution function is 1.80 Å. The dotted line in Fig. 5.18 (f) represents the experimental results of neutron diffraction. From the perspective of both peak position and shape, the radial distribution functions and experimental data of the simulated structures showed a good fit. Tab. 5.2 lists the structural parameters of computer simulations for a-D thin films with different densities. Fig. 5.18 (a)–(e) correspond to models A–E in the table, respectively.

All structures did not contain isolated atoms or atomic chains. The low-density (2.0 g/cm^3) structure was mainly composed of sp^2 atoms, some sp^3 atoms, and a small

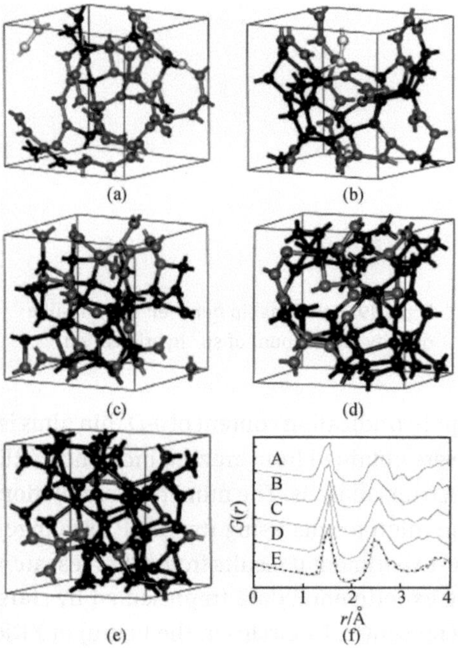

Fig. 5.18: Structural models and radial distribution functions of a-D thin films with different densities: (a)–(e) carbon network structures with densities of 2.0, 2.3, 2.6, 2.9, and 3.2 g/cm^3, respectively; black represents sp^3 carbon atoms, grey represents sp^2 carbon atoms, and white represents sp^1 carbon atoms; (f) atomic radial distribution functions.

Tab. 5.2: Structural parameters of computer simulations for a-D thin films with different densities.

Model	Density (g/cm^3)	Bond length (Å)	Bond angle (°)	Coordination number			
				2	3	4	Mean
A	2.0	1.51	116.0	1.7	78.1	17.2	3.1
B	2.3	1.51	112.4	6.2	50.0	43.8	3.4
C	2.6	1.54	111.7	0	50.0	50.0	3.5
D	2.9	1.56	109.5	0	28.1	71.9	3.7
E	3.2	1.53	109.5	0	15.6	84.4	3.8

amount of sp^1 atoms; the mean coordination number was 3.1. In this type of low-density carbon network structure, sp^2 carbon atoms formed long chains and macrocycles, and the structure was relatively loose. As density increased, the structure became more densified, macrocycles gradually decreased, and long-chain sp^2 carbon atoms were fragmented by sp^3 carbon atoms. When the sp^3 content was relatively high, sp^2 carbon chains were unable to penetrate the entire structure and could only appear in the form of short chains or π-bond pairs. When the density exceeded 2.6 g/cm^3, the structure was mainly composed of sp^3 carbon atoms, and sp^1 carbon atoms could not be observed. When the density was 3.2 g/cm^3, the sp^3 content reached 84.4 %, and the mean bond angle was 109.5°, which is very close to the bond angle of diamond. With increasing density, significant changes occurred in the structural properties of the thin film, such as a decreasing bond angle and increasing coordination number.

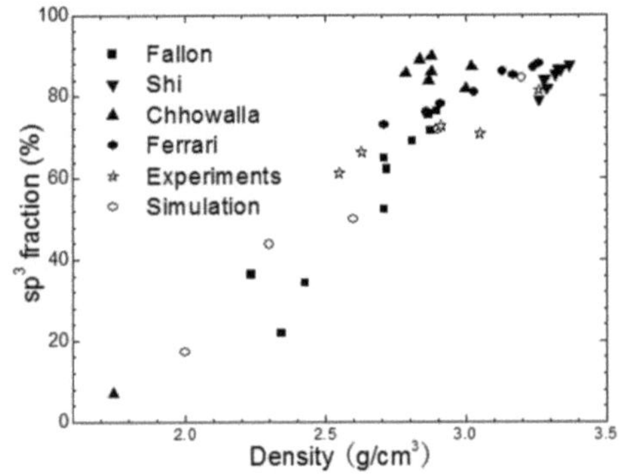

Fig. 5.19: Relationship between a-D thin-film density and amount of sp^3 hybridization.

The relationship between the density and sp^3 hybridization content of a-D thin films is shown in Fig. 5.19. The experimental data were obtained from grazing-incidence XRR density testing and electron energy loss spectrum analysis. The numerical simulation was based on C-P molecular dynamics simulation results using the OLYP exchange-correlation energy. The figure also shows the experimental results from other research groups as reference. It can be seen that the experimental data (represented by stars in the figure) and computer simulations (represented by circles in the figure) of XRR density tests showed a good fit, and that sp^3 hybridization essentially showed a linear increase with increasing thin-film density.

In summary, grazing-incidence x-ray reflectometry can be used to determine that the vertical cross section of thin-film growth from the surface to the substrate shows a low/high/low layered density distribution. The low-density surface constitutes obvious experimental evidence supporting the subplantation deposition mechanism of a-D thin films. Molecular dynamics simulation was then applied to establish the structural models of a-D thin films with different densities. The amount of sp^3 hybridization in the thin films shows an almost linear increase with increasing densities. In the low-density a-C network, the carbon network structure is looser, and sp^2 carbon atoms will form macrocycles and long chains that are distributed throughout the carbon network. As density gradually increases, the macrocycles will be disrupted by sp^3 carbon atoms but the sp^2 carbon atoms still exist in the form of long chains. When the density is higher, the carbon network structure will become more densified, and sp^2 carbon atoms can only exist in the form of short chains.

5.3.3 Surface morphology of thin films

The widespread application of atomic force microscopy has facilitated the investigation of the influence of deposition processes on a-D surface morphology. Fig. 5.20 shows the surface morphology of a-D thin films fabricated under different negative substrate biases.

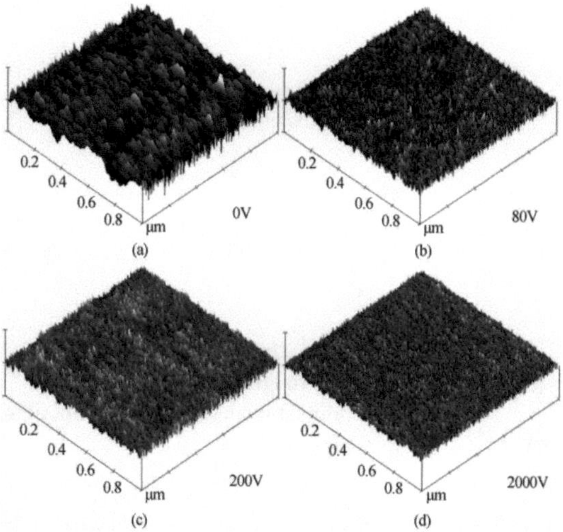

Fig. 5.20: Surface morphology of a-D thin films fabricated under different negative substrate biases (the scaling of all *z*-axes in the figure is 5 nm/div).

As mentioned previously, the optimal negative substrate bias for thin films with rich sp^3 hybridization is 80 V. C^+ ions with this energy are able to overcome the surface binding energy of monocrystalline Si substrates to be implanted a few atomic layers beneath the surface. Subsequently, under the action of the "atomic hammer", a large amount of ions will enter atomic gaps with a smaller volume than the deposited ions, which will cause an extremely high local compressive stress greater than 10 GPa. Based on the carbon stress-temperature phase diagram, the local shallow surface layer has conditions for sp^3 hybridization, as shown in Fig. 5.20 (b). This will lead to the formation of a-D thin films with a smooth surface and high amount of sp^3 hybridization. Without the addition of substrate bias, the energy of incident C^+ ions is insufficient to overcome the substrate surface; thus, they will be embedded on the surface. Under low temperature and low pressure, sp^2 hybridization is the stable phase, which will lead to the formation of a-D thin films with a rough surface and low amount of sp^3 hybridization (Fig. 5.20 (a)). If the deposition energy is slightly higher than the energy window for rich sp^3 hybridization, then when a substrate bias of 200 V is applied, C^+ ions will be implanted at the subsurface layer. A higher energy will result in the extension of the thermal stabilization phase, thereby causing some C^+ ions to migrate towards the surface and cluster. This, in turn, will lead to an increase in surface roughness while also causing a decrease in the amount of sp^3 hybridization in the thin film and the corresponding increase in the coupling coefficient. We can see from Fig. 5.20 (c) that there are rising ridges on the thin-film surface. However, it should be noted that when a high negative substrate bias of 2000 V was applied, the thin-film surface did not become rough; instead, it showed a smoother thin-film surface compared to when sp^3 hybridization was the highest, as shown in Fig. 5.20 (d).

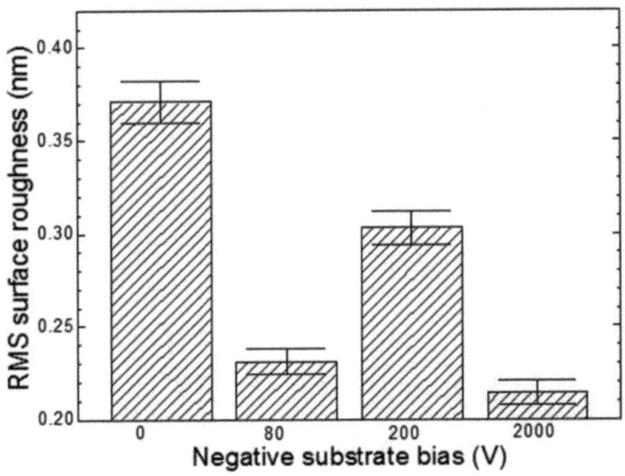

Fig. 5.21: Root-mean-square surface roughness of a-D thin films fabricated under different substrate biases.

In order to verify the experimental results further, the digital processing software package bundled with the atomic force microscopic system was used to compute the root-mean-square (RMS) roughness of the thin-film surface. The experimental results are shown in Fig. 5.21.

Thin-film RMS roughness was the lowest under a high negative substrate bias of 2000 V. This is because thin-film deposition is the dynamic equilibrium between surface growth and surface destruction. During the interactions between energetic ions and the substrate surface, the incident ions will collide with the surface atoms and be embedded onto the shallow surface layers. This is also accompanied by surface-growth processes of implanted ions under the effects of thermal pulses, including thermal vibrations, thermal diffusion, and surface migration. Furthermore, incident ions will have destructive effects on the surface atoms, such as sputtering, displacement, and defect formation. When the energy of incident ions was relatively low, the destructive effects on the surface were relatively weak, and were mainly manifested as thin-film growth processes. The application of subplantation growth mechanisms was able to provide a satisfactory explanation for the changes in surface morphology. A higher sp^3 content implied a lower RMS surface roughness. For a higher deposition energy, a relatively high incident energy led to a deeper implantation of deposited ions, which were embedded in atomic gaps smaller than the volume of the incident ions. This caused difficulties in the migration of deposited ions toward the surface, and preserved the smooth surface. On the other hand, the sputtering effects of incident ions on surface atoms were more significant, owing to the high energy, which led to preferential sputtering of asperities and thereby resulting in a smooth surface. However, when the energy was excessively high (e.g. 20 kV), the high-energy incident ions produced a large quantity of holes and other defects on the surface, which caused severe surface destruction, as well as an uneven thin-film surface. Beyond the energy window for rich sp^3 hybridization, the amount of sp^3 hybridization in the thin film decreased as energy gradually increased. Nevertheless, within a suitable energy range, the thin-film surface showed a trend towards greater smoothness due to the sputtering effects of incident ions. This is precisely the fact that has been ignored in previous research.

In summary, as the negative substrate bias increased, the a-D surface morphology showed a clear variation trend. At low energy, the thin-film surface roughness and sp^3 content were closely related. As the sp^3 content increased, the surface became smoother; this could be fully explained using the subplantation deposition mechanisms. However, under high-energy conditions, the sputtering of the surface by the deposited ions led to a smoothing effect. In fact, a surface even smoother than those of thin films with rich sp^3 hybridization could be obtained within a suitable energy range.

5.3.4 Surface composition of thin films

According to the subplantation growth theory of a-C, the penetration of the thin-film surface by incident ions is a necessary condition for local densification at the shallow subsurface layer. It can be seen from equation (5.15) that the critical penetration energy E_{p} is related to the surface binding energy E_{b} and the critical displacement energy E_{d}:

$$E_{\mathrm{p}} = E_{\mathrm{d}} + E_{\mathrm{b}}, \tag{5.15}$$

where the critical displacement energy of C^+ ions is approximately 25 eV, and the surface binding energy is approximately 7 eV; hence, the critical penetration energy is approximately 32 eV.

If the energy of deposited ions is less than the critical penetration energy, then the deposition of a-D thin films will be manifested as surface growth, and the original substrate component will be covered by the thin film. If the energy of the deposited ions exceeds the critical penetration energy, then the deposition of the a-D thin film will be manifested as subplantation growth, and the thin-film surface might retain components of the original substrate. In view of this fact, x-ray photoelectron spectroscopy was employed to detect the surface composition of an unsputtered a-D thin film, which is shown in Fig. 5.22. Without the addition of substrate bias, Si substrate components could not be detected in the deposited thin-film surface. In contrast, Si substrate components could be detected on the surfaces of thin films fabricated under other substrate biases. Only the XPS full-range scan spectra of thin films fabricated under a negative substrate bias of 80 V and without substrate bias are plotted in the figure. In practical terms, the investigation of the thin-film deposition mechanism from the perspective of surface composition is also highly speculative; hence, more in-depth research is required in this aspect.

In conclusion, there are shortcomings with the use of the subplantation deposition mechanism to explain certain experimental results, including the relationship between thin-film microstructure with different deposition patterns and the deposition temperature of a-D thin films fabricated using different processing methods, as

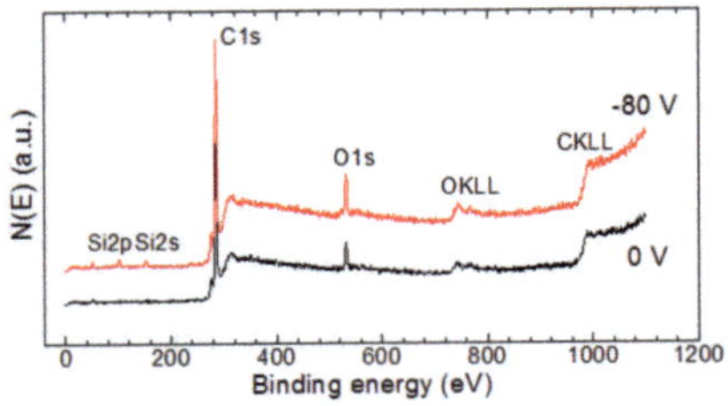

Fig. 5.22: XPS full-range scan spectra of thin films fabricated under a negative substrate bias of 80 V and without substrate bias.

well as the relationship between the growth speed and microstructure. Nevertheless, this mechanism is able to adequately explain the patterns of energy deposition in a-D thin films fabricated using the filtered arc process. By testing the layered density distribution of the growth cross sections, surface morphology, and surface composition of a-D thin-films, we were able to validate the use of the subplantation deposition mechanism, and provided the required theoretical and experimental basis for effectively controlling the filtered arc deposition process.

5.4 Process control rules of optical parameters of thin films

The refractive index, extinction coefficient, and optical band gap are crucial parameters for the characterization of thin-film optical properties. The investigation of the process control rules of these parameters not only facilitates the evaluation of microstructural characteristics of a-D thin films and a deeper understanding of the basic properties of the thin films, but also is a necessary foundation for the practical optical applications of a-D thin films. The widespread use of spectroscopic ellipsometers has provided a powerful means to characterise the optical properties of a-D thin films. This section involves the testing of optical constants in a-D thin films using spectroscopic ellipsometry in order to analyze the influence of processing conditions and investigate the optical properties of thin films.

5.4.1 Basic theory

The basic optical principles of ellipsometry are described in Fig. 5.23. The sample surface is illuminated with polarized light, and changes in polarization characteristics (amplitude ψ and phase Δ) are measured after reflection, based on which the properties of the thin film are derived. Since light waves are a type of transverse wave, the incident polarized light can be decomposed into the P wave component (parallel to the incident surface) and S wave component (perpendicular to the incident surface). The reflected and transmitted light can also be decomposed in a similar manner. The following relation based on Fresnel's law of reflection holds in this scenario:

$$\rho = \frac{R_P}{R_S} = \tan(\psi)e^{i\Delta}, \tag{5.16}$$

where ρ is the Fresnel reflection coefficient, and R_P and R_S are the reflectances of the P and S wave components of the incident polarized light, respectively. By applying the boundary conditions of different interfaces, Maxwell's equations can be used to obtain the mathematical expressions of R_P and R_S:

$$R_P = \frac{N_1 \sin\theta_0 - N_0 \sin\theta_1}{N_1 \cos\theta_0 + N_0 \cos\theta_1}, \tag{5.17}$$

$$R_S = \frac{N_0 \cos\theta_0 - N_1 \cos\theta_1}{N_0 \cos\theta_0 + N_1 \cos\theta_1}, \tag{5.18}$$

where θ_0 and θ_1 are the incident and refractive angles, respectively, and N_0 and N_1 are the complex refractive indices of air and the thin film, respectively.

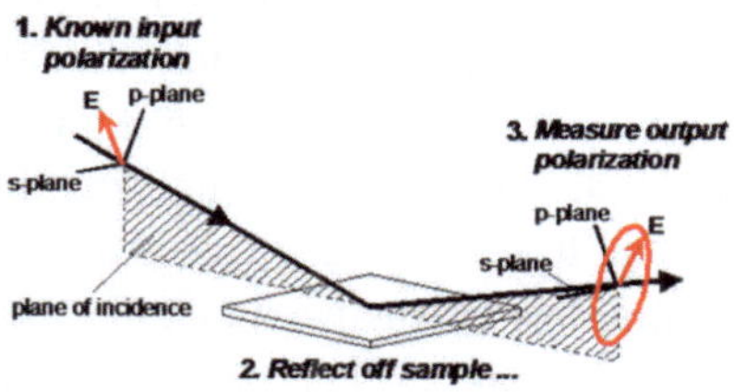

Fig. 5.23: Basic principles of ellipsometric measurements.

Based on Snell's law of refraction and the rule of equivalent complex numbers, the relation between thin-film optical constants (refractive index n and extinction coefficient k) and optical polarization parameters (amplitude ψ and phase Δ) can be established:

$$n = \left\{k^2 + n_0^2 \sin^2\theta_0 \left[1 + \frac{\tan^2\theta_0(\cos^2 2\psi - \sin^2 2\psi \sin^2\Delta)}{(1 + \sin 2\psi \cos\Delta)^2}\right]\right\}^{1/2}, \tag{5.19}$$

$$k = \frac{n_0^2 \sin^2\theta_0 \tan^2\theta_0 \sin 4\psi \sin\Delta}{2n(1 + \sin 2\psi \cos\Delta)^2}. \tag{5.20}$$

The optical properties of the medium can be expressed using complex refractive indices as follows:

$$N(\omega) = n(\omega) + ik(\omega), \tag{5.21}$$

where N, n, and k are real functions of the frequency of incident light; the real and imaginary parts of the complex refractive indices can be interconverted using the Kramers–Kronig dispersion relation.

As light is propagated deeper into the thin film, the energy of absorption loss increases. The degree of absorption can be characterized using the extinction coefficient k or absorption coefficient α, and will exhibit exponential decay according to Beer's law as follows:

$$I = I_0 e^{-\alpha x}, \quad \alpha = \frac{4\pi k}{\lambda}, \tag{5.22}$$

where I_0 is the initial light intensity, x is the propagation depth of incident light, and λ is the wavelength of incident light.

In terms of the optical band gap E_0, assume that the mathematical relation between the density of states and photon energy near the edge of the conduction band and valence band has a parabolic shape as follows:

$$\sqrt{\alpha h\omega} = B(h\omega - E_0), \tag{5.23}$$

where B is a constant related to the material property, h is Planck's constant, and E_0 is the optical band gap.

5.4.2 Refractive index

The dispersion relation the refractive index of the a-D thin film is shown in Fig. 5.24. When the incident light wavelength was relatively short, the thin-film refractive index increased with wavelength. However, in the infrared wave range, the refractive index essentially remained constant. Aside from drift potential, the refractive indices of thin films for the remaining substrate biases generally stabilized at 2.6–2.7. This is higher than the refractive index of diamond crystals or thin-films (refractive index of CVD-grown type I diamond is 2.42), and slightly higher than that of other a-D thin films. The refractive index reflects the degree of obstruction of materials for light waves. The propagation speed of light is the greatest in vacuum (3×10^8 m/s), and will decline in other media. Based on this implication, refractive index can be regarded as the "optical density" of materials, which is determined by the atomic density and polarization characteristics of the materials. In reality, the refractive index of a-D thin films is often higher than that of diamond. Evidently, it is not possible to explain this phenomenon from the perspective of atomic density. The atomic structure of a-D thin films is not a simple mixture of sp^2 and sp^3 hybridization. When light is propagated through this amorphous network and interacts with carbon atoms of different coordination numbers, it will naturally result in polarization phenomena that are different

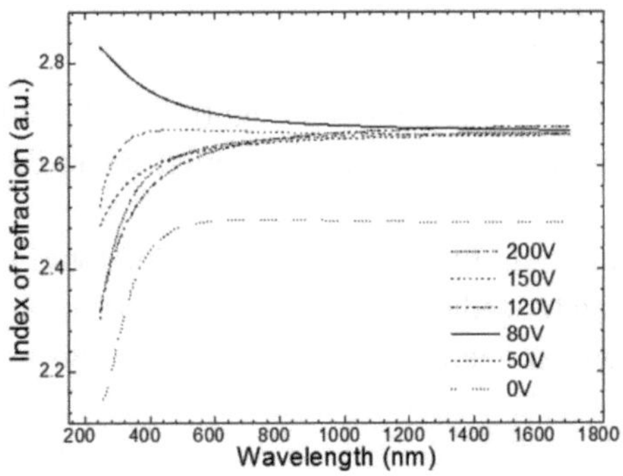

Fig. 5.24: Relationship between incident light wavelength and refractive indices of a-D thin films deposited under different substrate biases.

from the case of diamonds, which has a tetrahedral face-centred cubic structure. In diamond, the intermittent π-bond chains or rings embedded in the σ-bond parent structure play an extremely important role. Furthermore, the differences in sp^2 hybridization amount and distribution are precisely what cause the disparities in the optical properties of a-D thin films fabricated using different experimental methods. In addition, compared to hydrogenated amorphous carbon (a-C:H), the refractive index of a-D thin films is usually higher, but the adjustment range that can be achieved by modifying the fabrication processes is far narrower. Hydrogen atoms play a crucial role in stabilising the tetracoordinate bonds of a-C:H and controlling thin-film properties. The hydrogen content can vary in a relatively large range of 1–60 %, and the corresponding refractive index can be adjusted in a wide range of 1.6–2.6. The relative atomic mass of hydrogen is the lowest; thus, it also has the weakest dispersion effect on photons. Furthermore, the energy of the atomic group produced by the dissociation of hydrocarbon precursors is also difficult to control, thereby forming a looser film layer. Hence, the refractive index of a-C:H is generally less than that of a-D thin films.

It can clearly be seen from Fig. 5.25 that the trend of change in thin-film refractive index with negative substrate bias was an initial increase followed by a decrease, and the maximum was reached when the negative substrate bias was 80 V. When the wavelength of incident polarized light was relatively short, the disparities in the refractive indices of a-D thin films deposited using different processes were substantial. In the infrared wave range, aside from drift potential, the changes in refractive indices of thin films fabricated under the remaining conditions were not large. In other words, the adjustment range of refractive indices in the infrared wave region that can be achieved by altering the incident ionic energy is limited. The energy of incident photons will decrease with increasing wavelength. Since the photon energy is lower than the optical band gap in the infrared wave range, changes in the refractive index are mainly determined by the dispersion effect of the amorphous carbon network on photons. Since the results of CPMD computer simulation indicated that high-density a-D thin films have an amorphous network with smaller structural differences, the differences in the photon dispersion effect of the amorphous network are also relatively small. The energy of incident photons will increase with decreasing wavelength. Hence, in the visible light and ultraviolet wave range, photon energy approached or exceeded the optical band gap, which caused certain electrons to jump from the valence band to the conduction band. As tricoordinate π-bonds are the main bond type, which also determines the differences in sp^2 distribution and amount of varying optical band gaps, the refractive index will necessarily show a relatively large disparity when high-energy photons pass through the thin film. Based on the variations between the energy of deposited ions and refractive indices of a-D thin films, the maximum refractive index falls within the energy window for rich sp^3 hybridization. With increasing amount of sp^3 hybridization, the refractive index increases. This is inconsistent with the experimental results obtained by Lossy et al. and Chen et al. They suggested that the higher the ratio of

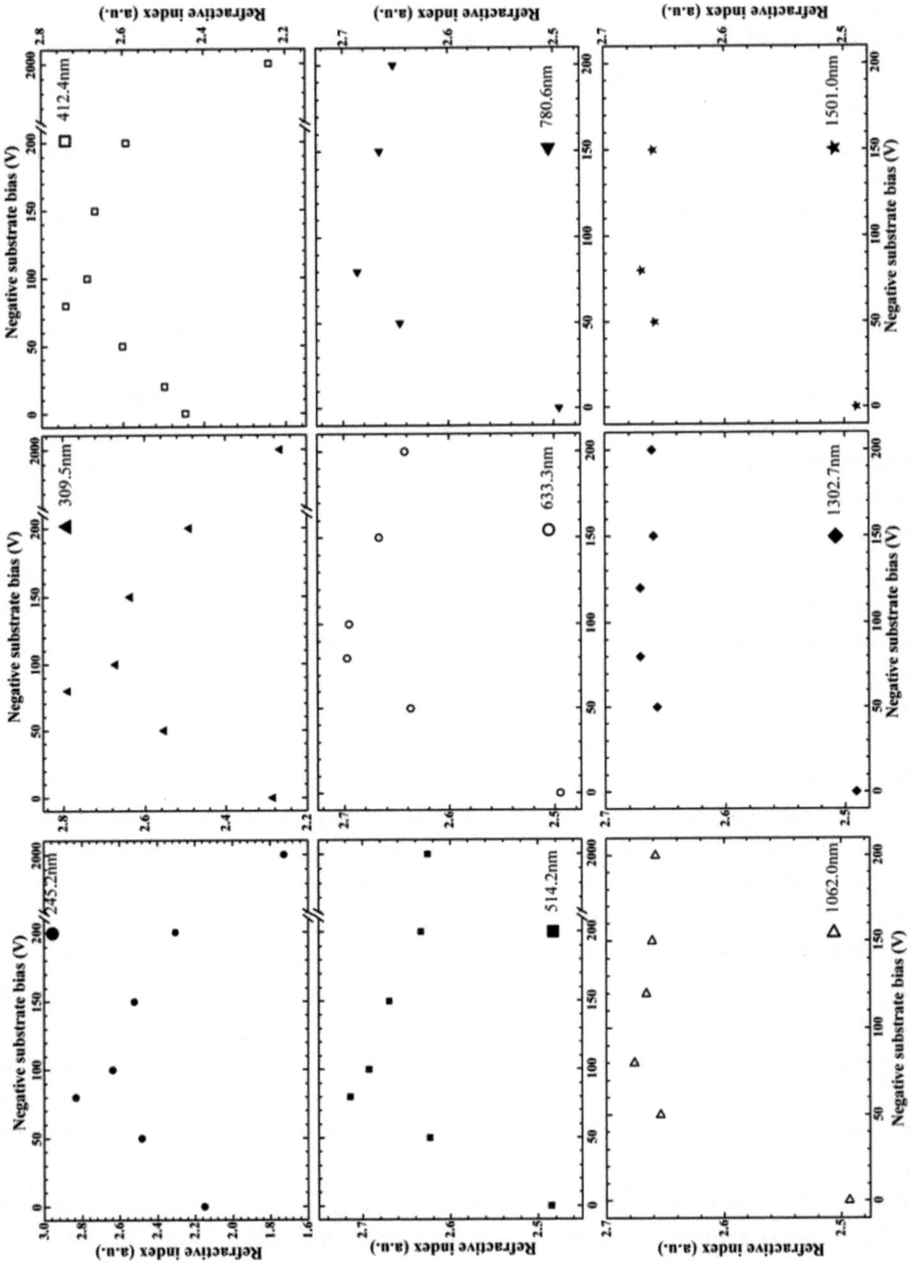

Fig. 5.25: Relationship between a-D thin-film refractive index and deposition energy under different wavelengths.

sp^3 hybridization in thin films, the closer their properties are to diamond crystals. As the refractive index of diamond is 2.42, the trend of variation in thin-film refractive index with the energy of deposited ions should show an initial increase followed by a decrease. Specific reasons for this discrepancy await further investigations.

5.4.3 Extinction coefficient

We can see from Fig. 5.26 that the extinction coefficient of a-D thin films decreased gradually with increasing wavelength of testing polarized light, and approached zero. This trend of change is often used as one of the criteria to determine the success or failure of ellipsometric data fitting. Within the ultraviolet wave range, the extinction coefficient was relatively high; hence, the thin film appeared opaque or semitransparent. Within the wave range of visible light, the extinction coefficient decreased sharply; hence, the thin film appeared semitransparent to transparent. As film thickness increased, the thin film presented different colors ranging from light brown to pale black owing to the coherence phenomena. As the incident light entered the infrared wave range, the extinction coefficient was already very small, and the thin film appeared transparent. By applying this property, a-D thin films can be used as antireflective and protective coatings for infrared optical devices.

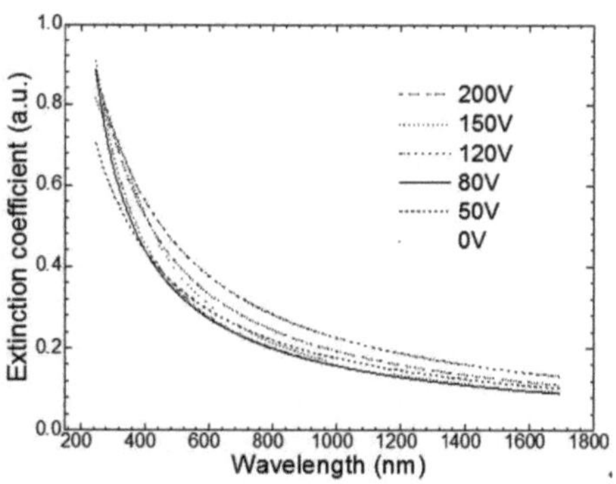

Fig. 5.26: Relationship between extinction coefficient and incident light wavelength of a-D thin films deposited under different substrate biases.

In order to investigate further the influence of deposition ion energy on extinction coefficient, the relationship between extinction coefficient and substrate bias under different wavelengths of incident polarized light is plotted in Fig. 5.27.

It can be seen from Fig. 5.27 that, as negative substrate bias increased gradually, the extinction coefficient of a-D thin films showed a U-shaped pattern of variation, and was the lowest when the negative substrate bias was 80 V. From the visible light to the infrared wave range, the degree of influence of processing conditions decreased as the extinction coefficient decreased. For example, when the incident wavelength was 412.4 nm, the difference in extinction coefficient under different substrate biases was

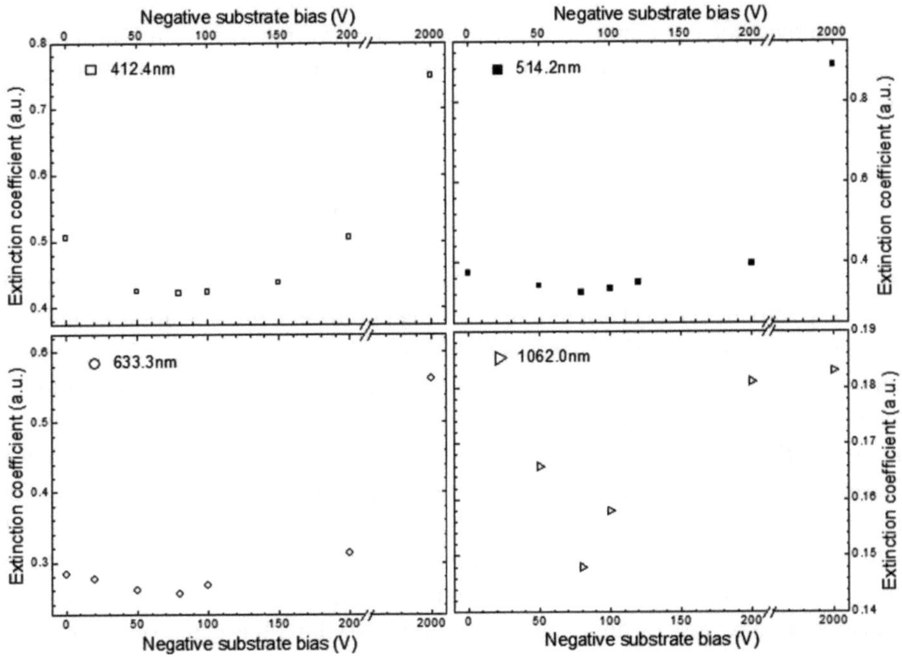

Fig. 5.27: Relationship between extinction coefficient and deposition energy of a-D thin films under different wavelength conditions.

up to 0.35, whereas when the incident wavelength was 1 μm, the difference in extinction coefficient under different substrate biases was less than 0.05. On the other hand, the difference in the internal stress of thin films under different processing conditions was very large, even exceeding 5 GPa. In engineering practices, a high internal stress is often the bottle neck restricting the widespread application a-D thin films. This fact has encouraged us to avoid a high-stress energy window for rich sp^3 hybridization in certain infrared optical applications while simultaneously obtaining coating layers with comparable optical properties, much lower internal stress, and the necessary optical thickness. In addition, within the ultraviolet wave range, the extinction coefficient did not show significant variations, because the interband state of a-D thin films is relatively complex, with a lack of obvious boundaries for localized-state or extended-state electron transitions caused by incident photons.

5.4.4 Optical band gap

As with crystalline semiconductor materials, many amorphous semiconductor materials exhibit intrinsic absorption processes due to the electron-hole pairs produced when electrons absorb photons and undergo transition from the valence band to the conduction band. The optical band gap is able to reflect the difficulty of intrinsic absorption processes. Under general circumstances, the complete absorption spectra of amorphous semiconductor materials should include three parts: a high-absorption region, an exponential-absorption region, and a weak-absorption region. The absorption spectrum of a-D thin films within the range of the test spectrum falls in the high-absorption region and exponential-absorption region. We can assume that the relationship between absorption coefficient and photon energy in the high-absorption region follows a parabolic function, and the Tauc band gap can be calculated using equation (5.23), as shown in Fig. 5.28.

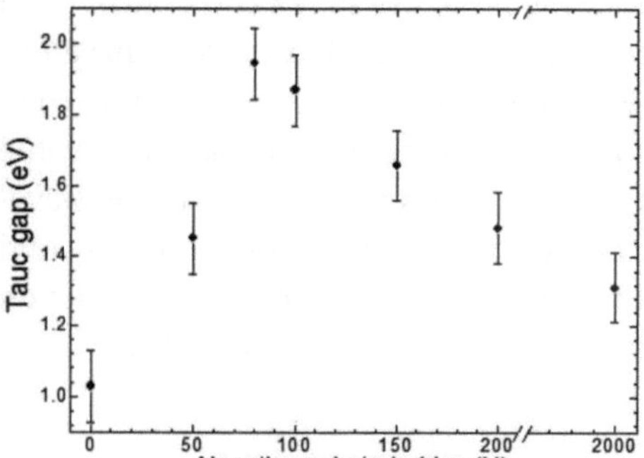

Fig. 5.28: Relationship between optical band gap and deposition energy of a-D thin films.

The relationship between the band gap of a-D thin films and the negative substrate bias shows good consistency with the relationships of deposition energy with mechanical properties, surface morphology, and so on. The relationship is also determined by the hybridization ratio in the thin films. Fig. 5.29 clearly shows that an almost linear increase occurs in the optical band gap E_0 as the thin-film sp^3 content increased, and that the thin-film became more transparent as the sp^3 content increased. When the negative substrate bias was 80 V, the thin film showed the maximum optical band gap and minimum extinction coefficient. At this point, the thin film also showed the

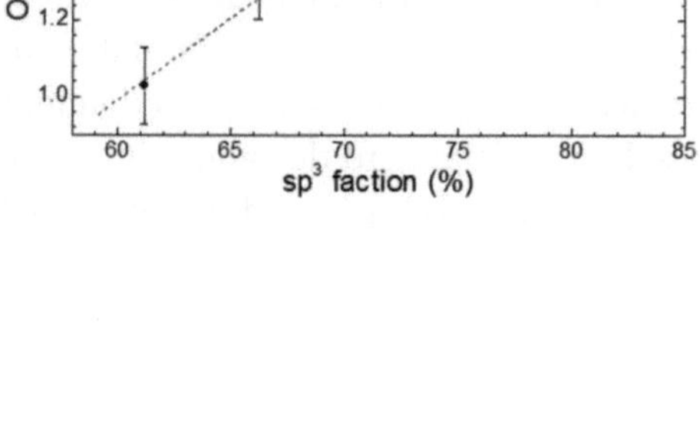

Fig. 5.29: Relationship between optical band gap and amount of sp^3 hybridization in a-D thin films.

maximum refractive index as the density was the highest. If the deposition condition deviated from the processing conditions for rich sp^3 hybridization, the optical band gap became narrower while the extinction coefficient showed a corresponding increase. Compared to σ-bonds, π-bonds are closer to the Fermi level. Hence, the optical properties of a-D thin films should be determined by the discrete sp^2 fields in the rigid tetra-coordinated network. The broadening of the optical band gap with the increase in sp^3 content should actually be attributed to the narrowing of the π-bond electron density of states and broadening of the π–π^* gap.

In conclusion, the refractive index, extinction coefficient, and optical band gap of a-D thin films show obvious trends of variation with the incident energy of deposited ions. As the negative substrate bias increased, the refractive index and optical band gap both showed an initial increase followed by a decrease, and the maximum values occurred when the negative substrate bias was 80 V; in contrast, the extinction coefficient showed an initial decrease followed by an increase, and the maximum value occurred when the negative substrate bias was 80 V. With the increase in the wavelength of testing polarized light, the extinction coefficient decreased gradually and approached zero. The adjustment range of thin-film refractive index and extinction coefficient achieved by changing the deposition energy decreased gradually.

5.5 Effects of thin-film thickness and stress

If a-D thin films are used as antireflective and protective coatings for infrared windows/domes, then according to the optical principles of thin films, it is necessary to adjust the optical thickness of the thin film in order to achieve antireflective effects. Under a given refractive index, the optical thickness is mainly determined by the geometric thickness of the thin film. There are currently no exact answers as to whether changes in film thickness will influence the mechanical properties and microstructure of a-D thin films. Therefore, this section will mainly focus on the effects of film-thickness variation on the mechanical properties and microstructure of a-D thin films.

5.5.1 Mechanical properties of amorphous diamond thin films with different thickness

During the testing process, the diamond indenter penetrated the thin film and entered the substrate in order to obtain the maximum value of thin-film hardness. Then, the TestWorks 4 software package and Analyst software were employed for data processing. As shown in Fig. 5.30, each point represents the mean value of 10 different indentation data, and the error bars are based on the standard deviation. It can be seen that, with increasing film thickness, the hardness and elastic modulus increased

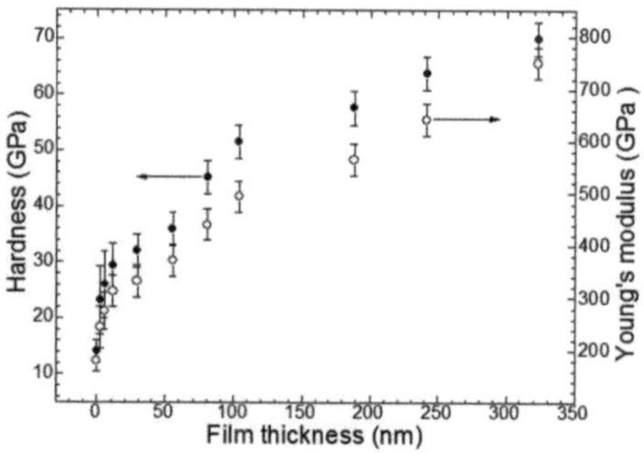

Fig. 5.30: Relationship of thin-film thickness with hardness and elastic modulus.

continuously. When the film thickness was approximately 300 nm, the hardness was almost 70 GPa, and the elastic modulus exceeded 750 GPa. The hardness test was performed using the Pharr–Oliver method, and hardness was calculated using the indentation load–displacement curve. In reality, this is the composite hardness of both the substrate and thin film; thus, the influence of substrate will become more significant with thinner films. The actual thin-film hardness and elastic modulus should be higher than the measured values in the nanoindentation test. In order to obtain the actual hardness of the thin film, the finite-element simulation method can be employed to remove the influence of the "soft" substrate in order to determine the yield strength of the thin film.

The critical scratch load is determined by the changes in the lateral scratch force and coefficient of friction between the diamond indenter and thin film. When the thin film and substrate are separated, a sudden change occurs in the lateral scratch force. Hence, the magnitude of the critical scratch load reflects the thin-film adhesion strength to a certain extent. However, it is also directly constrained by the experimental parameters and the characteristics of the thin film itself. The relationship between critical scratch load and thin-film thickness is shown in Fig. 5.31. Each point represents the mean value of 5 different scratch data, and the errors bars are based on the standard deviation. It can be seen that the critical scratch load decreased rapidly with increasing thin-film thickness. However, when the film thickness exceeds 100 nm, the thin film will possess the properties of diamond. The relatively high levels of hardness and elastic modulus attained will necessarily cause a higher normal indentation load, which manifests as an increase in critical scratch load once again. Some study demonstrated that the critical load increased continuously with increasing film thickness. However, some others were completely different, and suggested that film thickness and critical load have the following empirical relation: $L = 1.46 + 0.0059t - 3 \times 10^{-6}t^2$. The curve inflection points of the relation were all around 100 nm. This disparity might have been related to the measurement devices and experimental conditions, which also reflect the limitations of using the critical scratch load to evaluate thin-film adhesion performance.

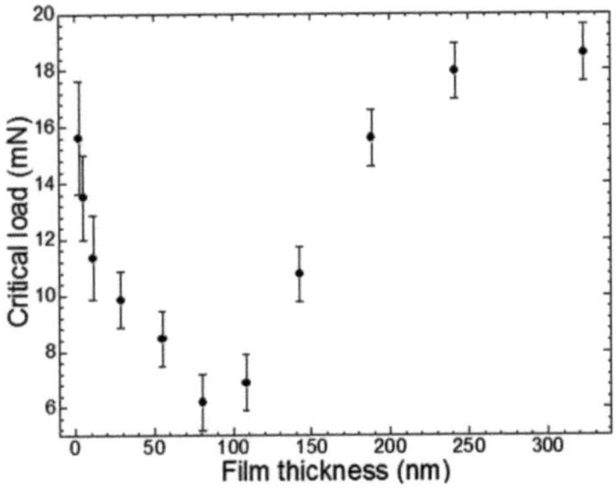

Fig. 5.31: Relationship between critical scratch load and thin-film thickness.

5.5.2 Raman characterization of amorphous diamond films in different thicknesses

Three main characteristics can be obtained from the Raman signals measured within the wavenumber range of 200–2000 cm^{-1}, which are shown in Fig. 5.32. At a wavenumber of 520 cm^{-1}, we can observe the first-order Raman spectral peak of the Si substrate, and this peak position is also used for system calibration. Within the range of 900–1000 cm^{-1}, the second-order signal of the Si substrate can also be observed, and this spectral peak is often used to determine the transparency of the thin film. With increasing film thickness, the intensity of both the first-order and second-order peaks of Si decreased. The profiles of the spectral peaks gradually faded into the background when the film thickness exceeded 200 nm for the first-order peak and 100 nm for the second-order peak. The first-order spectral peak of a-C mainly concentrated around 1300–1400 cm^{-1}, and was presented as an asymmetrical broad peak. The BWF function can be used to fit the Raman data in order to obtain important parameters such as maximum peak intensity, FWHM, and peak position.

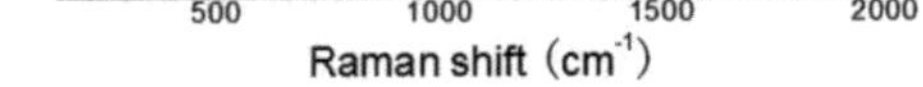

Fig. 5.32: Visible Raman spectrum of typical a-D thin film.

Fig. 5.33 shows the Raman spectra of a-D thin films with different thicknesses. It can be clearly seen in Fig. 5.33 (a) that, when the film thickness was 55.32 nm, the Raman spectral line profile of the a-D thin film was significantly more prominent than other spectral lines. Fig. 5.33 (b), e show that the Raman peak at this point had the maximum peak intensity and minimum FWHM. This indicates that, within the thickness range of 50–80 nm, the Raman peaks of a-D thin films were sharper and easier to analyse. If the film was too thin, the effects of the surface layer increased significantly, which led to biases in the analysis; furthermore, the low peak intensity caused the severe broadening of the FWHM, thereby increasing fitting error. If the film was too thick, the time spent on fabrication will be prolonged, and the thin film might be delaminated owing to instability. In addition, the increase in maximum peak intensity and FWHM showed near-periodic variations with increasing film thickness. This might have been the result of optical interference in the thin film caused by lasers.

Fig. 5.33 (d) shows that, when the film thickness exceeded 50 nm, the coupling coefficient remained stable. However, as the film thickness decreased, the Raman spectrum gradually became more inclined, and the absolute value of the coupling coefficient decreased. This is because the cross section of a-D thin films has a "sandwich-like" layered structure. The surface layer contains an ordered structure of graphite-like layers with relatively high amounts of sp^2 hybridization. When analyzed from the perspective of topology and defects, numerous surface atoms are mutually bonded as

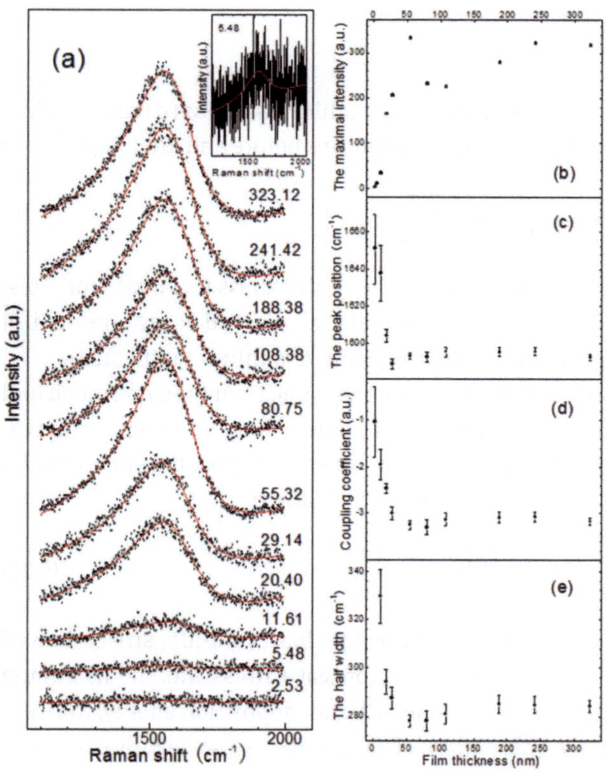

Fig. 5.33: Raman spectra of a-D thin films with different thickness: (a) BWF function fitting parameters; Relationship of thin-film thickness with (b) maximum peak intensity; (c) peak position; (d) coupling coefficient; (e) full width at half maximum.

planar rings or chains. Beneath the surface layer lies the amorphous system with rich sp^3 hybridization, followed by the amorphous mixed layer of carbon and substrate atoms. Due to the structural disparities, the surface layer often has different properties, and its thickness ranges from approximately a few Angstrom to more than 30 Å. When the film is relatively thin, the surface layer will account for a higher proportion and its influence will be stronger, causing the Raman spectra to have a steeper slope.

As can be seen from Fig. 5.33 (c), when the film thickness exceeded a certain level, the position of the broad peak was relatively stable. However, when the film was relatively thin (e.g. film thickness less than 30 nm), the peak position shifted toward the high-frequency end as the thin-film thickness decreased. It is generally believed that, with increasing amount of sp^2 hybridization, the G-peak position will shift toward the low-frequency end. As thin-film thickness decreased, the ratio of tricoordinate planar rings or chains in the surface layer increased, and the Raman peak position shifted toward the high-frequency end instead. This phenomenon is related to the thin-film stress.

5.5.3 Effect of thin-film stress

Based on Stoney's equation shown in equation (5.24), thin-film stress (σ) can be calculated using the change in the sample radius of curvatures before and after deposition:

$$\sigma = \frac{E_S}{6(1 - \nu_S)} \frac{t_S^2}{t_f} \left(\frac{1}{R} - \frac{1}{R_0} \right), \tag{5.24}$$

where E_S, t_S, and ν_S are the elastic modulus (180 GPa), thickness (0.5 mm), and Poisson's ratio (0.26), respectively, of the substrate material; R_0 and R are the radii of curvature before and after deposition, respectively; and t_f is thin-film thickness.

Considering that the substrate of all stress samples came from the same batch and had the same shape and thickness (fluctuations less than ±10 μm), the radius of curvature is approximated to infinity before deposition in order to simplify the experimental process. Thus, Stoney's equation can be simplified as follows:

$$\sigma = \frac{E_S t_S^2}{6(1 - \nu_S)} \times \frac{1}{t_f R}. \tag{5.25}$$

Fig. 5.34 shows that, with increasing negative substrate bias, thin-film stress showed a trend of increase followed by decrease; the compressive stress was the maximum when the negative substrate bias was 80 V. Under a higher deposition energy, the thin-film stress decreased significantly. This pattern of variation can also be fully explained by the subplantation growth mechanism of thin films. Compared to the a-D thin films fabricated by McKenzie, Polo, Fallon et al., the pulsed bias employed in this book to fabricate thin films had a narrower pulse width that significantly reduced thermal input, resulting in lower stress.

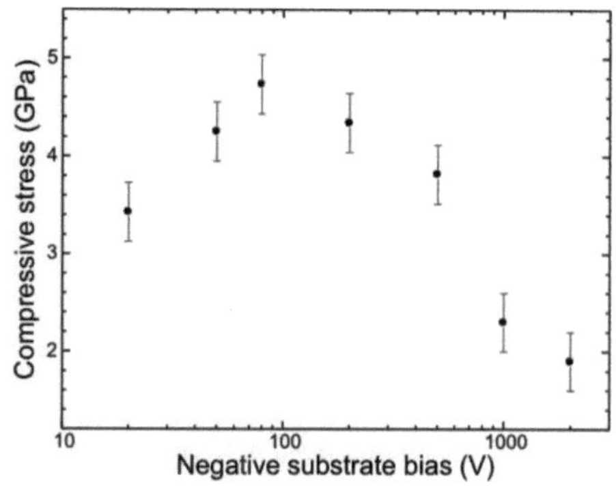

Fig. 5.34: Internal stress of a-D thin films fabricated under different substrate biases.

Fig. 5.35 shows that stress decreased continuously as film thickness increased. When the film was relatively thin (thickness less than 30 nm), the magnitude of decrease was relatively large, and it then leveled off. The experimentally measured stress is the sum of thin-film internal residual stress and surface stress. When the film is relatively thin, the effects of surface tension are significant. With increasing film thickness, the effects of surface tension will be rapidly diminished. This is consistent with the experimental fact that thinner films curl more easily when delaminated from the substrate. In addition, as the film thickness continues to increase, heat accumulation occurs in the thin-film growth process, and internal stress decreases owing to local stress relaxation.

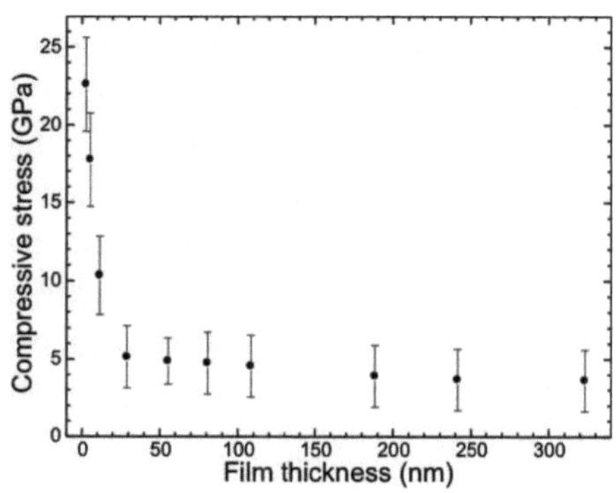

Fig. 5.35: Compressive stress of a-D thin films with different thicknesses.

J. W. Ager III compared the Raman spectra of a-D thin films before and after delamination. The study found a significant shift in the spectral peak position, and suggested that internal stress will promote the shift of the characteristic peak of Raman scattering to a higher frequency by nearly 20 cm^{-1}. Jin-Koog Shin performed the peak deconvolution of asymmetrical broad peaks and found that the high internal residual stress caused the G-peak to shift to higher frequency. As shown in Fig. 5.33 (c), stress decreased with increasing film thickness while the Raman spectral peak position grad-

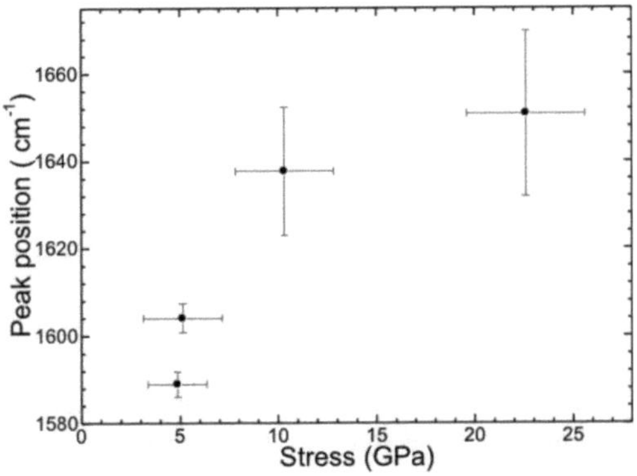

Fig. 5.36: Relationship between a-D thin-film stress and broad peak position.

ually shifted toward lower frequency; however, the magnitude of shift in the Raman peak position was relatively small.

From Fig. 5.36, we can estimate that the magnitude of stress-induced shift in broad peak position was approximately 3 cm^{-1}/GPa. This is the result of synergistic effects as thinner films have increased sp^2 ratio, which causes the peak position to shift to lower frequency.

In conclusion, with increasing film thickness, the hardness and elastic modulus of a-D thin films increase continuously. When the thickness of a-D coating on an Si substrate exceeds 300 nm, the hardness and elastic modulus will reach 70 GPa and 750 GPa, respectively. If the influence of the substrate is removed, its mechanical properties will be extremely similar to those of block diamond. When the film thickness is within the range of 50–80 nm, visible Raman spectroscopy indicated that the intensity of the asymmetrical broad peak is the highest and FWHM is the narrowest; thus, it is optimal for reflecting characteristic information. When the film is relatively thin, the high amounts of tricoordinate planar rings or chains in the surface layer will result in a higher coupling coefficient, whereas the increase in thin-film stress will cause the Raman peak position to shift significantly toward a higher frequency. Thin-film stress also showed a trend of initial increase followed by a decrease with increasing negative substrate bias. Furthermore, the thin-film stress was maximum under the deposition conditions for rich sp^3 hybridization.

5.6 Thermal stability

A substantial amount of research has been conducted on the thermal stability of a-D thin films. In terms of vacuum thermal stability, the focus of research is on how to perform the annealing of a-D thin films at 500–600 °C for the relaxation of internal stress while simultaneously ensuring that the hybridization ratio of the thin film remains unchanged. Regarding the thermal stability of thin films in air, the focus of research is on determining the temperature of oxidation loss, as well as the changes in thin-film properties and structure induced by increasing temperatures. Nevertheless, practical applications often require the use of a-D thin films with specified thicknesses that are fabricated under different deposition conditions. The service environment might even involve low temperatures below 0 °C (e.g. the use of a-D thin films as a wear-resistant protective layer against sliding friction in deployable satellite antennae). Research work in this area is still rarely reported. Therefore, this section will focus on not only the high-temperature thermal stability, but also the low-temperature thermal stability of thin films. We will also analyse and compare the thermal stability of a-D thin films fabricated under different substrate biases.

5.6.1 Thermal stability of thin films in air

a-D thin-films have already been applied as wear-resistant protective layers of certain tools. As these application processes might generate high increases in temperature, the effectiveness of thin films will be directly determined by its thermal stability. a-D can also be utilized as a coating layer for components operating in space or low-temperature conditions. Therefore, the structural and performance stability of a-D thin films under low-temperature conditions are equally important. Raman spectroscopy is an optimal and nondestructive characterization method to obtain information on the a-C film structure. The analysis of Raman spectra is a very convenient means to resolve the evolution pattern of thin-film microstructure, as well as to determine the clustering tendency and scale of ordering of thin films. It is an ideal method to investigate the thermal sensitivity of a-D thin films. This book will employ micro-Raman spectroscopy to investigate the thermal stability of a-D thin films in air. We will also observe the structural stability of thin films in real time during the cooling process from room temperature using liquid nitrogen.

5.6.1.1 Annealing experiments

In order to present clearly the evolution process of thin-film microstructure with the increase in annealing temperature, the Raman spectra of samples annealed under different temperatures in the range of 100–600 °C are presented as a stacked plot in Fig. 5.37 (a). Thin films deposited under a negative substrate bias of 2000 V were selected as the experimental sample. Although thin films deposited under a high substrate bias have relatively high amounts of sp^2 hybridization, which causes their thermal stability to be lower than that of thin films deposited under a low substrate bias, a high substrate bias will significantly reduce thin-film stress. Hence, this is widely applied to fabricate films in order to enhance the coating–substrate adhesion.

The Raman spectroscopy of a-D thin films is mainly characterized by an asymmetrical broad peak in the range of 1200–1800 cm^{-1}, which can be fitted using a monoclinal Lorentzian line shape described by the BWF function. As can be seen

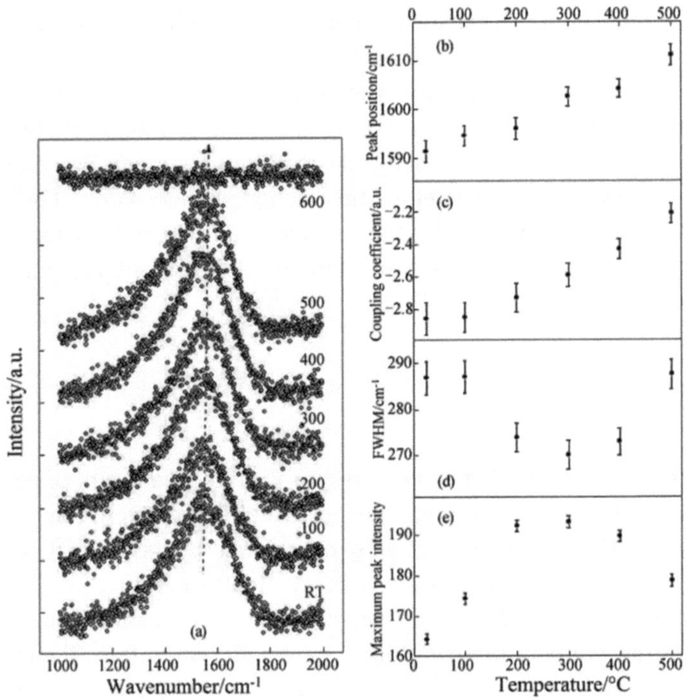

Fig. 5.37: Visible Raman spectroscopy of a-D thin films subjected to low-temperature annealing at different temperatures: (a) fitting parameters of BWF function; relationship of annealing temperature with (b) peak position; (c) coupling coefficient; (d) FWHM; (e) maximum peak intensity.

in Fig. 5.37 (a), from room temperature to 500 °C, the spectral line profile essentially remained stable, which indicates the thin film has relatively high thermal stability. Research has shown that in the range of 1300–1400 cm^{-1}, the changes in spectral line profile are closely related to the scale of ordering L_a of the sp^2 field in the continuous sp^3 network. The characteristic of significant sp^2 field clustering is the appearance of D shoulder peaks. The larger the ordering scale of sp^2 fields, the higher is the D-peak intensity. Unlike the performance of thin-film samples annealed using the same processes in vacuum, which showed obvious D shoulder peaks after annealing at 500 °C, these were not observed in thin-film samples in air. Tay et al. proposed that an aerobic atmosphere is able to increase the thermal sensitivity of a-D thin films, and their results are different from those obtained in this book. This might have been because the oxidation and volatilization on the thin-film surface in the experiment removed some heat, which delayed the sp^2 field clustering effect. However, as the temperature continued to increase, the oxidation loss of the film material increased significantly, such that after annealing at 600 °C, the asymmetrical broad peak disappeared completely. The experiment shows that the temperature at which complete oxidation loss will occur in a-D thin films under aerobic conditions is approximately 550 °C.

As the amount of sp^2 hybridization and its scale of ordering in a-D thin films were both too small, the D-peak intensity of the visible Raman scattering cross section was too weak. Hence, BWF single-peak fitting could be employed for analysis, and the fitting parameters are shown in Fig. 5.37 (b)–(e). As annealing temperature increased, the G-peak position shifted significantly toward higher frequencies, while the coupling coefficient showed an increasing trend. However, the changes in FWHM and maximum peak intensity were relatively more complicated. The appearance of the G-peak could be attributed to the E_{2g} symmetric mode in monocrystalline graphite, i.e. the stretching vibration mode of π-bonds. Due to the loss of long-range ordering in a-D thin films, significant broadening was observed in the G-peak. Furthermore, visible Raman spectroscopy is able to effectively measure depths far greater than the size of sp^2 fields in thin films. Under the influence of different quantities, distributions, morphologies, and environments of sp^2 fields, the Raman scattering cross section is actually the superimposition of different vibrational modes, which will also broaden the G-peak. As the temperature increased to 500 °C, the peak position shifted by almost 20 cm^{-1}.

The coupling coefficient is a measure of the inclination of the asymmetrical broad peak. A higher amount of sp^2 hybridization in the thin film implies a greater degree of slope and a smaller absolute value of the coupling coefficient. As can be seen from Fig. 5.37 (c), when the annealing temperature increased, the coupling coefficient increased, and the amount of sp^2 hybridization in the thin film increased as well. However, since the photon energy and π–π^* excitation energy are comparable, and π-bonds have long-range polarising effects, and visible Raman spectroscopy will reflect π–π^* resonant scattering. Therefore, in order to characterise the ratios of different hybridizations, we will also need to rely on electron energy loss spectroscopy, nuclear magnetic resonance, or ultraviolet Raman spectroscopy. Fig. 5.37 (b) shows that the peak position shifted significantly as temperature increased, but the spectral line profile did not change significantly. It can be seen that the peak position of the a-D thin film annealed in air is more sensitive to temperature changes.

As shown in Fig. 5.37 (d), the FWHM showed an initial decrease followed by an increase with increasing annealing temperature, and the minimum value occurred at approximately 300 °C. The thin-film samples deposited under different processing conditions also showed a similar trend of change. As a-D thin films are formed under room temperature by ion beam deposition, the deposition process can be explained using the subplantation growth mechanism, which can also be divided into three stages: collision, thermalization, and relaxation. The duration of the relaxation process is approximately of the order of 10^{-10} s. This will ensure that a large quantity of defects produced during the process will be preserved, and the experiment also verified that the thin-film defect density is extremely high. During the annealing process, accompanied by atomic diffusion and local rearrangements, the thin film had sufficient time for relaxation, which reduced the defect density. The changes in FWHM are related to the long-term relaxation caused by annealing, but their specific causes await further investigation. Fig. 5.37 (e) shows that as temperature increased, the maximum peak intensity showed an initial increase followed by a decrease. As the scattering cross-section of visible Raman spectroscopy performed on a-D thin films mainly reflected the vibrational information of sp^2 fields, and the amount of sp^2 hybridization increased continually with the rise in annealing temperature, this was manifested as the increase in maximum peak intensity. However, when the temperature continued to increase until 500 °C, the oxidation and volatilization of the constituents of the film material became more significant, which led to increased loss of film thickness, as well as a decrease in maximum peak intensity.

The hardness and Young's elastic modulus were measured using a nanoindenter. Each data point in Fig. 5.38 represents the mean value of multiple measurements. The figure shows that thin-film hardness and elastic modulus remained unchanged after annealing at 400 °C. They showed a slight decrease after annealing at 500 °C due to the loss of film thickness and increase in sp^2 ratio, but they still remained at a relatively high level. Under high-temperature conditions, the a-D thin film fabricated using the filtered arc method retained a certain level of hardness and elastic modulus. This thermal stability is extremely beneficial for applications involving temperature increases.

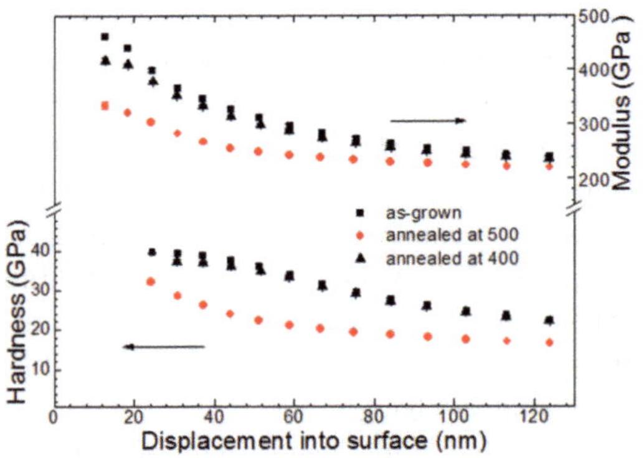

Fig. 5.38: Hardness and elastic modulus of a-D thin films before and after annealing at 400 °C and 500 °C.

5.6.1.2 Low-temperature experiments

Thin films deposited under a negative substrate bias of 80 V were selected as samples for low-temperature experiments because the negative substrate bias of 80 V is the optimal condition for depositing a-D thin films with rich sp^3 hybridization. This will ensure that the thin film will attain the highest level of mechanical and optical properties, and is often used to fabricate the working layer of coating systems. Fig. 5.39 (a) shows the real-time Raman spectra of a-D thin films during the process of cooling with liquid nitrogen.

We can see that there were virtually no changes in the spectral line profile from room temperature to −190 °C, and the peak position and coupling coefficient also remained constant. This implies that the thin-film structure remained stable under low-temperature conditions. The spectral peak intensity was relatively low at 0 °C owing to the strong scattering effect of the ice and water mixture on the testing laser, leading to the severe loss of incident photon energy and the significant reduction of received signals or maximum peak intensity. However, it can easily be observed from the relationship between spectral parameters and temperature that, at approximately −130 °C, the maximum peak intensity increased and FWHM decreased, whereas at −190 °C, the maximum peak intensity was significantly diminished. Under a state of low temperature, the thermal vibrations of atoms are significantly weaker, and the vibrational states induced by photons in the a-D thin-film lattice will necessarily exhibit corresponding changes, leading to the differences observed in spectral parameters.

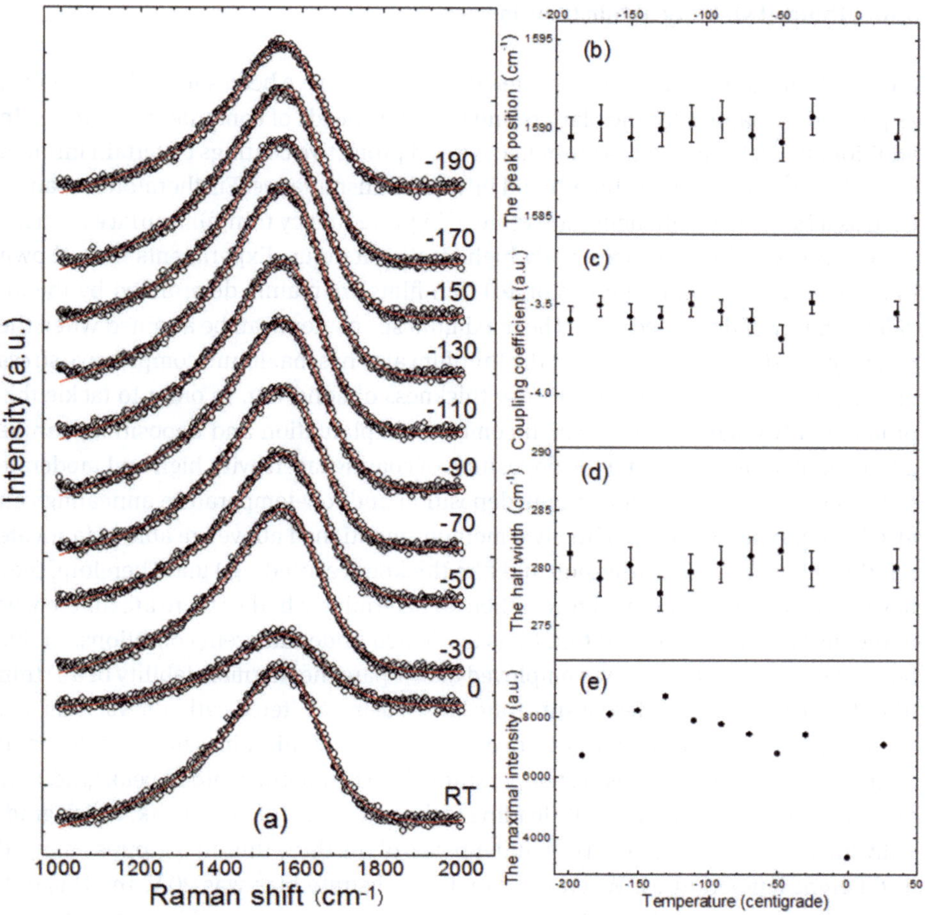

Fig. 5.39: Real-time visible Raman spectroscopy of a-D thin film cooled with liquid nitrogen: (a) fitting parameters of BWF function; relationship of temperature with (b) peak position; (c) coupling coefficient; (d) FWHM; and (e) maximum peak intensity.

Furthermore, the nanoindentation experiment indicated that thin-film hardness and elastic modulus before and after cooling were equal.

In summary, the a-D thin film fabricated using filtered cathodic vacuum arc had good thermal stability. After annealing at 400 °C in air, its hardness and elastic modulus essentially remained unchanged, and its structure was stable until 500 °C. However, at 600 °C, the thin film underwent rapid consumption due to oxidation. The visible Raman spectroscopy of a-D thin films indicated that the spectral peak position shifted toward higher frequencies as the temperature increased. During the cooling process at low temperatures, the thin film was not sensitive to temperature changes, and its structure remained stable.

5.6.2 Thermal stability of thin films in vacuum

a-D thin films fabricated using the filtered arc method have been successfully applied as protective coatings for the platters and magnetic heads of computer hard drives. In addition, it can be applied as antireflective and protective coatings of certain infrared optical devices owing to its broadband infrared transmittance. Furthermore, it can be used as a frequency-enhancing substrate of high-frequency thin-film surface acoustic wave sensors owing to its extremely high elastic modulus. Experiments have shown that the structure and properties of a-D thin films are mainly determined by the incident energy of deposited ions. The maximum sp^3 content can be attained when the negative substrate bias is 80 V, but the thin-film also has maximum compressive stress at this point, which limits the effective thickness of adhesion. In order to tackle this problem, a two-step process, which combines implantation and deposition, can be adopted, thereby alternating the deposition of coating layers with high and moderate sp^3 content. Processes that combine deposition and low-temperature annealing can also be implemented. Both of the two methods mentioned above are able to fabricate a-D thin-films with good adhesion and film thickness exceeding 1 µm. Therefore, they have attracted the attention of researchers to determine whether there are differences in the thermal stability of a-D thin films fabricated under different conditions. In this book, Raman spectroscopy was employed to compare the thermal stability of a-D thin films fabricated under different substrate biases using a filtered cathodic vacuum arc.

Fig. 5.40 shows the Raman spectra of three groups of thin-film samples before and after annealing. With increasing temperature, the asymmetrical broad peak gradually became more inclined, eventually leading to the emergence of the D-peak, which gradually increased. However, the D shoulder peaks of the three thin-film groups emerged at different temperatures. When the negative substrate bias was 20 V, the D shoulder peak only appeared at 700 °C. When the negative substrate bias was 80 V, the D shoulder peak appeared at 600 °C. When the negative substrate bias was 2000 V, the D shoulder peak appeared at 500 °C. The visible Raman spectra of a-D thin films consist mainly of two characteristics: G-peak and D-peak. The G-peak originates from the

E_{2g} symmetry of monocrystalline graphite. Owing to the loss of long-range ordering of the tetrahedral a-C thin-film microstructure, the peak width will broaden significantly and form an asymmetrical broad peak centred at approximately 1580 cm^{-1}. This broad peak is the most basic characteristic observed in the visible Raman spectra of tetrahedral a-C. D-peak will appear in the range of 1300–1400 cm^{-1}, and it originates from the breathing vibrations of sp^2 fields in polycrystalline graphite rings. As the size of sp^2 fields increases, the amount of sp^2 fields will grow, and D-peak intensity will be enhanced. Thus, the D-peak is an important basis for the description of sp^2 clustering tendency and thin-film stability in a-C. As shown in Fig. 5.40, the trends of change observed in the three groups of Raman spectra indicate that the thermal stability of a-D thin films will decrease with increasing deposition energy.

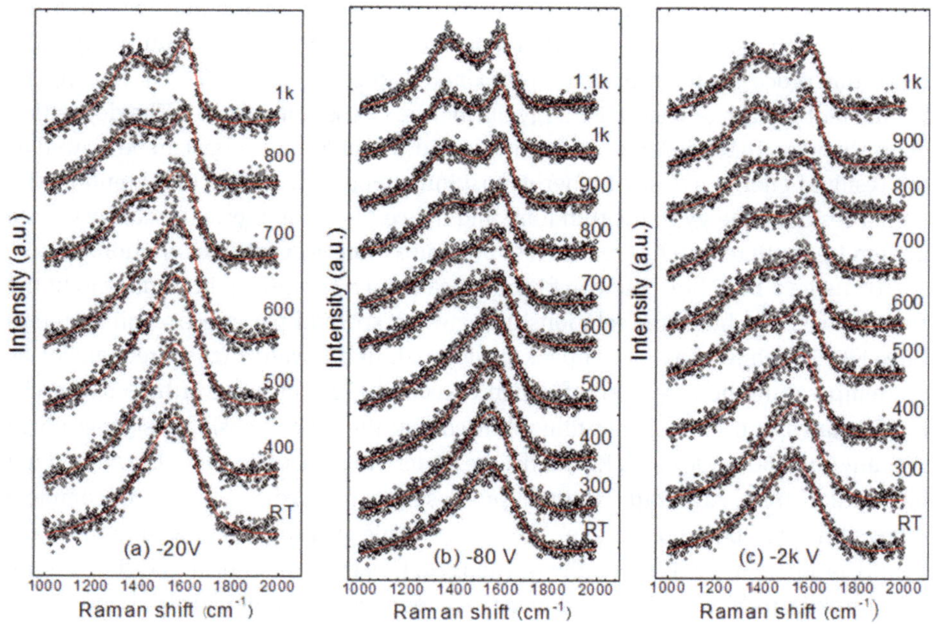

Fig. 5.40: Visible Raman spectra of a-D thin films fabricated under different negative substrate biases and subjected to vacuum annealing at temperatures in the range of 300–1100 °C.

The photon energy of visible light is comparable to the π–π^* excitation energy. Moreover, owing to the long-range polarising effects of π-bonds, the Raman cross section mainly reflects the vibrational information of sp^2 fields. Therefore, K-edge electron energy loss spectroscopy was performed to measure the hybridization ratios of thin films. The experiment showed that, at negative substrate biases of 20 V, 80 V, and 2 kV, the amounts of sp^3 hybridization were 70 %, 80 %, and 65 %, respectively. This indicates that thin films with the highest amount of sp^3 hybridization did not have the best thermal stability. However, it is generally believed that the thermal stability of a-D thin films will decrease with decreasing amount of sp^3 hybridization. According to the subplantation growth mechanism of a-C films, only deposited particles of a certain energy level are able to overcome the surface binding energy, and this energy is approximately 20–30 eV. With the rational optimization of deposition energy, a-D thin films with a maximum amount of sp^3 hybridization can be produced.

Experiments have shown that the optimal negative substrate bias for rich sp^3 hybridization is approximately 80 V. If the energy continues to increase, the amount of sp^3 hybridization will decrease once again. However, we have also discovered that, even under zero substrate bias (deposition energy of 20–30 eV), the thin film had a high amount of sp^3 hybridization (greater than 50 %). Although the subplantation growth mechanism provides a good explanation on the intrinsic relation between deposition energy and thin-film structure, there remain certain flaws that prevent the elucidation of all experimental phenomena. This book supposes that the structural relaxation and growth process of thin films are concomitant, but the duration of the relaxation process is too transient (approximately on the order of 10^{-10} s), causing the insufficient release of residual energy during the deposition process. The deposition energy beyond that needed to overcome the surface binding energy can be viewed as the residual energy. Thus, a higher deposition energy will lead to the retention of a higher residual energy in the thin film, and hence, thin-film thermal stability will be poorer. In-depth research is required to provide a more accurate explanation.

By using a BWF line shape to fit the G-peak and a Lorentzian line shape to fit the D-peak, we can obtain the relationship between spectral parameters and temperature. Fig. 5.41 shows that the G-peak position shifted toward higher frequencies as the temperature increased, all of which exceeded 1600 cm^{-1}. Nevertheless, the visible Raman spectra of a-D thin-films showed that the G-peak position is generally in the range of 1550–1600 cm^{-1}. Evidently, the underlying reason for this shift should not be internal stress. After rapid thermal processing at approximately 600 °C, complete relaxation could be achieved in a-D thin films while ensuring that the amount of sp^3 hybridization remained essentially unchanged. However, the G-peak position did not shift as expected toward lower frequencies after stress relaxation. This phenomenon can be attributed to the clustering of sp^2 hybridization in the thin film. As temperature increases, the intermittent π-bonds in the sp^3 hybridization network will gradually aggregate and grow. This will cause sp^2 fields to grow in size, increase in quantity, and decrease in spacing, which will cause the transformation from a short-range ordered state to a medium-range ordered state. The appearance of such medium-range ordered clusters with size exceeding 1 nm has already been empirically verified by fluctuation spectra and electron microscopy. If the temperature continues to increase, the tendency for graphitization will become more prominent and be manifested as the gradual growth of the D-peak.

Fig. 5.42 shows the relationship between the G-peak FWHM of thin films fabricated under different substrate biases and annealing temperature. The figure indicates that, when the negative substrate bias was 20 V, thin-film FWHM remains basically un-

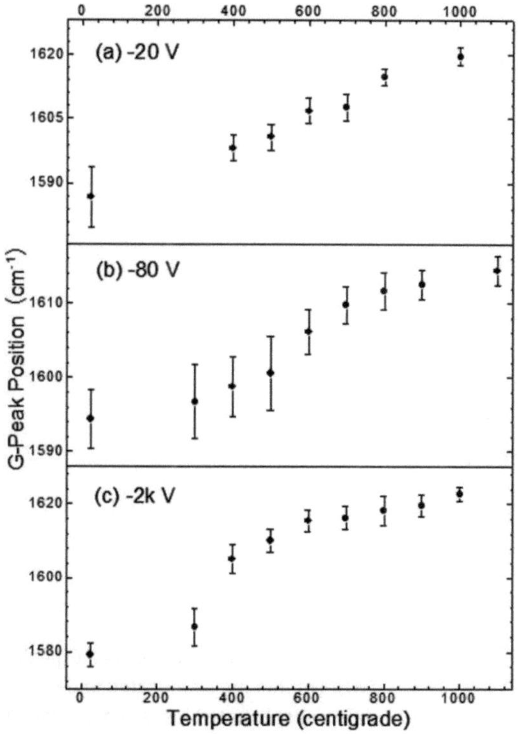

Fig. 5.41: Relationship between G-peak position of thin films fabricated under different substrate biases and annealing temperature.

changed below 600 °C; as the temperature increased, FWHM drastically decreased. When the negative substrate bias was 80 V, FWHM began to decrease drastically when the temperature exceeded 500 °C. When the negative substrate bias was 2000 V, the FWHM decreased continuously with increasing annealing temperature. The factors influencing G-peak FWHM mainly include the degree of crystalline state or ordering of the thin film, size and distribution of sp^2 fields, and thin-film internal stress. The a-D thin films fabricated under room temperature have a typical short-range order structure, with diverse distribution and size of sp^2 fields, which resulted in severe broadening of the G-peak. However, when the annealing temperature reached a certain level, the graphitization tendency of the thin film was significantly enhanced, while the size of sp^2 clusters increased continuously.

The photons of visible light will cause lattice vibrations. Changes in the wave vector are given by $\Delta k \propto 2\pi/d$, where d is the cluster size. As the degree of ordering becomes more significant, the size of sp^2 clusters will increase and FWHM will decrease. Sakata et al. reported that a linear increase occurs between G-peak FWHM and stress. Thus, it can be inferred that the reduction in G-peak FWHM is related to thin-film stress relaxation.

Fig. 5.43 shows the relationship between the D-peak FWHM of thin films fabricated under different substrate biases and annealing temperature. The figure clearly

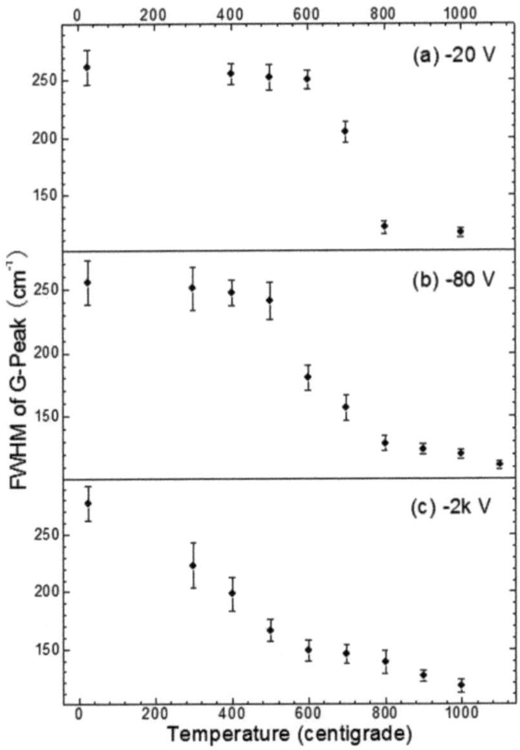

Fig. 5.42: Relationship between G-peak FWHM of thin films fabricated under different substrate biases and annealing temperature.

shows that the trend of change in D-peak FWHM was an initial increased followed by a decrease with increasing annealing temperature. The transition temperatures of the three groups of thin films fabricated under different deposition energies were 800 °C, 700 °C, and 600 °C, respectively, which were all higher than the D shoulder peak temperature of 100 °C. The increase in D-peak FWHM can be attributed to the significant aggregation of sp^2 fields that were originally intermittently distributed. However, as the temperature increased further, the graphitization tendency was enhanced significantly. As with the G-peak, the FWHM of the D-peak decreased drastically with increasing temperature.

In conclusion, thin films fabricated using the filtered cathodic vacuum arc method have good thermal stability. However, the deposition energy has a significant influence on thermal stability, and a high deposition energy will reduce the thermal stability of thin films. When the negative substrate bias is 20 V, the D shoulder peak only appears in the visible Raman spectrum after annealing at 700 °C. When the negative substrate bias is 80 V, the D shoulder peak appears in the visible Raman spectrum after annealing at 600 °C, despite having the highest amount of sp^3 hybridization. With increasing annealing temperature, the G-peak position shifts toward higher frequencies, its FWHM will decrease drastically after the appearance of the D shoulder peak,

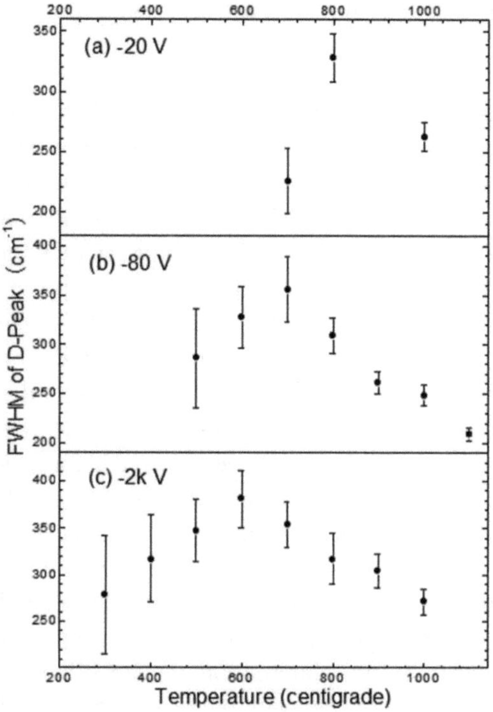

Fig. 5.43: Relationship between D-peak FWHM of thin films fabricated under different substrate biases and annealing temperature.

and the D-peak FWHM shows a trend of initial increase followed by a decrease. Visible Raman spectroscopy is a powerful method for characterising the evolution and clustering tendency of sp^2 fields in a-D thin films.

5.7 Adhesion with infrared window materials

The application of a-D thin films relies, to a large extent, on the adhesion between the thin film and substrate. However, the stress produced during the deposition process often limits the maximum adhesion thickness of a-D thin films on the substrate, which ultimately leads to the delamination of the thin film from the substrate. In view of this fact, researchers have employed several techniques, including implantation modification and the introduction of an intermediate layer, to improve the film-substrate interface adhesion. Good adhesion has already been achieved between the a-D thin film and various substrates, such as Si, glass, piezoelectric crystals, and aluminium alloy. However, the poor adhesion of a-D thin films with zinc sulphide (ZnS) and other infrared window materials remains an unresolved problem. This section will analyse the interface between a-D thin films and typical infrared window materials, including ZnS, magnesium fluoride (MgF_2) and germanium (Ge), to determine the technological pathway for improving the adhesion between a-D thin films and ZnS.

5.7.1 Interface properties of amorphous diamond thin films with zinc sulphide and magnesium fluoride

With regards to the direct deposition of a-D thin films on polished hot-pressed ZnS and MgF_2 substrates, if the deposition time is relatively short and the film is thin, then the thin film will be able to adhere to the substrate and withstand the test of alcohol wipes. However, with an increased deposition time and a film thickness greater than 80 nm on a ZnS substrate, the a-D thin film will delaminate naturally after exiting the vacuum chamber, regardless of the deposition conditions applied. Fig. 5.44 shows the appearance of a delaminated a-D thin film on a ZnS substrate. The thin-film surface has numerous irregular wrinkles and folds, and the interface separation was rapid. The initiation and propagation process of fractures at the interface can be observed under optical microscopy. If the deposition time is relatively long, the spread of fractures can be observed using the naked eye, and cracks might even have occurred before the thin film exited the vacuum chamber. Unprocessed a-D thin films after deposition still retain a relatively large amount of compressive stress, causing the gradual accumulation of elastic strain energy at the interface. If this exceeds the constraints of the weak interface adhesion strength, then thin-film delamination will occur. The relaxed wrinkles on the thin-film surface shown in the figure indicate the state resulting after the release of compressive stress. Following thin-film delamination, the ZnS substrate retained its greenish-white colour, indicating that the deposition process did not lead to effective adhesion between carbon and the substrate.

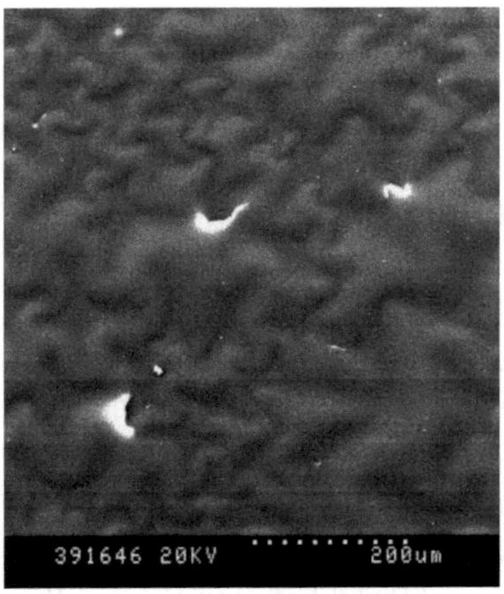

Fig. 5.44: Morphology of a-D thin film after delamination from ZnS substrate.

In order to explore the interface properties of a-D thin films with ZnS and MgF_2, depth profiling using x-ray photoelectron spectroscopy (XPS) was applied to analyse the composition distribution and chemical environment of the interface region. The XPS depth profiling of the a-D/ZnS interface is shown in Fig. 5.45. The atomic contents of zinc, sulphur, carbon and oxygen elements were calculated using the integrated area

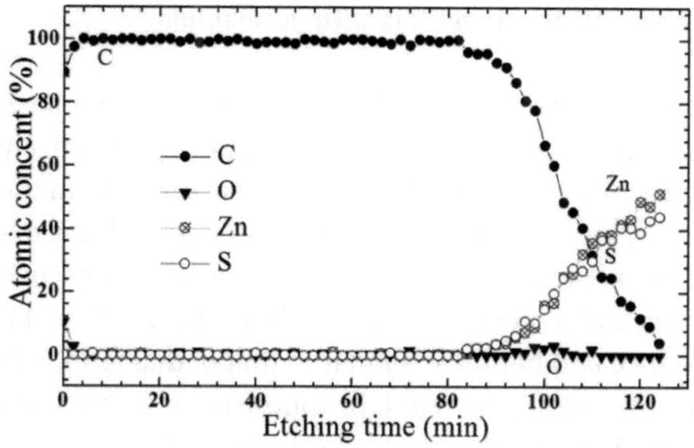

Fig. 5.45: XPS depth profiling of a-D/ZnS interface.

under the relevant spectral peaks of Zn 2p, S 2p, C 1s, and O 1s, respectively. The thickness of the interface zone was estimated to be approximately 8 nm based on the etching rate during the experiment. The thickness of the interface zone is related to substrate surface roughness. The relatively rough surface of hot-pressed crystals causes the component mixing zone to be relatively wide. A comparison of the C 1s peak position in the thin-film growth cross section indicated that no shift existed. Moreover, the analysis of S and Zn spectral peaks did not reveal any traces of compounds formed between the C and substrate components.

Similarly, the XPS depth profiling of the a-D/MgF_2 interface was performed by analysing the evolution of Mg 2p, F 1s, C 1s, and O 1s during the course of etching, as shown in Fig. 5.46. Traces of compounds formed between carbon and the substrate components still could not be detected. Therefore, it can be inferred that the a-D thin film is mainly interconnected with ZnS and MgF_2 via intermolecular forces (van der Waals forces), which only produces limited adhesion strength, resulting in an extremely fragile film-substrate interface.

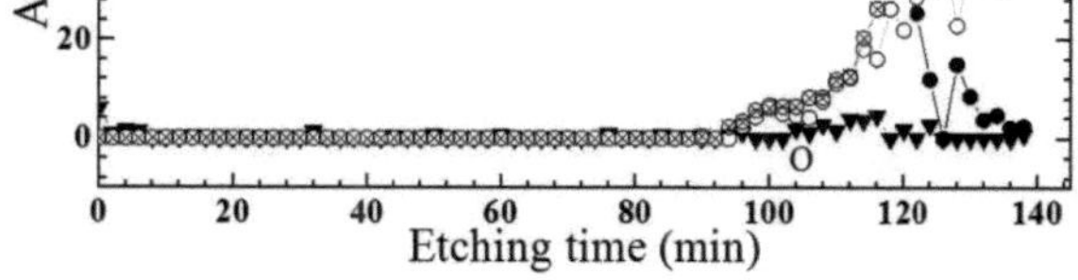

Fig. 5.46: XPS depth profiling of a-D/MgF_2 interface.

5.7.2 Interface properties of amorphous diamond films with germanium

As Ge has low scattering, small chromatic aberration, and other advantages, it is one of the most common materials used for infrared windows and lenses. Ge has a relatively high refractive index; in the absence of additional antireflective measures within the working wave range, its transmittance will usually be less than 50 %. Ge also has a relatively soft texture, and is unable to withstand harsh working environments. Hence, it is necessary to provide antireflective and protective measures. Since the a-D thin film has broadband infrared transmittance and hardness similar to that of diamond, it is a suitable antireflective and protective coating material. In contrast to a-D thin films on ZnS substrates, a-D thin films are able to form a good adhesion with Ge substrates. The following section will present an analysis of the a-D/Ge interface in order to determine the root cause for the differences in such adhesions and provide experimental evidence for the formation of compounds at the interface.

The XPS depth profiling of the a-D/Ge interface is shown in Fig. 5.47. In the figure, the atomic contents of germanium, carbon, and oxygen were calculated based on the integrated area under the spectral peaks of Ge 2p, C 1s, and O 1s. Based on an etching rate of 0.1 nm/min, the width of the interface zone is estimated to be approximately 5–8 nm. Since the RMS surface roughness of the monocrystalline Ge substrate is less than 1 nm, the interface zone of this composition gradient should still be related to atomic diffusion and the subplantation process of deposited C ions. When deposited C ions overcome the binding energy of the Ge surface and penetrate the substrate by a distance ranging from a few Angstroms to greater than 30 Å, the subsequent collision cascade of the deposited ions not only creates the conditions for different C hybridization, but also promotes the chances of mixing and interactions between C and substrate atoms. Experiments have shown that the width of the interface zone cannot determine film-substrate adhesion, but the interactive effect between C and substrate atoms is often a key factor that determines interface adhesion properties.

Fig. 5.48 clearly shows the variation of the C 1s peak position as etching progressed. At the initial stage of etching, the large reduction in the C 1s peak position is due to significant graphitization caused by the continuous bombardment of the a-D

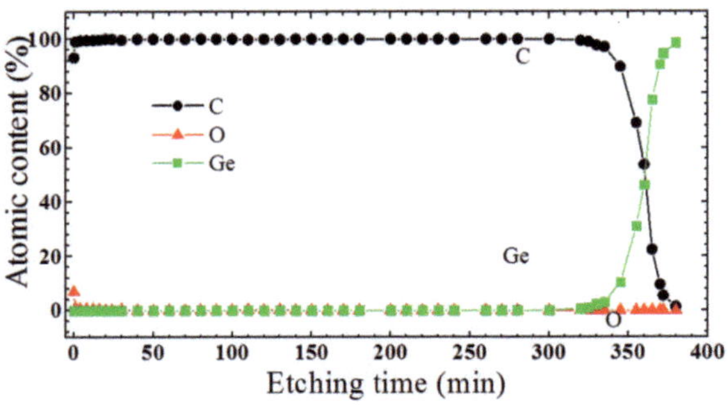

Fig. 5.47: XPS depth composition profiling of a-D/Ge interface.

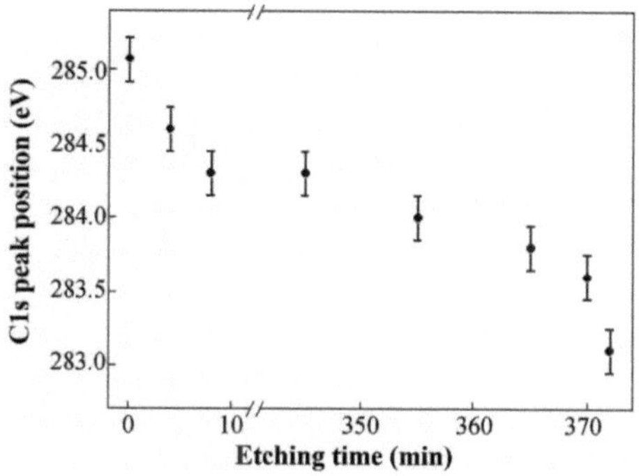

Fig. 5.48: Variation of C 1s peak position with increasing etching time.

thin film by argon (Ar) ions. This eventually stabilized at 284.3 eV, which is equivalent to the chemical binding energy of pyrolytic graphite. When etching progressed to the film-Ge substrate interface, the C 1s binding energy showed a significant reduction once again. Unlike the first decrease in the C spectral peak, which was due to the bombardment of noble gas ions, this decrease can be attributed to the formation of carbides.

In order to reveal the atomic compositions of different carbides at the initial stage of film growth, peak deconvolution was performed on the C optical emission spectra, which are plotted in Fig. 5.49. With reference to the C 1s binding energy of graphite (284.3 eV) and an interval of 0.8–1.0 eV between the sp^2 and sp^3 hybridization states, another C deconvoluted peak was detected below 284.0 eV. Furthermore, with the gradual removal of the a-D thin film, this component became increasingly stronger. According to relevant data, this component can be attributed to the appearance of germanium carbide at the interface zone. When the interface layer was covered by the a-D thin film, information from C compounds was flooded by signals from thin-film components. It was only when the thin-film was gradually removed that the spectral peaks of carbides were clearly observed.

As shown in Fig. 5.50 (a), the peak position of germanium carbide gradually shifted toward higher binding energies with increasing thin-film growth. The ratio of C atoms involved in the formation of germanium carbide can be obtained by computing the ratio between the integrated area under the deconvoluted peak of the carbide phase and the sum of the integrated area under the two deconvoluted peaks representing sp^2 and sp^3 hybridizations. Fig. 5.50 (b) indicates that, during the initial stage of thin-film deposition, a high proportion of incident C ions participated in the formation of carbides. Fig. 5.51 demonstrates that, at the interface zone, the peak position of the germanium-carbide phase gradually shifted toward higher binding energies with increasing C content. These results indicate that the carbides formed from the interactions between C ions and the Ge substrates are not uniform compounds with a fixed chemical composition. Hence, the C content in germanium carbides might increase as more C ions are deposited. During the initial stage of thin-film growth, the

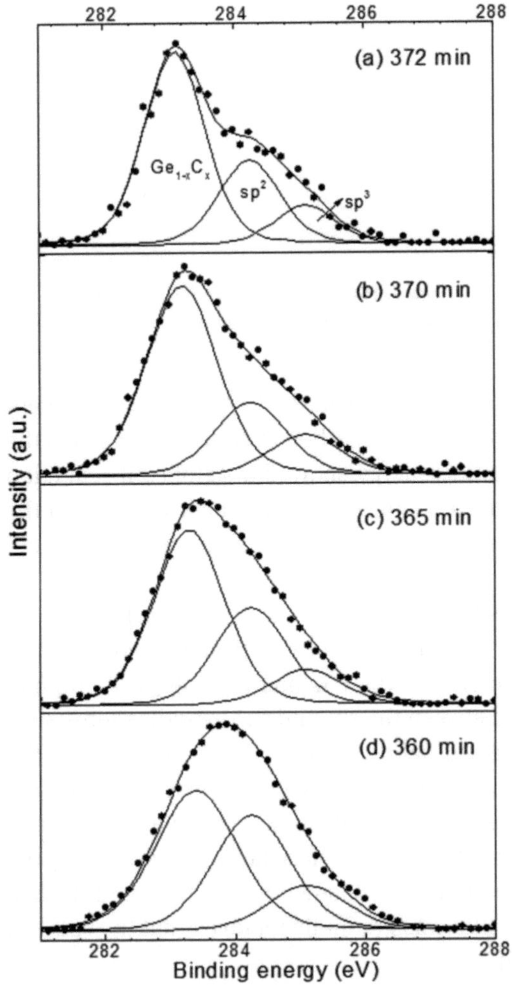

Fig. 5.49: C 1s spectral peaks for different etching times: (a) 372 min; (b) 370 min; (c) 365 min; (d) 360 min.

interface zone consists of a germanium-carbide phase that gradually changes from substrate components to thin-film components. This carbide transition layer composed of substrate and thin-film components is able to strengthen the adhesion of a-D thin films with Ge substrates.

In order to verify further the presence of C–Ge interfacial reactions, the spectral peaks of Ge 2p and Ge 3d were also measured, and the relationship between the Ge peak position and etching time is plotted in Fig. 5.52. The $2p^{3/2}$ and 3d spectral peaks of monocrystalline Ge were located at 1217 eV and 29 eV, respectively. However, the peak positions of interfacial Ge were higher than the corresponding peak position of monocrystalline Ge, and gradually shifted toward higher binding energies with thin-film growth. The variations in Ge peaks can also be viewed as evidence for the germanium-carbide phase produced in the interface zone.

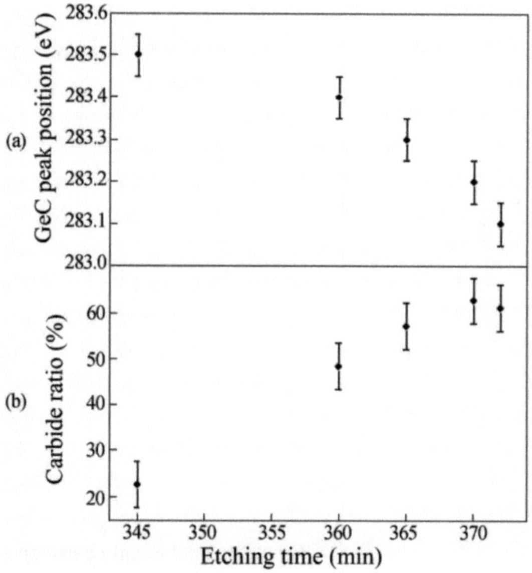

Fig. 5.50: Spectral characteristics of germanium carbide at the interface zone: (a) relationship between germanium-carbide peak position and etching time; (b) relationship between ratio of carbides among carbon atoms and etching time.

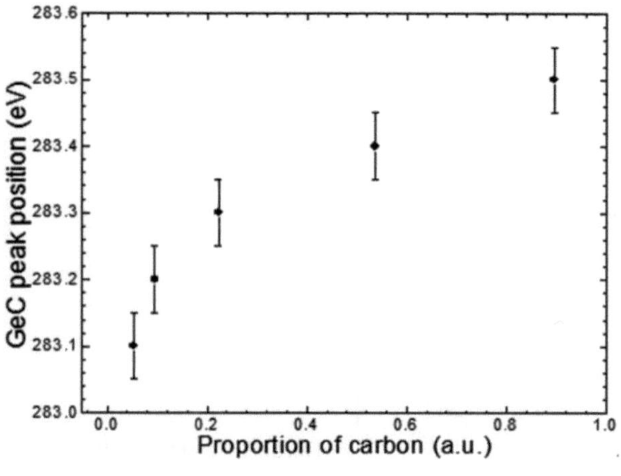

Fig. 5.51: Relationship between germanium-carbide peak position and carbon content at the interface zone.

In order to reflect the adhesion properties of a-D thin films on Ge substrates, a tape test was performed for qualitative assessment. The adhesive tape was adhered to the thin film and forcefully peeled off; this was repeated a number of times. It was found that the thin film showed good adhesion. Then, a nanoscratch test was performed to determine thin-film adhesion. During the experiment, a diamond indenter was used to scratch the thin film, and the critical scratch load was determined based on the changes in lateral friction. As shown in Fig. 5.53, the critical scratch load of the a-D thin film was nearly 100 mN, and the thin-film thickness measured using a step profiler was 292 nm. However, when cracks occurred in the thin film, the diamond indenter had already reached a depth of nearly 500 nm beneath the surface, and the damage

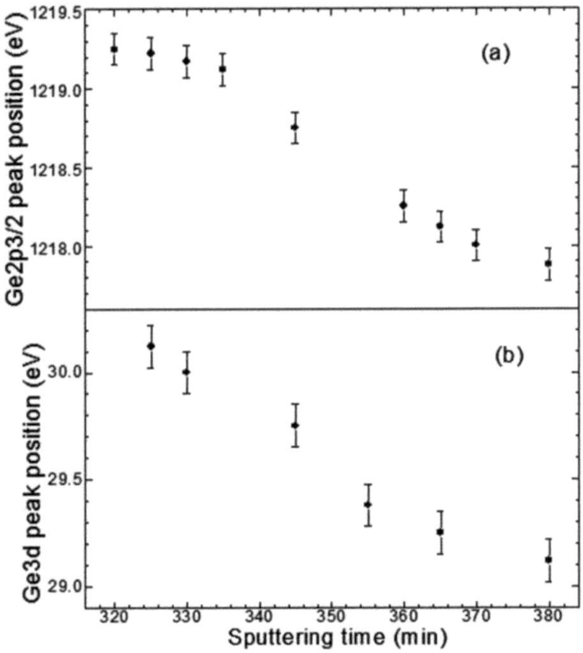

Fig. 5.52: Relationship between Ge 2p and Ge 3d peak positions at the interface zone and etching time.

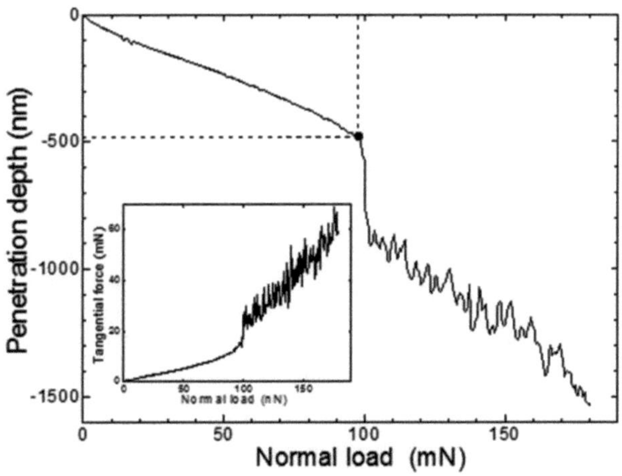

Fig. 5.53: Results of a-D/Ge scratch test; the inset shows the lateral friction of the nanoindenter.

occurred at the substrate, rather than at the interface. Hence, we can infer that the a-D/Ge interface has the higher adhesion strength.

In summary, measurements of the shifts in C and Ge XPS spectral peaks with etching time have verified the presence of a germanium-carbide phase at the interface between the a-D thin film and Ge substrate. This interfacial compound layer with graded changes in composition from the Ge substrate to the a-D thin film is beneficial to thin-film adhesion.

5.8 Structure and stress analysis of multilayered amorphous diamond films

Deposition using the filtered cathodic vacuum arc (FCVA) method enables the fabrication of high-performance ta-C thin films at high deposition speed near room temperature. Nevertheless, the rich sp^3 structure of thin films formed by the subplantation of deposited ions and local densification retains a high degree of compressive stress, which can easily lead to the delamination of the coating layer from the substrate. Furthermore, when applied as an infrared optical protective coating, at a reference wavelength of 10 μm, the geometric thickness of the thin film needs to be greater than 1 μm to satisfy optical antireflective requirements. Therefore, the relatively high internal compressive stress of the coating will severely limit the application of ta-C within the field of infrared protective coatings. By modifying the substrate bias, it is possible to use an sp^2-rich coating as the innermost layer that is bound to the substrate and an sp^3-rich coating as the outermost layer, thereby forming a graded or alternating multilayered coating structure. This will effectively reduce the internal stress of ta-C thin films, increase coating–substrate adhesion strength, increase the thickness of coating growth, and retain the DLC properties of the thin film. This section will investigate the microstructure and interface structure of multilayered ta-C films, as well as the stress changes in each sublayer after deposition.

5.8.1 Stress theory of multilayered films

With regards to the adhesion of n film layers on a substrate (Fig. 5.54), the thickness and elastic modulus of the i-th layer are t_i and $E_{f,i}$ ($i = 1, 2, \ldots, n$), respectively, while the thickness and elastic modulus of the substrate are t_s and E_s, respectively. Also,

$$\sum_{i=1}^{n} t_i = t_f \ll t_s, \quad \sum_{i=1}^{n} E_{f,i} t_i \ll E_s t_s.$$

The middle atomic layer of the substrate at $z = -t_s/2$ is taken as the unstressed neutral plane, and the strain of the ith film layer and substrate with respect to their flat states are $\varepsilon_{f,i}$ and ε_s, respectively. The average strain of the overall coating–substrate system is given by

$$\bar{\varepsilon} = \frac{E_s t_s \varepsilon_s + \sum_{i=1}^{n} E_{f,i} t_i \varepsilon_{f,i}}{E_s t_s + \sum_{i=1}^{n} E_{f,i} t_i} \simeq \varepsilon_s + \frac{\sum_{i=1}^{n} E_{f,i} t_i (\varepsilon_{f,i} - \varepsilon_s)}{E_s t_s}. \tag{5.26}$$

Therefore, the strain produced by substrate deformation due to the stress of the entire coating layer is given by

$$\varepsilon_s - \bar{\varepsilon} \simeq \frac{-\sum_{i=1}^{n} E_{f,i} t_i (\varepsilon_{f,i} - \varepsilon_s)}{E_s t_s}. \tag{5.27}$$

By performing a first-order approximation of equation (5.27), we obtain $\varepsilon_s = \bar{\varepsilon}$. The strain of the i-th layer with respect to the substrate, which is caused by internal film

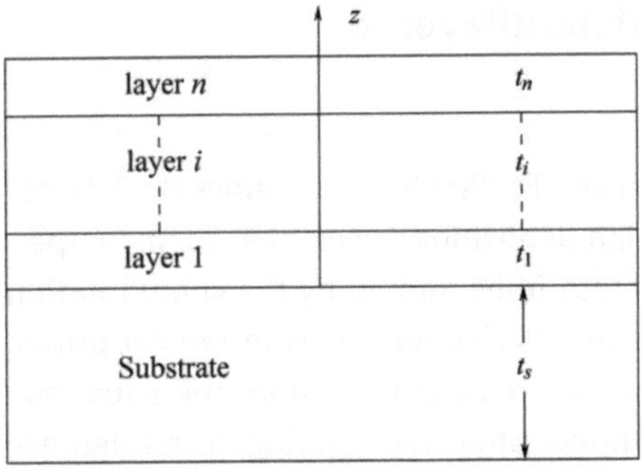

Fig. 5.54: Schematic diagram of the structure of a multilayered film adhered onto a substrate.

stress resulting from elastic mismatch, is given by

$$\varepsilon_{\mathrm{m},i} = \varepsilon_{\mathrm{f},i} - \bar{\varepsilon} \simeq \varepsilon_{\mathrm{f},i} - \varepsilon_{\mathrm{s}}. \tag{5.28}$$

Equation (5.28) shows that the strain in each layer caused by internal film stress is also influenced by the strain of the substrate itself. In the stress models and calculations discussed in this book, polished monocrystalline Si substrates with strictly the same surface morphology and size were used, and the coating layers were fabricated near room temperature. Hence, the effects of substrate deformation and thermal strain on the coating strain will not be considered for the time being, i.e. $\varepsilon_{\mathrm{m},i} \approx \varepsilon_{\mathrm{f},i}$.

Vilms et al. analyzed a simple stress model of multilayered thin films, which indicated that the stress of the sublayers is distributed almost uniformly in the coating, and the stress of each sublayer is equivalent to the linear approximation of stress in a monolayered film. Therefore, according to the theory of mechanical equilibrium, the stress-induced curvature of the neutral plane in the multilayered film-substrate system can be expressed as

$$\kappa = \frac{-6}{t_{\mathrm{s}}}(\varepsilon_{\mathrm{s}} - \bar{\varepsilon}) = \frac{6}{E_{\mathrm{s}} t_{\mathrm{s}}^2} \sum_{i=1}^{n} E_{\mathrm{f},i} t_i \varepsilon_{\mathrm{m},i} = \sum_{i=1}^{n} \kappa_i. \tag{5.29}$$

Based on the reciprocal relation between curvature and radius of curvature r, the following can be obtained:

$$\kappa = \frac{1}{r} = \sum_{i=1}^{n} \frac{1}{r_i}. \tag{5.30}$$

According to the stress-strain relation of first-order elasticity theory, $\sigma = E\varepsilon$, the stress of the i-th coating layer can be expressed as

$$\sigma_{\mathrm{f},i} = E_{\mathrm{f},i} \varepsilon_{\mathrm{m},i} = -E_{\mathrm{s}} t_{\mathrm{s}}^2 / 6 t_i r_i. \tag{5.31}$$

Therefore, in a $t_{\mathrm{f}} \ll t_{\mathrm{s}}$ multilayered film-substrate system, the total curvature is equal to the linear sum of the curvature of each sublayer, the average stress of the multilayered film is only related to the total radius of curvature, and the stress of each sublayer is only related to the component of radius of curvature r_i of that particular layer and not to the stress of adjacent layers or the stacking order of the film layers. Furthermore, the local interfacial shear stress at the substrate boundary will not influence the system curvature and the distribution of normal stress within the coating layer.

For a planar two-dimensional coating–substrate system, the elastic modulus E in the equation above can be substituted by the biaxial elastic modulus $E^* = E/(1 - \nu)$, where ν is the Poisson's ratio of the material. This will provide the Stoney stress expression, as well as theoretical evidence for the stress research of multilayered ta-C films.

5.8.2 Microstructure of multilayered amorphous diamond films

5.8.2.1 Raman spectrum analysis

The multilayered coating designs in this book all involve an sp^3-rich coating layer formed by a thin film deposited at −80 V substrate bias and an sp^2-rich coating layer formed by a thin film deposited at low substrate biases of −1000 to −2000 V. The visible Raman spectra of monolayered ta-C thin films deposited at −2000 and −80 V are shown in Fig. 5.55.

It can be seen from Fig. 5.55 that the spectral lines showed an asymmetrical broad peak within the region of 1200–1700 cm^{-1}, and the centre of the spectral line was located at approximately 1560 cm^{-1}. In the spectral profile of the thin film deposited at −2000 V, the broadening of the D-peak induced by cyclic sp^2 breathing vibrations is relatively significant, and the D shoulder peak appeared near 1360 cm^{-1}; however, D-peak scattering was masked in thin films deposited at −80 V. Therefore, peak deconvolution was performed on the spectral lines in this book, and the fitted curves are presented as solid lines in Fig. 5.55.

The structure of multilayered films was also analyzed using Raman spectroscopy, and Fig. 5.56 shows the visible Raman spectral lines of each sublayer in a graded, multilayered ta-C coating after deposition. The graded multilayered coating consisted of four sublayers, and the substrate biases of each sublayer were −1500, −1000, −500, and −80 V, respectively. The Raman spectra of all sublayers after deposition showed asymmetrical broad peaks within the region of 1200–1700 cm^{-1}, and the peak centre was located at approximately 1550 cm^{-1}. D-peaks, which reflect sp^2 breathing vibration, could not be observed between 1300–1400 cm^{-1}. The spectral lines had good symmetry. In addition, in the spectral line of the bottom layer A_1 ($\approx$ 20 nm), the second-order peak of the Si substrate appeared at 965 cm^{-1}.

Using a simple Gaussian function for the curve fitting of each spectral line, the variations in D-peak to G-peak intensity ratio ($I(D)/I(G)$) and G-peak FWHM with the gradual deposition of each coating layer can be obtained (see Fig. 5.57; the dotted lines represent the interface between each sublayer).

It can be seen from Fig. 5.57 that $I(D)/I(G)$ decreased with the addition of each sublayer, whereas the G-peak FWHM increased gradually. The $I(D)/I(G)$ value reflects the changes in sp^2 cluster size, and has an approximately negative linear proportional

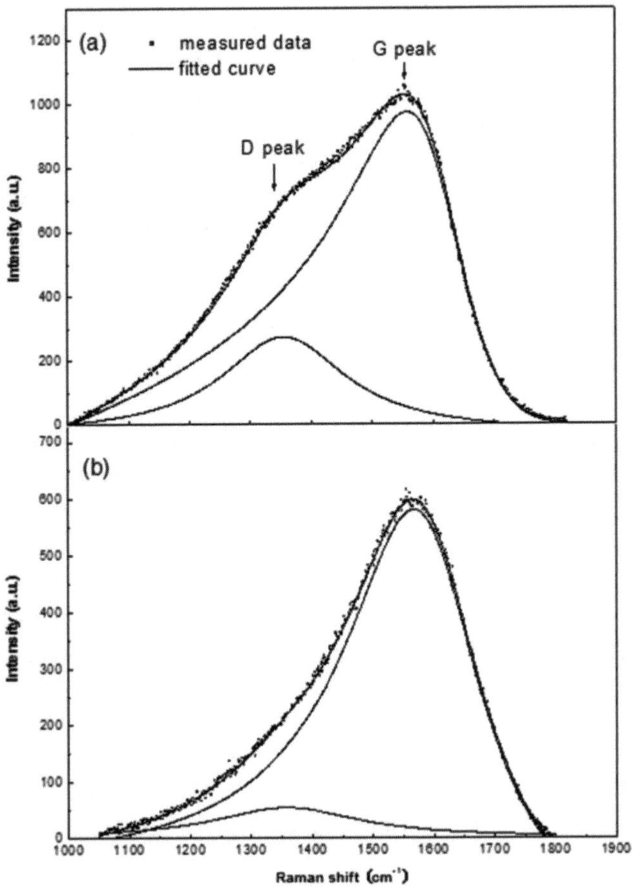

Fig. 5.55: Visible Raman curves and BWF-Lorentzian fitted curves of ta-C thin films deposited under different substrate biases.

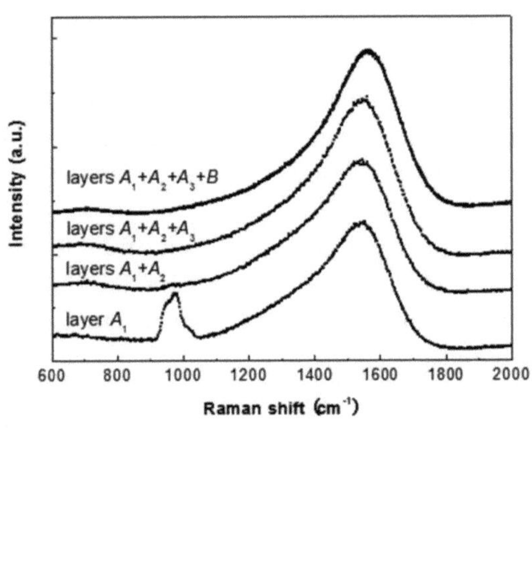

Fig. 5.56: Visible Raman spectral lines of graded multilayered ta-C thin film.

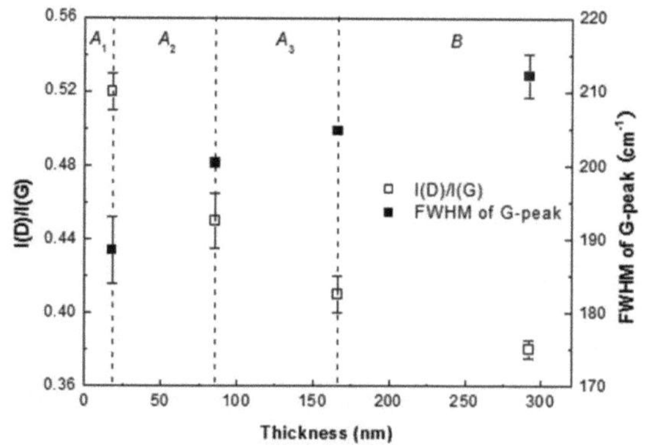

Fig. 5.57: Changes in Gaussian-fitted ratio of peak intensity $I(D)/I(G)$ and G-peak FWHM with the deposition of each sublayer.

relationship with the amount of sp^3 hybridization in ta-C thin films. Therefore, the Raman spectra reflect the variations in the average sp^3 content of the coating layer. The results of the analysis indicated that the FWHM of the G-peak originates from the stress-induced changes in C–C bond length and bond angle, and is able to reflect indirectly the variations of thin-film internal stress. Further, a positive linear proportional relationship exists between the two. Therefore, the increase in G-peak FWHM indicates that the internal stress of the graded multilayered ta-C coating will increase gradually with the deposition of each sublayer. Moreover, owing to the increase in thin-film stress, the G-peak position will also shift gradually toward the high-frequency region of Raman scattering. The changes in the internal compressive stress of the graded multilayered ta-C coating will be verified in the subsequent analysis of stress tests.

5.8.2.2 Analysis of grazing-incidence x-ray reflectometry

Fig. 5.58 shows the XRR measurement curves of alternating multilayered and bilayered ta-C coatings. Among them, the alternating multilayered coating is composed of two groups of A_i and B_i bilayers ($i = 1, 2$). As the alternating multilayered coating has more interfaces, the interference fringes of its reflectance curve are more significant, and it does not have a singular amplitude modulation period; specifically, the small amplitude modulation period corresponds to the overall thickness, and the large amplitude modulation period corresponds to the thickness of each sublayer.

Based on the Parratt algorithm, the Rfit2000 software was employed to fit the reflectance curves, and an optimization function was applied to find the global minimum, thereby obtaining the structural characteristics of each sublayer and interface in the thin film. Tab. 5.3 lists the XRR curve-fitting data of alternating multilayered ta-C films. The optimization function employed is expressed as follows:

$$\chi^2 = \frac{1}{N-M}\sum\left(\frac{R\exp(j) - KR\,\mathrm{cal}(j)}{\mathrm{d}R\exp(j)}\right)^2, \tag{5.32}$$

where j is the measurement point ($j = 1, 2, \ldots, N$), M is the number of variables, R is the reflection intensity, and K is the scale factor.

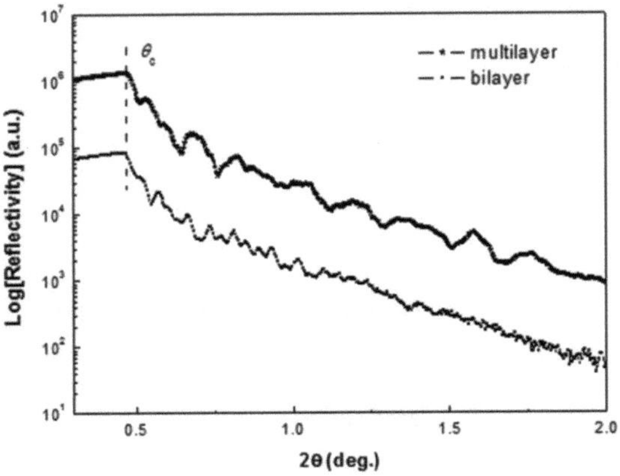

Fig. 5.58: XRR measurement curves of alternating multilayered and bilayered ta-C coatings.

Tab. 5.3: XRR curve-fitting data of alternating multilayered ta-C films.

Thin-film	Thickness (nm)	Density (g/cm^3)	Surface roughness (nm)
sp^3-rich film layer B_2	61.7	3.01	0.6
B_2/A_2 interface	2.5	2.55	0.8
sp^2-rich film layer A_2	56.5	2.38	1.0
A_2/B_2 interface	3.0	2.41	1.2
sp^3-rich film layer B_1	64.5	3.08	0.5
B_1/A_1 interface	2.5	2.52	0.8
sp^2-rich film layer A_1	63.3	2.40	1.0
B_1/Si interface	3.5	2.54	1.1
Si substrate	∞	2.33	0.3

It can be observed from Tab. 5.3 that the interface between the sp^2-rich film layer A_1 and the Si substrate have a relatively high density. This is because, during the deposition of film layer A at a substrate bias of −2000 V, high-energy C^+ ions will diffuse in the proximal substrate surface, causing a densified structure similar to SiC. Moreover, the high-energy ion bombardment of the sp^2-rich film layer during deposition will lead to the increased roughness of the A_i/B_i sublayer interface. It will also cause the sp^3 hybridization in the surface of the sp^3-rich B_i layer to convert to sp^2 hybridization, resulting in decreased density. Anders et al. employed transmission electron microscopy (TEM) to observe the interfacial structure of alternating multilayered a-C films, and found that the interface region of the sp^2-rich and sp^3-rich sublayers was relatively wide owing to high-energy ion bombardment when depositing the sp^2-rich film layer. A similar phenomenon was also observed in Tab. 5.3, as indicated by the greater thickness of the A_2/B_2 interface. Furthermore, as the local lattice mismatch was small, the interface thickness between the sublayers was less than that of the coating–substrate interface.

5.8.3 Stress analysis of multilayered amorphous diamond films

5.8.3.1 Gradient multilayered films

The Stoney formula was employed to compute the stress changes in each sublayer of a gradient multilayered ta-C film after deposition, which is shown in Fig. 5.59. The dotted lines in the figure represent the interface of each sublayer, and the error values of stress calculation are due to the measurement errors of film thickness and radius of curvature. During deposition, the substrate biases of sublayers $A_1 - B$ were −1500, −1000, −500, and −80 V, respectively. According to the analytical results of the microstructure of a graded multilayered ta-C film in this book, the amount of sp^3 hybridization gradually increased from layer A_1 to layer B; sp^2 hybridization accounted for a large proportion in layers A_1 and A_2, whereas layer B is an sp^3-rich film layer.

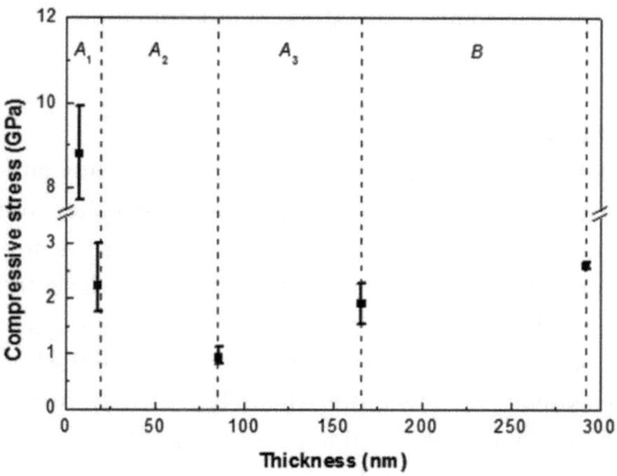

Fig. 5.59: Variations in compressive stress among the sublayers of a graded multilayered ta-C film.

During the deposition of film layers A_1 and A_2 at low substrate biases, the local thermal peak effects due to the high energy of C^+ ions will cause the migration of thin-film surface atoms, enabling the relaxation of compressive stress and resulting in a smaller stress value. However, we can observe from Fig. 5.59 that, during the deposition process of the thin bottom layer A_1, compressive stress was relatively high (8.79 ± 1.12 GPa), which is inconsistent with the claim that the sp^2-rich film layer should have low stress values. Patsalas et al. suggested that this was due to the implantation of high-energy C^+ ions into the Si substrate surface layer, which then diffused into the substrate and formed a mixed layer with a densified SiC-like structure at the coating–substrate interface. The lattice distortion between C atoms and Si atoms resulted in high additional stress, which is manifested at the thin bottom layer A_1, leading to the phenomenon of high stress. Mathioudakis et al. performed the molecular dynamics simulation of the multilayered a-C film structure, and found that a small amount of sp^3 hybridization in the bottommost sp^2-rich film layer was aggregated near the substrate interface, which would lead to high interface compressive stress as well. XRR results also showed that a densified interface layer was formed between the sp^2-rich bottom layer and the Si substrate, and its thickness was 3.5 nm, which is similar to the initial thickness of the A_1 film layer. Hence, high compressive stress was exhibited.

As the thickness of layer A_1 increased and layer A_2 was deposited, the influence of additional stress from the coating–substrate mismatch on thin-film stress gradually became weaker, and the level of compressive stress reverted to the low value already present in sp^2-rich thin films (0.93 ± 0.20 GPa). After layers A_3 and B were deposited, the thin-film sp^3 content began to increase gradually, and the points where they gathered led to the formation of compressive stress fields, which in turn caused another small increase in thin-film compressive stress. The average compressive stress of a graded multilayered ta-C film was 2.60 ± 0.06 GPa. This value is approximately 30 % smaller compared to the compressive stress value of 3.65 ± 0.02 GPa measured from a monolayered sp^3-rich thin film deposited at −80 V. Therefore, a graded multilayered film design is able to effectively reduce the compressive stress of ta-C thin films.

5.8.3.2 Alternating multilayered films

Measurements performed on the behaviour of stress changes in A_iB_i alternating multilayered ta-C films are shown in Fig. 5.60. A_i is the sp^2-rich film layer deposited at a negative substrate bias of 1000 V, and B_i is the sp^3-rich film layer deposited at a negative substrate bias of 80 V. The thickness ratio of the two layers d_{Ai}/d_{Bi} is designed to be approximately 1.0. The entire multilayered film is composed of three groups of A_iB_i bilayers ($i = 1, 2, 3$), and the total thickness is approximately 1 μm. After the bottom layer A_i is deposited, stress was relatively low (approximately 1.58 GPa). At this point, the film layer is relatively thick; hence, the additional stress from the coating–substrate interface mismatch layer does not significantly influence the thin-film stress. After layer B_i is deposited, the stress value substantially increased to 2.85 GPa. Subsequently, during the alternating deposition of layers A and B, the stress values of

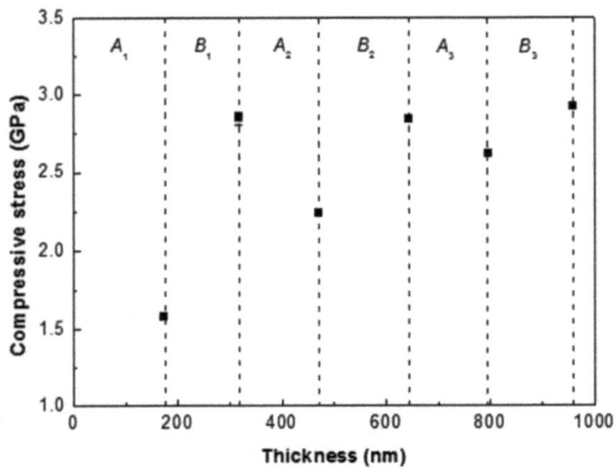

Fig. 5.60: Variations in compressive stress between each sublayer of an alternating multilayered ta-C film.

the coating layer showed jagged fluctuations, and the magnitude of change gradually decreased. The final average compressive stress obtained for the multilayered film was 2.92 ± 0.01 GPa, which is a 20 % decrease compared to the compressive stress of a 300-nm-thick monolayered sp^3-rich thin film. The trend of change in the sublayers of the alternating multilayered ta-C film is very similar to that observed in the study by Logothetidis et al. on alternating a-C films. Regarding the reasons behind the stress relaxation of alternating multilayered films, Logothetidis et al. proposed that the densified interface formed between the sublayers increased the system energy. In order to reduce the additional energy at these interfaces (surface energy + strain energy), bending occurs in the substrate, which produces macro tensile stress and offsets a part of the internal film stress. Furthermore, the more plastic sp^2-rich layer A can act as a buffering layer in the multilayered film, causing the relaxation of compressive stress in the entire thin film to a certain extent.

The results of microstructure and stress research on graded and alternating multilayered ta-C films have shown that the addition of sp^2-rich film layers can effectively relax the compressive stress of thin films. Moreover, the entire multilayered film retains a structure with rich sp^3 hybridization. A comparison of the two design methods indicates that the fabrication process of alternating multilayered films can yield a thicker ta-C thin film with thinner sublayers; this method is also simple and easily implemented. Hence, it has been widely researched and applied, and is also the main technique used in this book to fabricate multilayered ta-C films. The influence of the thickness parameters of each sublayer, such as the thickness ratio d_{Ai}/d_{Bi}, periodic thickness of A_iB_i bilayer, and other factors, on the mechanical properties of multilayered films will be discussed in subsequent chapters. However, more experiments and analyses are still required to understand the stress relaxation mechanisms of multilayered films.

5.9 Mechanical properties of multilayered amorphous diamond films

Owing to their excellent mechanical properties, multilayered ta-C films have been widely applied in the field of wear-resistant coatings. This section will explore the typical mechanical properties of multilayered ta-C films, including hardness, elastic modulus, fracture toughness, scratch-resistance, and adhesion, as well as the thermal stability of multilayered films. In the design of ta-C films, differences in sublayer structural parameters can have a substantial impact on their mechanical properties. Logothetidis et al. analyzed the impact of the thickness ratio of "soft"/"hard" film layers and periodic thickness on the properties of multilayered ta-C films, including the average compressive stress and hardness. They found that, as the thickness of "soft" (sp^2-rich) film layers increased, the average compressive stress of the multilayered film layer decreased gradually and the film hardness decreased as well. Moreover, a small periodic thickness helped enhance film hardness. Pujada et al. verified that the average stress in multilayered films decreases with decreasing periodic thickness. The study by Ager found that optimal wear resistance could be achieved when the ratio of "soft" (sp^2-rich) to "hard" (sp^3-rich) film layers was approximately 50 %. Therefore, this section will investigate the effects of sublayer thickness ratio and periodic thickness on the mechanical properties of multilayered ta-C films to optimise the fabrication process of multilayered ta-C films.

5.9.1 Hardness and Young's modulus

Fig. 5.61 shows the hardness and elastic-modulus data of graded and alternating multilayered ta-C films obtained using the continuous stiffness measurement (CSM) method in a nano-scratch test. The error of each data point originates from the standard deviation of multiple measurements. It can be seen that the data in the figure exhibits a very similar pattern of variation. In the initial stage, plastic deformation of the thin film was induced at the tip of the indenter to avoid abnormal stress. As the indentation displacement increased, the measured values of hardness and elastic modulus increased rapidly. As the indentation depth increased, the area of contact expanded gradually, the plastic deformation zone at the indenter tip was reduced, and the magnitude of increase in measured values was reduced. As the indentation displacement continued to increase, the plastic deformation zone expanded to the coating–substrate interface, the thin film began to exhibit elastic deformation, and the measured value reached the maximum and remained there for a certain distance. At this point, the indentation depth of the indenter is slightly greater than the contact depth. Subsequently, when the relatively softer Si substrate started to undergo plastic deformation, the measured values that were affected began to decrease, and they gradually reached the level of the Si substrate. The disparity in the degree of increase between hardness and elastic modulus at the initial stage is due to the difference of order of magnitude between their relationships with the indenter contact area. In addition, when the indentation displacement was lower, the plastic deformation zone was smaller, and the measured values were influenced by the interface of the outer sublayers. Therefore, small fluctuations occurred at the initial stage after the maximum hardness was attained.

It should be noted that in Fig. 5.61 (b), as the indentation displacement increased, the decrease in the hardness of the alternating multilayered ta-C film after reaching the maximum was not smooth and continuous but instead showed two gradient plateaus (indicated as a and b by arrows in the figure). A comparison of the indentation displacement and the thickness of each sublayer at the plateaus showed that the position of their appearance is within the thickness range of sublayer B_i in the multilayered film. This is because, during the process of continuous indentation in the multilayered film, the plastic deformation zone expanded to the hard B_i layer and was met

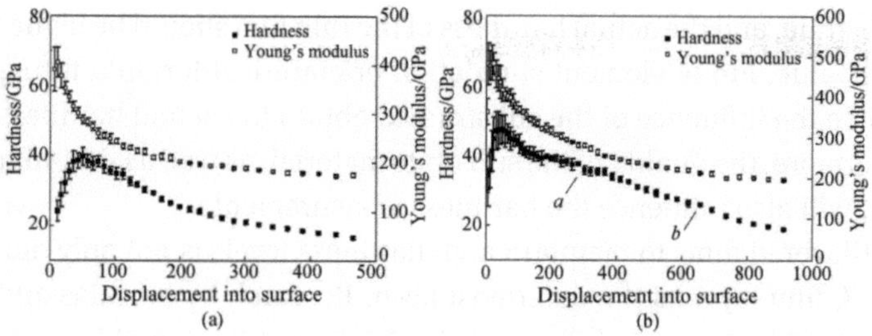

Fig. 5.61: Hardness and elastic-modulus data from the nanoindentation test of ta-C thin films.

with resistance. Thus, the elastic deformation of layer B_i delayed the expansion of the contact area, which caused the hardness values to show a recovery plateau. However, this phenomenon was not detected in the measured hardness curve of the graded multilayered ta-C film (Fig. 5.61 (a)). Ziebert et al. employed the small-angle cross-section (SACS) nanoindentation method to characterise the mechanical properties of a-D thin-film depth cross sections. The study provided a detailed description of the variations in the hardness of the sublayers and the interface zones between them within the measured indentation depth.

Fig. 5.62 shows the measured hardness and elastic-modulus values of monolayered and multilayered ta-C films. The monolayered thin film was deposited at −1000 V (sp^2-rich) and −80 V (sp^3-rich), and its film thickness was the same as that of the graded multilayered film at approximately 300 nm. It can be seen in the figure that the multilayered ta-C film retained the high hardness level of sp^3-rich thin films, especially the 1-µm-thick alternating multilayered films, which had hardness values of up to 47 GPa. Furthermore, the elastic modulus of multilayered thin films were all increased, indicating the enhanced elastic properties of the thin film. During the process of indentation load application, the substrate undergoes deformation simultaneously with thin-film deformation. Therefore, the measured value is the composite hardness

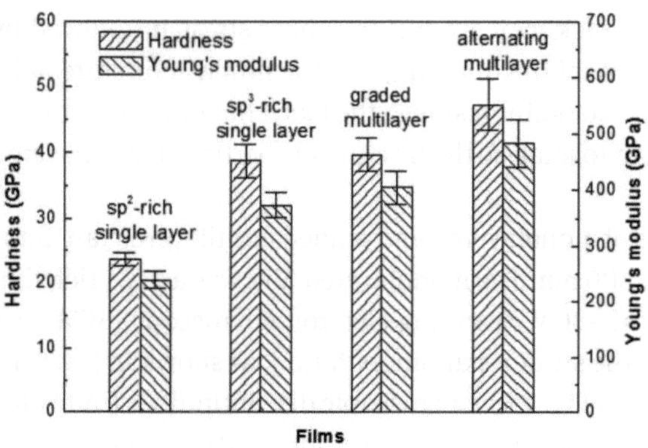

Fig. 5.62: Measured hardness and elastic-modulus values of monolayered and multilayered ta-C films.

of the thin film and substrate, and the actual hardness of the thin film should be higher than the experimental value. Finite-element simulation or empirical-formula fitting can be applied to remove the influence of the substrate to obtain the actual hardness of the thin films. Furthermore, the quality of the substrate material, as well as thin-film binding and adhesion, will also influence the hardness measurements.

The ability of multilayered films to maintain high hardness levels is not only due to the use of the hard B_i film layer as the outermost layer. Research by Patsalas and Logothetidis have indicated that a part of the sp^2 hybridization at the top of layer A_i is influenced by the compressive stress fields and converts to sp^3 hybridzation during the deposition of the adjacent layer B_i, thereby densifying the B_i/A_i interface layer. This phenomenon can also be verified by the analysis of XRR interface results, which showed a higher density at the B_i/A_i interface layer. Furthermore, in order to compensate for the lattice mismatch caused by structural transformation, the neighboring B_i and A_i layers will also produce coherency strain, which will have a hardening effect on the multilayered thin film. Simultaneously, this type of coherency strain will cause the crosslinking of the graphite planes in the interface zone, thereby increasing the elastic properties of the multilayered film. As there are several interface layers, the enhancement effect on the hardness and elastic properties is more significant in alternating multilayered ta-C films. In addition, Patsalas et al. performed observations using x-ray diffraction and high-resolution TEM, and found that after bombardment by high-energy (> 1000 eV) Ar^+ ions, the sp^3 positions in a-C films produced nanocrystalline diamond structures. During the deposition process of alternating multilayered ta-C films described in this book, the deposition ion energy for the sp^2-rich layer A has similar bombardment energy as Ar^+ ions; hence, thin-film hardness might have been enhanced owing to the formation of nanocrystalline structures.

5.9.2 Fracture properties

During the nanoindentation test, when the diamond indenter is pressed to a certain depth into the thin-film surface, a crack may occur if the shear stress introduced by the normal load exceeds the hardness of the thin film, and the film will fracture or be delaminated. By applying the continuous stiffness test, the load changes within the indentation displacement can be used to analyse the fracture properties of multilayered ta-C films.

Fig. 5.63 shows the load curves of monolayered and graded multilayered ta-C films with thicknesses of approximately 300 nm. The monolayered film was an sp^3-rich thin film deposited at a substrate bias of −80 V. Scanning electron microscopy (SEM) was used to observe the morphology of the surface nanoindentation, as shown in Fig. 5.64. As observed from the loading curve in Fig. 5.63, microscopic discontinuities (indicated by arrows as a and b) can be observed during load application, and changes occurred in the trend of variation after these discontinuities. This indicates that radial cracks had begun to form at the corresponding load, and the area enclosed by the loading curve and the indentation displacement represents the elastic deformation of the entire coating layer. The radial cracks observed in Fig. 5.63 (a) expanded in the angular direction of the triangular-pyramidal indenter. However, in Fig. 5.63 (b), the indentation on the graded multilayered film did not expand but was contained within the surface indentation. The discontinuities at *a* and *b* on the loading curves correspond to loads of 18.7 mN and 43.8 mN, respectively, and are known as the critical loads of crack formation. The relation between the critical load P_c and the critical compressive stress for coating–substrate interface fracture σ_c is given by

$$\sigma_{max} \propto (P_c E^2)^{1/3}. \tag{5.33}$$

Furthermore, at the point of contact on the loading axis, the critical shear force that will cause interface fractures $\tau_c = 0.31\sigma_{max}$. Therefore, graded multilayered ta-C films have a higher critical fracture stress than monolayered sp^3-rich thin films, thereby enhancing the stability of the coating layer. During the unloading process shown in Fig. 5.63(b), when the load was reduced to 13.4 mN, the indenter seemed to be pushed upwards suddenly, thereby forming an inflection point on the curve (indicated by an arrow at *c*). This phenomenon is caused by the plastic deformation during the loading compression process, and the compressive stress field formed within the contact deformation zone has a certain level of load-bearing capacity. However, during the unloading process, thin-film deformation is released, causing the formation of lateral cracks at the edge of the deformation zone, which expand to the film surface. The reaction force acting on the indenter will result in load displacement. Fig. 5.64 shows that lateral cracks appeared at the indentation boundary of the multilayered film surface, but the film layer showed no bulging or delamination. This indicates that the cracks only expanded within the thin film and did not affect the coating–substrate interface adhesion.

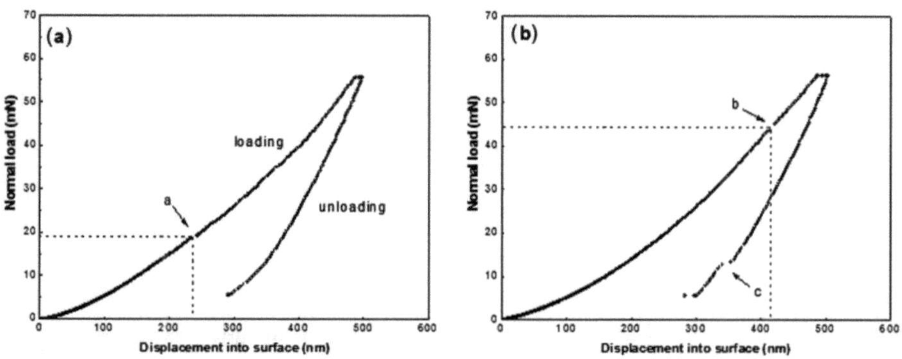

Fig. 5.63: Loading curves of ta-C thin films: (a) monolayered; (b) multilayered.

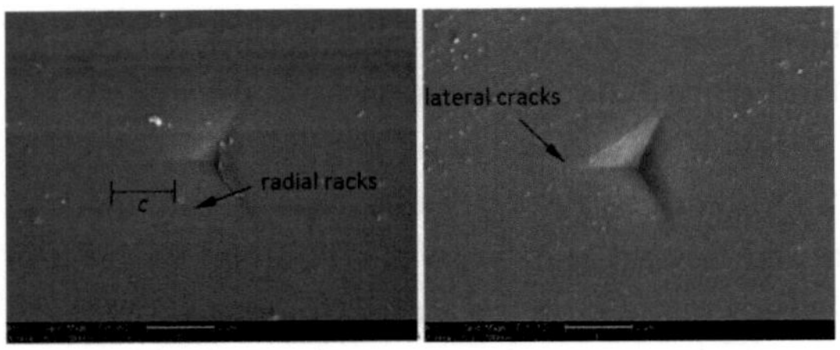

Fig. 5.64: SEM morphology of surface indentation on ta-C thin films: (a) monolayered; (b) graded multilayered.

In order to investigate the effects of loading conditions on the fracture properties of multilayered films, a different maximum indentation displacement was applied during the nanoindentation experiment on multilayered ta-C films, and the loading curves are shown in Fig. 5.65.

In Fig. 5.65, the multilayered ta-C film is formed by alternating layers A_i and B_i, the substrate biases of which are −1000 and −80 V, respectively. The total thickness is approximately 400 nm, and the thickness ratio of the sublayers $d_{Ai}/d_{Bi} \approx 1.0$ ($i = 1, 2, 3, 4$). With the increase in the maximum indentation displacement (500–1400 nm), the maximum indentation load increased (70–460 mN). In Fig. 5.65 (a),

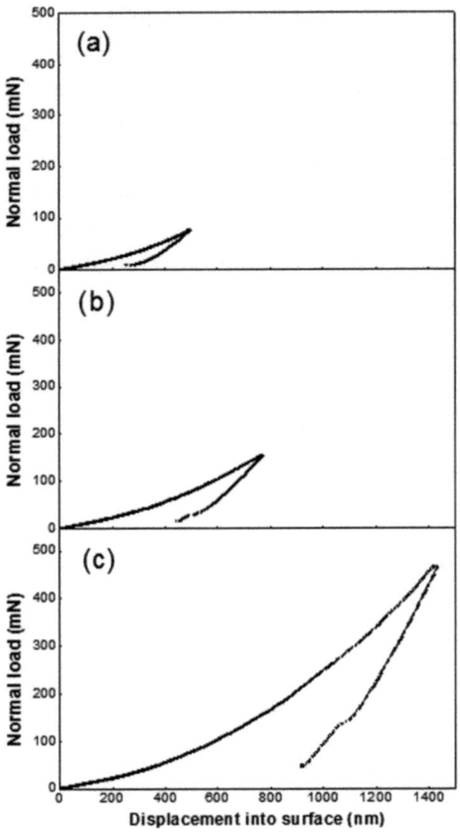

Fig. 5.65: Loading curves for different maximum indentation displacements: (a) indentation displacement of 500 nm; (b) indentation displacement of 800 nm; (c) indentation displacement of 1400 nm.

when the maximum indentation displacement was relatively low (500 nm), the load applied was small, load changes were continuous, the extent of recovery in the unloading curve was large, and the indentation was low (Fig. 5.66 (a)). In Fig. 5.65 (b), when the maximum indentation displacement was increased to 800 nm, an inflection point appeared in the unloading curve. This was more significant in Fig. 5.65 (c) when the maximum indentation displacement was 1400 nm. The indentation was deeper, radial cracks extended along the edges of the indentation, and lateral cracks emerged at the edges (Fig. 5.66 (b)). However, the substrate had a greater impact on load changes when deeper indentation displacement was applied. Therefore, measurements of shallow indentation displacement were used in this book when calculating thin-film fracture toughness.

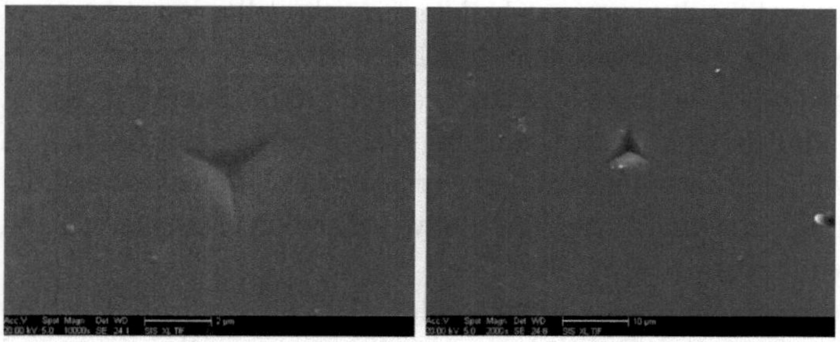

(a) (b)

Fig. 5.66: SEM morphology of indentations under different maximum indentation displacements.

Jungk et al. gave the following formula for calculating thin-film fracture toughness:

$$K_{IC} = \lambda \left(\frac{E_f}{H_f} \right)^{1/3} \frac{P_{max}}{t_f c^{1/2}}, \tag{5.34}$$

where E_f, H_f, and t_f are the elastic modulus, hardness, and thickness of the thin film, respectively, P_{max} is the maximum indentation load; c is the radial crack length, and λ is the indenter constant (for a Berkovich indenter, $\lambda = 0.016$).

Equation (5.34) can be used to calculate the fracture toughness of the thin film above, and the calculation results are listed in Tab. 5.4. It can be seen that the fracture toughness of multilayered ta-C films increased to different extents, indicating that the thin films had enhanced antifracture properties. Furthermore, the calculated values were all higher than the fracture toughness value of $(4.25 \pm 0.7)\ \mathrm{MPa \cdot m^{1/2}}$ for monolayered ta-C thin films (about 500 nm), as reported in the literature.

Tab. 5.4: Data on elastic modulus, hardness, and fracture toughness of monolayered and multilayered ta-C films.

Thin-film	t_f (nm)	E_f (GPa)	H_f (GPa)	K_{IC} (MPa·m$^{1/2}$)
sp^3-rich monolayered film	320	393.7	38.5	4.78 ± 0.3
Graded multilayered film	290	404.6	39.7	5.17 ± 0.4
Overlapping multilayered film	370	410.2	40.1	5.34 ± 0.6

5.9.3 Scratch resistance and adhesion performance

The adhesion performance of coating layers is one of the most important mechanical properties of multilayered films, and it is directly related to the practical applications of thin films. It is difficult to accurately determine the adhesion strength of the coating–substrate interface using the nanoscratch test. However, direct measurements can be performed on the changes in surface scratch profile, lateral force, and friction factor before, after, and during the scratching process through the use of high load accuracy. This will allow us to determine the critical load for cracking and delamination of the coating, providing a semiquantitative evaluation of the binding and adhesion properties of the coating.

Fig. 5.67 shows the curves of surface scratch profile and friction factor of monolayered and multilayered ta-C thin films within the scratch distance, as well as the typical scratch morphology under optical microscopy. The thin films tested included monolayered sp^3-rich, bilayered, graded multilayered, and alternating multilayered ta-C films. The thickness of the first three was approximately 300 nm, and that of the alternating multilayered film was approximately 1 μm. Within an overall scratch distance of 500 μm, the curves of the scratch surface profile were composed of three components: prescanning, scratching, and post-scanning. Pre- and postscanning were completed under an extremely small load (0.1 mN) in order to describe the surface morphology of the films before and after scratching. The normal scratch load increased linearly from 0 mN to 180 mN during the scratching process. When the profile curve is at zero, it indicates the thin-film surface; negative displacements in the Y direction indicate that the indenter has scratched into the thin film, while positive displacements indicate surface protrusions and the accumulation of chippings. The surface profile curves in Fig. 5.67 (a)–(c) are divided by two dotted lines into three zones, which correspond to the scratch photograph in Fig. 5.67 (e). At the initial stage of the first zone, only elastic deformation occurred in the coating layer. As the load increased, although the negative displacement of the scratching curve exceeded the thin-film thickness, the post-scanning curve only showed a small negative displacement, indicating that a high level of elastic recovery existed in the coating layer, which only led to an insignificant amount of plastic deformation. In Fig. 5.67 (e), the scratch at this point appeared relatively smooth and was shallow. As the scratch load increased, the scratch gradually became deeper, and the friction factor began to increase as well. When

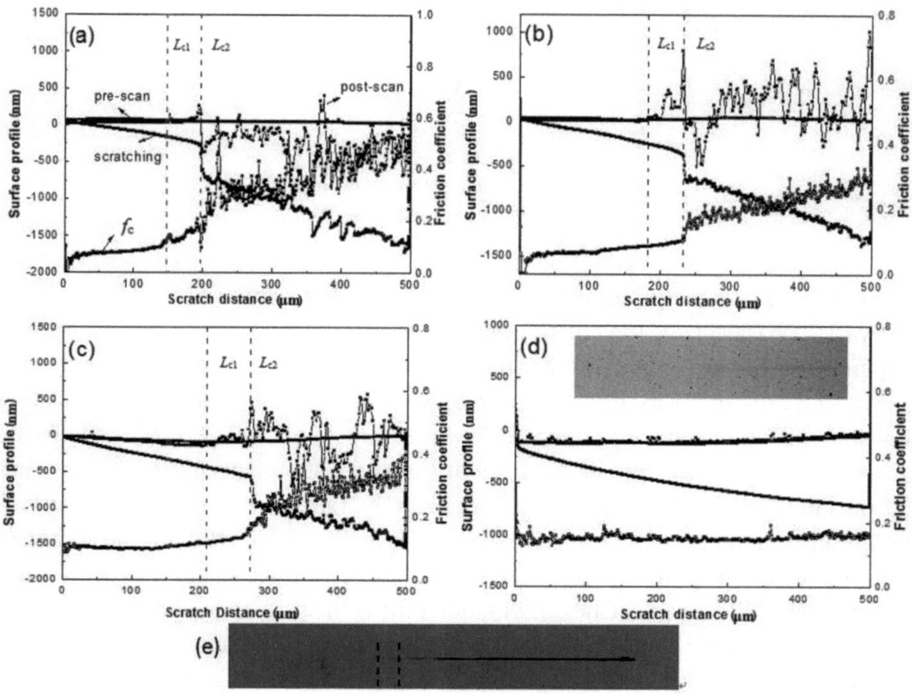

Fig. 5.67: Scratch surface profile and friction factor curves of different ta-C thin films, and typical scratch morphology under optical microscopy.

the indenter reached the interface of the second zone (the position of the first dotted line in the figure), the scratch stress exceeded the adhesion strength of the coating layer; hence, cracks and ridges were formed. Small fluctuations appeared in the post-scanning curve, and the rate of increase in the friction factor increased gradually.

The SEM photograph in Fig. 5.68 shows parallel cracks extending along the scratching direction and angular cracks at a certain angle to the scratching direction. The corresponding load at this point is known as the critical load L_{C1}, which characterises the critical point of crack formation. At the interface between the second and third zones (the position of the second dotted line in the figure), the friction factor, scratching, and post-scanning curves all showed a sudden step change, and the degree of fluctuation became very severe. In addition, the negative displacement of the scratching curve far exceeded the thin-film thickness. These phenomena indicate the complete fracture and delamination of the coating layer from the substrate, while the large fluctuations in the post-scanning curve are caused by the accumulation of coating chippings. In Fig. 5.68, coating chippings were distributed on both sides of the scratch. As the crack is formed because of the concentration and release of stress at the indenter tip, the cycle of stress concentration and release during the continuous process of scratch loading caused the coating chippings to exhibit a continuous undu-

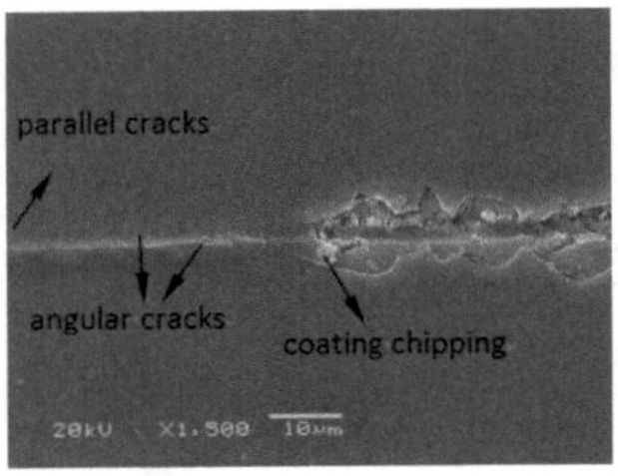

Fig. 5.68: SEM scratch morphology of graded multilayered ta-C film.

lating shape. The corresponding normal load at this point is known as the critical load L_{c2}, which characterises the critical point of coating delamination. Through energy-dispersive x-ray spectroscopy (EDX) analysis, it was found that between critical loads L_{c1} and L_{c2}, the intensity ratio of C and Si K_a peaks I_C/I_{Si} is 0.38 ± 0.01, indicating that the coating was not delaminated from the substrate. At L_{c2}, I_C/I_{Si} rapidly decreased to 0.02 ± 0.01, which is the level of the Si substrate surface, indicating that the thin film had been delaminated from the substrate. In Fig. 5.67 (d), the surface profile curve of the 1-μm-thick alternating multilayered ta-C film did not show sudden changes during the entire scratch loading process. This phenomenon indicates that the indenter only caused elastic deformation during the scratch loading process, and did not cause coating damage; the critical scratch load was greater than 180 mN. The optical micrograph of the scratch in the figure only showed a shallow scratch, with no indications of coating fracture or delamination. Nevertheless, the pre-scanning displacement showed a small magnitude of increase, which was due to the particles of impurities on the surface. In addition, a comparison of the friction factors for the various ta-C films before fracture shown in Fig. 5.67 revealed that the friction factors of multilayered thin films were slightly higher.

Aside from thin-film adhesion, the critical scratch load is also affected by various factors, including the experimental environment, indenter diameter, loading speed, film thickness, and other coating–substrate mechanical properties. Therefore, in terms of thermodynamics, the critical scratch load cannot be simply linked with thin film adhesion, but it can provide a semi-quantitative assessment of thin-film adhesion under the same conditions. Tab. 5.5 lists the measured values of critical scratch load for different ta-C films. Under the same thin-film thickness, multilayered films had higher critical loads, which amply demonstrates the improvement in thin-film scratch resistance and adhesion performance. Furthermore, it can be seen that in a bilayered ta-C film composed of an sp^2-rich bottom layer and an sp^3-rich layer, the increased energy of deposited C^+ ions resulted in enhanced thin-film adhesion, which in turn led to a corresponding increase in critical scratch load.

For thin-films adhered to rigid substrates, Attar and Johannesson gave the following expression for the critical load when the indenter scratched at a depth greater than twice the film thickness:

$$L_C = \frac{d_C}{\mu_C}\left(\frac{E_f}{1 - \nu_f^2} 2t_f W\right)^{1/2}, \tag{5.35}$$

Tab. 5.5: Measurement values of critical scratch load in different ta-C films.

Thin-film name	Thickness t_f (nm)	Hardness H (GPa)	L_{c1} (mN)	L_{c2} (mN)
sp^3-rich monolayered film	320	38.5	53.7 ± 2.1	71.0 ± 1.5
Bilayered film r	310	38.2	65.1 ± 1.8	83.7 ± 1.0
Graded multilayered film	290	39.7	76.7 ± 2.5	98.4 ± 1.2
Overlapping multilayered film	960	47.3		

where d_c is the scratch width at the critical scratch load, μ_c is the critical friction factor, E_f and t_f are the thin-film elastic modulus and thickness, v is the thin-film Poisson's ratio (for a ta-C film, $v = 0.25$), and W is the work of adhesion.

The work of adhesion W is equal to the release rate of interfacial fracture energy per unit area G_c, i.e. the adhesion energy per unit area, which can be given by

$$W = G_c = \frac{\sigma_c^2 t(1-v)}{2E_f}, \tag{5.36}$$

where σ_c is the critical compressive stress needed to cause coating–substrate interfacial failure. The relation between fracture toughness K_{IC} and fracture energy release rate G_c is given by

$$K_{IC} = \left(G_c \frac{E_f}{1-v_f^2}\right)^{1/2}. \tag{5.37}$$

Using equations (5.35)–(5.37), the work of adhesion, critical stress, and fracture toughness of the different ta-C films above can be calculated, and the values are listed in Tab. 5.6. As can be seen, the multilayered thin films had higher critical stress for interfacial failure and fracture toughness, which is consistent with the pattern of fracture properties observed in the nanoindentation test. However, during the scratch loading process, there is a relatively high shear stress in the scratching direction acting on the interface fracture in addition to the normal compressive stress. Thus, the calculated work of adhesion and critical stress values are higher than the actual values, and the fracture toughness obtained from the scratch test is slightly higher than that obtained from the nanoindentation test.

Tab. 5.6: Work of adhesion, critical stress, and fracture toughness of different ta-C thin films.

Thin-film name	Work of adhesion W or G_c (J/m²)	Critical stress σ_c (GPa)	Fracture toughness K_{IC} (MPa·m$^{1/2}$)
sp^3-rich monolayered film	61.9 ± 1.0	14.3 ± 1.0	5.1 ± 0.2
Bilayered film	64.2 ± 2.0	14.8 ± 2.0	5.2 ± 0.3
Graded multilayered film	73.9 ± 2.2	16.6 ± 2.1	5.6 ± 0.2

During the scratch loading process of multilayered films, the different mechanical properties of each sublayer enable the coating interfaces to store a proportion of the elastic deformation energy, which then reduces the energy at the coating–substrate interface and diminishes the interfacial stress introduced by the same scratch load. Moreover, the firm bonding of the sp^2-rich bottom layer with the substrate will also enhance the adhesion of the coating layer. In addition, the multilayered structure is also able to overcome the weakening of bond strength caused by impurities and macro-particles in the thin film.

5.9.4 Effect of sublayer thickness on the mechanical properties of multilayered films

The ability of alternating multilayered ta-C films to maintain excellent mechanical properties is primarily due to the preservation of the compressive stress and densified structure in the sp^3-rich film layer. However, if the sp^2-rich layer completely introduces tensile stress to the sp^3-rich layer through lattice relaxation, the multilayered film will become a monostructural sp^2-rich film layer, and the mechanical properties of the thin film will deteriorate. Due to the disparities between the mechanical properties of sp^2-rich and sp^3-rich films, the relative thickness ratio of the sublayers and the periodic thickness are crucial design parameters for multilayered ta-C films. The mechanical properties of multilayered films can also be controlled by optimising these two parameters. This section will discuss the effects of sublayer thickness ratio and periodic thickness on the average compressive stress, hardness, elastic modulus, scratch resistance, and adhesion of multilayered ta-C films.

5.9.4.1 Thickness ratio of sp^2-rich/sp^3-rich sublayers

The use of an sp^2-rich film as the buffering layer can reduce the average compressive stress of multilayered films. The stress values of each sublayer after deposition showed a jagged fluctuation, and low-stress hard ta-C films with a thickness of approximately 1 µm were finally obtained. With the same film thickness (approximately 1 µm) and periodic thickness (approximately 300 nm), the effects of the thickness ratio between the sp^2-rich (A_i) and sp^3-rich (B_i) sublayers (d_{Ai}/d_{Bi} = 0.2–2.0) on the average compressive stress can be observed in Fig. 5.69. The figure shows that, as the thickness ratio of the sp^2-rich layer increased, the average compressive stress of the multilayered film decreased gradually. When d_{Ai}/d_{Bi} = 2.0, the average compressive stress of the multilayered film decreased to 2.66 ± 0.01 GPa, which is a 27 % reduction compared to the average compressive stress of the 300-nm-thick monolayered sp^3-rich film. Therefore, in the design of multilayered films, thicker sp^2-rich layers can enhance the relaxation of internal film stress.

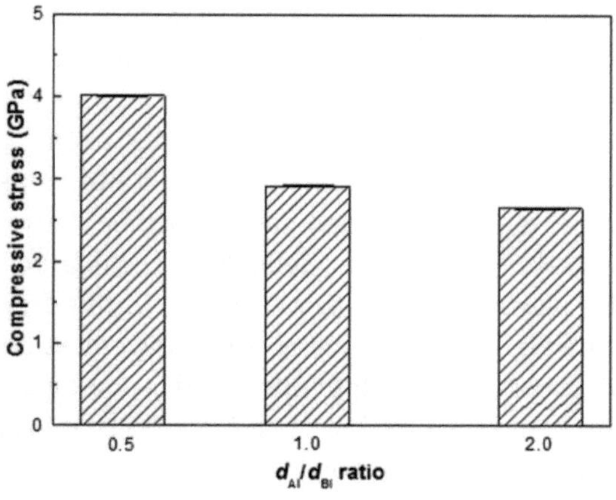

Fig. 5.69: Effect of sublayer thickness ratio d_{Ai}/d_{Bi} on the average compressive stress of multilayered ta-C films.

Owing to the coherency stress of the densified interface layer between the sublayers, the hardness and elastic modulus of multilayered films are enhanced. With the same interfacial structure and number of layers, the nanoindentation hardness and elastic modulus of multilayered ta-C films with different d_{Ai}/d_{Bi} are shown in Fig. 5.70. In the figure, when d_{Ai}/d_{Bi} was small, i.e. when the "hard" sp^3-rich layers were thicker, the hardness and elastic modulus of the multilayered films were higher at 51 GPa and 506 GPa, respectively. As d_{Ai}/d_{Bi} increased, when the "soft" sp^2-rich layers were thicker, the hardness and elastic modulus of the multilayered films decreased, but the hardness still remained at a high level of 45 GPa.

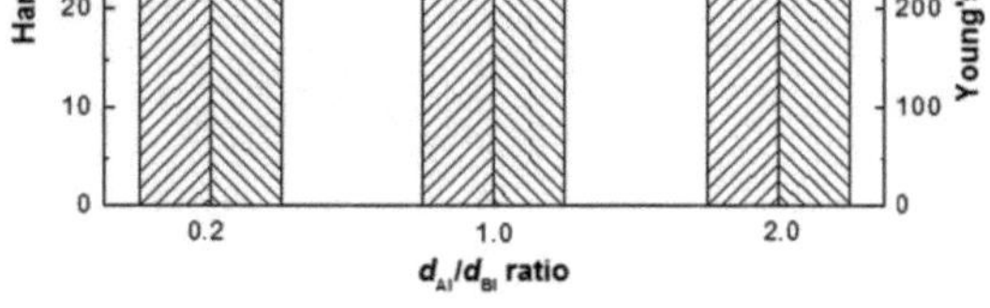

Fig. 5.70: Hardness and elastic modulus of multilayered ta-C films with different sublayer thickness ratios d_{Ai}/d_{Bi}.

The experimental results above indicate that a high thickness ratio of sp^2-rich layers can greatly reduce the compressive stress of multilayered ta-C films while retaining a high level of hardness. Therefore, during the fabrication process of ta-C films, is it more beneficial for the mechanical properties of the thin film to maximise the thickness of sp^2-rich layers? In response to this question, the effects of different d_{Ai}/d_{Bi} thickness ratios on the scratch resistance and adhesion of multilayered films were investigated. Fig. 5.71 shows the surface profile and friction factor curves of each multilayered film from the nanoscratch test, as well as the optical micrograph of the scratch.

In the surface profile curves in the figure, the overlap between the post-scanning and pre-scanning curves indicate that only elastic deformation occurred during the scratching process, and complete elastic recovery occurred after unloading. Microscopic negative displacements were observed in the post-scanning curve, which indicate that a small amount of irreversible plastic deformation occurred in the thin film. Subsequently, large fluctuations occurred in the post-scanning curve, and cracks occurred in the thin film; the corresponding normal load is the critical load L_{c1}. As for the scratching curve, the gradual increase in negative displacement indicates that load caused the indenter to scratch into the thin film. The point at which negative displacement increased suddenly indicates the occurrence of delamination of the thin film from the substrate, and the corresponding normal load is the critical load L_{c2}.

In Fig. 5.71 (a), the critical loads L_{c1} and L_{c2} of the multilayered film coincided ($L_{c1} = L_{c2} = 160.4 \pm 0.1$ mN) when $d_{Ai}/d_{Bi} = 0.2$. This is due to the higher ratio of hard sp^3-rich layers, which enhanced the scratch-resistance capacity of the thin film.

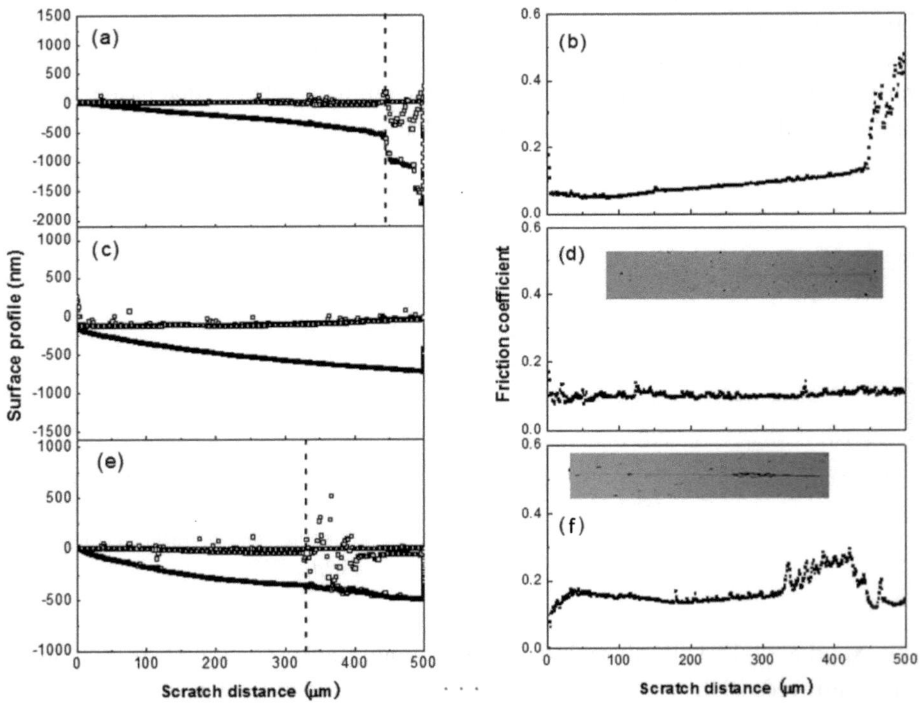

Fig. 5.71: Scratch profile and friction factor curves of multilayered ta-C films with different sublayer thickness ratios d_{Ai}/d_{Bi}.

Hence, the stress introduced by loads prior to L_{c1} were insufficient to cause the expansion of cracks in the thin film. However, as the load increased to L_{c2}, the stress concentration at the indenter tip exceeded the load-bearing capacity of the thin film and the adhesion strength of the substrate-coating interface. This caused the coating layer to fracture and led to the sudden increase in friction factor, as shown in Fig. 5.71 (b). When d_{Ai}/d_{Bi} increased to 1.0, sudden changes were not observed in the profile curves and friction factor in Fig. 5.71 (c) and (d). Furthermore, the optical micrograph in Fig. 5.71 (d) showed that the scratch on the film surface was extremely shallow. This indicates that cracks and delamination did not occur in the thin film during the scratching process, and the microscopic fluctuations were caused by defects and impurity particles on the thin-film surface. When d_{Ai}/d_{Bi} increased to 2.0, Fig. 5.71 (e) shows that there were only large fluctuations caused by thin-film cracks in the post-scanning curve, and the corresponding critical load $L_{c1} = 119.8 \pm 0.2$ mN. However, the sudden negative displacement in the scratching curve caused by the critical load L_{c2}, which will lead to thin-film delamination, was not observed. The primary reason for this phenomenon is the large ratio of "soft" sp^2-rich layers in the multilayered film, which decreased the load-bearing capacity and scratch resistance of the thin films, and increased the susceptibility to cracking. The fluctuations of the friction factor in Fig. 5.71 (f) are mainly due to coating chippings. Nevertheless, owing to the increased thickness of the sp^2-rich layer bound to the substrate, the thin film had greater adhesion. Hence, the scratch load did not reach the critical conditions for the coating–substrate interfacial adhesion, and the coating layer was not delaminated from the substrate. In addition, as the ratio of sp^2-rich layers increased, the average compressive stress of the multilayered film decreased. Hence, the stress value introduced by the small scratch load was lower and thin-film adhesion was enhanced.

5.9.4.2 Periodic film thickness

In the design of alternating multilayered ta-C films, the thickness of each bilayer A_iB_i is known as the periodic thickness of the multilayered film. Variations in the periodic thickness also imply changes in the number of sublayers or interface layers. According to the analyses above on multilayered ta-C films, the interfacial structure of sublayers is extremely important to their mechanical properties. Alternating multilayered ta-C films and bilayered films with periodic thicknesses of 80 nm and 160 nm, respectively, were fabricated by alternating between substrate biases of −2000 V (A_i) and −80 V (B_i). The sublayer thickness ratio d_{Ai}/d_{Bi} was 1.0, and the total film thickness was approximately 320 nm. The hardness and elastic modulus of the different layers measured using the nanoindentation test are shown in Fig. 5.72. Continuous stiffness measurement was performed during the nanoindentation test. The indentation depth was 1400 nm, and the peak load was 460 mN. Fig. 5.73 shows the SEM indentation morphology of the different films layers.

When the periodic thicknesses were 80 nm and 160 nm, the hardness value of the multilayered film increased slightly (approximately 36 GPa). The bilayered ta-C thin film maintained a high level of hardness (approximately 35 GPa) owing to the thick sp^3-rich layer.

During the deposition of the sp^2-rich layer A_i, the increase in the C^+ ion energy will cause thermal peak effects, which will increase the sp^2 content and graphitization of the film layer while reducing the hardness of the entire multilayered film. The relation between multilayered ta-C films and periodic thickness can be simulated using the Koehler model as follows:

$$H_f = H_A + \frac{k}{\Lambda}, \tag{5.38}$$

where H_A is the hardness of sp^2-rich layers, k is a coefficient related to the thickness of sp^3-rich layers and amount of sp^3 hybridization, and Λ is the periodic thickness.

The relation above is not clearly shown in the measurement results of hardness, which might have been due to the large periodic thickness of the design. Moreover, the hardness of the sp^3-rich layers might have weakened the hardness enhancements from this effect. Logothetidis selected a periodic thickness of 10–50 nm, and the resulting hardness of sp^3-rich layers was only 24 GPa.

In Fig. 5.72, the increase in thin-film elastic modulus with decreasing periodic thickness is significant. Specifically, the elastic modulus was 390 GPa when the periodic thickness was 160 nm, which was an increase of almost 40 GPa compared to the elastic modulus of the bilayered film. In the studies by Koehler and Pickett on nanomultilayered Cu/Ni films, the supermodulus effect was observed. They believe that it was because the multilayered structure segmented the overall continuity, thereby increasing the free energy in the system. In terms of noncoherent interfaces, the increase in free energy is directly proportional to the interface area per unit volume, and is inversely proportional to sublayer thickness. According to the coherent composite model, two film layers with similar lattice constants and thermal diffusivities but disparate elastic moduli will produce coherency strain, which will cause the elastic modulus to increase. Moreover, this supermodulus effect is more significant in thin films with smaller periodic thickness. Cammarata used the coherency strain model of nanomultilayered thin films and found that there exists a certain critical periodic thickness Λ_c below which the formation of coherent interfaces is facilitated, causing the stress and compressive strain of noncoherent lattices to disappear. This will result in decreased hardness and elastic modulus. The coherent interface strain can also be used to explain the variations in the hardness and elastic modulus of ta-C films with changes in periodic thickness.

A comparison of nanoindentation morphology formed under similar experimental conditions (Fig. 5.73) revealed that, when the periodic thickness of the multilayered film was 160 nm, the size of indentation in Fig. 5.73 (b) was relatively small, and the elastic recovery of the thin film was large. When the periodic thickness was reduced to 80 nm, Fig. 5.73 (a) shows that the distance of expansion in radial cracks increased

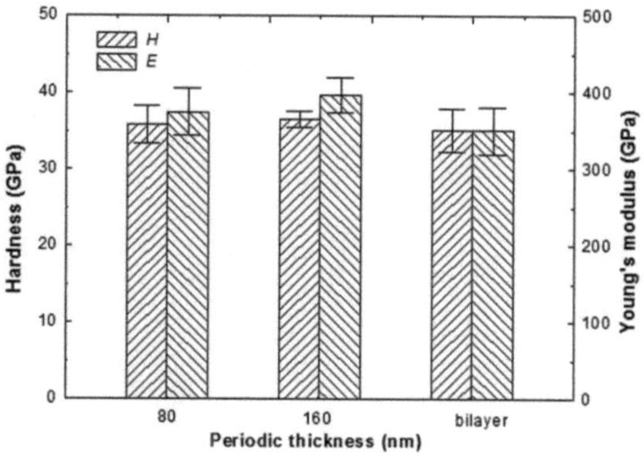

Fig. 5.72: Hardness and elastic modulus of multilayered ta-C films with different periodic thicknesses.

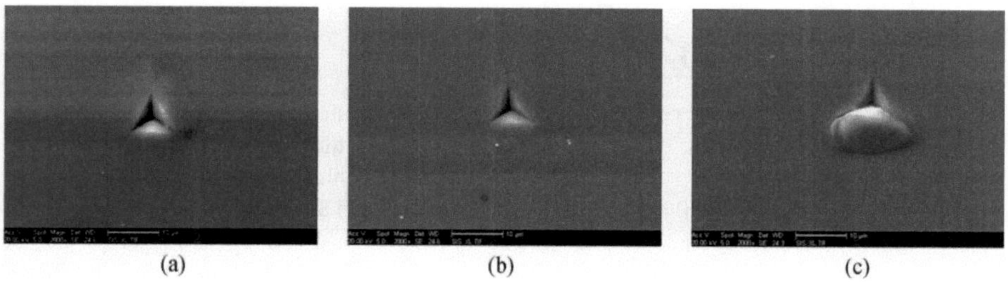

Fig. 5.73: SEM morphology of multilayered ta-C films with different periodic thicknesses: (a) 80 nm; (b) 160 nm; (c) bilayered film.

and a protrusion occurred at the indentation edge. This phenomenon was particularly obvious in the bilayered film (Fig. 5.73 (c)), which indicates the occurrence of fracture at the coating–substrate interface and the delamination of the thin-film from the substrate.

Fig. 5.74 and 5.75 show the surface profile curves and SEM scratch morphology, respectively, of multilayered ta-C films with different periodic thicknesses.

For the bilayered film, parallel cracks and angular cracks formed along the boundary of the scratch at the critical load L_{c1} (114.2 ± 0.5 mN), as shown in Fig. 5.75 (a). As the scratch load increased, the angular cracks caused the coating to fracture, and the post-scanning curve showed large fluctuations. Fig. 5.75 (b) shows that angular cracks were formed in the scratch, which had a continuous distribution and were oriented at 120° to the scratching direction. The density of cracks had increased, and a part of the coating was curled and delaminated. The formation of these cracks is due to the tensile force formed at the back of the indenter, and the triangular-pyramidal shape of the indenter determined the orientation of the angular cracks. When the load increased to the critical load L_{c2} (173 ± 0.3 mN), the scratching stress exceeded the adhesion strength of the coating–substrate interface, and a semi-circular crack was formed on the scratch (Fig. 5.75 (c)). Subsequently, the coating was delaminated

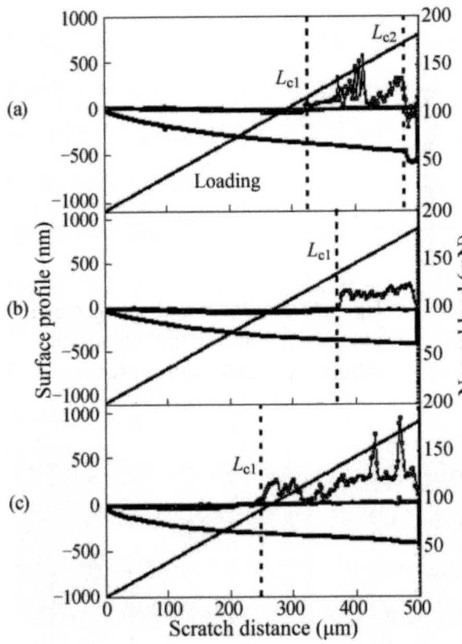

Fig. 5.74: Scratch surface profiles and loading curves of multilayered ta-C films under different periodic thicknesses: (a) bilayered film; (b) 160 nm; (c) 80 nm.

and formed chippings, and the surface of the Si substrate was exposed. A sudden change occurred in the negative displacement of the scratching curve. However, in the scratch profiles of multilayered ta-C films with small periodic thicknesses of 160 nm and 80 nm, only the critical loads L_{c1} leading to crack formation were observed, which were 134.8 ± 0.3 mN and 84.8 ± 0.2 mN, respectively. In contrast, the critical loads leading to thin-film delamination, L_{c2}, could not be observed. A possible reason is the enhanced thin-film adhesion strength due to the multilayered design. Furthermore, the deposition of sp^2-rich layers involved high C$^+$ ion energy, which enhanced the coating–substrate adhesion strength. This phenomenon was also observed in the relatively high critical load L_{c2} of the bilayered film.

In comparison, as can be seen in Fig. 5.75 (d)–(f), the scratch was shallow in the multilayered film with a periodic thickness of 160 nm, the size of cracks was small, and the thin film did not undergo delamination. Thus, the thin film had excellent scratch resistance and adhesion properties. When the periodic thickness was reduced to 80 nm, as shown in Fig. 5.75 (g)–(i), the size and density of cracks increased, and the cracks gradually converged. Thus, the scratch resistance of the thin film was poorer. Therefore, periodic thickness has a significant impact on the scratch resistance of multilayered ta-C films.

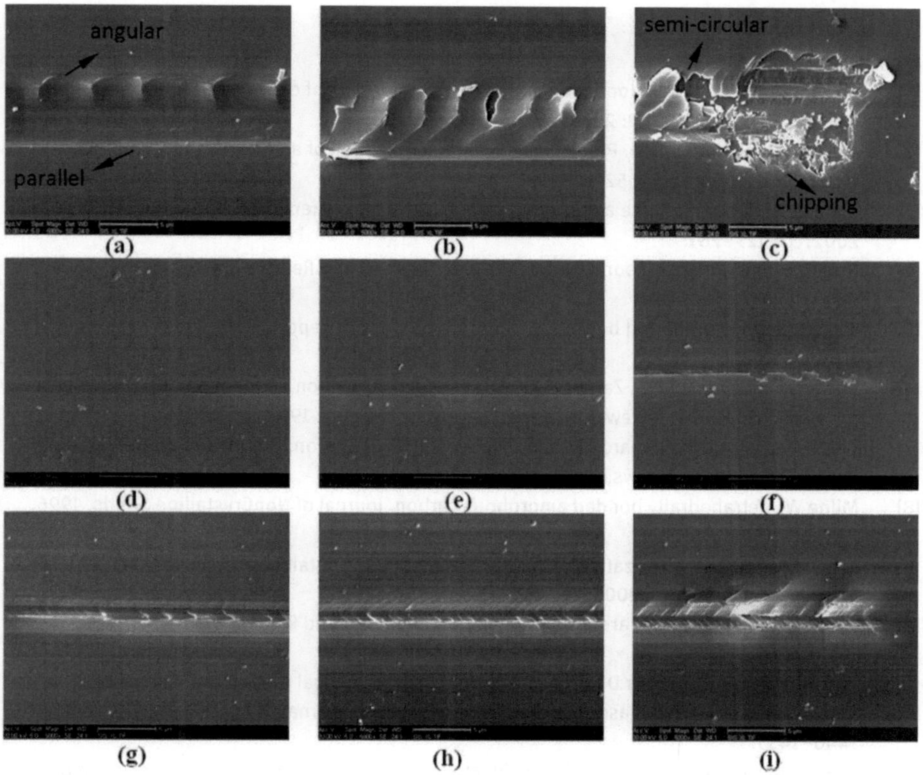

Fig. 5.75: SEM observations of scratches on multilayered ta-C films with the same periodic thickness (a)–(c) bilayered film; (d)–(f) 160 nm; (g)–(i) 80 nm.

In conclusion, the reduction of periodic thickness can increase the hardness and elastic modulus of multilayered ta-C films. However, the nanoscratch test revealed that the formation of micro-cracks originates from the application of external load. This will first lead to the yield deformation of the substrate surface, gradually followed by extension toward the thin-film interface and inside the substrate. When yield deformation reaches the film interface, deflection will occur and cracks will be produced. If the number of film interfaces is too high, the sp^3-rich layers will be too thin, which will reduce the load-bearing capacity of the thin film and increase the probability of crack formation and expansion. This will cause the decrease in the scratch resistance of the thin film. The results have shown that alternating multilayered ta-C films with a periodic thickness of 160 nm had the best scratch resistance. Research by Zhang et al. also revealed that multilayered DLC thin films with a periodic thickness of 200 nm had the best wear resistance. However, as the periodic thickness decreased, the rate of wear increased and the protective properties of the thin film were diminished. These results are very similar to the research results obtained in this book.

References

[1] Aissenberg S, Chabot R. Ion-beam deposition of thin-films of diamond-like carbon. Journal of Applied Physics, 1971, 42: 2953–2958.
[2] Aissenberg S, Kimock FM. Properties and characterization of amorphous carbon films. Materials Science Forum, 1990, 52-52: 1–40.
[3] Robertson J. Diamond-like amorphous carbon. Materials Science & Engineering: R: Reports, 2002, 37: 129–281.
[4] Grill A. Diamond-like carbon: State of the art. Diamond and Related Materials, 1999, 8: 428–434.
[5] McKenzie DR. Tetrahedral bonding in amorphous carbon. Reports in Progress Physics, 1996, 59: 1611–1664.
[6] Voevodin AA, Donley MS, Zabinski JS. Pulsed laser deposition of diamond-like carbon wear protective coating: A review. Surface & Coating Technology, 1997, 92: 42–49.
[7] Wei Q, Narayan J. Superhard diamondlike carbon: Preparation, theory and properties. International Materials Reviews, 2000, 45: 133–154.
[8] Milne WI. Tetrahedrally bonded amorphous carbon. Journal of NonCrystalline Solids, 1996, 198–200: 605–610.
[9] Chu PK, Li LH. Characterization of amorphous and nanocrystalline carbon films. Materials Chemistry and Physics, 2006, 96: 253–277.
[10] Lifshitz Y. Diamond-like carbon-present status. Diamond and Related Materials, 1999, 8: 1659–1676.
[11] Davanloo F, Lee TJ, Jander DR, et al. Adhesion and mechanical properties of amorphic diamond films prepared by laser plasma discharge source. Journal of Applied Physics, 1992, 71: 1446–1453.
[12] Veerasamy VS, Amaratunga AJ, Milne WI, et al. Optical and electronic properties of amorphous diamond. Diamond and Related Materials, 1993, 2: 782–787.
[13] A Division of Morgan Advanced Ceramics. http://www.diamonex.com/abouttech_dlc.htm.
[14] Xu NS, Chen J, Deng SZ. Physics origin of nonlinearity in the Fowler-Nordheim plot of field-induced emission from amorphous diamond films: Thermionic emission to field emission. Applied Physics Letters, 2000, 76: 2463–2465.
[15] Coll BF, Chhowalla M. Amorphous diamond film by enhanced arc deposition. Surface & Coating Technology, 1996, 79: 76–86.
[16] Xiong FL, Wang YY, Chang RPH. Complex dielectric function of amorphous diamond films deposited by pulsed-excimer-laser ablation of graphite. Physical Review B, 1993, 48: 8016–8023.
[17] Hakovirta M, Salo J, Lappalainen R, et al. Correlation of carbon ion energy with sp^2/sp^3 ratio in amorphous diamond films produced with a mass-separated ion beam. Physics Letters A, 1995, 205: 287–289.
[18] Wei A X, Chen DH, Peng SQ, et al. Optical and electrical characteristics of amorphous diamond films. Diamond and Related Materials, 1997, 6: 983–986.
[19] Jacob W, Möller W. On the structure of thin hydrocarbon films. Applied Physics Letters, 1993, 63: 1771–1773.
[20] Shi X, Tay BK, Tan HS, et al. Properties of carbon ion deposited tetrahedral amorphous carbon films as a function of ion energy. Physics Letters A, 1996, 79(9): 7234–7240.
[21] Chhowalla M, Robertson J, Chen CW, et al. Influence of ion energy and substrate temperature on the optical and electronic properties of tetrahedral amorphous carbon (ta-C) films. Journal of Applied Physics, 1997, 81: 139–145.

[22] Lossy R, Pappas DL, Roy RA, et al. Properties of amorphous diamond films prepared by a filtered cathodic arc. Journal of Applied Physics, 1995, 77: 4750–4756.

[23] Martin PJ, Filipczuk SW, Netterfield RP, et al. Structure and hardness of diamond-like carbon films prepared by arc evaporation. Journal of Materials Science Letters, 1988, 7: 410–412.

[24] Malzbender J, den Toonder JMJ, Balkenende AR, et al. Measuring mechanical properties of coatings: A methodology applied to nano-particle-filled sol-gel coatings on glass. Materials Science & Engineering: R: Reports, 2002, 36: 47–103.

[25] Zhang YB, Lau SP, Sheeja D, et al. Study of mechanical properties and stress of tetrahedral amorphous carbon films prepared by pulsed biasing. Surface & Coating Technology, 2005, 195: 338–343.

[26] Weiler M, Sattel S, Giessen T, et al. Preparation and properties of highly tetrahedral hydrogenated amorphous carbon. Physics Review B, 1996, 53(3): 1594–1608.

[27] Shi X, Flynn D, Tay BK, et al. Mechanical properties and Raman spectra of tetrahedral amorphous carbon films with high sp^3 fraction deposited using a filtered cathodic arc. Philosophical Magazine, 1997, 76(3): 351–361.

[28] Schneider D, Tucker MD. Nondestructive characterization and evaluation of thin-films by laser-induced ultrasonic surface waves. Thin Solid Films, 1996, 290–291: 305–311.

[29] Teo KBK, Rodil SE, Tsai JTH, et al. Effect of graphitic inclusions on the optical gap of tetrahedral amorphous carbon films. Journal of Applied Physics, 2001, 89(7): 3706–3710.

[30] Dasgupta D, Demichelis F, Pirri CF, et al. π bands and gap states from optical absorption and electron-spin-resonance studies on amorphous carbon and amorphous hydrogenated carbon films. Physical Review B, 1991, 43: 2131–2135.

[31] Ferrari AC, Robertson J. Interpretation of Raman spectra of disordered and amorphous carbon. Physical Review B, 2000, 61: 14095–14107.

[32] Shroder RE, Nemanich RJ, Glass JT. Analysis of the composite structures in diamond thin films by Raman spectroscopy. Physical Review B, 1990, 41: 3738–3745.

[33] Merkulov VI, Lannin JS, Munro C H, et al. UV studies of tetrahedral bonding in diamond-like amorphous carbon. Physical Review Letters, 1997, 78: 4869–4872.

[34] Gilkes KWR, Prawer S, Nugent KW, et al. Direct quantitative detection of the sp^3 bonding in diamond-like carbon films using ultraviolet and visible Raman spectroscopy. Journal of Applied Physics, 2000, 87: 7283–7289.

[35] Tamor MA, Vassell WC. Raman “finger printing” of amorphous carbon films. Journal of Applied Physics, 1994, 76: 3823–3830.

[36] Schwan J, Ulrich S, Batori V, et al. Raman spectroscopy on amorphous carbon films, 1996, 80: 440–447.

[37] Tay BK, Shi X, Tan HS, et al. Raman studies of tetrahedral amorphous carbon films deposited by filtered cathodic vacuum arc. Surface & Coating Technology, 1998, 105: 155–158.

[38] Hu A, Alkhesho I, Zhou H, et al. Optical and microstructural properties of diamond-like carbon films grown by pulsed laser deposition. Diamond and Related Materials, 2007, 16: 149–154.

[39] Prawer S, Nugent KW, Lifshitz Y, et al. Systematic variation of the Raman spectra of DLC films as a function of sp^2:sp^3 composition. Diamond and Related Materials, 1996, 5: 433–438.

[40] Yu Khriachtchev L, Lappalainen R, Hakovirta M, et al. Raman diagnostics of amorphous diamond-like carbon films produced with a mass-separated ion beam. Diamond and Related Materials, 1997, 6: 694–699.

[41] Stanishevsky AV, Khriachtchev LY, Lappalainen R, et al. On correlation between the shape of Raman spectra and short-range order structure of hydrogen-free amorphous carbon films. Diamond and Related Materials, 1997, 6: 1026–1030.

[42] Profeta M, Mauri F. Theory of resonant scattering of tetrahedral amorphous carbon. Physical Review B, 2001, 63: 245415.

[43] Chen ZY, Zhao JP, Yano T, et al. Observation of sp^3 bonding in tetrahedral amorphous carbon using visible Raman spectroscopy. Journal of Applied Physics, 2000, 88: 2305–2308.
[44] Masek K, Fabik S, Maskova A, et al. Photoelectron spectroscopy characterization of diamond-like carbon films. Applied Spectroscopy, 2006, 60(8): 936–940.
[45] Calliari L. AES and core level photoemission in the study of a-C and a-C:H. Diamond and Related Materials, 2005, 14(8): 1232–1240.
[46] Reinke P, Garnier MF, Oelhafen P. In situ photoelectron spectroscopy analysis of tetrahedral amorphous carbon films. Journal of Electron Spectroscopy and Related Phenomena, 2004, 136(3): 239–245.
[47] Diaz J, Paolicelli G, Ferrer S, Comin F. Separation of the sp^3 and sp^2 components in the C1s photoemission spectra of amorphous carbon films. Physical Review B, 1996, 54: 8064–8069.
[48] Brühwiler PA, Maxwell AJ, Puglia C, et al. π^* and σ^* excitons in C1s absorption of graphite. Physical Review Letters, 1995, 74: 614–617.
[49] Morar JF, Himpsel FJ, Hollinger G, et al. C1s excitation studies of diamond (111). I. Surface core levels. Physical Review B, 1986, 33: 1340–1345.
[50] Bubert H, Jenett H. Surface and Thin Film Analysis: Principles, Instrumentation, Applications. Wiley-VCH VerlaggmbH, 2002: 6–30.
[51] Ilie A, Hart A, Flewitt AJ, et al. Effect of work function and surface microstructure on field emission of tetrahedral amorphous carbon. Journal of Applied Physics, 2000, 88: 6002–6010.
[52] Cooper NC, Fagan MS, Goringe CM, et al. Surface structure and sputtering in amorphous carbon thin films: A tight-binding study of film deposition. Journal of Physics-Condensed Matter, 2002, 14: 723–730.
[53] Zhao JP, Chen ZY. Sandwich atomic structure in tetrahedral amorphous carbon: Evidence of subplantation model for film growth from hyperthermal species. Physical Review B, 2001, 63: 115318.
[54] Kelires PC. Surface properties of diamond-like amorphous carbon. Journal of NonCrystalline Solids, 1998, 227–230: 597–601.
[55] Titantah JT, Lamoen D. sp^3/sp^2 characterization of carbon materials from first-principles calculations: x-ray photoelectron versus high energy electron energy-loss spectroscopy techniques. Carbon, 2005, 43: 1311–1316.
[56] Turgeon S, Paynter RW. On the determination of carbon sp^2/sp^3 ratios in polystyrene-polyethylene copolymers by photoelectron spectroscopy. Thin Solid Films, 2001, 394: 44–48.
[57] Bagus PS, Illas F, Pacchioni G, et al. Mechanisms responsible for chemical shifts of core-level binding energies and their relationship to chemical bonding. Journal of Electron Spectroscopy and Related Phenomena, 1999, 100: 215–236.
[58] Haerle R, Riedo E, Pasquarello A, et al. sp^2/sp^3 hybridization ratio in amorphous carbon from C1s core-level shifts: x-ray photoelectron spectroscopy and first-principles calculation. Physical Review B, 2002, 65: 045101.
[59] Frauenheim T, Jungnickel G, Stephan U, et al. Atomic-scale structure and electronic properties of highly tetrahedral hydrogenated amorphous carbon. Physical Review B, 1994, 50: 7940–7945.
[60] Marks NA, McKenzie DR, Pailthorpe BA, et al. Microscopic structure of tetrahedral amorphous carbon. Physical Review Letters, 1996, 76: 768–771.
[61] Kulik J, Lifshitz Y, Lempert GD, et al. Electron-energy-loss spectroscopy of mass-selected ion-beam-deposited diamondlike carbon. Journal of Applied Physics, 1994, 76: 5063–5069.
[62] Papworth AJ, Kiely CJ, Burden AP, et al. Electron-energy-loss spectroscopy characterization of the sp^2 bonding fraction within carbon thin films. Physical Review B, 2000, 62: 12628–12631.

[63] Ferrari AC, Libassi A, Tanner BK, et al. Density, sp^3 fraction, and cross-sectional structure of amorphous carbon films determined by x-ray reflectivity and electron-loss spectroscopy. Physical Review B, 2000, 62: 11089–11103.

[64] Wan D, Komvopoulos K. Probabilistic analysis of tetrahedral carbon hybridization in amorphous carbon films. Applied Physics Letters, 2006, 88: 221908.

[65] McKenzie DR, Muller D, Pailthorpe BA. Compress-stress-induced formation of thin-film tetrahedral amorphous carbon. Physical Review Letters, 1991, 67(6): 773–776.

[66] McKenzie DR. Generation and applications of compressive stress induced by low energy ion beam bombardment. Journal of Vacuum Science & Technology B, 1993, 11: 1928–1935.

[67] Ferrari AC, Kleinsorge B, Morrison NA, et al. Stress reduction and bond stability during thermal annealing of tetrahedral amorphous carbon. Journal of Applied Physics, 1999, 85: 7191–7197.

[68] Alam TM, Friedmann TA, Schultz PA, et al. Low temperature annealing in tetrahedral amorphous carbon thin films observed by ^{13}CNMR spectroscopy. Physical Review B, 2003, 67: 254309.

[69] Kelires PC. Intrinsic stress and local rigidity in tetrahedral amorphous carbon. Physical Review B, 2000, 62: 15686-15694.

[70] Lifshitz Y, Kasi SR, Rabalais JW, et al. Subplantation model for film growth from hyperthermal species. Physical Review B, 1990, 41: 10468–10480.

[71] Davis CA. A simple model for the formation of compressive stress in thin films by ion bombardment. Thin Solid Films, 1993, 226: 30–34.

[72] Robertson J. The deposition mechanism of diamond-like a-C and a-C:H. Diamond and Related Materials, 1994, 3: 361–368.

[73] Marks NA. Evidence for subpicosecond thermal spike in the formation of tetrahedral amorphous carbon. Physical Review B, 1997, 56: 2441–2446.

[74] Uhlmann S, Frauenheim T, Lifshitz Y. Molecular-dynamics study of the fundamental processes involved in subplantation of diamond-like carbon. Physical Review Letters, 1998, 81: 641–644.

[75] Ronning C. Ion-beam synthesis and growth mechanism of diamond-like materials. Applied Physics A, 2003, 77: 39–53.

[76] Fallon PJ, Veerasamy VS, Davis CA, et al. Properties of filtered-ion-beam-deposited diamond-like carbon as a function of ion energy. Physical Review B, 1993, 48(7): 4777–4782.

[77] Davis CA. A simple model for the formation of compressive stress in thin films by bombardment. Thin Solid Films, 1993, 226: 30–34.

[78] Marks NA. Evidence for subpicosecond thermal spike in the formation of tetrahedral amorphous carbon. Physical Review B, 1997, 56: 2441–2446.

[79] Zhang Q, Yoon SF, Ahn RJ, et al. Study of hydrogenated diamond-like carbon films using x-ray reflectivity. Journal of Applied Physics, 1999, 86: 289–296.

[80] Logothetidis S, Stergioudis G. Studies of density and surface roughness of ultrathin amorphous carbon films with regards to thickness with x-ray reflectometry and spectroscopic ellipsometry. Applied Physics Letters, 1997, 71: 2463–2465.

[81] Attwood DT. Soft x-ray and Extreme Ultraviolet Radiation: Principles and Applications, Translated by Zhang J et al., Beijing Science Press, 2003.

[82] Marks NA, McKenzie DR, Pailthorpe BA, et al. Ab initio simulations of tetrahedral amorphous carbon. Physical Review B, 1996, 54: 9703–9714.

[83] Frauenheim T, Blaudeck P, Stephan U, et al. Atomic structure and physical properties of amorphous carbon and its hydrogenated analogs. Physical Review B, 1993, 48: 4823–4834.

[84] Car R, Parrinello M. Unified approach for molecular dynamics and density-functional theory. Physical Review Letters, 1985, 55: 2471-2474.

[85] Nosé S. A unified formation of the constant temperature molecular dynamics methods. Journal of Chemical Physics, 1984, 81: 511-519.

[86] Gilkes KWR, Gaskell PH, Robertson J. Comparison of neutron-scattering data for tetrahedral amorphous carbon with structural models. Physical Review B, 1995, 51: 12303–12312.

[87] Lifshitz Y. Hydrogen-free amorphous carbon films: Correlation between growth conditions and properties. Diamond and Related Materials, 1996, 5: 388–400.

[88] Woollam JA, Johs B, Herzinger CM, et al. Overview of Variable Angle Spectroscopic Ellipsometry (VASE), Part I. Basic Theory and Typical Applications from Optical Metrology. SPIE, 1999, CR72: 1–28.

[89] Johs B, Woollam JA, Herzinger CM, et al. Overview of Variable Angle Spectroscopic Ellipsometry (VASE), Part II, Advanced Applications. 1999, CR72: 29–58.

[90] Liao YB. Polarization Optics. Beijing, Science Press, 2003.

[91] Musienko T, Rudakov V, Solov'ev L. On the application of Kramers-Kronig relations to media with spatial dispersion. Journal of Physics- Condensed Matter, 1989, 1: 6745.

[92] Frederick W. Optical Properties of Solids. New York, Academic Press, 1972.

[93] Tauc J, Menth A. States in the gap. Journal of NonCrystalline Solids, 1972, 8–10: 569.

[94] Tropf WJ, Harris TJ, Thomas ME. Optical materials: Visible and infrared, in: Waynant RW, Ediger MN (eds.), Electro-Optics Handbook, 2nd edn. Columbus, McGraw-Hill Press, 2000.

[95] Lifshitz Y, Lempert GD, Grossman E, et al. Optical and photoemission studies of DLC films prepared with a systematic variation of the sp^3:sp^2 composition. Diamond and Related Materials, 1997, 6: 687–693.

[96] Zhang XW, Cheung WY, Ke N, et al. Optical properties of nitrogenated tetrahedral amorphous carbon films. Journal of Applied Physics, 2002, 92: 1242–1247.

[97] Canillas A, Polo MC, Andujar JL, et al. Spectroscopic ellipsometric study of the tetrahedral amorphous carbon films optical properties and modeling. Diamond and Related Materials, 2001, 10: 1132–1136.

[98] Knoll J, Geiger J. Optical constants of arc-evaporated amorphous carbon in the far-infrared spectral region. Physical Review B, 1984, 29: 5651–5655.

[99] Chen ZY, Zhao JP. Optical constants of tetrahedral amorphous carbon films in the infrared region and at a wavelength of 633 nm. Journal of Applied Physics, 2000, 87: 4268–4273.

[100] De Souza DR, Soares LL, Cescato L, et al. Measurement of optical constants of a thin a-C:H films. Microelectronics Journal, 2000, 31: 251–254.

[101] Smith FW. Optical constants of a hydrogenated amorphous carbon film. Journal of Applied Physics, 1984, 55: 764–771.

[102] Godet C, Conway NMJ, Boureé JE, et al. Structural and electronic properties of electron cyclotron resonance plasma deposited hydrogenated amorphous carbon and carbon nitride films. Journal of Applied Physics, 2002, 91: 4154–4162.

[103] Forouhi AR, Bloomer I. Optical dispersion relations for amorphous semiconductors and amorphous dielectrics. Physical Review B, 1986, 36: 7018–7026.

[104] Lettington AH. Applications of diamond-like carbon thin films. Carbon, 1998, 36: 555–560.

[105] Miller AJ, Reece DM, Hudson MD, et al. Diamond coatings for IR windows applications. Diamond and Related Materials, 1997, 6: 386–389.

[106] Polo MC, Andu JL, Hart A, et al. Preparation of tetrahedral amorphous carbon films by filtered cathodic vacuum arc deposition. Diamond and Related Materials, 2000, 9: 663–667.

[107] Liu EK, Zhu BS, Luo JS, et al. Semiconductor Physics. Beijing, National Defence Industry Press, 2002.

[108] Robertson J. Recombination and photoluminescence mechanism in hydrogenated amorphous carbon. Physical Review B, 1996, 53: 16302–16305.

[109] Savvides N. Fourfold to threefold transition in diamond-like amorphous carbon films: A study of optical and electrical properties. Journal of Applied Physics, 1985, 58: 518–521.
[110] Oppedisano C, Tagliaferro A. Relationship between sp^2 carbon content and E_{04} optical gap in amorphous carbon-based materials. Applied Physics Letters, 1999, 75: 3650–3652.
[111] Oliver WO, Pharr GM. An improved technique for determining hardness and elastic modulus using load and displacement sensing indentation experiments. Journal of Materials Research, 1992, 7: 1564–1583.
[112] Knapp JA, Follstaedt DM, Barbour JC, et al. Finite-element modeling of nanoindentation for determining the mechanical properties of implanted layers and thin films. Nuclear Instruments & Methods in Physics Research Section B, Beam Interactions with Materials and Atoms, 1997, 127–128: 935–939.
[113] Qi J, Chan CY, Bello I, et al. Film thickness effects on mechanical and tribological properties of nitrogenated diamond-like carbon films. Surface & Coating Technology, 2001, 145: 38–43.
[114] Sheeja D, Tay B K, Lau S P, et al. An empirical relation for critical load of DLC coatings prepared on silicon substrates. International Journal of Modern Physics B, 2002, 16: 958-962.
[115] Friedmann TA, Sullivan JP, Knapp JA, et al. Thick stress-free amorphous-tetrahedral carbon films with hardness near that of diamond. Applied Physics Letters, 1997, 71: 3820–3822.
[116] Dong JJ, Drabold DA. Ring formation and the structural and electronic properties of tetrahedral amorphous carbon. Physical Review B, 1998, 57: 15591–15598.
[117] Gerstner EG, McKenzie DR, Puchert MK, et al. Adherent carbon film deposition by cathodic arc with implantation. Journal of Vacuum Science & Technology B, 1995, 13(2): 406–411.
[118] Waidmann S, Knupfer M, Fink J, et al. High-resolution electron energy-loss spectroscopy of undoped and nitrogen-doped tetrahedral amorphous carbon films. Diamond and Related Materials, 2000, 9: 722–727.
[119] Cheng GX. Raman and Brillouin Scattering – Principles and Applications. Beijing, Science Press, 2001.
[120] Michel MD, Muhlen LV, Achete CA, et al. Fracture toughness, hardness and elastic modulus of hydrogenated amorphous carbon films deposited by chemical vapor deposition. Thin Solid Films, 2006, 496: 481–488.
[121] Gupta BK, Bhushan B. Micromechanical properties of amorphous carbon coatings deposited by different deposition techniques. Thin Solid Films, 1995, 270: 391–398.
[122] Sheeja D, Tay BK, Leong KW, et al. Effect of film thickness on the stress and adhesion of diamond-like carbon coatings. Diamond and Related Materials, 2002, 11: 1643–1647.
[123] Roth TH, Kloos KH, Broszeit E. Structure, internal stress, adhesion and wear resistance of sputtered alumina coatings. Thin Solid Films, 1987, 153: 123–133.
[124] Ager III JW, Anders S, Anders A, et al. Effect of intrinsic growth stress on the Raman spectra of vacuum-arc-deposited amorphous carbon films. Applied Physical Letters, 1995, 66: 3444–3446.
[125] Shin JK, Lee CS, Lee KR, et al. Effect of residual stress on the Raman-spectrum analysis of tetrahedral amorphous carbon films. Applied Physics Letters, 2001, 78: 631–633.
[126] Anders S, Díaz J, Ager III JW, et al. Thermal stability of amorphous hard carbon films produced by cathodic arc deposition. Applied Physics Letters, 1997, 71: 3367–3369.
[127] Monteiro OR, Ager III JW, Lee DH, et al. Annealing of nonhydrogenated amorphous carbon films prepared by filtered cathodic arc deposition. Journal of Applied Physics, 2000, 88: 2395–2399.
[128] Bilek MMM, McKenzie DR. Comprehensive model of stress generation and relief processes in thin films deposited with energetic ions. Surface & Coating Technology, 2006, 200: 4345–4354.

[129] Czaplewski DA, Sulivan JP, Friedmann TA, et al. Temperature dependence of the mechanical properties of tetrahedrally coordinated amorphous carbon thin films. Applied Physics Letters, 2005, 87: 161915.
[130] Nakazawa H, Yamagata Y, Suemitsu M, et al. Thermal effects on structural properties of diamond-like carbon films prepared by pulsed laser deposition. Thin Solid Films, 2004, 467: 98–103.
[131] Tay BK, Sheeja D, Lau SP, et al. Time and temperature-dependent changes in the structural properties of tetrahedral amorphous carbon films. Surface & Coating Technology, 2000, 130: 248–251.
[132] Tarrant RN, Warschkow O, McKenzie DR. Raman spectra of partially oriented sp^2 carbon films: Experimental and modelled. Vibrational Spectroscopy, 2006, 41: 232–239.
[133] Anders S, Ager III JW, Pharr GM, et al. Heat treatment of cathodic arc deposited amorphous hard carbon films. Thin Solid Films, 1997, 308–309: 186–190.
[134] Anders A. Energetic deposition using filtered cathodic arc plasma. Vacuum, 2002, 67: 673–686.
[135] Bilek MM, McKenzie DR, Moeller W. Use of low energy and high frequency PBII during thin film deposition to achieve relief of intrinsic stress and microstructural changes. Surface & Coating Technology, 2004, 186: 21–28.
[136] Tarrant RN, Montross CS, McKenzie DR. Combined deposition and implantation in the cathodic arc for thick film preparation. Surface & Coating Technology, 2001, 136: 188–191.
[137] Sheeja D, Tay BK, Lau SP, et al. Characterization of ta-C films prepared by a two-step filtered vacuum arc deposition technique. Surface & Coating Technology, 2000, 127: 246–250.
[138] Friedmann TA, McCarty KF, Barbour JC, et al. Thermal stability of amorphous carbon films grown by pulsed laser deposition. Applied Physics Letters, 1996, 68: 1643–1645.
[139] Rey S, Antoni F, Prevot B, et al. Thermal stability of amorphous carbon films deposited by pulsed laser ablation. Applied Physics A, 2000, 71: 433–439.
[140] Tay BK, Shi X, Liu EJ, et al. Heat treatment of tetrahedral amorphous carbon films grown by filtered cathodic vacuum-arc technique. Diamond and Related Materials, 1999, 8: 1328–1332.
[141] Bhargava S, Bist HD, Narlikar AV, et al. Effect of substrate temperature and heat treatment on the microstructure of diamond-like carbon films. Journal of Applied Physics, 1996, 79: 1917–1925.
[142] Yoshikawa M, Katagiri G, Ishida H, et al. Raman spectra of diamond-like amorphous carbon films. Journal of Applied Physics, 1988, 64: 6464–6468.
[143] Britton DT, Härting M, Hempel M, et al. Defect characterization in amorphous diamond-like carbon coatings. Applied Surface Science, 1999, 149: 130–134.
[144] Shi X, Hu YH, Hu L. Tetrahedral amorphous carbon (ta-C) ultrathin films for slider overcoat application. International Journal of Modern Physics B, 2002, 16: 963–967.
[145] McCulloch DG, Peng JL, McKenzie DR, et al. Mechanisms for the behaviour of carbon films during annealing. Physical Review B, 2004, 70: 85406.
[146] Chen XD, Sullivan JP, Friedmann TA, et al. Fluctuation microscopy studies of medium-range ordering in amorphous diamond-like carbon films. Applied Physics Letters, 2004, 84: 2823–2825.
[147] Orwa JO, Andrienko I, Peng JL, et al. Thermal induced sp^2 clustering in tetrahedral amorphous carbon (ta-C) films. Journal of Applied Physics, 2004, 96: 6286–6297.
[148] Shroder R, Nemanich R, Glass J. Analysis of the composite structures in diamond thin films by Raman spectroscopy. Physical Review B, 1990, 41: 3738–3745.
[149] Sakata H, Dresselhaus G, Dresselhaus MS, Endo M. Effect of uniaxial stress on the Raman spectra of graphite fibers. Journal of Applied Physics, 1988, 63: 2769–2772.

[150] Zehnder T, Balmer J, Lüthy W, Weber HP. Determination of limits in deposition of adhering a-C films on silicon produced by pulsed laser deposition. Thin Solid Films. 1995, 263: 198–202.
[151] Chhowalla M, Amaratunga GAJ. Strongly adhering and thick highly tetrahedral amorphous (ta-c) thin films via surface modification by implantation. Journal of Materials Research, 2001, 16: 5–8.
[152] Leistner AJ, Frischer-Cripps AC, Bennett JM. Indentation hardness and modulus of the surface of a large super-polished single crystal silicon sphere in Goodman WA (ed.), Optical Materials and Structures Technologies. Proceedings of SPIE 2003, 5179: 215–222.
[153] Yoshinori F, Kaoru A, Haruyuki Y, Tadaaki S. Adhesion strength of DLC films on glass with mixing layer prepared by IBAD. Surface & Coating Technology, 2000, 128–129: 308–312.
[154] Zhgoon SA, Zhang Q, Yoon SF, Ahn J, Rusli. Modification of piezoelectrics surface by diamond-like carbon films and application to SAW devices. Journal of Electroceramics, 2001, 7: 13–16.
[155] Anttila A, Lappalainen R, Hakovirta M, Salo J. Improvement of adhesion of diamond-like using whisker-composite techniques. Surface & Coating Technology, 1996, 82: 130–132.
[156] Davanloo F, Park H, You JH, Collins CB. Adhesion properties of amorphous diamond films deposited on zinc-sulfide substrates. Journal of Adhesion Science and Technology, 1995, 9: 681–694.
[157] Savage JA. Infrared Optical Materials and Their Antireflective Coatings. Adam Hilger Ltd, Bristol and Boston. 1985: 59–76.
[158] Tropf WJ, Thomas ME, Frazer R K. Windows and Domes: Past, Present, and Future. Proceedings of SPIE 2003: 5078: 80–89.
[159] Yoshinori F, Kaoru A, Haruyuki Y, Tadaaki S. Adhesion strength of DLC Films on glass with mixing layer prepared by IBAD. Surface & Coating Technology, 2000, 128–129: 308–312.
[160] Reinke P, Francz G, Oelhafen P. Structural changes in diamond and amorphous carbon induced by low-energy ion irradiation. Physical Review B, 1996, 54: 7067–7073.
[161] Mackowski JM, Cimma B, Pignard R. Rain Erosion Behavior of Germanium Carbide (GeC) Films Grown on ZnS Substrates [C]. Proceedings of SPIE 1992, 1760: 201–209.
[162] Moulder JF, Stickle WF, Sobol PE, Bomben KD. Handbook of x-ray Photoelectron Spectroscopy. Minnesota, Physical Electronics, Inc. 1995, p. 92–93.
[163] Klein CA. Multi-layered Optical Windows: Strains, Stresses, and Curvature. Proceedings of SPIE 1999, 3705: 250–265.
[164] Klein CA. Comment on "The multilayer-modified Stoney's formula for laminated polymer composites on a silicon substrate". Journal of Applied Physics, 86, 5474 (1999). Journal of Applied Physics, 2000, 88: 5499–5500.
[165] Vilms J, Kerps D. Simple stress formula for multilayered thin films on a thick substrate. Journal of Applied Physics, 1982, 53: 1536–1537.
[166] Kim JS, Paik KW, Oh SH. The multilayer-modified Stoney's formula for laminated polymer composites on a silicon substrate. Journal of Applied Physics, 1999, 86(10): 5474–5479
[167] Hsueh CH. Modeling of elastic deformation of multilayers due to residual stresses and external bending. Journal of Applied Physics, 2002, 91: 9652–9656.
[168] Parratt LG, Hempstead CF. Anomalous dispersion and scattering of X-rays. Physics Review, 1954, 95: 359–363.
[169] Patsalas P, Logothetidis S, Kelires PC. Surface and interface morphology and structure of amorphous carbon thin and multilayer films. Diamond and Related Materials, 2005, 14: 1241–1254.
[170] Anders S, Callahan DL, Pharr GM, Tsui TY, Bharia CS. Multilayer of amorphous carbon prepared by cathodic arc deposition. Surface & Coating Technology, 1997, 94–95: 189–194.

[171] Baranov AM, Varfolomeev AE, Fanchenko SS, Nefedov AA., Calliari L, Speranza G, Laidani N. Investigation of the structural properties of thin amorphous carbon films and bilayer structures. Surface & Coating Technology, 2001, 137: 52–59.
[172] Mathioudakis C, Kelires PC, Panagiotatos Y, Patsalas P, Charitidis C, Logothetidis S. Nanomechanical properties of multilayered amorphous carbon structures. Physical Review B, 2002, 65: 205203.
[173] Logothetidis S, Gioti M, Charitidis C, Patsalas P, Arvanitidis J, Stoemenos J. Stability, enhancement of elastic properties and structure of multilayered amorphous carbon films. Applied Surface Science, 1999, 138–139: 244–249.
[174] Logothetidis S, Charitides C, Gioti M, Panayiotatos Y, Handrea M, Kautek W. Comprehensive study on the properties of multilayered amorphous carbon films. Diamond and Related Materials, 2000, 9: 756–760.
[175] Logothetidis S, Kassavetis S, Charitidis C, Panayiotatos Y, Laskarakis A. Nanoindentation studies of multilayer amorphous carbon films. Carbon, 2004, 42: 1133–1136.
[176] Pujada BR, Tichelaar FD, Janssen GCAM. Stress in tungsten carbide-diamond like carbon multilayer coatings. Applied Physics Letters, 2007, 90: 021013.
[177] Ager III JW, Anders S, Brown IG, Nastasi M, Walter KC. Multilayer hard carbon films with low wear rates. Surface & Coating Technology, 1997, 91: 91–94.
[178] Qi J, Lai KH, Lee CS, Bello I, Lee ST, Luo JB, Wen SZ. Mechanical properties of a-C:H multilayer films. Diamond and Related Materials, 2001, 10: 1833–1838.
[179] Charitidis C, Logothetidis S, Douka P. Nanoindentation and nanoscratching studies of amorphous carbon films. Diamond and Related Materials, 1999, 8: 558–562.
[180] Ziebert C, Bauer C, Stüber M, Ulrich S, Holleck H. Characterization of the interface region in stepwise bias-graded layers of DLC films by a high-resolution depth profiling method. Thin Solid Films, 2005, 482: 63–68.
[181] Knapp JA, Follstaedt DM, Myers SM, Barbour JC, Friedmann TA. Finite-element modeling of nanoindentation. Journal of Applied Physics, 1999, 85: 1460–1474.
[182] Hou QR, Gao J, Li SJ. Adhesion and its influence on micro-hardness of DLC and SiC. European Physical Journal B, 1999, 8: 493–496.
[183] Li XD, Bhushan B. A review of nanoindentation continuous stiffness measurement technique and its applications. Materials Characterization, 2002, 48: 11–36.
[184] Michel MD, Muhlen LV, Achete CA, Lepienski CM. Fracture toughness, hardness and elastic modulus of hydrogenated amorphous carbon films deposited by chemical vapor deposition. Thin Solid Films, 2006, 496: 481–488.
[185] Cáceres D, Vergara I, González R, Monroy E, Calle F, Muñoz E, Omnès F. Nanoindentation on ALGaN thin films. Journal of Applied Physics, 1999, 15: 6773–6778.
[186] Karimi A, Wang Y, Cselle T, Morstein M. Fracture mechanisms in nanoscale layered hard thin films. Thin Solid Films, 2002, 420–421: 275–280.
[187] Jungk JM, Boyce BL, Buchheit TE, Friedmann TA, Yang D, Gerbrerich WW. Indentation fracture toughness and acoustic energy release in tetrahedral amorphous carbon diamond-like thin films. Acta Materialia, 2006, 54: 4043–4052.
[188] Jonnalagadda K, Cho SW, Chasiotis I, Friedmann T, Sullivan J. Effect of intrinsic stress gradient on the effect mode-I fracture toughness of amorphous diamond-like carbon films for MEMS. Journal of the Mechanics and Physics of Solids, 2008, 56: 388–401.
[189] Quinn JP, Lemoine P, Maguire P, Mclaughlin JA. Ultra-thin tetrahedral amorphous carbon films with strong adhesion, as measured by nanoscratch testing. Diamond and Related Materials, 2004, 13: 1385–1390.

[190] Blees MH, Winkelman GB, Balkenende AR, den Toonder JMJ. The effect of friction on scratch adhesion testing: Application to a sol-gel coating on polypropylene. Thin Solid Films, 2000, 359: 1–13.
[191] Attar F, Johannesson T. Adhesion evaluation of thin ceramic coatings on tool steel using the scratch testing technique. Surface & Coating Technology, 1996, 78: 87–102.
[192] Laugier MT. An energy approach to the adhesion of coatings using the scratch test. Thin Solid Films, 1981, 76: 289–295.
[193] Laygier MT. Adhesion of TiC and TiN coatings prepared by chemical vapor deposition on WC-Co-based cemented carbides. Journal of Materials Science, 1986, 21: 2269–2272.
[194] Chang SY, Huang YC. Analyses of interface adhesion between porous SiO_2 low-k film and SiC/SiN layers by nanoindentation and nanoscratch tests. Microelectronic Engineering, 2007, 84: 319–327.
[195] Mathioudakis C, Kelires PC, Panagiotatos Y, Patsalas P, Charitidis C, Logothetidis S. Nanomechanical properties of multilayered amorphous carbon structures. Physical Review B, 2002, 65: 205203.
[196] Koehler JS. Attempt to design a strong solid. Physical Review B, 1970, 2: 547–551.
[197] Pickett WE. Relationship between the electronic structure of coherent composition modulated alloys and the supermodulus effect. Journal of Physics F: Metal Physics, 1982, 12: 2195–2204.
[198] Cammarata RC, Sieradzki K. Effects of surface stress on the elastic moduli of thin films and superlattices. Physical Review Letters, 1989, 62: 2005–2008.
[199] Liu Y, Meletis EI. Tribological behavior of DLC coatings with functionally gradient interfaces. Surface & Coating Technology, 2002, 153: 178–183.
[200] Zhang W, Tanaka A, Xu BS, Koga Y. Study on the diamond-like carbon multilayer films for tribological application. Diamond and Related Materials, 2005, 14: 1361–1367.

6 Germanium-carbide film

6.1 Overview

Germanium-carbide coatings have many special properties, making it highly promising for future applications.

Firstly, it is an ideal infrared antireflective/protective coating with the following distinct advantages:

(1) Germanium carbide has a sphalerite-like structure and good compatibility with most infrared window materials such as zinc sulphide and germanium, making it appropriate as the intermediate transition layer.

(2) It shows small absorption in the mid- and far-infrared wavebands, as well as low stress; thus, it can be fabricated into thick coatings.

(3) Germanium-carbide coatings show refractive-index variation within the range of 2–4 with change in the content of germanium and carbon atoms, which is highly beneficial to the design and fabrication of multi-layer coatings.

Germanium carbide is also a promising semiconductor coating material owing to the good adjustability of its optical band gap within a relatively wide range. This has a great significance for photovoltaic applications.

When fabricating nonhydrogen germanium-carbide coatings, the raw gaseous material cannot contain hydrogen. At present, there are mainly three methods to fabricate germanium carbide without hydrogen: ion implantation, molecular beam epitaxy, and magnetron co-sputtering. Germanium-carbide coatings fabricated using ion implantation shows a nonuniform thickness distribution. Germanium-carbide coatings fabricated using molecular beam epitaxy have a relatively low carbon content and extremely low coating-growth speed. In contrast, magnetron co-sputtering has the advantages of simple equipment demand, rapid deposition, homogeneous composition, and wide-range adjustability; thus, it is adopted as the preferred coating fabrication scheme here. There exist two approaches of magnetron co-sputter deposition: magnetron sputtering with a single composite target and multiple-target magnetron sputtering. Compared with the former, the latter has the simpler fabrication process and provides better control of coating composition. Therefore, we will introduce the double-target co-sputtering method for fabricating nonhydrogen germanium-carbide coatings in the following subsections.

https://doi.org/10.1515/9783110489514-009

6.2 Deposition rate of germanium-carbide film prepared by magnetron co-sputtering

6.2.1 Preparation of germanium-carbide film by magnetron co-sputtering

Since Grove discovered the sputtering phenomenon in the 1850s, sputter deposition has become an effective coating deposition method that is widely applied in the fields of electronics, optics, and surface processing. As a sputter deposition technique, the working principle of magnetron co-sputtering is identical to that of magnetron sputtering. This method can constrain electrons on the target surface through an applied magnetic field controlling electron motions, and extend the moving paths of electrons, which is beneficial to the ionization of working gas and thereby increases the sputtering efficiency. The magnetron co-sputtering technique is most commonly used in the fabrication of alloy coatings and multiple-layer coatings that are expected to meet various requirements of composition and performance. Because of multiple factors, the desired effects often cannot be achieved using conventional alloy targets or composite targets. For instance, different elements have different selective sputtering properties, film resputter yields, and adhesive forces. However, owing to the mutual independence between targets, multiple-target sputtering can separately control the sputtering with a superior performance in the fabrication of multiple-layer coatings and hybrid coatings.

Fig. 6.1 shows the structural schematic of the magnetron co-sputtering system. Discharging gas is fed into the vacuum chamber from the side wall and extracted out of the vacuum chamber by the vacuum pumping system, which consists of a mechanical pump and molecular pump. The two magnetron sputtering targets are configured with their own control power supplies as well as the water-cooling system to cool the targets through water circulation. The substrate is placed above the rotatable specimen stage and heated by the quartz lamp array beneath the stage through light irradiation. When the two magnetron targets are too close to each other, the glow will be inhomogeneously distributed owing to the repulsive effect of magnetic lines of force from the two magnetic cylinders. Therefore, the uniformity of thin-film composition and thickness can be improved by rotating the specimen stage and adopting a perforated plate during the fabrication process of germanium-carbide coatings.

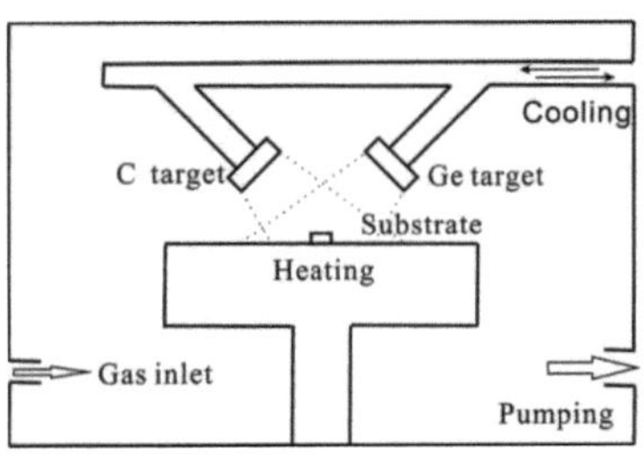

Fig. 6.1: Structural schematic of the magnetron co-sputtering system.

6.2.2 Effect of power on deposition rate

Here, we introduce the magnetron co-sputtering method used for fabricating germanium-carbide coatings in an Ar atmosphere. During the test process, the sputtering power of the germanium target directly affects the sputtering efficiency of the germanium target and ionization degree of discharge gas, thereby influencing the growth speed of coating. In our test, the DC power of the carbon target was set to 100 W, and the flow rate and working pressure of the Ar gas were maintained at 25 sccm and 1.0 Pa, respectively. The substrate temperature was controlled at 200 °C using the electric heater, and the germanium target power was set in the range of 40–160 W by using the controller of the radio-frequency (RF) power supply.

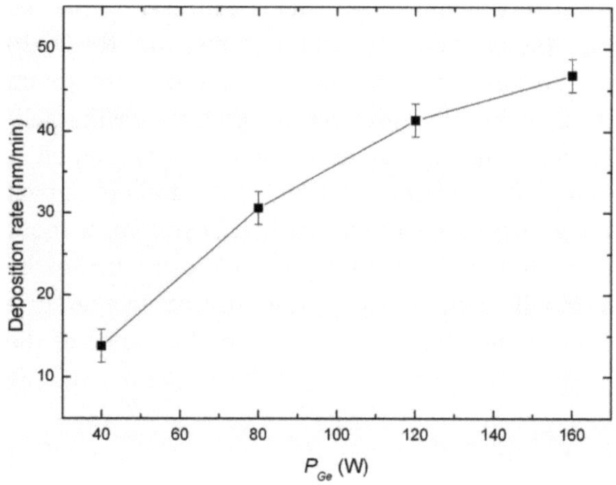

Fig. 6.2: Relationship between the deposition rate of the germanium-carbide coating and the sputtering power of the germanium target.

Fig. 6.2 illustrates the relationship between the growth speed of the germanium-carbide coating and the sputtering power of the germanium target. It can be observed that, with the increase of sputtering power of the germanium target from 40 W to 160 W, the deposition rate gradually increases. As is known, the power density on the target surface can directly influence the plasma intensity and the sputtering efficiency of RF sputtering. When the RF power increases, automatic bias and ion current will simultaneously increase. The automatic bias increase will cause more Ge ions to be sputtered from the target per unit time; increases in the ion current and the ion quantity participating in sputtering. Therefore, an increase in the RF power results in the sputtering of a greater number of Ge ions, leading to an increase in the growth speed of the germanium-carbide coating.

6.2.3 Effect of substrate temperature on deposition rate

Substrate temperature, as a crucial parameter in the coating deposition process, has a significant influence on the thin-film microstructure and plays a critical role in the modification of multiple performance parameters for the coating. In our test, the DC power of the carbon target was set to 100 W, and the RF power of the monocrystalline germanium target was set to 140 W. The inlet flowrate and working pressure of the Ar gas were set to 25 sccm and 1.0 Pa, respectively. The substrate temperature was controlled between the ambient temperature and 600 °C by using an electric heater.

Fig. 6.3 shows the relationship between the growth rate of germanium carbide and the substrate temperature. It can be seen that, when the substrate temperature increases from the ambient temperature to 600 °C, the deposition rate of the germanium-carbide coating does not vary significantly. This is because, during the deposition process, neither the sputtering speed of carbon atoms from the high-purity graphite target nor that of Ge ions from the monocrystalline germanium target is affected by the substrate temperature. However, an increase in the substrate temperature during the fabrication process can enhance the migration ability of deposited particles along the substrate surface, which will substantially impact the microstructural variation of the germanium-carbide coating. Therefore, without changing the growth rate of the germanium-carbide coating, the microstructure of the germanium-carbide coating can be modified by controlling the substrate temperature for the purpose of adjusting the mechanical/optical/electrical performance of the germanium-carbide coating.

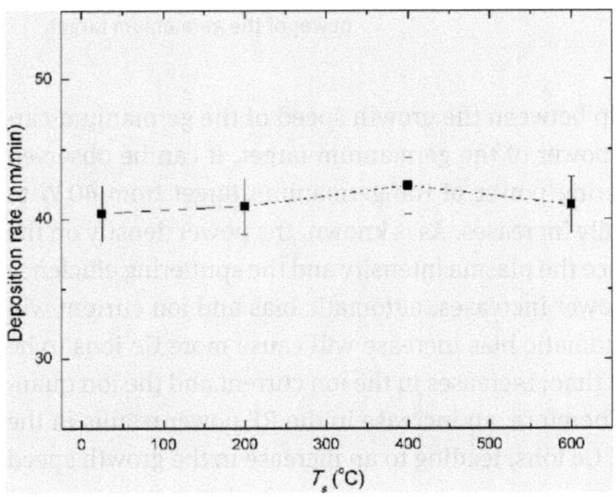

Fig. 6.3: Relationship between the deposition rate of the germanium-carbide coating and the substrate temperature.

6.3 Surface morphology, crystal structure and composition

6.3.1 Surface morphology

For analysing the surface structure and morphology of a coating material using an atomic force microscope (AFM), there is no special requirement on the intrinsic physical properties of the coating material or a need for particular processing methods such as metal/carbon spraying; thus, the surface morphology of the coating material can be well preserved. Hence, AFM is superior for direct imaging on an insulated plane.

Fig. 6.4 shows the surface morphology of germanium-carbide coatings fabricated under different values of germanium-target sputtering power, which indicates that the sputtering power of the germanium target has a great impact on the surface morphology of germanium coatings. Fig. 6.5 displays the surface-roughness variation of the germanium-carbide coating against the sputtering power of the germanium target. It can be seen that the coating surface roughness decreases when the sputtering power of the germanium target increases from 40 W to 80 W, and it then increases with the further increase of sputtering power up to 160 W. The reason for this variation can be explained as follows. When the sputtering power of the germanium target is 40 W, the kinetic energy of film-forming particles is relatively low, making it difficult for the particles to diffuse and leading to a relatively large surface roughness; when the power reaches 80 W and continues to increase, the coating surface roughness increases because of the gradual acceleration of the coating deposition rate.

Fig. 6.6 demonstrates the AFM surface morphology of germanium-carbide coatings fabricated at different substrate temperatures. It can be observed that the substrate temperature significantly affects the surface morphology. Fig. 6.7 shows the variation of surface roughness of germanium-carbide coatings with the substrate temper-

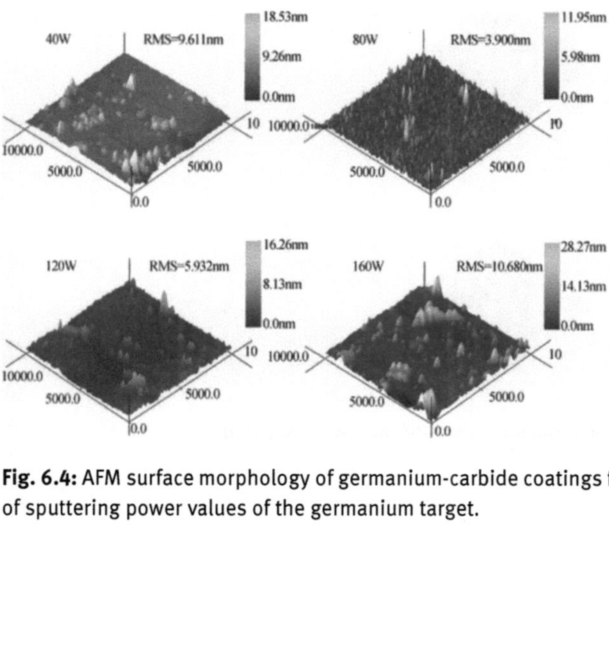

Fig. 6.4: AFM surface morphology of germanium-carbide coatings fabricated under different values of sputtering power values of the germanium target.

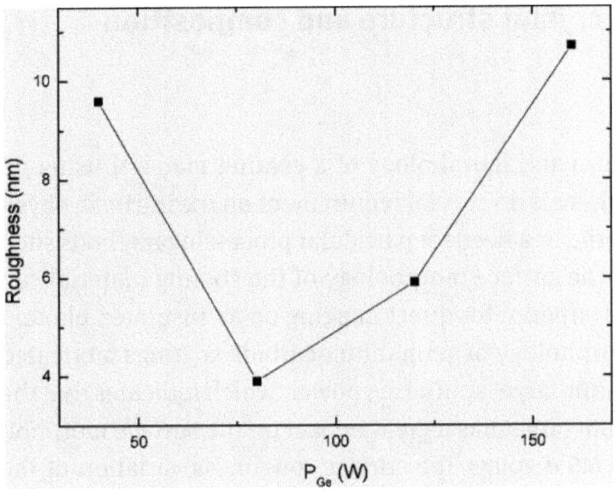

Fig. 6.5: Relationship between the surface roughness of germanium-carbide coatings and the sputtering power of the germanium target.

ature. It can be seen from the figure that the coating surface roughness decreases as the substrate temperature increases. That is, the higher the substrate temperature is, the smoother the coating surface will be. This illustrates that the surface diffusion mechanism of film-forming particles has an important effect on the coating surface roughness. During the deposition process, when the substrate temperature is relatively low, particles will have relatively low kinetic energy after reaching the substrate surface. As the substrate temperature increases, particles acquire a higher kinetic energy subsequent to arriving at the substrate surface and thereby stronger diffusion ability along the substrate surface, leading to a reduction of coating surface roughness.

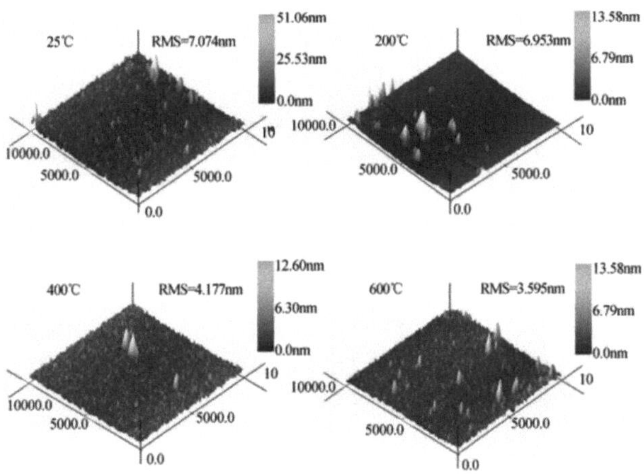

Fig. 6.6: AFM surface morphology of germanium-carbide coatings fabricated at different temperatures.

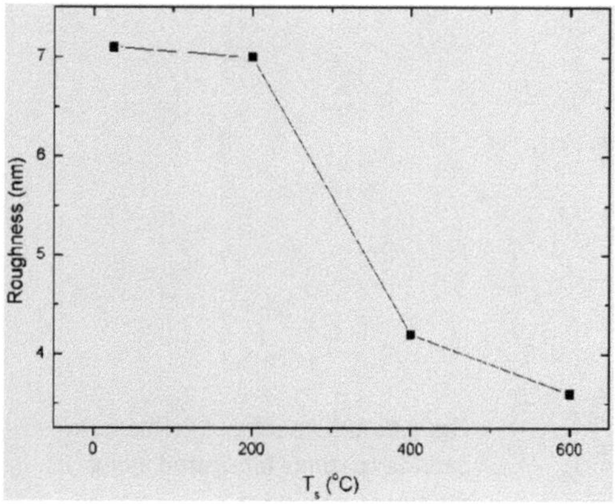

Fig. 6.7: Relationship between the surface roughness of the germanium-carbide coating and the substrate temperature.

6.3.2 Crystal structure

Owing to the periodicity of atomic arrangement in a crystal, when an x-ray is incident on a crystal or crystalline coating, it will be scattered by the atoms' electrons, driving each scattered wave to produce coherent scattering with a wavelength identical to that of the incident x-ray. The intensity of the coherent scattering undergoes interference strengthening in some particular crystal face orientations. This phenomenon is referred to as x-ray diffraction. The corresponding incident angle is defined as the diffraction angle, and the interplanar spacing of diffraction can be determined by Bragg's law. According to Bragg's law, the diffraction condition can be described as

$$2d\sin\theta = n\lambda, \tag{6.1}$$

where λ is the wavelength of the x-ray and θ is the diffraction angle.

Fig. 6.8 shows a schematic of the x-ray powder diffraction (XRD) of germanium-carbide coatings fabricated using different techniques. All the fabricated germanium-carbide coatings shown are in the amorphous state. From the perspective of thermodynamics, the crystalline state is the steady state of the atomic system, whereas the amorphous state is a metastable state. Based on the thin-film growth dynamics, rapid energy variation can lead to the formation of a metastable amorphous structure. When the driving forces of the crystalline state and amorphous state are comparable, as long as the metastable phase promotes nucleation and fast growth, a remarkable variation of free energy occurs in the system and the metastable state will be maintained in the initial growth stage. Therefore, amorphous coatings are likely to be formed under a relatively high sputtering power of the germanium target. Generally, for a germanium-carbide coating fabricated using magnetron sputtering, when the substrate temperature reaches 450 °C, crystalline grains are formed in the coating. However, in the figure, the germanium diffraction peak cannot be observed even

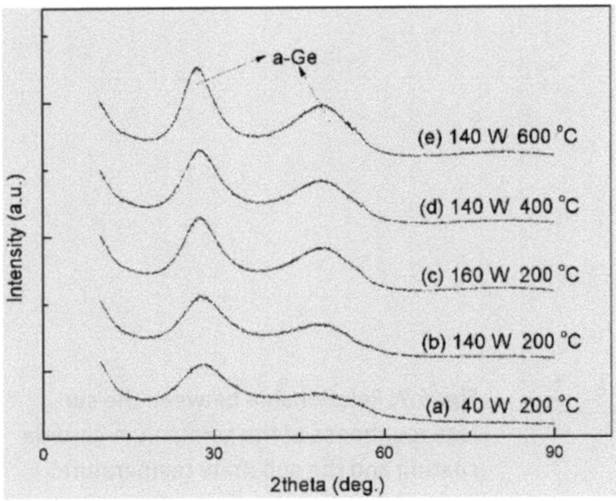

Fig. 6.8: XRD spectra of germanium-carbide coatings fabricated using different techniques.

when the substrate temperature increases to 600 °C, indicating that the existence of carbon has an inhibiting effect on germanium crystallization.

6.3.3 Composition analysis

It is difficult to measure the absolute content of germanium and carbon atoms in a thin film. Therefore, the relative content of germanium and carbon atoms is calculated instead of the absolute values. In our calculation, the atomic distribution of the coating at a depth of 10 nm (from the surface) is assumed to be homogenous. On this basis, the relative content of germanium atoms and carbon atoms in the coating can be calculated using the sensitivity factor method with the following expressions:

$$C_{\mathrm{Ge}} = \frac{\frac{I_{\mathrm{Ge\,3d}}}{S_{\mathrm{Ge}}}}{\frac{I_{\mathrm{Ge\,3d}}}{S_{\mathrm{Ge}}} + \frac{I_{\mathrm{C\,1s}}}{S_{\mathrm{C}}}}, \tag{6.2}$$

$$C_{\mathrm{C}} = \frac{\frac{I_{\mathrm{C\,1s}}}{S_{\mathrm{C}}}}{\frac{I_{\mathrm{Ge\,3d}}}{S_{\mathrm{Ge}}} + \frac{I_{\mathrm{C\,1s}}}{S_{\mathrm{C}}}}, \tag{6.3}$$

where C_{Ge} denotes the relative atomic concentration of germanium, C_{C} the relative atomic concentration of carbon, $I_{\mathrm{Ge\,3d}}$ the intensity of Ge 3d peak, $I_{\mathrm{C\,1s}}$ the intensity of C/S peak, S_{Ge} the sensitivity factor of germanium, and S_{C} the sensitivity factor of carbon.

Fig. 6.9 shows the full-scan XPS spectrum of a germanium-carbide coating fabricated under a germanium-target sputtering power of 140 W, carbon-target sputtering power of 100 W, working pressure of 1.0 Pa, and substrate temperature of 200 °C ((a) argon ions before sputtering, (b) argon ions after sputtering). It can be seen that the sputtering significantly changes the spectrum of argon ions at the O 1s peak. The peak strength of O 1s is high before sputtering but nearly disappears after sputtering, indicating that the oxygen atoms on the coating surface are basically those absorbed

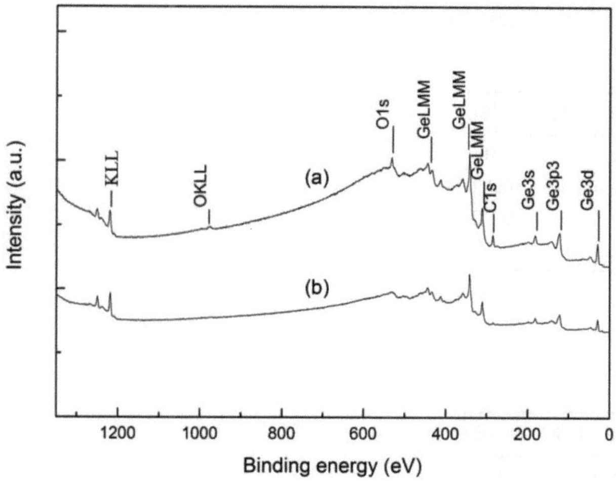

Fig. 6.9: Typical full-scan XPS spectrum of germanium-carbide coating.

by the surface from the surrounding air to which the specimen is exposed. Apart from the O 1s peak, C 1s and Ge 3d peaks can also be observed, which shows that the main composition of the coating is exactly the desired germanium carbide.

Here, we introduce the fabrication of a germanium-carbide coating in an argon atmosphere by using magnetron sputtering. In order to investigate the effects of RF power of the germanium target and substrate temperature on the coating composition, two groups of control-variable tests were performed:

(1) the DC power of the carbon target was set to 100 W and argon inlet/working pressure to 25 sccm/1.0 Pa; the substrate temperature was controlled at 200 °C using heating current, and the germanium-target power was controlled in the range of 40–160 W by using the controller on the RF power supply;
(2) the DC power of the germanium target was set to 100 W, RF power of the monocrystalline germanium target to 140 W, and argon inlet/working pressure to 25 sccm/1.0 Pa; the substrate temperature was controlled between the ambient temperature and 600 °C by varying the heating current.

Tab. 6.1 and 6.2 list the measured results of coating composition, from which the following conclusions can be drawn:

(1) The composition of the germanium-carbide coating is closely related to the power that is loaded on the magnetron sputtering targets. If the DC power of the high-purity graphite target is fixed, then with the increase of RF power loaded to the monocrystalline germanium target, the germanium atom content in the germanium-carbide coating correspondingly increases.
(2) The composition of the germanium-carbide coating has a weak relationship with the substrate temperature. In other words, the substrate temperature cannot affect the sputtering speed of magnetron sputtering targets.

Tab. 6.1: Composition of germanium-carbide coatings fabricated under different values of germanium target power.

Specimen	C (at. %)	Ge (at. %)	C/Ge
a-$Ge_{1-x}C_x$-40 W	44.9	55.1	0.81
a-$Ge_{1-x}C_x$-80 W	29.6	70.4	0.42
a-$Ge_{1-x}C_x$-120 W	23.2	76.8	0.30
a-$Ge_{1-x}C_x$-160 W	15.0	85.0	0.18

Tab. 6.2: Composition of germanium-carbide coatings fabricated under different substrate temperatures.

Specimen	C (at. %)	Ge (at. %)	C/Ge
a-$Ge_{1-x}C_x$-RT	20.7	79.3	0.26
a-$Ge_{1-x}C_x$-200 °C	22.1	77.9	0.28
a-$Ge_{1-x}C_x$-400 °C	22.7	77.3	0.29
a-$Ge_{1-x}C_x$-600 °C	20.7	79.3	0.26

6.4 Bonding structure and bonding mechanism

6.4.1 FTIR spectra of germanium-carbide thin films

Infrared spectrometry is an effective technique for the structure analysis of organic materials. Because different bonding structures exhibit different absorption peaks in infrared absorption spectra, the bonding-structure variation of the specimen can be analyzed through the peak-position changes in the infrared absorption spectra. Similarly, Fourier transform infrared spectroscopy (FTIR) can also be applied to analyse the bonding-structure variation of inorganic materials.

Fig. 6.10 shows the FTIR spectrum of a germanium-carbide coating fabricated under a germanium-target sputtering power of 140 W, carbon-target sputtering power of 100 W, working pressure of 1.0 Pa, and substrate temperature of 200 °C.

It can be seen from Fig. 6.10 that there exist four obvious infrared absorption bands in the infrared spectrogram with the peak positions of 610 cm^{-1}, 750 cm^{-1}, 1100 cm^{-1}, and 1500 cm^{-1}, respectively. The absorption peak located at 610 cm^{-1} is caused by the stretching vibration of the Ge–C bond, indicating that a portion of the germanium and carbon atoms underwent chemical reactions when the germanium-carbide thin film is fabricated using magnetron co-sputtering. The absorption peak at 750 cm^{-1} is caused by the Ge–C–C–C–Ge structure vibration, where germanium atoms force the sp^2 C atoms to generate bending vibration in the plane. The absorption peak at 1100 cm^{-1} is produced by the stretching vibration of Si–O bonds. Since the surface treatment has already been conducted using HF acid before fabricating

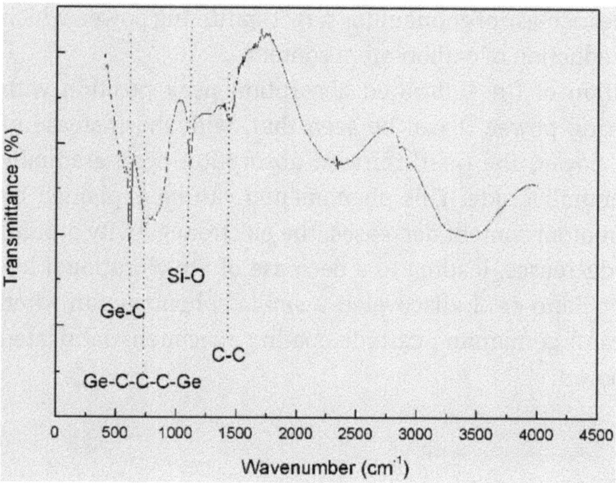

Fig. 6.10: FTIR spectrum of a typical germanium-carbide coating.

the germanium-carbide coating, the occurrence of these Si–O bonds is due to the oxygen atoms in the gaps of the substrate. The absorption peak at 1500 cm^{-1} is caused by the C–C bond vibration. As we all know, the C–C bond has no absorption peak in the infrared spectrum. However, when the C–C bond is impacted by the surrounding germanium atoms, its symmetry is damaged whereas its infrared activity is initiated, leading to the appearance of the absorption peak of the C–C bond in the FTIR spectrum.

Fig. 6.11 demonstrates the FTIR spectra of germanium-carbide coatings fabricated under different values of germanium-target sputtering power. It can be observed that, with the increase of germanium-target sputtering power, the intensity of the Ge–C peak at 610 cm^{-1} is slightly reduced, which shows that a high value of germanium-target sputtering power can decrease the carbon-atom content in the coating and thereby reduce the Ge–C bond content. Similarly, the peak strength of the C–C bond

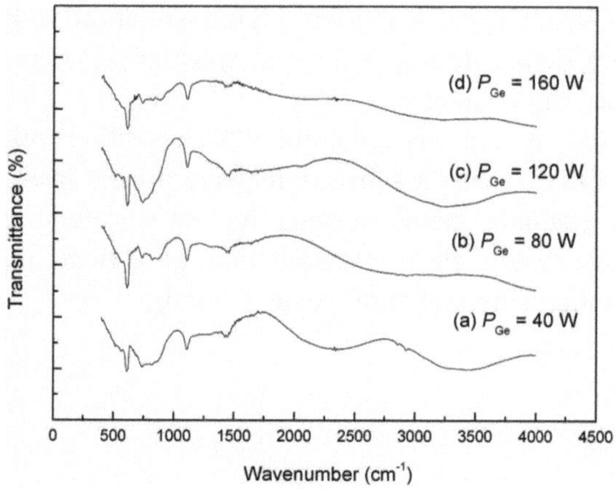

Fig. 6.11: FTIR spectra of germanium-carbide coatings fabricated under different values of germanium-target sputtering power.

at 1500 cm^{-1} decreases with the increase of germanium-target sputtering power, which can also be explained by the reduction of carbon-atom content.

Fig. 6.12 shows the variation of Ge–C infrared absorption peak position with the germanium-target sputtering power. It can be seen that, with the increase of germanium-target sputtering power, the Ge–C infrared absorption peak gradually moves towards the low-wavenumber side. This phenomenon can be explained by the fact that, when the carbon-atom content decreases, the electronegativity around germanium atoms gradually decreases, leading to a decrease of the vibrational frequency of the Ge–C bonds. N. Saito et al. discovered a similar phenomenon when investigating the FTIR spectra of germanium-carbide coating specimens fabricated under different values of RF power.

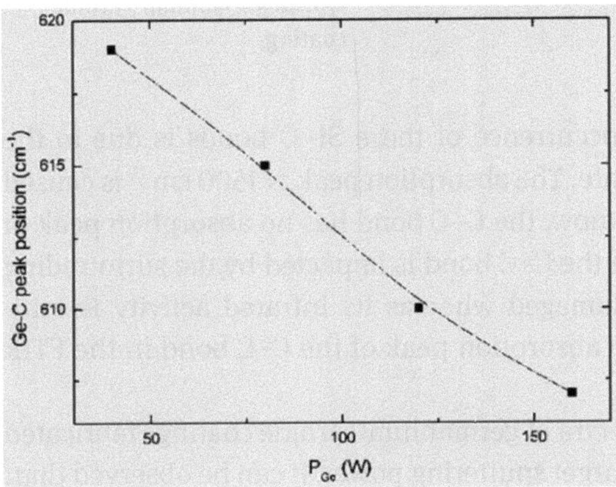

Fig. 6.12: Variation of the Ge–C absorption peak position with the germanium-target sputtering power.

Fig. 6.13 shows the FTIR spectra of germanium-carbide coatings fabricated under different substrate temperatures. It can be seen that, with the gradual increase of substrate temperature, the intensity of the Ge–C peak at 610 cm^{-1} is slightly enhanced, suggesting that a high substrate temperature has a positive effect on the formation of the Ge–C bond. The intensity of the C–C bond peak at 1500 cm^{-1} is also enhanced with the increase of substrate temperature, demonstrating that a high substrate temperature can promote the binding between carbon atoms.

Fig. 6.14 shows the variation of the Ge–C infrared absorption peak position with the substrate temperature. We can see that, when the substrate temperature increases, the Ge–C infrared absorption peak gradually moves towards the low-wavenumber side. This is because the gradual decrease of electronegativity around germanium atoms leads to a decrease of the vibrational frequency of the Ge–C bonds.

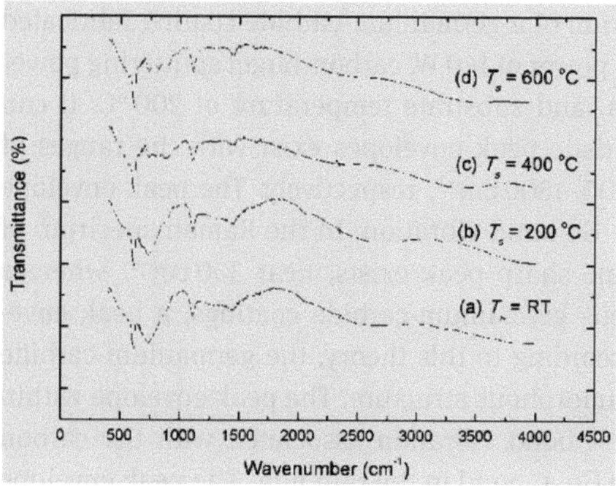

Fig. 6.13: FTIR spectra of germanium-carbide coatings fabricated under different substrate temperatures.

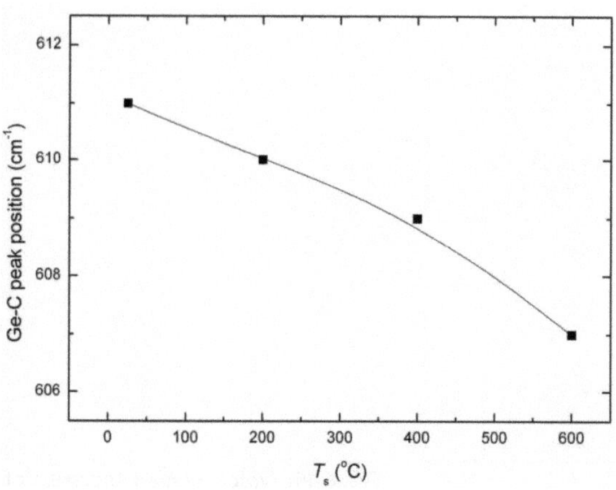

Fig. 6.14: Variation of Ge–C absorption peak position with the germanium-target sputtering power.

6.4.2 Raman spectra of germanium-carbide thin films

FTIR spectroscopy is an effective method for the study of asymmetric bonding structures. For symmetric bonding structures, however, Raman spectroscopy is superior. Raman spectroscopy is the most convenient and effective approach to analyse carbon structures. For example, as the structural vibrational frequencies of graphite (1580 cm^{-1}) and diamond (1350 cm^{-1}) in Raman spectra are significantly different from each other, it is quite simple to identify the structure of a carbon thin film. For the analysis of amorphous carbon coatings, visible Raman spectroscopy is generally adopted. Within the visible range, the sp^2-C Raman scattering cross section is far greater than the sp^3-C Raman scattering cross section; consequently, this method is only able to identify the sp^2-C structure. The Raman spectrum of a carbon coating usually has two characteristic peaks, namely the G peak of typical graphite near 1580 cm^{-1} and the D peak of disordered graphite near 1350 cm^{-1}.

Fig. 6.15 shows the Raman spectrum of a germanium-carbide coating fabricated under a germanium-target sputtering power of 140 W, carbon-target sputtering power of 100 W, working pressure of 1.0 Pa, and substrate temperature of 200 °C. It can be seen that three distinct characteristic peak envelopes exist with the ranges of 100–350 cm^{-1}, 500–700 cm^{-1}, and 1200–1800 cm^{-1}, respectively. The peak envelope within 100–350 cm^{-1} denotes the Ge–Ge bond vibration. In the Raman spectrum of monocrystalline germanium, only one sharp peak exists, near 300 cm^{-1}, whereas in the Raman spectrum of amorphous germanium-carbide coatings, a peak envelope occurs within 100–350 cm^{-1}. According to this theory, the germanium-carbide coating prepared in our test has the amorphous structure. The peak envelope within 500–700 cm^{-1}, attributed to the Ge–C bond vibration associated with the carbon atoms, illustrates the existence of the Ge–C bond in the thin film. The peak envelope within 1200–1800 cm^{-1} is generated by the C–C bond vibration.

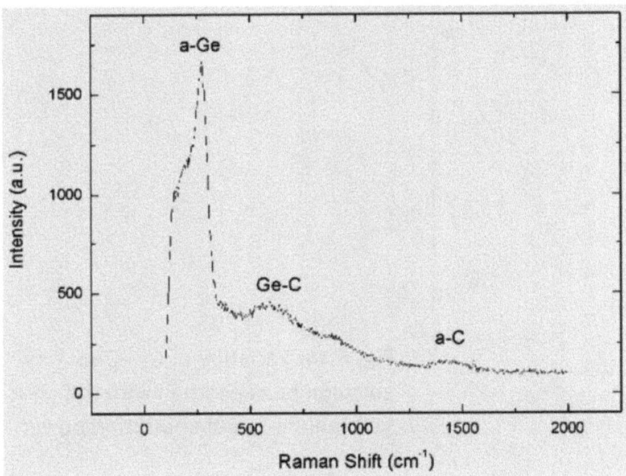

Fig. 6.15: Typical Raman spectrum of a germanium-carbide coating

Fig. 6.16 shows the Raman spectrum of a germanium-carbide coating within 100–350 cm^{-1} and 1200–1800 cm^{-1} and the corresponding fitting results of the germanium characteristic peak within 100–350 cm^{-1} and carbon characteristic peak within 1200–1800 cm^{-1}. The germanium characteristic peak within 100–350 cm^{-1} can be divided into four Gaussian peaks centred at 130 cm^{-1}, 160 cm^{-1}, 220 cm^{-1}, and 275 cm^{-1}, which correspond to the four vibration modes of germanium, i.e., TA, LA, LO, and TO, respectively. The carbon characteristic peak within 1200–1800 cm^{-1} can be divided into three Gaussian peaks centred at 1370 cm^{-1}, 1495 cm^{-1}, and 1560 cm^{-1}, which are caused by the vibration of disordered graphene nanoribbons (D ribbons), the carbon sp^2 vibration related to germanium, and the vibration of graphene nanoribbons (G ribbons), respectively.

Fig. 6.17 shows the Raman spectra in the range of 100–2000 cm^{-1} for germanium-carbide coatings fabricated under different germanium-target sputtering powers; the Raman spectrum of a diamond-like carbon (DLC) thin film is shown for comparison.

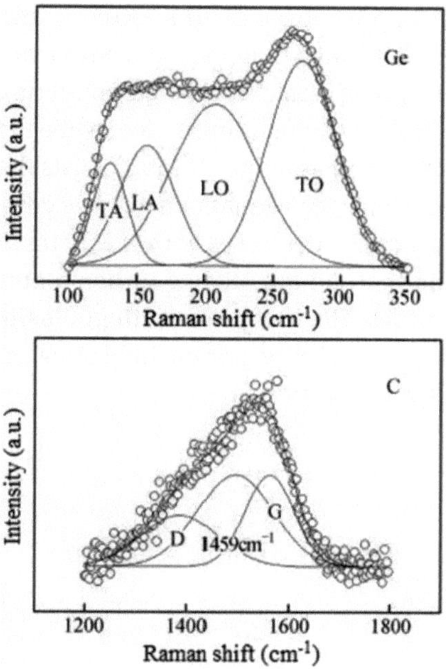

Fig. 6.16: Typical Raman spectral fitting of a germanium-carbide coating.

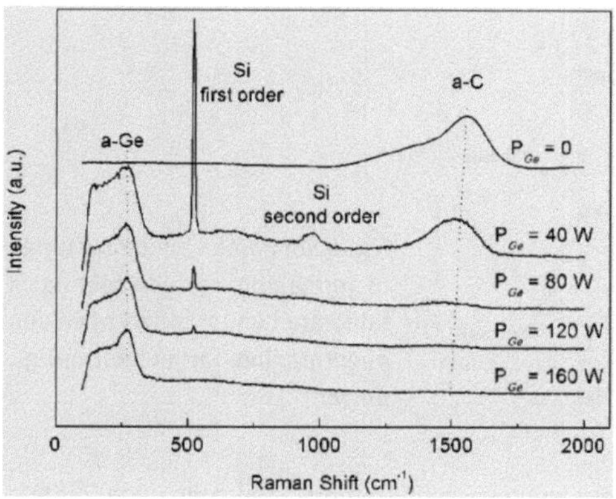

Fig. 6.17: Raman spectra of germanium-carbide coatings fabricated under different values of germanium-target sputtering power.

Since the fabricated coatings are relatively thin, the laser used in Raman spectroscopy can traverse some coatings and reflect the silicon information of the substrate in the Raman spectra. Especially in the case of 40 W germanium-target power, the silicon information of the substrate is substantial in the Raman spectrum. It can be seen from the Raman spectral variation that, when the germanium-target power is greater than 80 W, the information of amorphous germanium exists in the Raman spectrum, but not the characteristic peak of carbon; when the germanium-target power decreases to 40 W, characteristic peaks of amorphous graphite clearly appear in the Raman spectrum, demonstrating a large amount of sp^2 hybridised carbon in the coatings.

Fig. 6.18 shows the Raman spectral fitting in the ranges of 100–350 cm^{-1} and 1200–1800 cm^{-1} for germanium-carbide coatings fabricated under various power values. It is known from the figure that, with the increase of germanium-target sputtering power, the Ge-Ge TO vibration mode around 275 cm^{-1} is gradually enhanced, while the intensity of carbon Raman peaks within 1200–1800 cm^{-1} is reduced. Tab. 6.3 lists the parameters of Gaussian fitting within 100–350 cm^{-1} for the germanium-carbide coating. The peak position of the Ge–Ge TO mode gradually decreases with the increase of germanium-target sputtering power, showing that the quantity of carbon atoms that are adjacent to the germanium atoms decreases. The half-peak width gradually decreases with the increase of germanium-target sputtering power, indicating an increase in the orderliness of germanium atoms in the coatings.

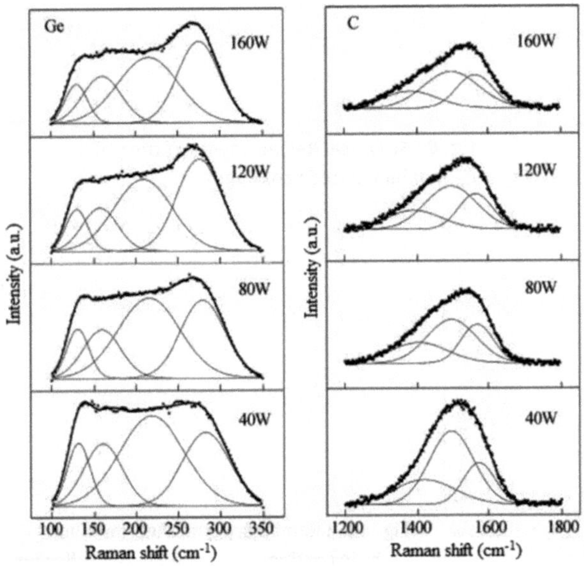

Fig. 6.18: Raman spectral fitting of germanium-carbide coatings fabricated under different values of germanium-target sputtering power.

Tab. 6.3: Raman spectral fitting parameters of germanium-carbide coatings fabricated under different values of germanium-target sputtering power.

P_{Ge} (W)	ω_{TA} (cm^{-1})	Γ_{TA} (cm^{-1})	ω_{LA} (cm^{-1})	Γ_{LA} (cm^{-1})	ω_{LO} (cm^{-1})	Γ_{LO} (cm^{-1})	ω_{TO} (cm^{-1})	Γ_{TO} (cm^{-1})
40	132	31	160	53	218	90	282	65
80	131	31	160	52	215	87	279	62
120	130	30	158	49	213	81	276	62
160	130	31	160	53	210	85	275	58

Gaussian linear fitting is commonly adopted to analyse the Raman spectra of carbon films. The variations of the fitting parameters are shown in Fig. 6.19. It can be seen that, when the germanium-target sputtering power increases from 40 W to 160 W, the relative-intensity ratio of D ribbon and G ribbon, I_D/I_G, gradually decreases from 1.17 to 0.75, and the position of G ribbon decreases from 1572 cm^{-1} to 1563 cm^{-1} with its half-peak width increasing from 102 cm^{-1} to 117 cm^{-1}. All these changes consistently illustrate that, with the increase of germanium-target sputtering power, the sp^2 carbon content in the germanium-carbide coating decreases gradually.

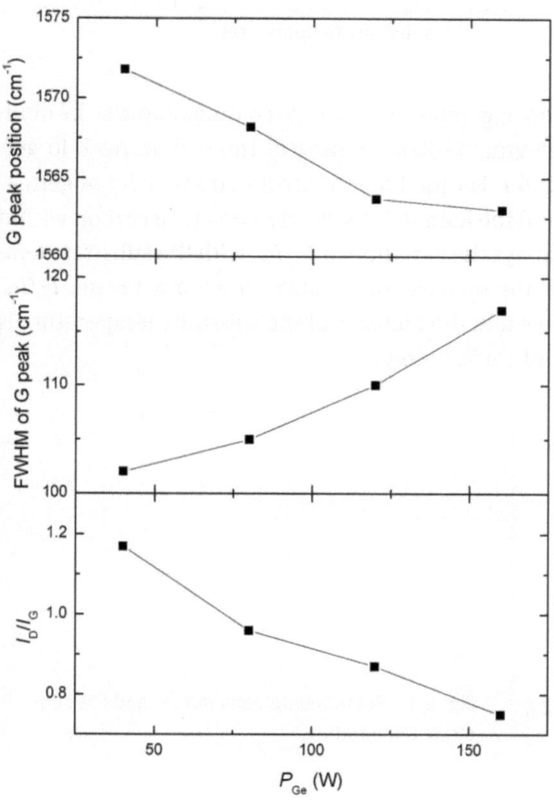

Fig. 6.19: Raman spectral fitting parameters of germanium-carbide coatings fabricated under different values of germanium-target sputtering power.

Fig. 6.20 shows the Raman spectra in the range of 100–2000 cm^{-1} for germanium-carbide coatings fabricated at different substrate temperatures. From this group of Raman spectra, no obvious change can be observed in the wide range of substrate temperature from room temperature to 600 °C, showing that in a germanium-carbide coating mainly composed of germanium atoms, the germanium atoms can induce a large scale of sp^3 hybridization of carbon atoms but little sp^2 hybridization.

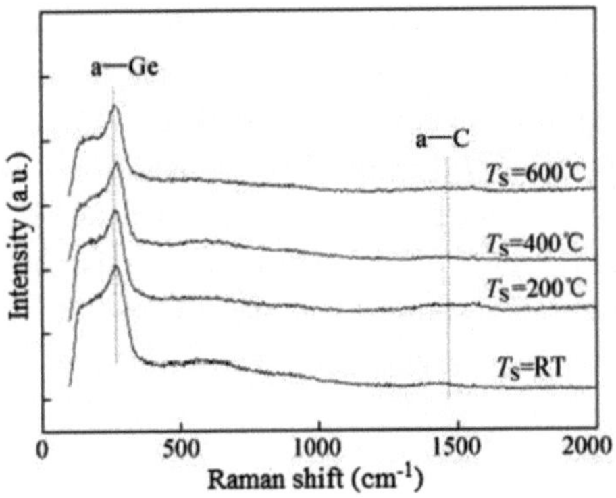

Fig. 6.20: Raman spectra of germanium-carbide coatings fabricated under different substrate temperatures.

The content of sp^2 hybridised carbon in germanium-carbide coatings can also be qualitatively analyzed by using the integrated intensity ratio of the carbon peak to germanium peak in the Raman spectrum. Let the Raman integrated intensity of germanium within 100–350 cm^{-1} be I_{Ge} and the Raman integrated intensity of carbon within 1200–1600 cm^{-1} be I_C. Fig. 6.21 displays the variation of I_C/I_{Ge} with the substrate temperature. It can be seen that, with the increase of the substrate temperature, I_C/I_{Ge} gradually increases, which indicates that the increase of the substrate temperature is beneficial to the sp^2 hybridization of carbon atoms.

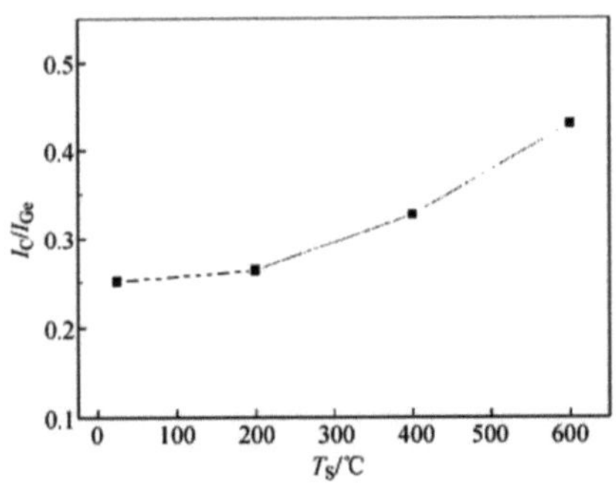

Fig. 6.21: Relationship between $\frac{I_C}{I_{Ge}}$ and the substrate temperature.

Fig. 6.22 shows the Raman spectral fitting curves within 100–350 cm^{-1} and 1200–1800 cm^{-1} for the germanium-carbide coatings fabricated under different substrate temperatures. The Gaussian fitting parameters of the germanium Raman peak in the range of 100–350 cm^{-1} are listed in Tab. 6.4, where the half-peak width of the Ge–Ge TO mode gradually decreases with the increase of substrate temperature, indicating an increase in the orderliness of germanium atoms in the coatings.

The variation of Raman peak-fitting parameters for carbon within 1200–1800 cm^{-1} is illustrated in Fig. 6.23. It can be seen that, when the substrate temperature increases from room temperature to 600 °C, the relative intensity ratio of D ribbon and G ribbon, I_D/I_G, gradually increases from 0.80 to 1.31, and the position of G ribbon in-

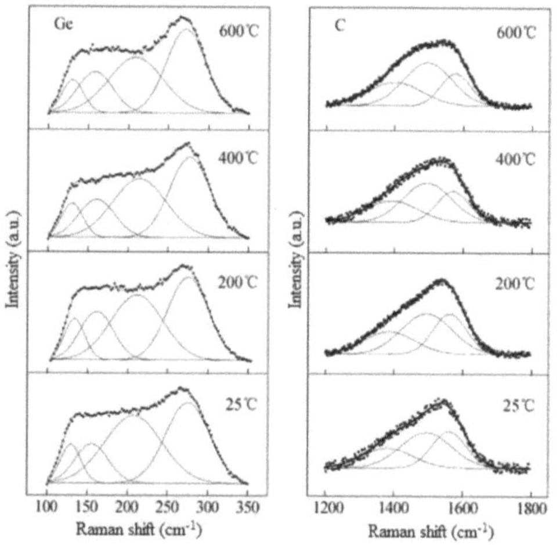

Fig. 6.22: Raman spectral fitting curves of germanium-carbide coatings fabricated under different substrate temperatures.

Tab. 6.4: Raman spectral fitting parameters of germanium-carbide coatings fabricated under different substrate temperatures.

T_s (°C)	ω_{TA} (cm^{-1})	Γ_{TA} (cm^{-1})	ω_{LA} (cm^{-1})	Γ_{LA} (cm^{-1})	ω_{LO} (cm^{-1})	Γ_{LO} (cm^{-1})	ω_{TO} (cm^{-1})	Γ_{TO} (cm^{-1})
RT*	130	30	156	48	208	84	275	64
200	130	30	157	49	207	76	275	60
400	130	31	160	52	213	78	276	59
600	130	31	159	51	209	79	277	58

Note: ω_{TA}, ω_{LA}, ω_{TO}, ω_{LO} represent the peak positions of Ge-Ge vibrational modes, i.e., TA, LA, LO, and TO, respectively; Γ_{TA}, Γ_{LA}, Γ_{LO}, Γ_{TO} represent the half-peak widths of Ge–Ge vibrational modes, i.e., TA, LA, LO, and TO, respectively.
* Room temperature

creases from 1561 cm^{-1} to 1576 cm^{-1} with its half-peak width decreasing from 117 cm^{-1} to 103 cm^{-1}. All these changes consistently indicate that the sp^2 carbon content in the germanium-carbide coatings gradually increases with the increase of the substrate temperature.

6.4.3 X-ray photoelectron spectra of germanium-carbide films

Fig. 6.24 shows the Ge 3d and C 1s core level spectra of a germanium-carbide coating fabricated under a germanium-target sputtering power of 140 W, carbon-target sputtering power of 100 W, working pressure of 1.0 Pa, and substrate temperature of 200 °C. It can be observed from the analytical results of FTIR and Raman spectroscopy that Ge–Ge, Ge–C, sp^3 C–C, and sp^2 C–C bonds exist in the thin film.

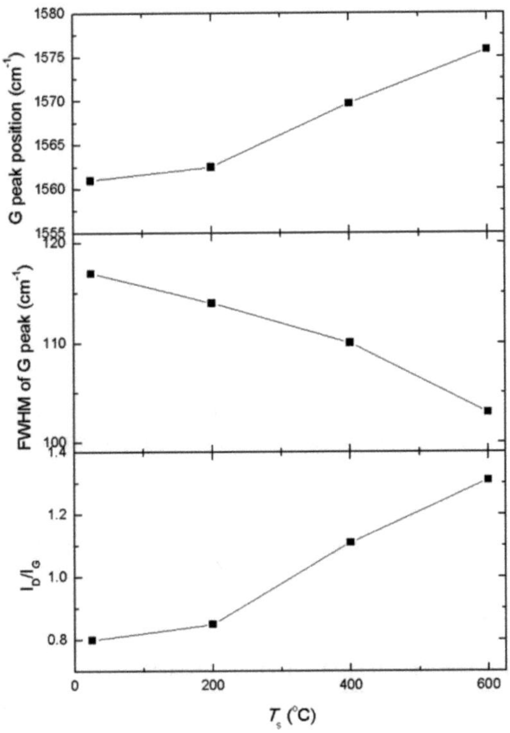

Fig. 6.23: Raman spectral fitting parameters of Ge_xC_{1-x} coatings fabricated under different substrate temperatures.

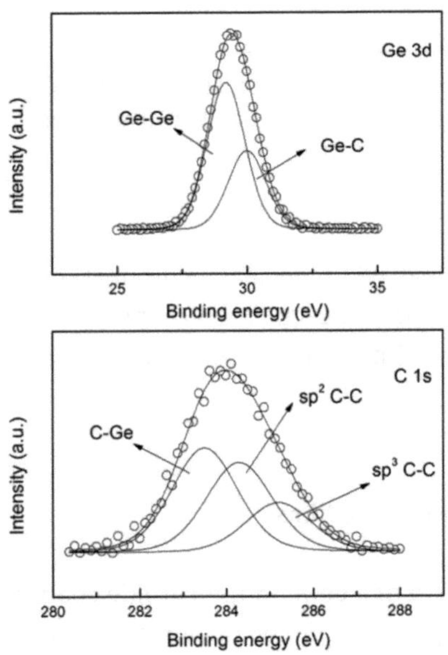

Fig. 6.24: Typical Ge 3d and C 1s core level spectra of germanium-carbide coating.

It can be seen from Fig. 6.24 that the half-peak width of the Ge 3d core level spectrum is relatively high and its spectral line can be fitted by two peaks: a Ge–Ge peak at 29.2 eV and a Ge–C peak at 29.9 eV. It can also be seen that the half-peak width of the C 1s spectrum of the germanium-carbide coating is approximately 2.4 eV, which far greater than those of diamond (1.0 eV) and graphite (0.6 eV). According to the FTIR and Raman spectral analysis, the C 1s spectrum can be divided into three peaks of Ge–C, sp^2 C–C, and sp^3 C–C at 283.5 eV, 284.3 eV, and 285.2 eV, respectively.

Fig. 6.25 displays the C 1s core level spectra of germanium-carbide coatings fabricated under different values of germanium-target sputtering power. It can be seen that, with the increase of the germanium-target sputtering power, the profile of the C 1s core level spectrum changes significantly, which indicates that the chemical environment around the carbon atoms in the thin films undergoes a great change on varying the germanium-target sputtering power.

Fig. 6.26 shows the relationship between the relative integrated intensity of each peak in the C 1s core level spectral fitting and the germanium-target sputtering power. It can be seen from the figure that an increase in the germanium-target sputtering power causes a rapid reduction of sp^2 C–C in the coating, which shows an agree-

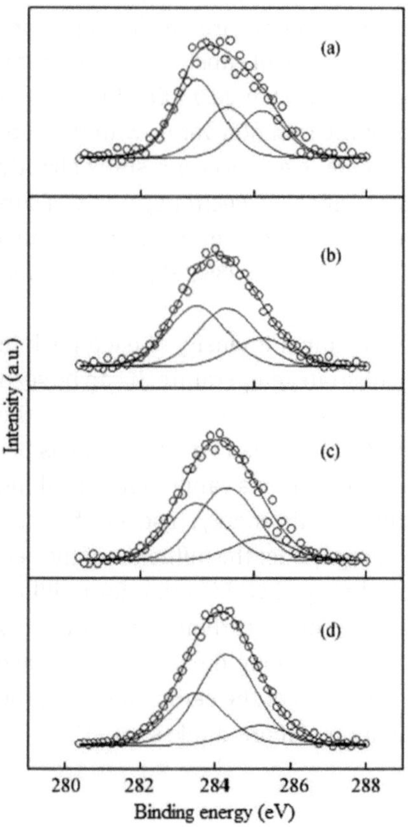

Fig. 6.25: XPS C 1s fitting curve of Ge_xC_{1-x} coatings fabricated under different values of germanium-target sputtering power: (a) 160 W; (b) 120 W; (c) 80 W; (d) 40 W.

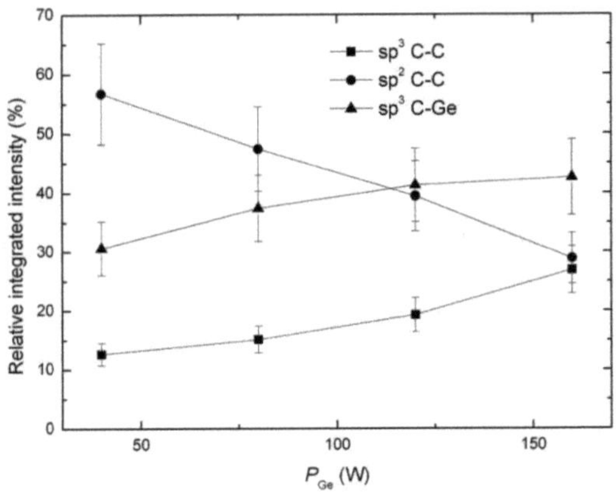

Fig. 6.26: Relationship between the relative integrated intensity of each peak after fitting and the germanium-target sputtering power.

ment with the results of Raman spectral analysis. Moreover, the relative content of sp^3 C–C and Ge–C in the coating increases owing to the enhancement of the sp^3 hybridization of carbon atoms by the increasing content of germanium atoms.

Fig. 6.27 shows the C 1s core level spectra of germanium-carbide coatings fabricated at different substrate temperatures. With the increase of substrate temperature, the profile of the C 1s core level spectrum undergoes some change, indicating that the chemical environment around the carbon atoms of the thin film varies with the substrate temperature. The peak position of the C 1s core level spectrum shifts slightly downward, which is mainly caused by the weakened effect of electronegativity of surrounding germanium atoms on the carbon atoms. The C 1s core level spectra can also be fitted by the three peaks of sp^3 C–C, sp^2 C–C, and Ge–C at 285.2 eV, 284.3 eV, and 283.5 eV, respectively.

Fig. 6.28 shows the variation of the relative integrated intensity of each peak in the C 1s core level spectrum after fitting with the substrate temperature. It can be seen that, with the increase of substrate temperature, the relative sp^3 C–C content in the coating is rapidly reduced but the relative sp^2 C–C content is increased. This is because, at a relatively low substrate temperature, carbon atoms cannot migrate along the substrate surface owing to the lack of sufficient kinetic energy; instead, they can only be distributed on the germanium-carbide coating. Under the influence of the germanium atoms, the carbon atoms mainly exist in the sp^3 hybridization form. When the substrate temperature increases, the kinetic energy of the carbon atoms gradually increases, enabling them to move for a certain distance along the substrate surface for a sufficiently long period of time. Therefore, a portion of the carbon atoms in the coating approach each other, promoting the formation of more stable sp^2 C–C.

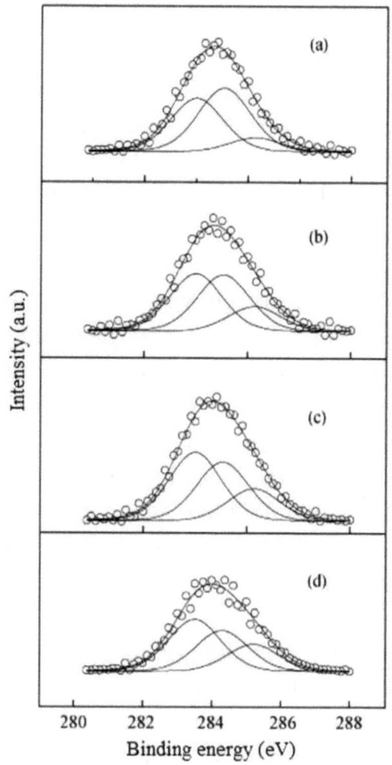

Fig. 6.27: XPS C 1s fitting curves of Ge_xC_{1-x} coatings fabricated under different substrate temperatures: (a) 600 °C; (b) 400 °C; (c) 200 °C; (d) room temperature.

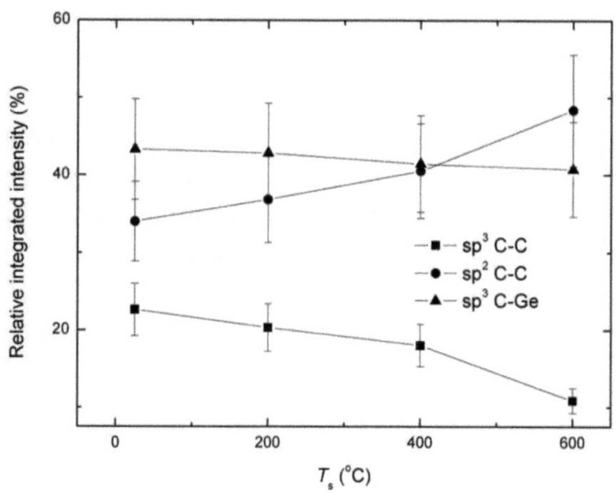

Fig. 6.28: Relationship between the relative integrated intensity of each peak after fitting and the substrate temperature.

6.4.4 Bonding mechanism and rules

The research mentioned above has clarified that the germanium-carbide coating fabricated using the magnetron co-sputtering method has an amorphous structure, in which the atomic arrangement exhibits a short-range orderliness and long-range disorder with various defects. Therefore, the coating properties are mainly influenced by the bonding structures formed between adjacent germanium and carbon atoms. However, the bond type/quantity and degree of atomic disorder are influenced by the deposition conditions. The influencing factors can be divided into two categories: kinetic factors and thermodynamic factors. When particles are attached to the substrate surface without sufficient response and transport time, kinetic factors play the dominant role. In the opposite scenario, thermodynamic factors play the major role.

When a germanium coating is fabricated under a relatively low germanium-target sputtering power, film-forming particles absorbed on the substrate surface have low kinetic energy and, thus, poor diffusivity along the surface, and the germanium and carbon atoms are distributed randomly, indicating an inferior arrangement with low orderliness and more defects. In this case, the carbon content in the coating and the content of corresponding Ge–C, sp^2 C–C, and sp^3 C–C are high. With the increase of germanium-target sputtering power, the kinetic energy of the film-forming particles absorbed on the substrate surface gradually increases, improving their diffusivity along the substrate surface, enhancing the arrangement order of germanium atoms, and reducing the defects. Simultaneously, the carbon content in the coating decreases, and thereby, the content of Ge–C, sp^2 C–C, and sp^3 C–C decreases. Under the influence of germanium atoms, carbon atoms will mainly exist in the sp^3 hybridization form. Similarly, when the substrate temperature is relatively low, film-forming particles absorbed on the substrate surface show weak kinetic energy and poor diffusivity on the substrate surface, and the germanium and carbon atoms demonstrate random distributions with a low level of arrangement order and large number of defects. At this point, carbon atoms in the coating mainly exist in the sp^3 hybridization form. With the increase of substrate temperature, the particles absorbed on the surface acquire increasing kinetic energy and diffusivity along the substrate surface, leading to an improved arrangement order of germanium atoms and a decreased number of defects. Moreover, the carbon atoms in the coating gain sufficient kinetic energy to move for a certain distance along the surface and form stable sp^2 C–C bonds with other carbon atoms, which can consequently decrease the sp^3 C–C content but increase the sp^2 C–C content in the coating.

6.5 Film density

Fig. 6.29 shows the C 1s and Ge 3d core-level spectra of a germanium-carbide coating before being etched by argon ions. According to the analysis in Chapter 5, the C 1s core-level spectrum can be divided into three Gaussian peaks C^1, C^2, and C^3 at 283.3 eV, 285.0 eV, and 286.5 eV, which correspond to the bonding energy of Ge–C, C–C, and C–O bonds, respectively. Similarly, the Ge 3d core-level spectrum can be divided into three Gaussian peaks Ge^1, Ge^2, and Ge^3 at 29.3 eV, 30.2 eV, and 32.2 eV, corresponding to the bonding energy of Ge–Ge, Ge–C, and Ge–O, respectively. This indicates that,

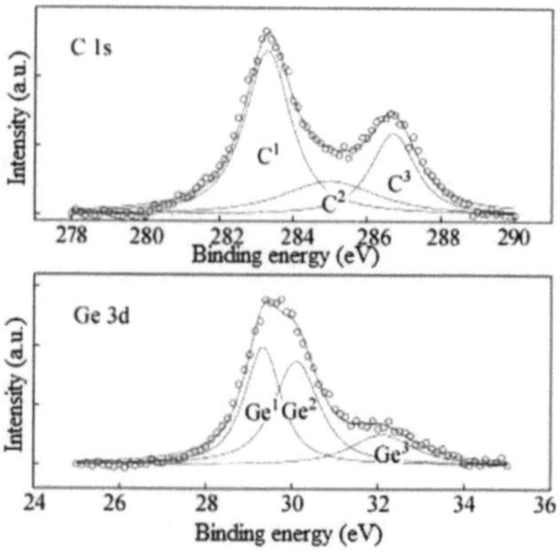

Fig. 6.29: C 1s and Ge 3d core-level spectra of germanium-carbide coating before being etched by argon ions.

while the film surface absorbs gas when the fabricated coating is exposed to air, the germanium atoms with dangling bonds can also combine with oxygen atoms to form oxides, making the coating surface state complicated. Therefore, the coating will have a superficial layer with structural and component complexity. Based on the analysis above, a fitting model of superficial layer/germanium carbide layer/quartz substrate can be established for performing fitting on the reflectance curve in order to obtain the coating density.

According to Parratt's algorithm, the reflectance curve can be fitted using the Rfit2000 software. The film density can be then obtained by searching for the global minimum using the optimization function. The adopted optimization function can be expressed as

$$x^2 = \frac{1}{N-M} \sum \left(\frac{R\exp(j) - KR\,\mathrm{cal}(j)}{\mathrm{d}R\exp(j)} \right)^2, \tag{6.4}$$

where j denotes the measurement point ($j = 1, 2, \ldots, N$), M the variable number, R the reflection strength, and K the scaling factor.

Fig. 6.30 shows the XRR measured curve, fitting results, and fitting model of a germanium-carbide coating, where solid lines and dots represent the measured curve and fitting points, respectively. The measurement shows that the critical angle θ_C of germanium-carbide coatings fabricated under different preparation technologies is always less than 0.3°, and the selected three-layer structure, i.e., superficial layer/ germanium carbide layer/quartz substrate, is able to fit the measured data perfectly.

Fig. 6.31 illustrates the variation of germanium-carbide coating density with the germanium-target sputtering power, where we can see that the film density increases with the increase of germanium-target sputtering power. The primary reason for the film-density increase is the increase of germanium-target sputtering power, germanium atom content in the coating, and the mean atomic weight of atoms in the coating. This trend has also been discovered in other studies. Fig. 6.32 shows the densities of germanium-carbide coatings fabricated in different laboratories vs the C/Ge atomic ratio. Benzi et al. fabricated germanium-carbide coatings through the decomposition

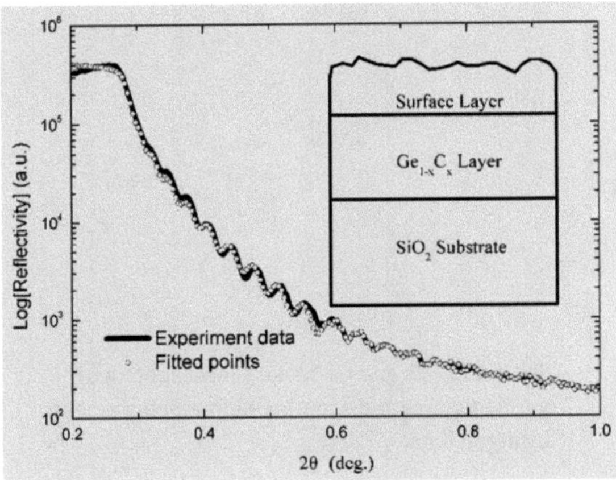

Fig. 6.30: XRR measured curve of the germanium-carbide coating.

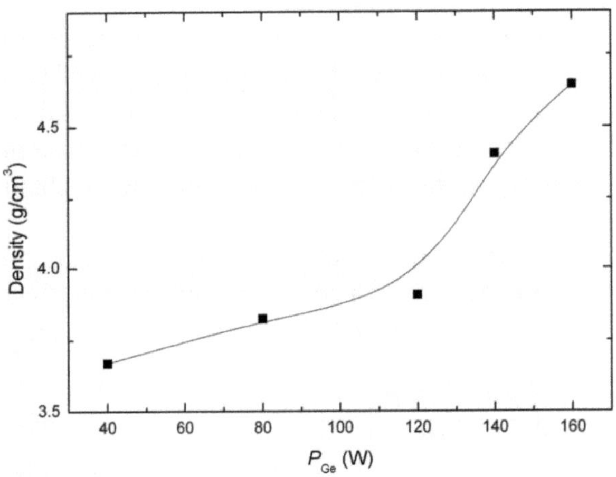

Fig. 6.31: Density variation of a germanium-carbide coating with the sputtering power.

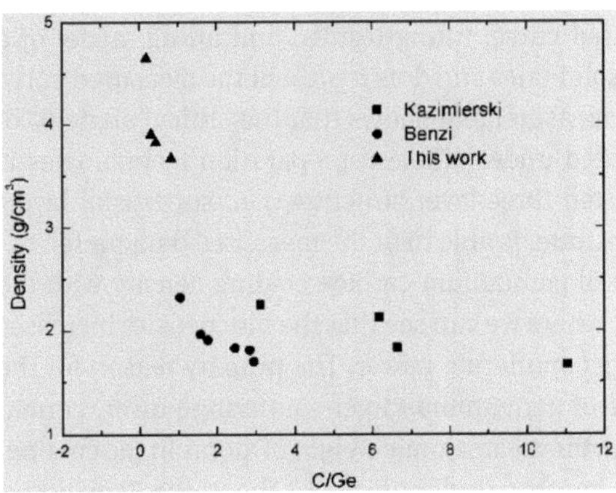

Fig. 6.32: Densities of germanium-carbide coatings versus the C/Ge atomic ratio.

reaction of germane and acetylene gas using x-ray activated chemical vapor deposition. The experimental results demonstrated that the coating density decreased with the increase of carbon atom content in the coating. Kamzimierski et al. fabricated amorphous germanium-carbide coatings using plasma-enhanced chemical vapor deposition. By decomposing organic germanium compounds, they discovered the trend that the film density increased with the increase of germanium content in the thin film. In comparison with the results of these studies, the nonhydrogen germanium-carbide coatings fabricated in our study have higher densities. This is mainly caused by the following three aspects:

(1) compared with chemical vapor deposition, magnetron sputter deposition can fabricate germanium-carbide coatings with superior compactness;
(2) since the precursor used for the thin-film fabrication with chemical vapor deposition contains hydrogen, the large quantity of hydrogen atoms in the coating makes the average molecular weight increase;
(3) germanium-carbide coatings fabricated in the study in this book contain relatively high volumes of germanium atoms.

Fig. 6.33 illustrates the density variation of germanium-carbide coating with the substrate temperature, which suggests that the coating density increases with increasing substrate temperature. There are mainly two reasons for this phenomenon:

(1) with the increase of substrate temperature, the gas between the specimen stage and magnetron sputtering targets expands when heated, leading to a relatively low pressure, a small loss in kinetic energy when film-forming particles fly towards the substrate, and eventually a compact thin film;
(2) film-forming particles gain kinetic energy from a high-temperature substrate, which promotes the shift of the particles along the substrate surface and helps generate thin films with fewer defects, and higher densities.

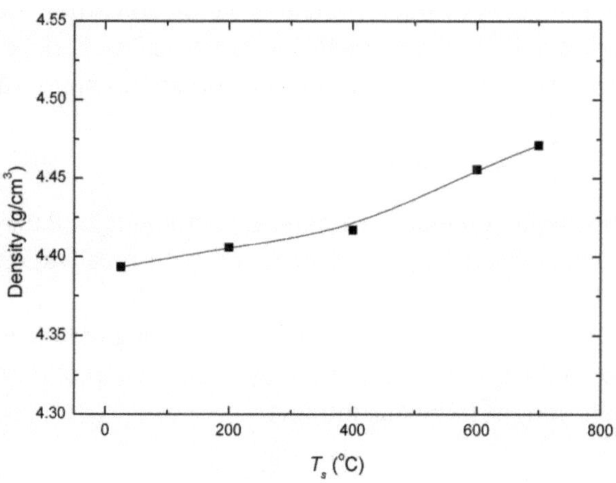

Fig. 6.33: Density variation of a germanium-carbide coating with the substrate temperature.

It is known from the descriptions above that nonhydrogen germanium-carbide coatings fabricated using magnetron co-sputtering have a higher density than hydrogen-containing germanium-carbide coatings. This is due to the fact that the hydrogen atom dramatically reduces the mean molecular weight of the coating as well as the fact that the structure of the hydrogen-containing germanium-carbide coating is not quite dense. Moreover, we also obtained the general trend that the density of germanium-carbide coatings increases with the increase of the Ge/C atomic ratio in the coating. The density of germanium-carbide coatings is primarily affected by two crucial technological parameters, namely the germanium-target sputtering power and substrate temperature. The coating density is increased when the germanium-target sputtering power increases, mostly because the increase of the germanium-target sputtering power can cause the germanium content as well as the average atomic weight of the coating to increase. The coating density is also increased with the increase of substrate temperature; this is because a higher substrate temperature increases the kinetic energy of film-forming particles, which promotes the migration of the particles along the substrate surface and generates coatings with less defects and higher densities.

6.6 Optical properties of thin films

6.6.1 Visible optical properties

The optical parameters of thin-film materials can be measured using multiple approaches, among which ellipsometry is extensively used for measuring the optical constants of coatings owing to its advantages of precise, noncontact, nonintrusive, and sensitive measurement. When a polarized light beam traverses a germanium-carbide coating, under the influence of the atoms in the coating, the polarization state of the light will be changed. Therefore, the optical constants of the germanium-carbide coating can be obtained by measuring the polarization-state variation and fitting the results with an appropriate model. When a planar wave traverses a specimen surface, the ratio of horizontal (ρ)/vertical (s) Fresnel reflection coefficients can be expressed as

$$\rho = \frac{r_{\mathrm{P}}}{r_{\mathrm{S}}} = \tan\phi \exp(\mathrm{i}\Delta), \tag{6.5}$$

where ϕ denotes the polarization angle, Δ the phase difference between light P and light S, r_{P} the polarized component of reflected light in plane P, and r_{S} the polarization component of reflected light in plane S.

In ellipsometry, different values of ϕ and Δ can be acquired by varying the wavelength of incident light, λ. Based on the material properties, by using an appropriate fitting model of optical parameters and the following formula

$$\langle\varepsilon\rangle = \langle\varepsilon_1\rangle + \mathrm{i}\langle\varepsilon_2\rangle = \langle n\rangle^2 = (\langle n\rangle + \mathrm{i}\langle k\rangle)^2 = \sin(\phi)^2 \cdot \left[1 + \frac{\tan^2\phi(1+\rho)}{(1+\rho)^2}\right], \tag{6.6}$$

one can obtain the optical constants and thickness of thin films. It is more difficult to analyse the optical properties of absorbing thin films than those of transparent thin films by using an ellipsometer. This is because the variation of optical constants (n, k) with wavelength is unknown prior to measurement. Therefore, the appropriate construction of the fitting model has critical significance for the validity of the fitting results.

6.6.1.1 Ellipsometry modeling

An appropriate fitting model includes two parts:
(1) a rational physical structure model, and
(2) a suitable dispersion relation.

For a germanium-carbide coating based on a silicon substrate, two physical structure models are adopted in our work for fitting:
(1) Model I: silicon substrate/germanium-carbide coating/effective medium layer (rough layer);
(2) Model II: silicon substrate/silicon-germanium interface layer/germanium-carbide coating/effective medium layer (rough layer).

The fitting results obtained using these two models are listed in Tab. 6.5. It can be seen that the introduction of a silicon-germanium interface layer between the germanium coating and silicon substrate can reduce the mean square error (MSE) by 14 %, significantly decrease the difference between the fitting values and the measured values, and greatly improve the simulation results.

For the constructed physical structure model, when the characteristic parameters of thin films are measured using ellipsometry, the following two types of optical dispersion models can be applied to conduct fitting:
(1) Tauc–Lorentz models. They are suitable for amorphous materials; E, ε_1, ε_2 are usually used to describe the light-material interaction through the expression

$$\varepsilon_2(E) = \frac{AE_0C(E-E_g)^2}{(E^2-E_0^2)^2+C^2E^2}\cdot\frac{1}{E} \quad (E > E_g,\ \text{or}\ E < E_g,\ \varepsilon = 0), \tag{6.7}$$

where A represents the amplitude, E_0 the peak shift energy, C the broadening term of the peak, and E_g the optical band gap.

Tab. 6.5: Fitting results of the two types of models.

Model	t_1, $Ge_{1-x}C_x$ (nm)	t_2, interface (nm)	Roughness (nm)	n (λ = 632.8 nm)	MSE
Model I	133.78 ± 0.470	–	8.38 ± 0.145	4.230	1.065
Model II	132.96 ± 0.454	3.040 ± 0.360	7.46±0.179	4.230	0.914

(2) Effective medium models. Generally, there exist gaps and impurities between the fabricated coating surface and the interface, which will cause a significant difference of optical properties between the surface/interface and the coating material. Equivalent medium models can be applied for the fitting of optical constants on the material surface and the interface. The three most commonly used equivalent medium models are the Lorentz–Lorenz effective medium model, Maxwell–Gamett effective medium model, and Bruggeman effective medium model.

In our test, the Levenberg–Marquardt algorithm is used for the fitting of thin-film parameters. By changing the coating thickness and other related coefficients, the practically measured curve is approached by the fitting curve. The mean square error (MSE) can be calculated using equation (6.8) to evaluate the fitting degree. A smaller MSE denotes a better fitting:

$$\mathrm{MSE} = \sqrt{\frac{1}{3n-m}\sum_{i=1}^{n}\left[(N_{\mathrm{E},i}-N_{\mathrm{G},i})^2+(C_{\mathrm{E},i}-C_{\mathrm{G},i})^2+(S_{\mathrm{E},i}-S_{\mathrm{G},i})^2\right]}\times 1000, \qquad (6.8)$$

where n is the number of measurements; m is the number of variable parameters in the model; $N = \cos(2\phi)$, $C = \sin(2\phi)\cos(\Delta)$, $S = \sin(2\phi)\sin(\Delta)$; symbols with subscript E denote the experimental measured values; and symbols with subscript G represent the theoretical computational values.

In order to verify the validity of the selected model, data measured using the ellipsometer from the germanium-carbide coating fabricated under a germanium-target sputtering power of 140 W and carbon-target sputtering power of 100 W are fitted. Fig. 6.34 illustrates the measured data and fitting results of a germanium-carbide coating. It can be seen that the theoretically calculated data shows a good agreement with measured ones, indicating that model II is highly accurate.

Fig. 6.35 shows the variations of refractive index and extinction coefficient of a germanium-carbide coating with the wavelength. It can be observed that the extinction coefficient of the germanium-carbide coating gradually decreases with the increase of wavelength in the range of 400–900 nm. However, the extinction coefficient is still fairly large (0.3–2.2) in the visible range, showing that the germanium-carbide

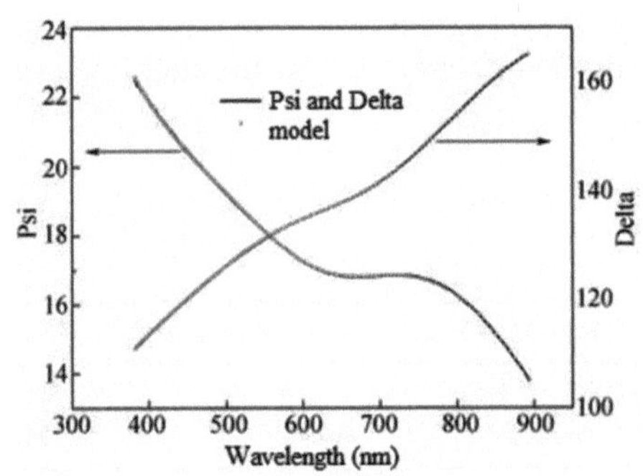

Fig. 6.34: Data measured using an ellipsometer from and fitting results of a germanium-carbide coating.

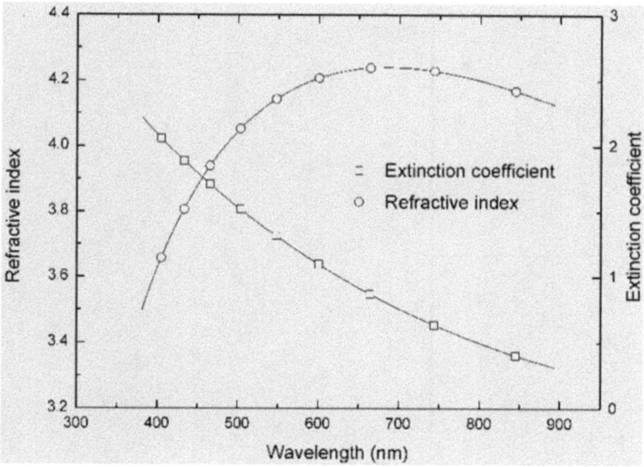

Fig. 6.35: Variations of refractive index and extinction coefficient of a germanium-carbide coating with the wavelength.

coating is a light-absorbing material. It can also be observed from the figure that, with the increase of wavelength, the refractive index rapidly increases before slowly decreasing, which is a trend similar to that shown by amorphous germanium thin films.

6.6.1.2 Optical constants

In order to investigate the influence of technological parameters on the optical constants, the refractive index and extinction coefficient in the case of $\lambda = 632.8$ nm are studied. Fig. 6.36 shows the relationship between the refractive index & extinction coefficient of a germanium-carbide coating and the germanium-target sputtering power.

It can be seen from Fig. 6.36 that, when the germanium-target sputtering power increases from 40 W to 160 W, the refractive index increases from 3.0 to 4.5 and extinction coefficient increases from 0.12 to 1.15. This phenomenon can be explained by considering the fact that the germanium content increases but the carbon content decreases when the germanium-target sputtering power is increased. As we know, the refractive index of amorphous germanium thin films is greater than that of amorphous carbon thin films. Therefore, an increasing germanium content in the germanium-carbide coating makes the coating structure closer to the amorphous germanium structure and increases the refractive index. Similarly, the increase of extinction coefficient with increasing germanium-target sputtering power is also due to the increase of germanium content in the coating.

Fig. 6.37 shows the variations of refractive index and extinction coefficient of a germanium-carbide coating with the substrate temperature. It can be observed that, when the substrate temperature increases from room temperature to 600 °C, the refractive index of the germanium-carbide coating increases from 4.1 to 4.4, and the

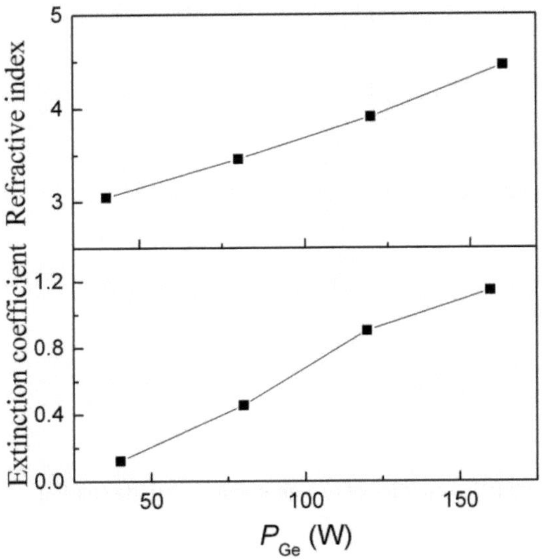

Fig. 6.36: Variation of refractive index and extinction coefficient of a germanium-carbide coating with the germanium-target sputtering power.

extinction coefficient increases from 1.10 to 1.44. This phenomenon can be understood if we consider the fact that, with the increase of substrate temperature, the defect quantity in the coating decreases and the coating density increases. It is known that the refractive index of amorphous coatings represents the overall performance of the scattering effect on photons caused by the atoms in the coating. Therefore, the larger the quantity of atoms per unit volume, the higher is the refractive index of thin films. There are two reasons for the increase of extinction coefficient with the substrate temperature:

(1) the increase of coating density enhances the light absorption of the coating;
(2) the volume of sp^2 carbon atoms in the coating increases with the increase of substrate temperature, leading to the increase of extinction coefficient.

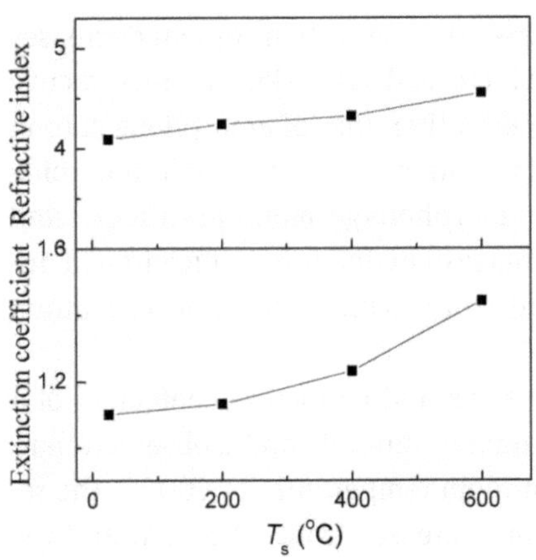

Fig. 6.37: Variations of refractive index and extinction coefficient of a germanium-carbide coating with the substrate temperature.

6.6.1.3 Optical band gap

Fig. 6.38 shows the calculation of the optical band gap of a germanium-carbide coating, where the absorptivity α can be solved based on the extinction coefficient k with the equation $\alpha = 4\pi k/\lambda$. The experimental measurement points are calculated using the extinction coefficient that is fitted by the ellipsometer approach mentioned above. By extrapolating the linear section of measurement points, when $\alpha h\nu = 0$, the value of photon energy is exactly the optical band gap of the germanium-carbide coating, E_g.

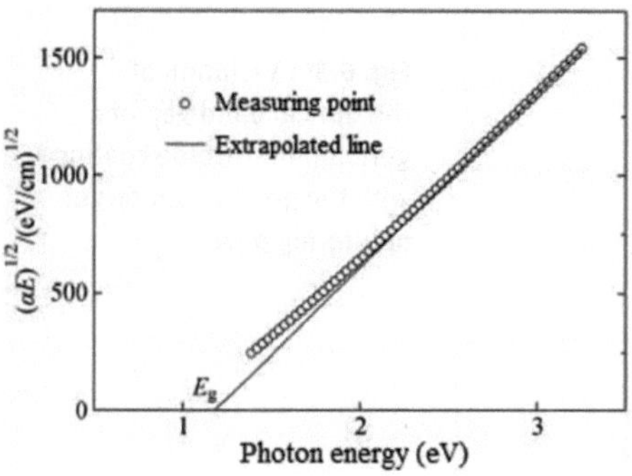

Fig. 6.38: Calculation of the optical band gap of a germanium-carbide coating.

Fig. 6.39 shows the relationship between the optical band gap of a germanium-carbide coating and the germanium-target sputtering power. We can see that, when the germanium-target sputtering power increases from 40 W to 160 W, the optical band gap of the germanium-carbide coating decreases from 1.55 eV to 1.05 eV. This indicates that the optical band gap of the germanium-carbide coating can vary with the change of germanium-target sputtering power in a relatively wide range. This is because, with the increase of the germanium-target sputtering power, the germanium/carbon atomic ratio in the coating gradually increases. The optical band gap of amorphous germanium coatings is only in the range of 0.6–0.8 eV, which is less than that of amorphous carbon coatings. When the germanium content is dominant in a germanium-carbide coating, its structure will be close to the amorphous germanium structure and, therefore, the optical band gap will be narrower.

Fig. 6.40 shows the relationship between the optical band gap of a germanium-carbide coating and the substrate temperature. It can be seen that, when the substrate temperature increases from room temperature to 600 °C, the optical band gap of the germanium-carbide coating slightly increases to 1.14 eV and then drops to 1.12 eV without a significant overall change. The trend of change of the optical band gap of the germanium-carbide coating is due to the variation of Ge–C and C–C bonds in the coating with the substrate temperature. In the case where the content of germanium is similar to that of carbon atoms in the coating, the optical band gap of the germanium-carbide coating is mainly dependent on the spatial distribution of π bonds in sp^2 hybridised carbon. Although the quantities of both Ge–C and C–C bonds increase

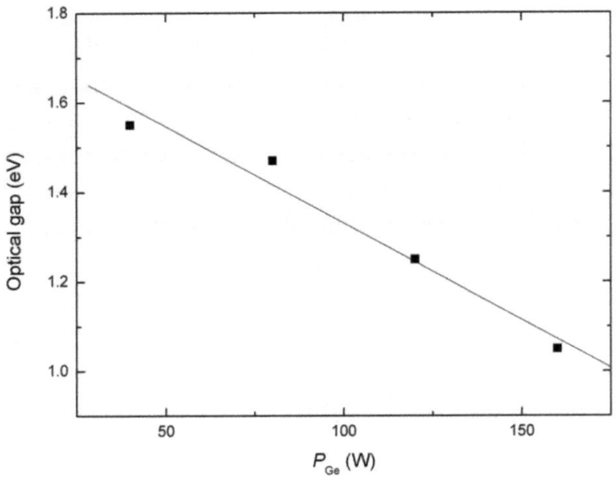

Fig. 6.39: Variation of the optical band gap of a germanium-carbide coating with the germanium-target sputtering power.

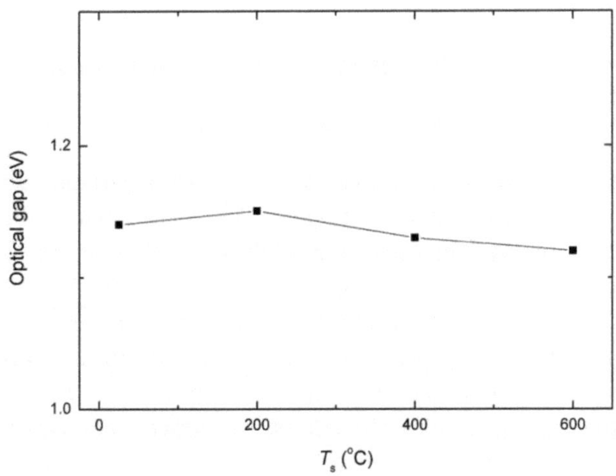

Fig. 6.40: Variation of the optical band gap of a germanium-carbide coating with the substrate temperature.

when the substrate temperature increases, growth rate of Ge–C bonds is higher than that of C–C bonds at a relatively low temperature, leading to an increase of the optical band gap in this range. However, when the temperature is high, Ge–C bonds grow significantly more slowly than C–C bonds, which will result in a large number of sp^2 carbon atoms and eventually reduce the optical band gap of the coating.

6.6.2 Infrared optical properties

6.6.2.1 Refractive index

For infrared antireflective/protective coating materials, the refractive index within the infrared range is a critical parameter in coating design. In the 1970s, Manifacier proposed a relatively simple method called the envelope method for calculating the optical constants of coatings. This method calculates the optical constants of coatings according to the envelope curves that are plotted based on the transmission curve measured in a certain range of wavelength. In the 1980s, Swanepoel modified this method.

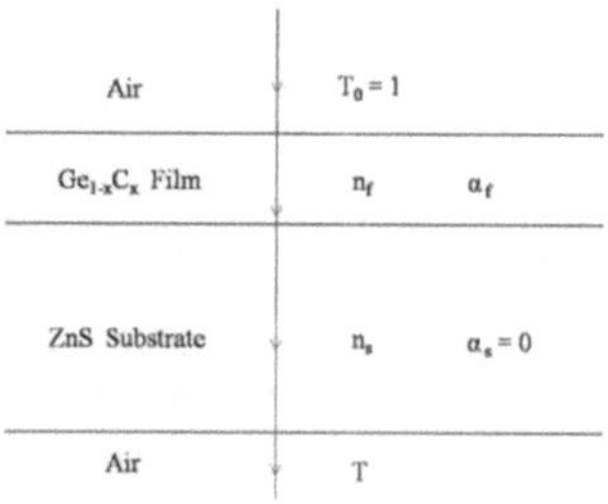

Fig. 6.41: Transmission schematic of the system consisting of a germanium-carbide coating and a zinc-sulphide substrate.

For the system consisting of a germanium-carbide coating and a zinc-sulphide substrate shown in Fig. 6.41, the thickness, refractive index, and absorption coefficient of the germanium-carbide coating are d, n_f, and α_f, respectively. The zinc-sulphide substrate has a thickness that is far greater than that of the germanium-carbide coating, a refractive index of n_s, and an absorption coefficient of 0. The transmission of this system can be obtained using the following equation:

$$T = \frac{Ax}{B - Cx\cos\varphi + Dx^2}, \tag{6.9}$$

where

$$A = 16n_s(n_f^2 + k^2); \quad B = [(n_f + 1)^2 + k^2][(n_f + 1)(n_f + n_s^2) + k^2];$$
$$C = 2\cos\varphi[(n_f^2 - 1 + k^2)(n_f^2 - n_s^2 + k^2) - 2k^2(n_s^2 + 1)]$$
$$- 2k\sin\varphi[2(n_f^2 - n_s^2 + k^2) + (n_s^2 + 1)(n_f^2 - 1 + k^2)];$$
$$D = [(n_f - 1) + k^2][(n_f - 1)(n_f - n_s^2) + k^2];$$
$$\varphi = 4\pi n_f d/\lambda; \quad x = \exp(-\alpha_f d); \quad \alpha_f = 4\pi k/\lambda.$$

When $k = 0$ and m is an integer, according to the optical interference equation,

$$2n_f d = m\lambda, \tag{6.10}$$

the following relationship exists at the extreme points of the coherent stripes:

$$T_M = \frac{Ax}{B - Cx + Dx^2}, \tag{6.11}$$

$$T_m = \frac{Ax}{B + Cx + Dx^2}, \tag{6.12}$$

$$\frac{1}{T_m} - \frac{1}{T_M} = \frac{2C}{A}. \tag{6.13}$$

According to equation (6.14), the refractive index of the coating can be obtained using the equation below:

$$n_f = \left[N + (N^2 - n_s^2)^{1/2}\right]^{1/2}, \tag{6.14}$$

where

$$N = 2n_s\left(\frac{1}{T_m} - \frac{1}{T_M}\right) + \frac{n_s^2 + 1}{2}.$$

Fig. 6.42 illustrates the transmission envelope curves of the germanium-carbide coating and zinc-sulphide substrate system. The process of solving for the refractive index of the coating using the envelope curves is described below:

(1) draw the practically measured transmission curve and find the extreme points on the curve;
(2) fit the envelope curves using the polynomial fitting approach based on the extreme points, and adjust the envelope curves;
(3) calculate the interference orders of refractive index, thickness, and extreme points, and round off the interference orders;
(4) if the interference orders are consecutive integers, recalculate the thickness using equation (6.14), calculate the refractive index of the extreme point based on modified data, and fit the refractive index curve; if not, readjust the envelope curves and then conduct the calculation.

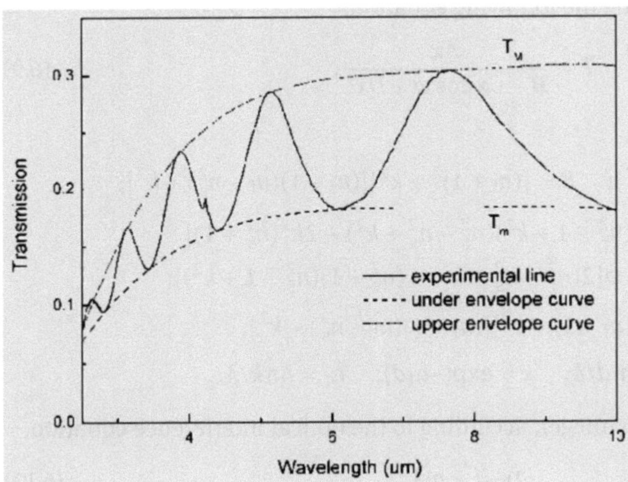

Fig. 6.42: Transmission envelope curves of the system of a germanium-carbide coating and zinc-sulphide substrate.

Here, we report a study on the variation of the refractive index at 9 μm with the technological parameters. Fig. 6.43 shows the variation of the refractive index of a germanium-carbide coating with the germanium-target sputtering power. It can be observed that, when the germanium-target sputtering power increases from 40 W to 160 W, the refractive index of the coating increases from 2.8 to 4.1. This phenomenon of refractive-index change can be explained by the fact that, when the germanium-target

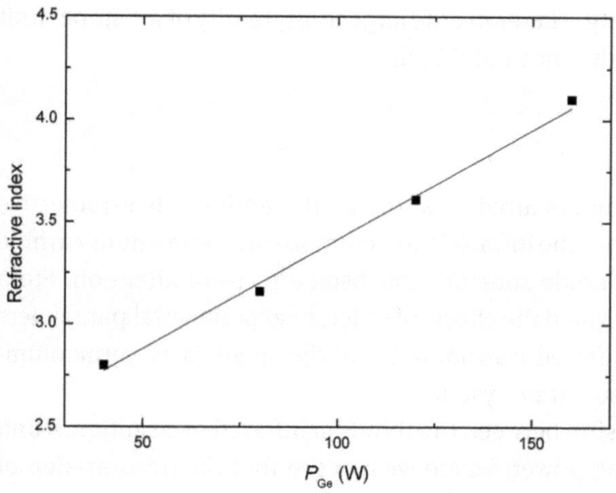

Fig. 6.43: Variation of the refractive index of a germanium-carbide coating with the germanium-target sputtering power.

sputtering power increases, the germanium content in the coating increases but the carbon content decreases. The refractive index of amorphous germanium coatings is higher than that of amorphous carbon coatings. When the germanium content in the germanium-carbide coating increases, the thin-film structure becomes similar to the amorphous germanium structure and, therefore, the refractive index increases.

Fig. 6.44 illustrates the relationship between the refractive index of a germanium-carbide coating and the substrate temperature. It can be seen that, when the substrate temperature increases from room temperature to 600 °C, the refractive index of the germanium-carbide coating increases from 3.9 to 4.1. This refractive-index change occurs because, with the increase of substrate temperature, the number of defects in the coating decreases and the coating density increases. The refractive index of an amorphous coating represents the overall performance of the scattering effect on photons

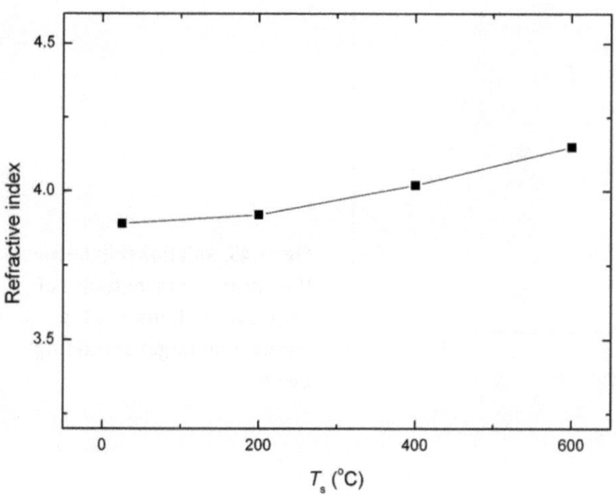

Fig. 6.44: Variation of refractive index of a germanium-carbide coating with the substrate temperature.

caused by the atoms in the coating. Therefore, the larger the quantity of atoms per unit volume, the higher is the refractive index of thin films.

6.6.2.2 Infrared transmittance

In the application of a germanium-carbide coating as the antireflective/protective coating of a zinc-sulphide window, the infrared transmittance of a germanium-carbide coating deposited on a zinc-sulphide substrate has been a focus of attention. From this perspective, we have investigated the effects of different experimental parameters on the infrared transmittanceinfrared transmittance of the single-layer germanium-carbide coating/zinc-sulphide substrate system.

Fig. 6.45 shows the relationship between the thin-film infrared transmittance and the germanium-target sputtering power, where we can see that the transmission of a zinc-sulphide substrate on which a germanium-carbide coating is deposited is less than that of an uncoated zinc-sulphide substrate. This is because the refractive index of the fabricated germanium-carbide coating is greater than that of the zinc-sulphide substrate. According to the optical thin-film theory, when the thin-film refractive index is greater than that of the substrate, the coating/substrate system will have a reflection enhancement effect that can reduce the transmission. By performing integration over the transmission curve, the mean transmission of each film system within the waveband of 8–12 µm can be obtained. Fig. 6.46 shows the relationship between the mean transmission within the band of 8–12 µm and the germanium-target sputtering power. It can be observed that, with the increase of germanium-target power, the mean transmission of films gradually decreases.

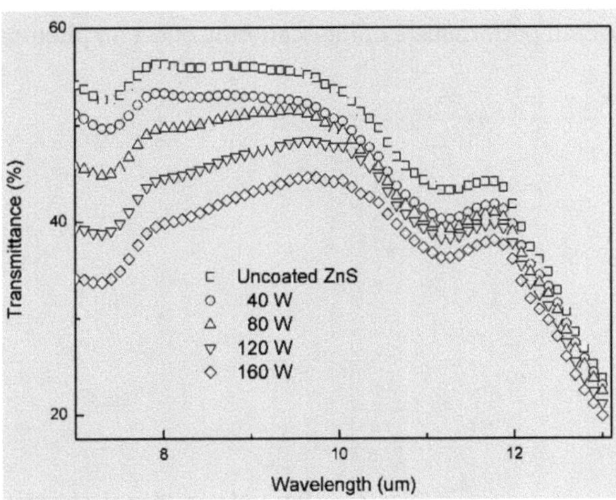

Fig. 6.45: Relationship between the infrared transmittance of Ge_xC_{1-x}/ZnS films and the germanium-target sputtering power.

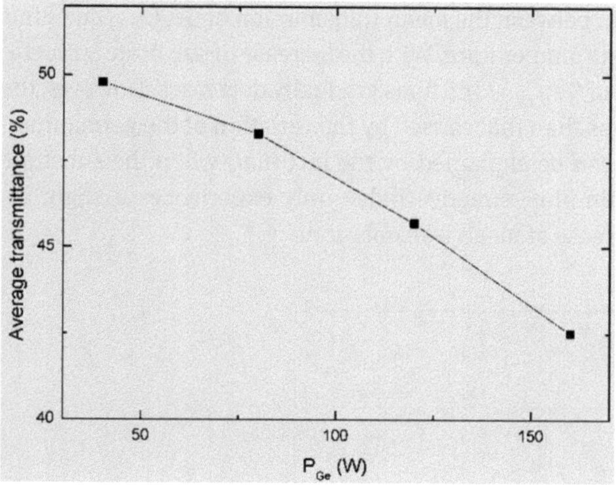

Fig. 6.46: Relationship between the mean transmission of Ge_xC_{1-x}/ZnS films within 8–12 µm and the germanium-target sputtering power.

Fig. 6.47 shows the relationship between the infrared transmittance of Ge_xC_{1-x}/ZnS films and the substrate temperature, indicating that the infrared transmittance of a zinc-sulphide substrate on which a germanium-carbide coating is deposited is less than that of an uncoated zinc-sulphide substrate. This is because the refractive index of the fabricated germanium-carbide coating is greater than that of the zinc-sulphide substrate.

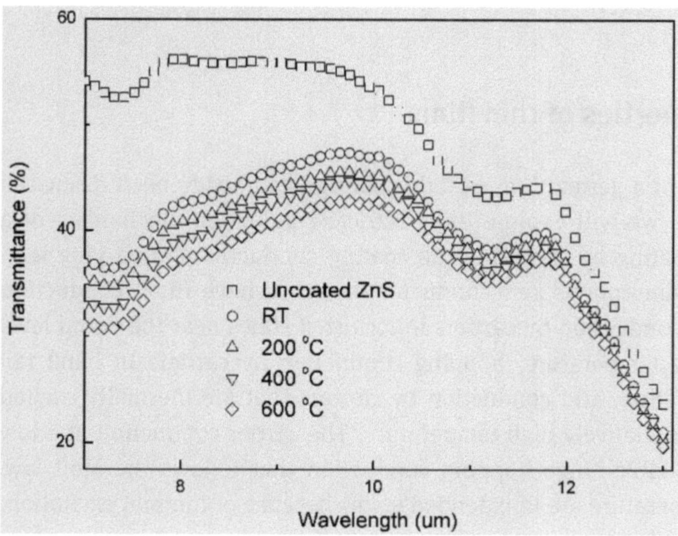

Fig. 6.47: Relationship between the infrared transmittance of Ge_xC_{1-x}/ZnS films and the substrate temperature.

Fig. 6.48 shows the relationship between the mean transmission of Ge_xC_{1-x}/ZnS films within 8–12 μm and the substrate temperature. With the increase of substrate temperature, the average transmission of Ge_xC_{1-x}/ZnS films gradually decreases. However, the magnitude of decrease is far less than that caused by the variation of the germanium-target sputtering power. This can be explained by the fact that, when the substrate temperature increases, the thin-film refractive index only experiences a slight increase, leading to a minor decrease of mean transmission.

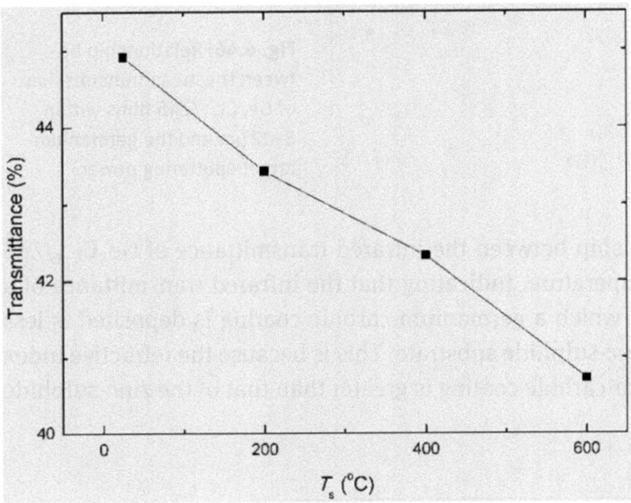

Fig. 6.48: Relationship between the mean transmittance of Ge_xC_{1-x}/ZnS films within 8–12 μm and the substrate temperature

6.7 Electrical properties of thin films

The optical band gap of a germanium-carbide coating has already been discussed above. In this section, we will explore the electrical conduction mechanism of a germanium-carbide coating by analysing the coating conductivity at varying temperatures. Generally, amorphous semiconductor materials have three conduction mechanisms, namely conduction by carriers in localized states near the Fermi level under a relatively low temperature, hopping conduction by carriers in band tail states at room temperature, and conduction by carriers that are thermally excited to extended states at a relatively high temperature. The carrier conduction at a low temperature is the variable-range hopping conduction that follows the Mott law. Carriers in a high temperature are in extended states because of thermal excitation, following the Arrhenius law:

$$\sigma(T) = \sigma_0 \exp\left(\frac{E_{act}}{kT}\right), \tag{6.15}$$

where σ_0 denotes the pre-exponential factor, E_{act} the activation energy, and k the Boltzmann constant.

Fig. 6.49 shows the variation of conductivity of germanium-carbide coatings fabricated under different values of germanium-target sputtering power. We can see that $\lg \sigma(T)$ is not always directly proportional to $1/T$ in the entire temperature range, indicative of at least two conduction mechanisms in the studied temperature range.

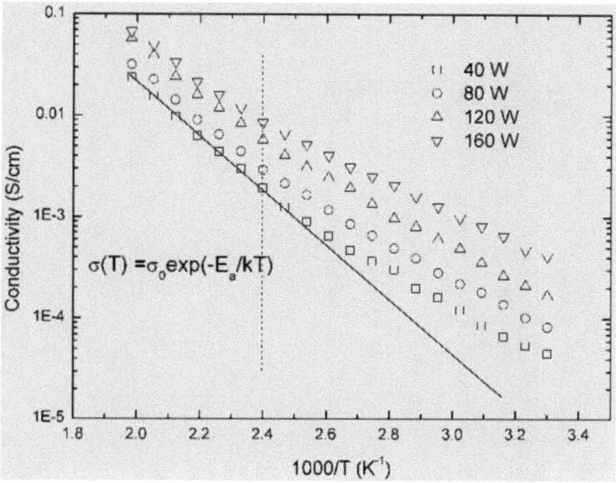

Fig. 6.49: Variation of Arrhenius conductivity of germanium-carbide coatings fabricated under different values of germanium-target sputtering power with the reciprocal of temperature.

Based on the Arrhenius equation, the values of σ_0 and E_{act} of a coating at a temperature greater than 400 K are listed in Tab. 6.6. It is observed from the table that, when the germanium-target sputtering power increases from 40 W to 160 W, σ_0 decreases from 3211 S/cm to 1543 S/cm. Moreover, with the increase of germanium-target sputtering power, E_{act} of the germanium-carbide coating gradually decreases. This is because, with the increase of germanium-target power, the germanium content in the coating gradually increases and carrier quantity participating in the hopping conduction rises, leading to an increase of conductivity and a decrease of activation energy.

Tab. 6.6: Fitting parameters of σ_0 and E_{act} for germanium-carbide coatings fabricated under different values of germanium-target sputtering power.

Specimen	Factor σ_0 (S/cm)	Activation energy E_{act} (eV)
a-$Ge_{1-x}C_x$-40 W	3211	0.516
a-$Ge_{1-x}C_x$-80 W	3128	0.499
a-$Ge_{1-x}C_x$-120 W	3290	0.458
a-$Ge_{1-x}C_x$-160 W	1543	0.407

Fig. 6.50 shows the variation of conduction of germanium-carbide coatings fabricated under different substrate temperatures with the temperature. It can be observed that the conductivity of the germanium-carbide coating slightly increases when the substrate temperature increases. However, the curve shape shows no significant change, demonstrating that the effect of substrate temperature on the thin-film electrical properties is relatively weak.

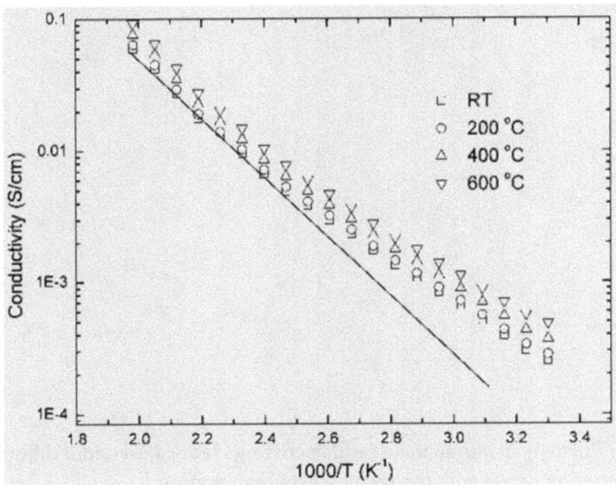

Fig. 6.50: Variation of Arrhenius conductivity of germanium-carbide coatings fabricated under different substrate temperatures with the reciprocal of temperature.

Similarly, σ_0 and E_{act} of the coatings are calculated according to the Arrhenius equation and listed in Tab. 6.7. It can be observed that, when σ_0 increases from 2094 S/cm to 2562 S/cm, E_{act} of the germanium-carbide coatings decreases slightly. At temperatures greater than 400 K, the computational values of σ_0 are all greater than 1000 S/cm. Therefore, conduction by carriers thermally excited to extended states is the conduction mode of these coatings. It can also be observed that, with the increase of substrate temperature, E_{act} of the germanium-carbide coating gradually decreases. This can be explained by the fact that, when the substrate temperature increases, the content of sp^2 carbon in the coating increases, the optical band gap decreases, and the excitation of carriers into localized states of the conduction band becomes easier, leading to an enhancement of conductivity and a reduction of activity energy of the coatings.

Based on the study of the electrical properties of germanium-carbide coatings described above, the conduction process of the thin film can be summarized as follows: when the temperature is less than 400 K, the carriers in the germanium-carbide coating are easily excited to localized states of the conduction band. At this time, the thin-film conductivity is substantially increased by the hopping conduction mode.

Tab. 6.7: Fitting parameters of σ_0 and E_{act} for germanium-carbide coatings fabricated under different substrate temperatures.

Specimen	Factor σ_0 (S/cm)	Activation energy E_{act} (eV)
a-$Ge_{1-x}C_x$-RT	2094	0.449
a-$Ge_{1-x}C_x$-200 °C	2186	0.446
a-$Ge_{1-x}C_x$-400 °C	2380	0.436
a-$Ge_{1-x}C_x$-600 °C	2562	0.429

When the temperature is greater than 400 K, carriers in the coating are thermally excited to extended states. At this point, the thin-film conductivity is increased by the thermal activation mode.

6.8 Mechanical properties of thin films

6.8.1 Hardness and Young's modulus

Nano-indentation is the ideal approach to measure the mechanical parameters such as hardness (H) and Young's modulus (E), which can be calculated using the Oliver–Pharr method based on the measured load–displacement (P–h) curve. However, this method can only provide a single value of hardness and Young's modulus from one indentation. In practice, the hardness and Young's modulus of a general coating–substrate system are not constants; they gradually change with the increase of indentation depth. The hardness-displacement (H–h) relationship can be acquired in a single-shot measurement by obtaining the contact stiffness-displacement (S–h) curve in the indentation process with the CSM dynamic testing technique. In conventional theoretical models, when an indenter is pressed into the specimen surface, elasto-plastic deformation is generated and triangular-pyramid hardness indentation is formed. After unloading, only a portion of the elastic displacement can be recovered. Fig. 6.51 shows a typical load–displacement curve of a germanium-carbide coating.

According to the theory of elastoplastic deformation, hardness can be expressed as

$$H = \frac{P_{max}}{A}, \tag{6.16}$$

where A is the projected contact area.

The contact stiffness S is the initial slope of the unloading curve, $\mathrm{d}P/\mathrm{d}h$, which is associated with the projected contact area and elastic modulus, and

$$S = \frac{\mathrm{d}P}{\mathrm{d}h} = 2\beta\sqrt{\frac{A}{\pi}}E_r, \tag{6.17}$$

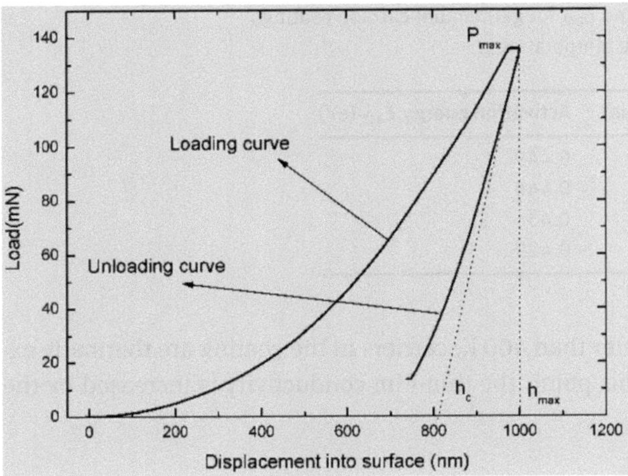

Fig. 6.51: Typical load–displacement curve of a germanium-carbide coating.

where β (1.034) represents the geometric constant of the Berkovich indenter, and E_r the equivalent Young's modulus over the indentation depth, including elastic deformation of both the indenter and the specimen, which can be expressed as

$$\frac{1}{E_r} = \frac{1 - \upsilon^2}{E} + \frac{1 - \upsilon_i^2}{E_i}, \tag{6.18}$$

where E and υ denote the Young's modulus and Possion's ratio of test specimen, respectively, and E_i and υ_i the Young's modulus (1141 GPa) and Possion's ratio (0.07) of diamond indenter, respectively.

Therefore, the hardness H and elastic modulus E of the thin-film material can be obtained by measuring the maximum load P_{max}, projected contact area A, and contact stiffness S from the loading curve.

Fig. 6.52 shows the variation of hardness and elastic modulus of a germanium-carbide coating with the germanium-target sputtering power. It can be observed that both the hardness and elastic modulus of the germanium-carbide coating increase with the increase of the germanium-target sputtering power. As discussed for germanium-carbide coating structures, in a germanium-carbide coating fabricated under a high germanium-target sputtering power, the content of sp^2 carbon atoms is relatively low but the content of sp^3 carbon atoms is high. In this case, the hardness of the germanium-carbide coating changes with its microstructural variation. On the other hand, the increase of carbon content in the thin film may increase the number of defects in the coating and lead to hardness reduction.

Fig. 6.53 shows the variation of hardness and elastic modulus of the germanium-carbide coating with the substrate temperature. Both the hardness and elastic modulus of the germanium-carbide coating increase with the increase of substrate temperature. As discussed for germanium-carbide coating structures, with the increase of

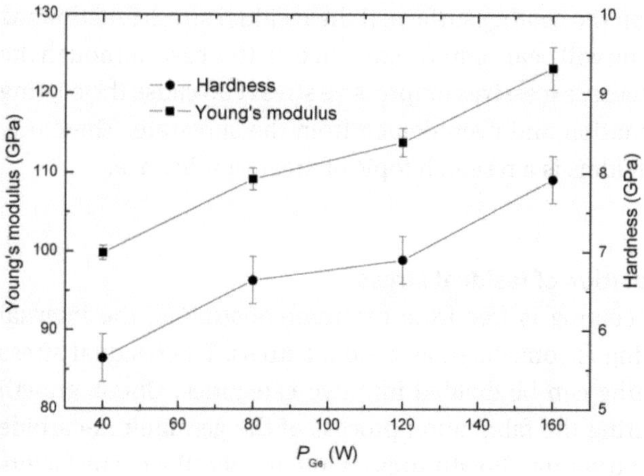

Fig. 6.52: Variation of hardness and elastic modulus of a germanium-carbide coating with the germanium-target sputtering power.

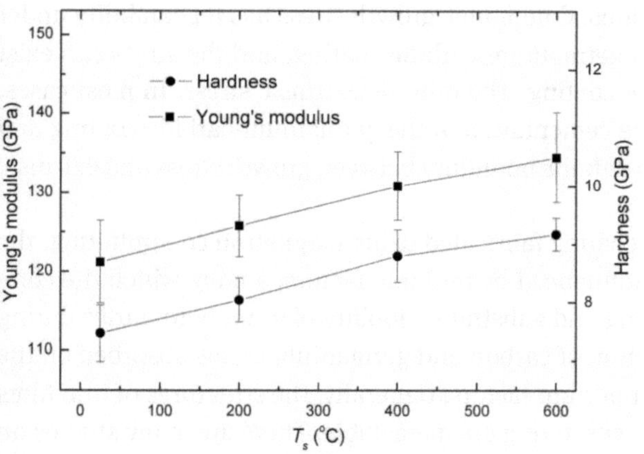

Fig. 6.53: Variation of hardness and elastic modulus of a germanium-carbide coating with the substrate temperature.

substrate temperature at high substrate temperatures, although the sp^2 carbon atom content in the fabricated germanium-carbide coating is increased, the coating is more compact with fewer defects and the coordination number of germanium atoms increases, leading to an enhancement in thin-film hardness.

6.8.2 Residual stress

Generally speaking, stress will inevitably occur in thin films during the film-growing process. The stress significantly impacts the structure and properties of the thin film. In terms of the force direction, residual stress can be divided into two categories: tensile stress and compressive stress. If the residual stress behaves as a tensile stress in the coating, the coating will bear the force that drives it to stretch. When the tensile stress is sufficiently large, cracking occurs between the coating and the substrate or even surface cracks are generated on the coating surface. If the residual stress functions as a compressive stress, the coating will bear compressive force. In this case, although the coating has superior compactness, excessive compressive stress can cause the coating to experience warping deformation and even detach from the substrate. Therefore, residual stress existing in thin films is a research topic of great significance.

6.8.2.1 Generation and calculation of residual stress

When a germanium-carbide coating is free from external constraints, the internal stress distributed in the coating is considered as residual stress. The residual stress of a germanium-carbide coating can be divided into two categories. One is growth stress, which is generated during the fabrication process of the germanium-carbide coating using magnetron co-sputtering. Growth stress is intensively affected by factors including carbon/germanium content in the coating, substrate temperature during deposition, and growth conditions. Commonly, growth stress has repeatability under a given growing condition of a germanium-carbide coating, and the stress can exist for a long period of time in the coating. The other is extrinsic stress. In most cases, extrinsic stress is caused by the cementation of the germanium-carbide coating and substrate. However, there is no definite boundary between growth stress and extrinsic stress.

For a germanium-carbide coating fabricated using magnetron co-sputtering, the variation of intrinsic stress is influenced by multiple factors, among which the combination condition of the coating and substrate, mobility of grain boundaries during the coating growth, and migration of carbon and germanium atoms absorbed on the coating surface are the most important factors. Generally, the structures of thin films fabricated using magnetron co-sputtering are metastable. There are many studies on the generation mechanisms of germanium-carbide coatings fabricated using the magnetron co-sputtering method. However, complete models for estimating the stress in films have not yet been proposed.

Owing to the impact of stress during the fabrication process of germanium-carbide coatings, the substrate tends to suffer from bending deformation. The residual stress in the coating can be calculated by measuring the radius of curvature for the bending deformation of the substrate. Coatings of different thicknesses have different residual stresses. Therefore, the thickness of the specimen fabricated in our study for the calculation of the stress of germanium-carbide coatings is set to 1 μm. Stoney proposed the correlation between the curvature radius of bending deflection of the substrate and the thin-film residual stress, as

$$\sigma_{\mathrm{f}} = \left(\frac{E}{1-\gamma}\right)_{\mathrm{s}} \frac{t_{\mathrm{s}}^2}{6rt_{\mathrm{f}}}, \tag{6.19}$$

where t_{f} denotes the film thickness, r the curvature radius of the substrate, E the elastic modulus of the substrate, and γ the Poisson's ratio of the substrate.

Germanium-carbide coatings in our test are fabricated at different temperatures. Owing to the difference in the coefficient of thermal expansion between the germanium-carbide coating and the silicon substrate, when the fabricated coating is in the process of reaching room temperature, a relatively large thermal mismatch is generated between the coating and the substrate. This generated stress is named thermal stress, which can be calculated using the following equation:

$$\sigma_{\text{therm}} = E_{\text{f}}(\alpha_{\text{f}} - \alpha_{\text{s}})(T_{\text{p}} - T_{\text{m}}), \tag{6.20}$$

where E_{f} is the elastic modulus of the coating; α_{f} and α_{s} the coefficients of thermal expansion of the coating and the substrate, respectively; and T_{p} and T_{m} the temperatures of deposition and measurement, respectively.

Tab. 6.8: Residual stress, thermal stress, and intrinsic stress of germanium-carbide coatings fabricated using different techniques.

Specimen	Residual stress (MPa)	Thermal stress (MPa)	Intrinsic stress (MPa)
a-$Ge_{1-x}C_x$-40 W	94.83	99.57	−4.74
a-$Ge_{1-x}C_x$-80 W	66.48	108.95	−42.47
a-$Ge_{1-x}C_x$-120 W	60.39	113.58	−53.19
a-$Ge_{1-x}C_x$-160 W	51.44	122.95	−71.51
a-$Ge_{1-x}C_x$-RT	−50.30	0	−50.30
a-$Ge_{1-x}C_x$-200 °C	52.94	125.45	−72.51
a-$Ge_{1-x}C_x$-400 °C	62.06	279.72	−217.66
a-$Ge_{1-x}C_x$-600 °C	109.33	437.52	−328.19

Tab. 6.8 lists the residual stress, thermal stress, and intrinsic stress of germanium-carbide coatings fabricated using different techniques, where positive values and negative values represent tensile stress and compressive stress, respectively. Residual stress can be calculated using equation (6.19) based on the radius of curvature measured from the bending deflection of the substrate. Thermal stress can be solved for using equation (6.20), where the coefficient of thermal expansion of the silicon substrate is 2.3×10^{-6}/°C and the coefficient of thermal expansion of the germanium-carbide coating is approximately 8.0×10^{-6}/°C. Considering the influence of the relative magnitudes of thermal expansion coefficients of the coating and substrate materials on the thermal stress, in order to investigate the effect of the structure on the coating stress, only intrinsic stress is discussed in our work. The computational formula is

$$\sigma = \sigma_{\text{f}} - \sigma_{\text{therm}}. \tag{6.21}$$

6.8.2.2 Experimental results

Fig. 6.54 shows the intrinsic stress of germanium-carbide coatings fabricated under different values of germanium-target sputtering power. The intrinsic stresses measured in experiments all behave as compressive stresses with relatively small magnitudes. It can be observed that the stress of germanium-carbide coatings is at the MPa level. It should be noted that the intrinsic stress of a germanium-carbide coating fabricated under a germanium-target sputtering power of 40 W is almost zero. With the increase of the germanium-target sputtering power, the intrinsic stress of the coating gradually increased from 5 MPa to 70 MPa. The results indicate that, when the sputtering power decreases to a certain degree, the internal stress of the germanium-carbide coating can be essentially eliminated.

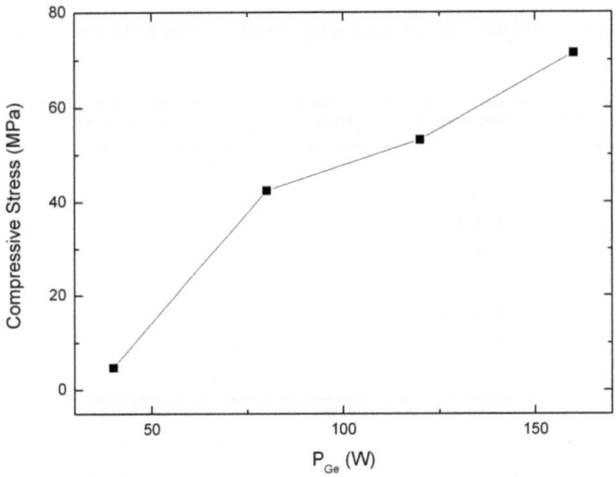

Fig. 6.54: Intrinsic stress of germanium-carbide coatings fabricated under different values of germanium-target sputtering power.

The low internal stress of germanium-carbide coatings can be explained from the perspective of atomic stress, which reflects the nonuniformity of the local microstructure of the material. The macroscopic stress measured in the test is the sum of local stresses. Carbon atoms in germanium-carbide coatings have two forms of hybridization, i.e. sp^2 and sp^3. The sp^2 atoms help release the local stress in the coating because the sp^2 C–C bond is relatively short at 0.142 nm, whereas the bond length of sp^3 C–C is 0.154 nm. When the carbon atoms in germanium-carbide coatings mainly exist in the form of sp^2 hybridization, the coating stress is relatively low. However, as germanium atoms only have sp^3 hybridization, they are likely to promote the transformation of the carbon atoms in the coating into the sp^3 hybridization form. The size of the sp^3 hybridization zone mainly based on germanium atoms will affect the size of local stress in the coating. Therefore, the phenomenon of decrease of stress in germanium-carbide coatings with the decrease of the germanium-target sputtering power can be explained as follows: when the germanium content in the thin film decreases and the carbon content increases, the quantity of sp^2 C–C bonds increases to release the local stress in the film and further reduce the macroscopic stress.

Previously, we analyzed the microstructural variation of germanium-carbide coatings fabricated under different values of the germanium-target sputtering power by using Raman spectroscopy, where I_C/I_{Ge} of Raman spectra reflected the content of sp^2 carbon atoms in the thin films. To establish the correlation between microscopic results and macroscopic stress, Fig. 6.55 shows the variation of I_C/I_{Ge} of Raman spectra for germanium-carbide coatings fabricated under different values of germanium-target sputtering power with the intrinsic stress.

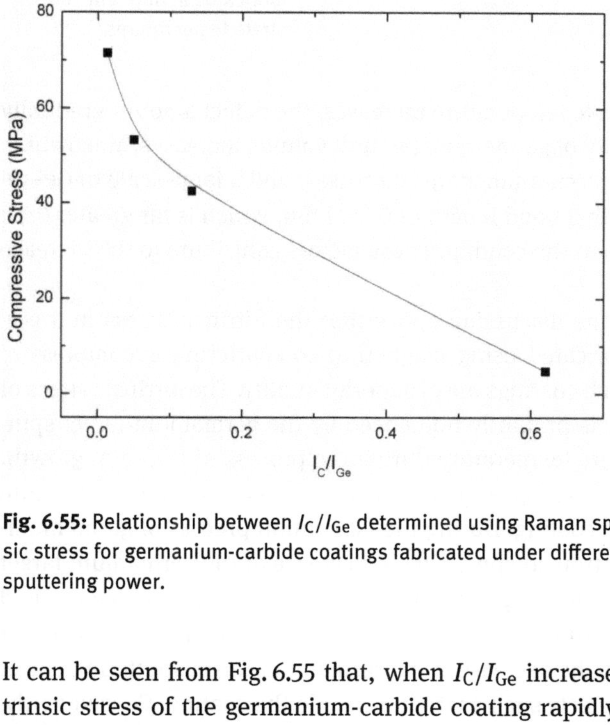

Fig. 6.55: Relationship between I_C/I_{Ge} determined using Raman spectroscopy and the intrinsic stress for germanium-carbide coatings fabricated under different values of germanium-target sputtering power.

It can be seen from Fig. 6.55 that, when I_C/I_{Ge} increases from 0.013 to 0.05, the intrinsic stress of the germanium-carbide coating rapidly decreases from 71.5 MPa to 53.2 MPa; when I_C/I_{Ge} continues to increase to 0.62, the intrinsic stress decreases to 4.7 MPa, almost linearly with I_C/I_{Ge}. This reflects that the decrease of macroscopic stress is caused by the decrease of local valence-bond structural variation in the coating, and that germanium atoms strongly influence the bonding of carbon atoms.

Fig. 6.56 shows the intrinsic stress of germanium-carbide coatings fabricated at different substrate temperatures. It can be observed that, with the increase of substrate temperature, the thin-film intrinsic stress increases from 50 MPa to 330 MPa. It is known from previous analyses that, in germanium-carbide coatings that are rich in

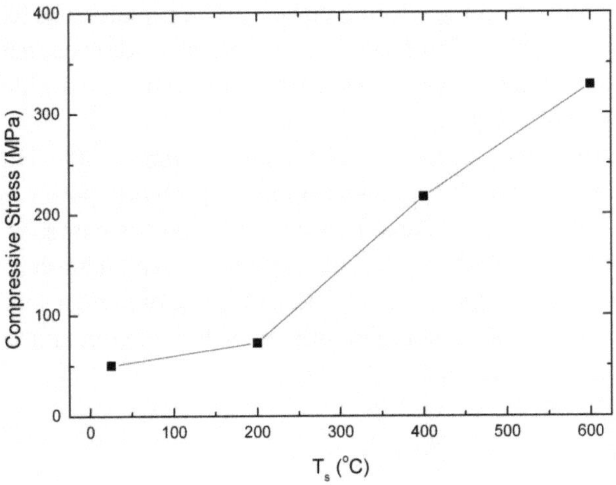

Fig. 6.56: Intrinsic stress of germanium-carbide coatings fabricated at different substrate temperatures.

germanium, when the substrate temperature increases, the defect amount gradually decreases, and atomic quantity of germanium per unit volume increases. Meanwhile, the coordination number of germanium atoms increases, and a large scale of Ge–Ge bonds is newly generated with a bond length of 0.2447 nm, which is far greater than that of other chemical bonds in the coating. These factors contribute to the increase of coating stress.

It can be inferred from the discussion above that the intrinsic stress in a germanium-carbide coating fabricated using magnetron co-sputtering is compressive stress, which ensures that such coatings are of superior quality. The intrinsic stress of germanium-carbide coatings is primarily influenced by the germanium-target sputtering power and the substrate temperature during the process of thin-film growth. That is, these two crucial technological parameters can affect the atomic structure and number of defects in the coating. During the fabrication process of germanium-carbide coating using magnetron co-sputtering, a decrease of the germanium-target sputtering power decreases the energy of particles accelerated by an electric field from the plasma generated through the glow discharge when the particles arrive at the substrate surface; further, the decrease of the germanium-target sputtering power promotes the formation of defects such as cavities inside the coating. Consequently, the volume of the coating is increased. Moreover, the decrease of the germanium-target sputtering power increases the number of carbon atoms in the coating, leading to the increase of sp^2 hybridised carbon atoms and the macroscopic stress decrease in the coating. The increase of the substrate temperature can decrease the number of defects inside the germanium-carbide coating that is rich in germanium. On the other hand, the formation of a large scale of Ge–Ge bonds will aggravate the local deformation, increase the stress, and consequently increase the macroscopic stress of the coating.

6.9 Thermal stability

Because of its superior properties, germanium-carbide coatings have great application prospects in the fields of infrared antireflective/protective coatings and photovoltaics. In these applications, germanium-carbide coatings may be exposed to a high-temperature environment (e.g. used as the antireflective/protective coating of the dome of a high-speed flying missile) or suffer from local high temperatures during operation. Therefore, the thermal stability of germanium-carbide coatings is a critical property determining its performance and service life, and it needs to be systematically studied. Researchers have adopted multiple approaches to fabricate germanium-carbide coatings around the world. They have also studied the associated properties from the optical, electrical, and structural aspects. However, there are few reports on the thermal stability of these films. For example, the effects of annealing on the bonding structures and mechanical properties of the thin film are not yet clearly understood. Considering the situations mentioned above, magnetron co-sputtering is used in our work to fabricate germanium-carbide coatings, and the coatings are annealed in vacuum. On this basis, the variation of the bonding structures and mechanical properties of the thin film with the annealing temperature is systematically studied.

6.9.1 Analysis of bonding structure

Fig. 6.57 shows the Raman spectra of germanium-carbide coatings fabricated at different annealing temperatures.

It can be observed that, with the increase of annealing temperature to 700 °C, the vibrational peak of the germanium TO model near 290 cm^{-1} gradually moves towards the position of 300 cm^{-1}, indicating an increase of orderliness in the coating, as well as the possibility of occurrence of microcrystalline germanium. With a further increase of

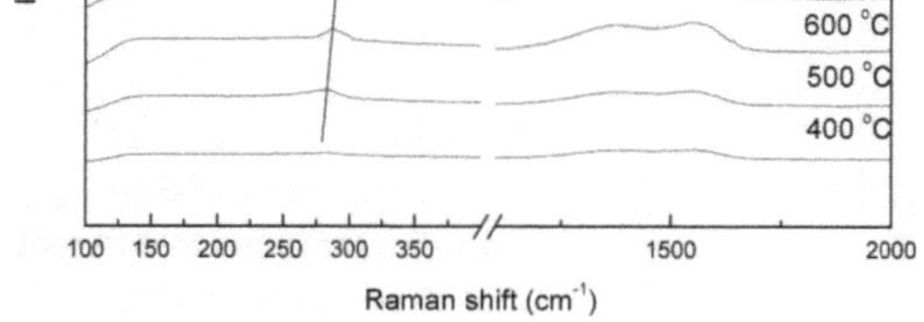

Fig. 6.57: Raman spectra of germanium-carbide coatings fabricated at different annealing temperatures.

temperature, the vibrational peak of the germanium TO model is gradually weakened, suggesting that the germanium atoms may have evaporated from the thin film. Experiments verify that, in the thermodynamic equilibrium state of the germanium-carbon system, germanium atoms will evaporate when the temperature increases to 963 °C. It is also observed from the figure that, with the increase of annealing temperature, the intensity of the wide peak within 1000–1800 cm^{-1} is gradually enhanced and clear peaks of carbon G ribbon and D ribbon appear.

The variation of Raman spectral fitting parameters for carbon in the range of 1200–1800 cm^{-1} is shown in Fig. 6.58. When the annealing temperature increases from 400 °C to 700 °C, the relative intensity ratio of D ribbon to G ribbon, I_D/I_G, slightly increases from 1.10 to 1.20, and the G ribbon position shifts from 1559 cm^{-1} to 1569 cm^{-1} with its half-peak width decreasing from 126 cm^{-1} to 120 cm^{-1}. When the annealing temperature increases beyond 700 °C, the relative intensity ratio of D ribbon to G ribbon, I_D/I_G, dramatically increases from 1.20 to 1.80, and the G ribbon position shifts from 1569 cm^{-1} to 1596 cm^{-1} with its half-peak width decreasing from 120 cm^{-1} to 75 cm^{-1}. This variation demonstrates that, with the increase of annealing temperature, the content of sp^2 carbon atoms in germanium-carbide coatings is

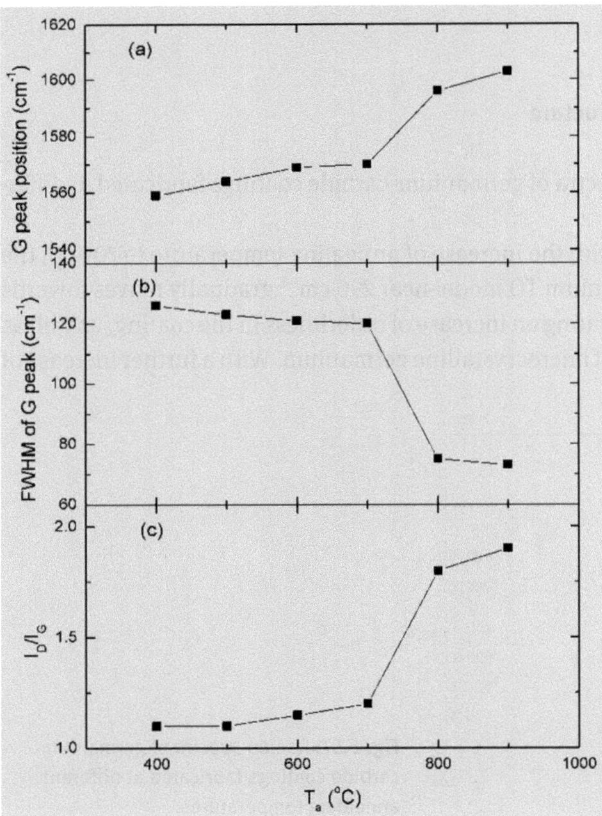

Fig. 6.58: Raman spectral fitting parameters of Ge_xC_{1-x} coatings fabricated at different annealing temperatures.

slightly and gradually increased. This change indicates that, when the annealing temperature is less than 700 °C, the germanium-carbide coating has a superior structural thermal stability; however, when the annealing temperature is greater than 700 °C, the germanium atoms in the coating evaporate, making the carbon atoms connect with each other in the coating to generate stable sp^2 hybridization.

Fig. 6.59 shows the C 1s core-level spectra of germanium-carbide coatings after annealing at different temperatures, from which we can see that the characteristics of the C 1s spectral curves are closely related to the hybridization state of carbon atoms.

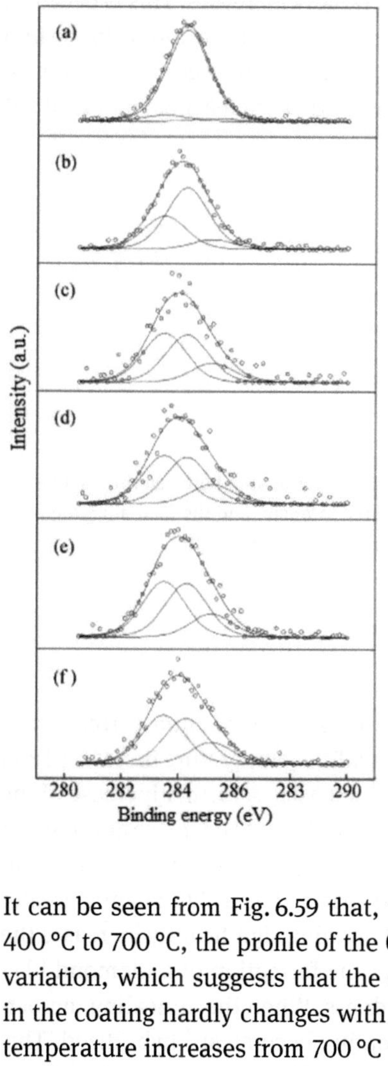

Fig. 6.59: XPS C 1s fitting curve of germanium-carbide coating after annealing at different temperatures: (a) T_a = 900 °C; (b) T_a = 800 °C; (c) T_a = 700 °C; (d) T_a = 600 °C; (e) T_a = 500 °C; (f) T_a = 400 °C.

It can be seen from Fig. 6.59 that, when the annealing temperature increases from 400 °C to 700 °C, the profile of the C 1s core-level spectrum undergoes no significant variation, which suggests that the chemical environment around the carbon atoms in the coating hardly changes with the annealing temperature. When the annealing temperature increases from 700 °C to 900 °C, the C 1s core-level spectral variation is substantial with the half-peak width narrowing and the peak position shifting up. This indicates an obvious change of the chemical environment around the carbon atoms with further increase of the annealing temperature due to the effect of electronegativity of the germanium atoms around the carbon atoms. The C 1s spectrum can be fitted using the three peaks of sp^3 C–C, sp^2 C–C, and Ge–C bonds at 285.2 eV, 284.3 eV, and 283.5 eV, respectively.

Fig. 6.60 shows the relative integrated intensity of each peak in the C 1s core-level spectrum prior to fitting versus the annealing temperature. When the annealing temperature increases from 400 °C to 700 °C, the relative integrated intensity of sp^3 C–C, sp^2 C–C, and Ge–C bonds essentially remains constant. However, with the further increase of the annealing temperature, the quantities of sp^3 C–C and Ge–C bonds rapidly decrease, but the relative content of sp^2 C–C bonds rapidly increases. This is because, with this further increase of annealing temperature, the germanium atoms in the film begin to evaporate, producing a higher chance of combination between carbon atoms, a larger tendency of graphitization of carbon atoms, and an increasing volume of sp^2 C–C bonds.

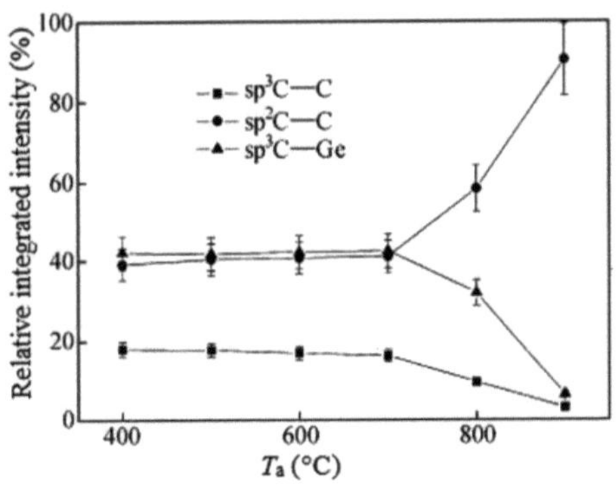

Fig. 6.60: Relationship between the relative integrated intensity and the annealing temperature of each peak after fitting.

6.9.2 Change of hardness

Fig. 6.61 shows the relationship between the hardness of germanium-carbide coatings and the annealing temperature. When the annealing temperature increases from 400 °C to 700 °C, the thin-film hardness gradually increases. A further increase of the annealing temperature, however, causes the coating hardness to decrease gradually. It is known from the analysis in the previous subsection that, prior to annealing at a temperature less than 700 °C, the germanium-carbide coating has a significantly greater number of Ge–Ge bonds. This change of chemical bonding significantly increases the atomic coordination number in the net, promotes the formation of a diamond-like-carbon structure, and enhances the thin-film hardness. When the annealing temperature exceeds 700 °C, the carbon atoms in the coating become sp^2 hybridized. This graphitization is able to decrease the coating hardness. Moreover, germanium atoms in the coating begin to evaporate with increasing temperature. As the annealing tem-

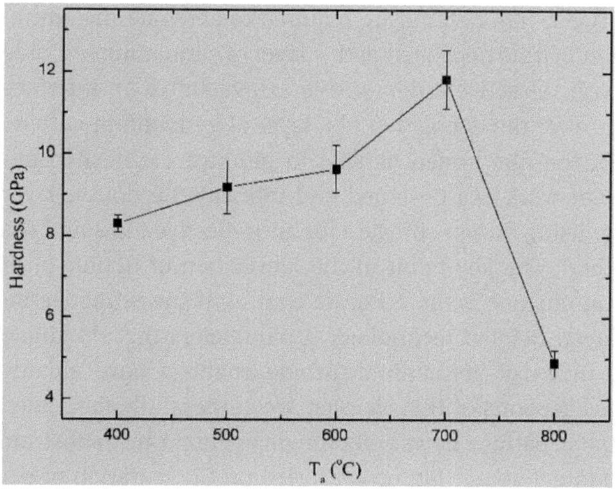

Fig. 6.61: Variation of hardness of a germanium-carbide coating with the annealing temperature.

perature increases to 800 °C, the number of germanium atoms in the coating dramatically decreases and the number of defects increases. When the annealing temperature increases to 900 °C, the annealing residue of the germanium-carbide coating is transformed into vitreous carbon, causing the hardness to decrease rapidly.

6.10 Germanium-carbide composite films

6.10.1 Germanium carbide/germanium-carbide composite films

A germanium-carbide coating mainly functions as the infrared antireflective/protective film in infrared applications. Infrared antireflective/protective coating refers to the thin film deposited on an infrared substrate to increase the infrared transmittance and hardness. At present, the most commonly used infrared window material in the range of 8–12 µm is zinc sulphide. However, neither the zinc sulphide fabricated using hot-pressed sintering nor multispectral zinc sulphide can satisfy the technological requirement of modern military and space applications in terms of intrinsic mechanical/optical properties. Generally, the hardness of zinc sulphide is approximately 3 GPa. During high-speed flight (*Ma* 3–5), the zinc-sulphide window is hardly free from the impact of rain drops and sand particles, making the external surface vulnerable. Infrared imaging is seriously affected as a result. Moreover, zinc sulphide has a refractive index of approximately 2.2 and a normal transmission of up to 72 %, which cannot meet the transmission requirement of infrared imaging. Hence, in order to satisfy the requirements of practical application, it is necessary to deposit an antireflective/protective coating on the zinc-sulphide substrate.

A single-layer germanium-carbide coating cannot produce the antireflective effect on zinc sulphide because its refractive index is greater than that of the zinc-sulphide substrate. Therefore, at least two layers of films are required to generate the antireflective effect. A zinc-sulphide substrate deposited with a layer of germanium-carbide coating (optical thickness: $\lambda_0/4$; refractive index: 4.1) is equivalent to an interface with a refractive index of 7.64. After the deposition of a layer of germanium-carbide coating (refractive index: 2.8), the film system is able to produce the desired antireflective effect. In the present work, we designed and fabricated a double-layer germanium-carbide coating by using design software for antireflective films and the magnetron co-sputtering method. The key point in the fabrication of double-layer germanium-carbide coatings in our test is the accurate control of the refractive index and thickness of every layer. Related technological parameters that dominate the thickness and refractive index of germanium-carbide coatings have already been discussed in the previous sections of this chapter. We successfully fabricated double-layer germanium-carbide coatings by varying the deposition time, substrate temperature, and germanium-target sputtering power based on the variation of deposition speed and refractive index to control the technological parameters of the germanium-carbide coating. The refractive index and thickness of the fabricated double-layer films are $n_1 = 4.1$, $d_1 = 580\,\text{nm}$, $n_2 = 2.8$, and $d_2 = 850\,\text{nm}$, respectively.

Fig. 6.62 shows a scanning electron microscopy (SEM) image of a zinc-sulphide substrate deposited with a double-layer germanium-carbide coating. It can be observed that the thin film is relatively compact with a smooth surface and no cracks. The interface between the zinc-sulphide substrate and the double-layer germanium-carbide coating is clear, whereas the interface between the two layers of the germanium-carbide coating is not obvious, suggesting a superior combination of the two layers without any mutual exfoliation.

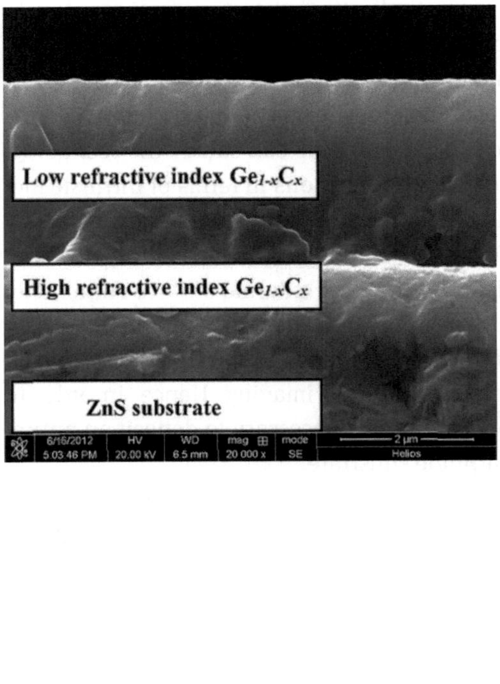

Fig. 6.62: SEM image of the zinc-sulphide substrate deposited with the double-layer germanium-carbide coating.

Fig. 6.63 shows the variation of infrared transmittance of the zinc-sulphide substrate before/after deposition of a double-layer germanium-carbide coating on one side and after annealing treatment at 400 °C with the wavelength. It can be seen that the transmission increases by 8 % at a wavelength of 9.6 μm after the deposition of the double-layer germanium-carbide coating on one side, but no significant change after the annealing process is observed.

Fig. 6.64 illustrates the hardness comparison of the zinc-sulphide substrate before/after the deposition of a germanium-carbide coating and after annealing treatment at 400 °C. We can see that the hardness substantially increases after the sub-

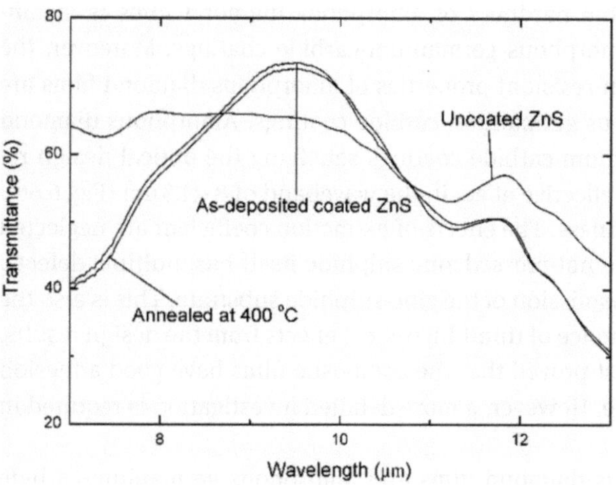

Fig. 6.63: Transmission variation of the zinc-sulphide substrate before/after the deposition and after the annealing treatment at 400 °C.

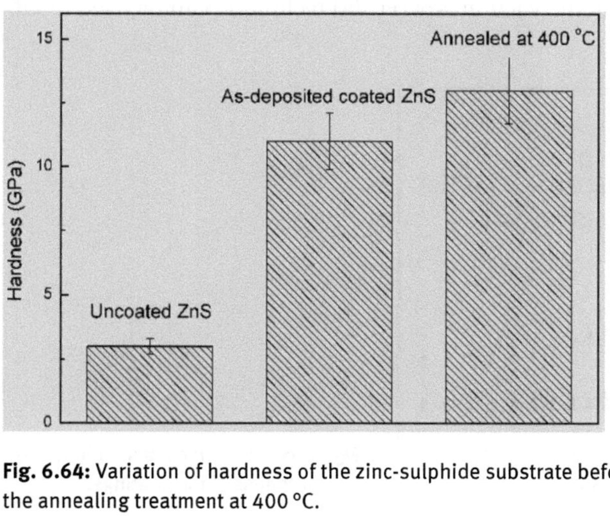

Fig. 6.64: Variation of hardness of the zinc-sulphide substrate before/after deposition and after the annealing treatment at 400 °C.

strate is deposited with the double-layer germanium-carbide coating, and it increases further after the 400 °C annealing process. By testing the infrared transmittance, hardness, and thermal stability, it was found that nonhydrogen germanium-carbide coating is a superior infrared antireflective/protective coating material, especially because of its excellent thermal stability, making suitable for application in high-temperature (e.g., 400 °C) environments.

6.10.2 Amorphous diamond films/germanium-carbide composite films

Fig. 6.65 clearly shows that the hardness of amorphous diamond films is significantly higher than that of amorphous germanium-carbide coatings. Moreover, the sand-impact- and rain-erosion-resistant properties of amorphous diamond films are superior to those of amorphous germanium-carbide coatings. Amorphous diamond films and amorphous germanium-carbide coatings satisfying the optical design requirements yield a good antireflective effect in the waveband of 8–11.5 μm (Fig. 6.66), exhibiting a high level of hardness. The effects of extinction coefficient are neglected in the design process, and the hot-pressed zinc sulphide itself has multiple defects, leading to a low practical transmission of the zinc-sulphide substrate. This is also the dominant reason for the difference of thin-film optical effects from the design results. A tape and regular scratch test proved that the composite films have good adhesion on the zinc-sulphide substrate. However, a more detailed investigation is required in the future.

In conclusion, amorphous diamond films and amorphous germanium-carbide coatings can serve as the antireflective/protective coating of infrared optical components such as zinc-sulphide substrates. The antireflective design can be developed and the optical parameters of films can be obtained based on the optical principles of thin films. The filtered cathodic vacuum arc (FCVA) technique can be used to fab-

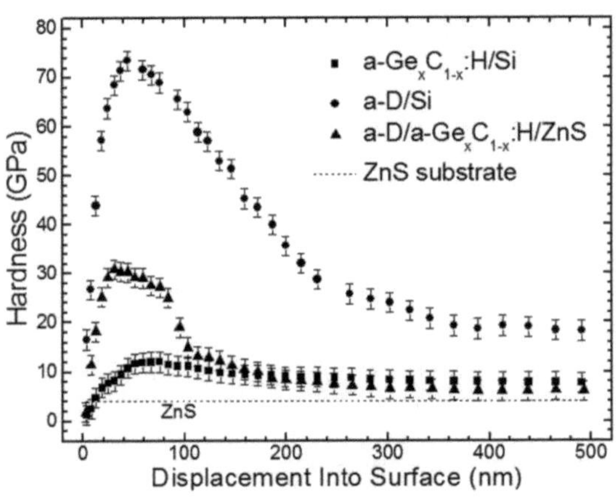

Fig. 6.65: Test curves of nanoindentation hardness of the coatings.

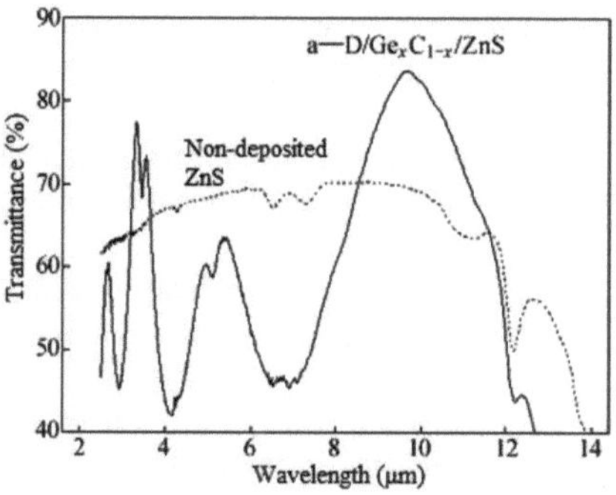

Fig. 6.66: Infrared transmittance curve of ta-C/a-$Ge_{1-x}C_x$ composite films.

ricate amorphous diamond coatings by changing the bias voltage of the substrate and controlling the refractive index/thickness of the coating during deposition. In the fabrication of amorphous germanium-carbide coatings using magnetron sputtering, the refractive index of the coating can be adjusted by changing the velocity ratio of methane, and the thickness can be controlled based on the velocity ratio and deposition time.

References

[1] Mackowski JM, Cimma B, Lacuve J, et al. Optical and Mechanical Behavior of GeC and BP Antireflection Coatings Under Rain Erosion Tests. Proceedings of SPIE, 1994: 552–560.

[2] Martin PM, Johnston JW, Bennett WD. Properties of Reactively Deposited SiC and GeC Alloys. Proceedings of SPIE, 1990: 291-307.

[3] Moure-Flores D, Quinones-Galvan JG, Hernandez-Hernandez A, et al. Structural, Optical and Electrical Properties of Cd Doped SnO_2 Thin Films Grown by RF Reactive Magnetron Co-sputtering. Applied Surface Science, 2012, 258(7): 2459–2463.

[4] Liu SJ, Chen LY, Liu CY, et al. Physical Properties of Polycrystalline Cr Doped SnO_2 Films Grown on Glasses Using Reactive DC Magnetron Co-sputtering Technique. Applied Surface Science, 2011, 257(6): 2254–2258.

[5] Sanchez JE, Sanchez OM, Ipaz L, et al. Mechanical, Tribological, and Electrochemical Behavior of $Cr_{1-x}Al_xN$ Coatings Deposited by RF Reactive Magnetron Co-sputtering Method. Applied Surface Science, 2010, 256(8): 2380–2387.

[6] Chung CK, Nautiyal A, Chen TS. Low Temperature Resistivity and Microstructure of Reactive Magnetron Co-sputtered Ta-Si-N Thin Films. Journal of Physics D: Applied Physics, 2010, 43(29): 295406.

[7] Jing Q, Xu Y, Zhang XY, et al. Zr-Cu Amorphous Films Prepared by Magnetron Co-sputtering Deposition of Pure Zr and Cu. Chinese Physics Letters, 2009, 26(8): 086109.

[8] Gonzalez CM, Rittby L, Graham WRM. FTIR Identification of the v4(σu) and v6(πu) Modes of LineargeC3Ge Trapped in Solid Ar. Journal of Physical Chemistry A, 2008, 112(43): 10831–10837.

[9] Kazimierski P, Jozwiak L. Transition from Amorphous Semiconductor to Amorphous Insulator in Hydrogenated Carbon-Germanium Films Investigated by IR Spectroscopy. Journal of NonCrystalline Solids, 2009, 355(4–5): 280–286.

[10] Kumeda M, Masuda A, Shimizu T. Structural Studies on Hydrogenated Amorphous Germanium-Carbon Films Prepared by RF Sputtering. Japanese Journal of Applied Physics, 1998, 37: 1754–1759.
[11] Saito N. Influence of RF Power on the Properties of Hydrogenated Amorphous Silicon Carbon Alloy Films Prepared by Magnetron Sputtering of Silicon in Methane-Argon Gas Mixtures. Journal of Applied Physics, 1985, 58(9): 3504–3507.
[12] Jacobsohn LG, Freire FL, Mariotto G. Investigation on the Chemical, Structural and Mechanical Properties of Carbon-Germanium Films Deposited by DC Magnetron Sputtering. Diamond and Related Materials, 1998, 7(2–5): 440-443.
[13] Mariotto G, Vinegoni C, Jacobsohn LG, et al. Raman Spectroscopy and Scanning Electron Microscopy Investigation of Annealed Amorphous Carbon-Germanium Films Deposited by DC Magnetron Sputtering. Diamond and Related Materials, 1999, 8(2–5): 668–672.
[14] Kanzawa Y, Katayama K, Nozawa K, et al. Preparation of $Ge_{1-y}C_y$ Alloys by C Implantation into Ge Crystal and Their Raman Spectra. Japanese Journal of Applied Physics. 2001, 40(10): 5880–5884.
[15] Choi S, Kim HW, Kim HJ, et al. Effect of Interstitial C Incorporation on the Raman Scattering of $Si_{1-x-y}Ge_xC_y$ Epitaxial Layer. Applied Physics Letters, 2008, 92(6): 061906.
[16] Kazimierski P, Tyczkowski J, Kozanecki M, et al. Transition from Amorphous Semiconductor to Amorphous Insulator in Hydrogenated Carbon-Germanium Films Investigated by Raman Spectroscopy. Chemistry of Materials, 2002, 14(11): 4694–4701.
[17] Kumeda M, Masuda A, Shimizu T. Structural Studies on Hydrogenated Amorphous Germanium-Carbon Films Prepared by RF Sputtering. Japanese Journal of Applied Physics, 1998, 37: 1754–1759.
[18] Benzi P, Bottizzo E, Demaria C, et al. Amorphous Nonstoichiometric $Ge_{1-x}C_x$:H Compounds Obtained by Radiolysis Chemical Vapor Deposition of Germane/Ethyne or Germane/Allene Systems: A Bonding and Microstructure Investigation Performed by X-ray Photoelectron Spectroscopy and Raman Spectroscopy. Journal of Applied Physics, 2007, 101(12): 124906.
[19] Inoue Y, Nakashima S, Mitsuishi S. Raman Spectra of Amorphous SiC. Solid State Communications, 1983, 48(12): 1071–1075.
[20] Vilcarromero J, Marques FC, Freire FL. Optoelectronic and Structural Properties of a-$Ge_{1-x}C_x$:H Prepared by RF Reactive Cosputtering. Journal of Applied Physics, 1998, 84(1): 174–180.
[21] Benzi P, Bottizzo E, Operti L, et al. Characterization and Properties of Amorphous Nonstoichiometricge$_{1-x}C_x$:H Compounds Obtained by Radiolysis CVD of Germane/Ethyne Systems. Chemistry of Materials, 2004, 16(6): 1068–1074.
[22] Kazimierski P, Tyczkowski J, Delamar M, et al. Transition from Amorphous Insulator to Amorphous Semiconductor in Hydrogenated Carbon-Germanium Films – Investigations in Submicrometer Scale. Journal of NonCrystalline Solids, 1998, 227: 422–426.
[23] Soukup RJ, Huguenin-Love JL, Ianno NJ, et al. Experimental Studies of $Ge_{1-x}C_x$ and $Ge_{1-x-y}C_xAl_y$ Thin Films. Journal of Vacuum Science and Technology A. 2008, 26(1): 17–22.
[24] Fang ZJ, Xi YB, Wang LJ, et al. Data Analysis of Optical Properties of Diamond Films by Infrared Spectroscopic Ellipsometry. Acta Optica Sinica, 2003, 23(12): 1507–1512.
[25] Li Q, Hang LX, Xu JQ. Ellipsometric analysis of optical constants for diamond-like carbon films deposited by UBMS. Journal of Applied Optics, 2009, 30(1): 105–109.
[26] Manifacier JC, Gasiot J, Fillard JP. A Simple Method for the Determination of the Optical Constants n, k and the Thickness of a Weakly Absorbing Thin Film. Journal of Physics E: Scientific Instruments, 1976, 9: 1002–1004.
[27] Swanepoel R. Determination of the Thickness and Optical Constants of Amorphous Silicon. Journal of Physics E: Scientific Instruments, 1983, 16: 1214–1222.

[28] Stoney G. The Tension of Metallic Films Deposited by Electrolysis. Proceedings of the Royal Society of London. Series A, Containing Papers of a Mathematical and Physical Character, 1909, 82(553): 172–175.
[29] Hoffmann L, Bach JC, Bech Nielsen B, et al. Substitutional Carbon in Germanium. Physical Review B, 1997, 55(17): 11167–11173.
[30] Scace RI, Slack GA. Solubility of Carbon in Silicon and Germanium. Journal of Chemical Physics, 1959, 30(6): 1551–1555.
[31] Seward CR, Coad EJ, Pickles CSJ, Field JE. The Rain Erosion Resistance of Diamond and Other Window Materials. Proceedings of SPIE, 1994, 2286: 285–300.
[32] Telling RH, Jibert GH, Field JE. The Erosion of Aerospace Materials by Solid Particle Impact. Proceedings of SPIE, 1997, 3060: 56–67.

7 Boron phosphide thin films

7.1 Overview

Boron phosphide thin film is the preferred material for infrared antireflective and protective window coatings. In infrared detection technology, windows and domes are at the front of infrared photoelectric systems, and they deteriorate continuously within the service environment of high-speed flight. There are very high performance requirements in the optical, thermal, mechanical, and other aspects of infrared window materials. However, a material that can fully satisfy all the requirements has yet to be reported. In particular, long-wave infrared optical materials such as zinc sulphide (ZnS), which are widely applied today, have low mechanical strength and fracture toughness and high brittleness. Hence, they are prone to surface damage due to the impact of raindrops and dust during high-speed flight. The decrease in infrared transmittance due to these damages will cause infrared photoelectric systems to be ineffective. The fabrication of infrared antireflective and protective coatings on infrared optical elements is an effective approach to satisfy the window-material requirements of high-speed infrared systems under harsh conditions. Not only do infrared antireflective and protective coatings need to be transparent to infrared light and have a low absorption coefficient, they should also have wear resistance, heat resistance, good substrate adhesion, and the ability to protect infrared elements against the impact of rain erosion. Therefore, the selection and fabrication techniques of infrared antireflective and protective coating materials are extremely important. Thus far, several types of infrared antireflective and protective coating materials have been researched and developed. Among them, the most ideal materials include germanium carbide (Ge_xC_{1-x}), diamond, diamond-like carbon (DLC), and boron phosphide.

Research has shown that boron phosphide films have good light transmittance within a broad wavelength range and the best rain erosion resistance among the same class of thin films, which maximally increases the damage threshold velocity (DTV) of the substrate. Moreover, boron phosphide shows better mechanical performance compared to germanium carbide. Its Knoop hardness is 47 GPa, which is more than twice that of sapphire, and its elastic modulus is up to 270 GPa, which is the highest among binary covalent compounds with a zinc-blende structure. Boron phosphide coatings show better adhesion and are more conducive to large-area deposition compared to diamond. Furthermore, boron phosphide has lower internal stress and better adhesion with various substrate materials compared to DLC, and there are no restrictions with boron phosphide in terms of deposition-film thickness. Therefore, boron phosphide has become the preferred material for the antireflective and protective coating of infrared windows in a service environment of high-speed flight.

Aside from applications as infrared antireflective and protective coatings for high-speed flight, the protective effects of boron phosphide films for components under

https://doi.org/10.1515/9783110489514-010

other harsh environments have received increasing attention from researchers. Boron phosphide is one of the ideal materials for the corrosion-resistant films of the infrared windows of submarine periscopes. Once a submarine has been deployed, its periscope windows will be submerged in a marine environment. Hence, not only do the windows need to ensure high transmittance, they also need to resist the corrosion of chemical compounds in seawater in order to maintain the sensitivity and resolution of the photoelectric system. However, the window material of periscopes, single-crystalline germanium (Ge), is highly susceptible to seawater corrosion, and the resulting pinhole pitting corrosion will greatly diminish its optical performance. Boron phosphide has excellent chemical stability and corrosion resistance. Hence, it is able to protect infrared windows against the erosion of seawater and preserve the original optical performance of infrared windows.

In addition, boron phosphide thin films have high hardness, good chemical stability, and excellent wear resistance. Hence, it is possible that they can be applied as surface protective coatings of cutting tools as well as wear-resistant and corrosion-resistant parts. The high thermal stability of boron phosphide also implies that it could be a potential thermoelectric material.

7.2 Research progress on the structure and properties of boron phosphide thin films

7.2.1 Structural features of boron phosphide thin films

7.2.1.1 Crystal structure

Boron phosphide is a group III–V compound composed of phosphorus (P) and boron (B). Its main polymorph is a cubic crystal, denoted as c-BP, which has a zinc-blende structure. The structure of c-BP is similar to that of diamond, as shown in Fig. 7.1 (a). In the zinc-blende structure of BP crystals, each B atom is surrounded by four P atoms. The four nearest P atoms are distributed on the vertices of a regular tetrahedron, wherein the P atoms and the central B atom each contribute a valence electron to form a covalent bond, as shown in Fig. 7.1 (b). Within the tetrahedral structure of the covalent crystal, the four covalent bonds form sp^3 hybrid orbitals based on the linear combination of S-state and P-state wave functions. These four bonds have an identical bond angle of 109°23′ and bond length of 0.1667 nm, with lattice constant $a = 4.538$ Å.

The cubic crystal system of c-BP remains stable up to 2500 °C under high pressure but dissociates at approximately 1100 °C in vacuum. c-BP undergoes a partial loss of P and forms the hexagonal polymorph $B_{12}P_2$, which has a rhombohedral structure. The rhombohedral structure of boron phosphide is similar to that of α-B. As shown in Fig. 7.1 (c), this structure is formed via the bonding of icosahedral B_{12} with cross-linking P–P diatomic chains.

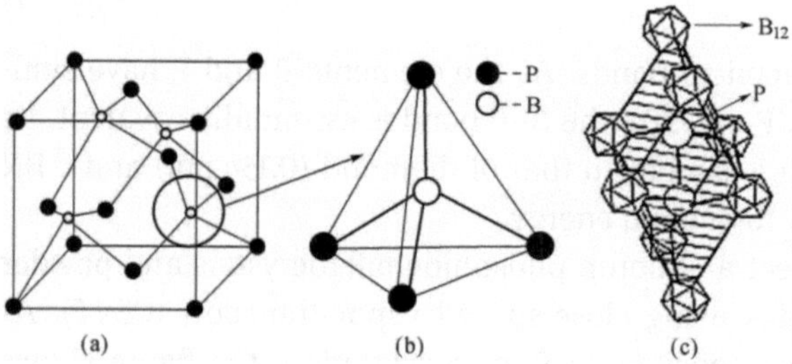

Fig. 7.1: Schematic diagram of cubic BP structure and rhombohedral $B_{12}P_2$ structure.

Kumashiro et al. reported that large-area monocrystalline BP thin films can be deposited on Si substrates using chemical vapor deposition (CVD). However, subsequent literature related to the fabrication of monocrystalline boron phosphide could not be found. This group of researchers discovered that changes in deposition temperature can cause transformations in crystal structure. Boron phosphide deposited using thermal CVD at 600 °C was amorphous, but when the temperature increased to 800 °C, the thin film changed into a polycrystalline structure, which showed multiple preferred orientations of (111), (200), and (220). Due to the differences in deposition processes and conditions, it is difficult to fabricate ideal single crystals. In fact, under most circumstances, boron phosphide thin films with polycrystalline and amorphous structures will be obtained.

7.2.1.2 Stoichiometric ratio

The stoichiometric ratio of cubic BP is 1 : 1. However, the thin films obtained in actual studies are generally nonstoichiometric. Kumashiro et al. employed low-pressure CVD (LPCVD) to deposit boron phosphide thin films, which showed a boron-rich state; the atomic ratio B/P analyzed using x-ray photoelectron spectroscopy (XPS) was 77/23. They suggested that one of the reasons for the boron-rich state was the tendency of B_2H_6 in the thin film toward self-aggregation into B_3H_x and entering B_{12} clusters. Another reason for the boron-rich state might have been that the P–P bonds are more difficult to form than B-B bonds. Lewis et al. obtained phosphide-rich boron phosphide thin films using plasma-assisted CVD (PACVD), which can be expressed as BP_x, where $x = 4–8$. The 2p electron binding energy of phosphorus was 129 eV, which is close to that of red phosphorus. This strongly indicates that boron phosphide thin films are polymeric with the phosphorous networks being stabilized by the boron atoms. Dalui et al. employed the co-evaporation technique to produce boron phosphide thin films that were boron-rich, phosphide-rich, and stoichiometric (B/P = 1). The studies above indicate that the chemical composition of boron phosphide thin films varies widely according to the fabrication method.

7.2.1.3 Chemical bonds

Boron phosphide has unique chemical bonds. As the elements B and P have similar electronegativities (B = 2.04, P = 1.90), the B–P bond is essentially covalent. Its bond length is 0.1667 nm, which is similar to that of diamond (0.154 nm) and c-BN (0.156 nm), and it has a relatively high bond energy.

Fig. 7.2 shows the Raman spectra of boron phosphide microcrystals and powder grown on Al_2O_3 and Si substrates using close-spaced vapor transport (CSVT). As highly-symmetrical vibrations are sensitive to Raman scattering, the figure shows that a shoulder peak and strong peak appeared at 800 and 827 cm^{-1}, respectively, which correspond to the transverse optical mode (TO mode) and longitudinal optical mode (LO mode) in the Brillouin zone of typical c-BP crystals.

Infrared absorption spectra can also reveal the vibrational mode of bonding atoms. Fig. 7.3 shows the infrared absorption spectrum of a typical boron phosphide thin film. Our analysis showed that the vibrational absorption of boron phosphide thin films in the infrared spectrum is mainly determined by its extrinsic absorption. The stretching vibrational absorption at 4.2 μm is induced by P–H bonds, the stretching vibrational absorption at 4.5 μm is induced by B–H bonds, and the bending vibra-

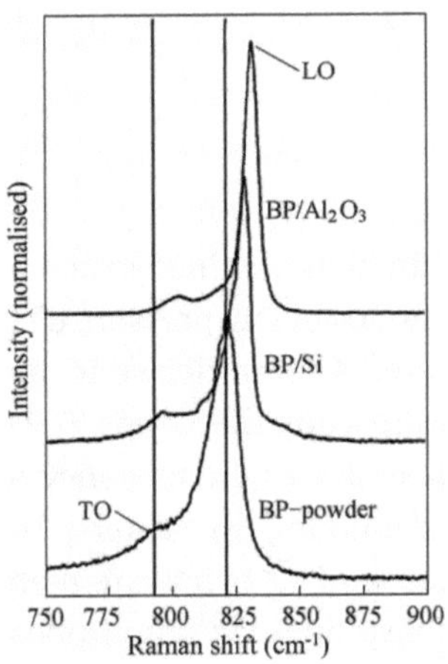

Fig. 7.2: Raman spectra of boron phosphide crystals.

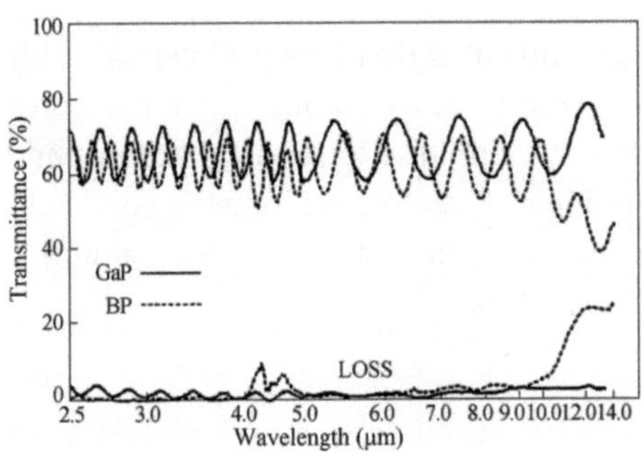

Fig. 7.3: Infrared transmission absorption spectra of boron phosphide thin film.

tional absorption at 13 μm is induced by B–H–B bonds. Therefore, we can see that the main impurity leading to infrared vibrational absorption is H, which originates predominantly from the residual H in the reactive gas. Hence, the intensity of absorption peaks is influenced by deposition conditions.

7.2.2 Performance features of boron phosphide thin films

7.2.2.1 Mechanical performance

Boron phosphide thin films have excellent mechanical performance. Tab. 7.1 compares the mechanical properties of common infrared optical materials.

Tab. 7.1: Mechanical properties of common infrared optical materials.

Material	Knoop hardness (GPa)	Elastic modulus (GPa)
BP	47	270
SiC	25.4	465
GaP	9	142
Ge	8.3	103
Si	11.5	131
ZnS (CVD)	2.5	75
Al_2O_3	20	335
Diamond	88	1050

As summarized in Tab. 7.1, the high hardness level of boron phosphide thin films is second only to that of diamond. Takenaka et al. reported that the Vickers hardness of BP single crystals obtained by epitaxial growth reached 4700 kg·mm^{-2}, which is more than twice the level of sapphire. Moreover, among all binary covalent compounds with a zinc-blende structure, boron phosphide has the highest elastic modulus of up to 270 GPa. Tab. 7.1 lists the hardness of a monocrystalline boron phosphide thin film. Nevertheless, under most circumstances, boron phosphide thin films with polycrystalline and amorphous structures will be obtained during the fabrication process, and their hardness will be less than that of the monocrystalline material. Boron phosphide thin films have low internal stress and good adhesion with various substrate materials; hence, they are not restricted in terms of the deposition thickness. The maximum thickness of a boron phosphide thin film reported in the literature thus far is 200 μm, and the thin film was fabricated by Kumashiro et al. using thermal decomposition on a Si substrate.

7.2.2.2 Optical performance

The optical properties of boron phosphide thin films are mainly determined by extrinsic absorption processes, and the main impurities are H and O. During the fabrication process of boron phosphide, a hydrogen-containing gas will generally be used as the precursor. Therefore, the residual H in the thin films significantly impacts infrared transmission. Fig. 7.3 shows the infrared transmission absorption spectra of boron phosphide thin film.

As can be seen from Fig. 7.3, the main impurity absorption originates from the B–H and P–H stretching vibrations at 4.2 and 4.5 µm, respectively, whereas the absorption at 13 µm is induced by the bending vibrations of B–H–B bonds. Furthermore, the presence of oxides in the thin film is disadvantageous to its optical properties. For example, BO_3 will lead to absorption bands at 7.9, 13.2, and 15.8 µm; P=O and P–O particles will lead to absorption bands at 8.3 and 13.2 µm, respectively. The intensity and width of absorption peaks will vary under different fabrication conditions; thus, the optimization of processing conditions can reduce the intensity of impurity absorption peaks. The decrease in the transmittance of boron phosphide thin films within the infrared wave range is due to its relatively high refractive index, which is 2.9–3.1 at 10 µm.

7.2.2.3 Resistance to rain erosion

The harsh service environment of high-speed infrared windows often involves erosion by sand, raindrop, and dust, among which erosion by raindrop impact is the most common type of damage. By performing rain erosion experiments, qualitative and quantitative evaluations can be conducted on the rain-erosion resistance of window materials and their protective coatings. These experiments also serve as an integrated assessment of their mechanical performance.

The main evaluation criteria for measuring the rain-erosion resistance of materials are the damage threshold velocity (DTV). It is defined as the maximum velocity that can be achieved by a single water drop or waterjet impacting the material surface at high speed such that no surface damage can be observed.

Fig. 7.4 shows the DTV curves of boron phosphide thin films with different thicknesses deposited on Ge substrates. The DTV curves of boron phosphide films with different thicknesses were obtained using the multiple impact jet apparatus (MIJA) in the Cavendish Laboratory at the University of Cambridge. It can be seen from the figure that the DTV curves show an increasing trend with increasing boron phosphide film thickness, such that under heavy rainfall (number of impacts > 100), the DTV of the uncoated Ge substrate increased significantly from 110 m/s to 300 m/s.

Similarly, Fig. 7.5 shows the variations in DTV with the number of impacts for different coating materials, which were also measured using MIJA in the Cavendish Laboratory. Several research teams have modified the raindrop speed, raindrop diameter, and other experimental conditions, and all reached the following conclusion. Com-

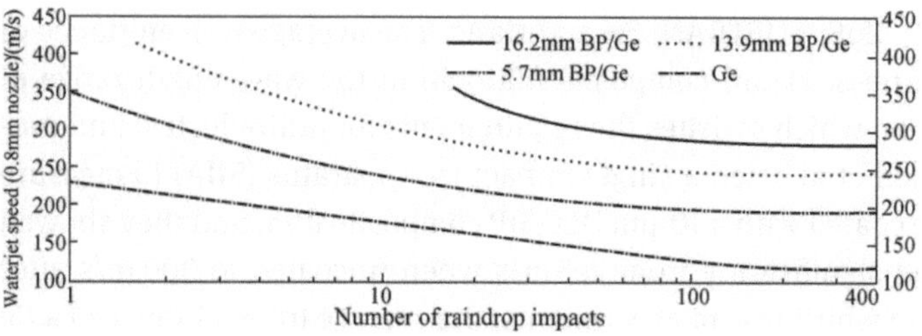

Fig. 7.4: Damage threshold velocity curves of boron phosphide thin films of different thicknesses.

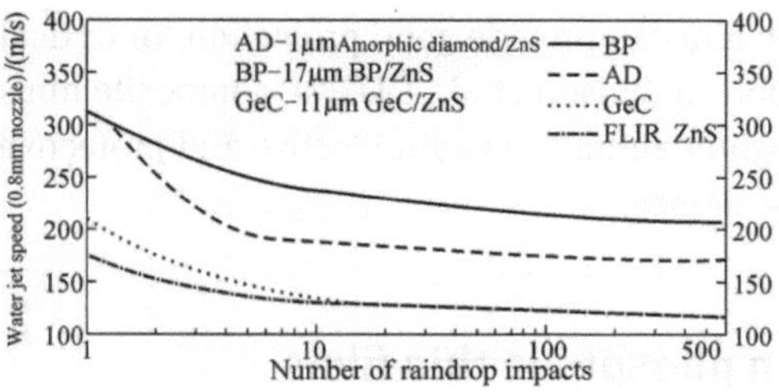

Fig. 7.5: Damage threshold velocity curves of different thin films deposited on ZnS substrates.

pared to similar antireflective and protective coatings (e.g. gallium phosphide, germanium carbide, etc.), boron phosphide led to the highest increase in substrate DTV, and thus showed the best performance in experiments on rain-erosion resistance.

7.2.3 Composite films

Based on the experiments on rain-erosion resistance, the DTV increased with increasing boron phosphide film thickness, and the requirements for resistance to rain and sand erosion were satisfied when the film thickness approached 10 μm. However, boron phosphide is unable to fulfil the requirements for both rain erosion resistance and antireflection simultaneously. This is because boron phosphide has a high refractive index ($n \approx 3$) and a relatively large absorption ratio, as a result of which the increase in film thickness will lead to diminishing optical performance. This implies that there always has to be a compromise between infrared transmission and mechanical performance. To compensate for the shortcomings of boron phosphide thin films, some researchers have designed composite coating systems for antireflective and protective films. The aim is to ensure that the thickness and refractive index of each layer are mutually complementary, ultimately achieving an optimal match between the antireflective and protective effects of the coating system. Currently, composite films that involve boron phosphide mainly include DLC/BP, BP/GaP, and BP/Ge.

The most common composite film is composed of DLC and boron phosphide. The DLC/BP composite film can be applied on a series of infrared window materials, including Ge, FLIR, ZnS, TUFTRAN, Si, and GaAs. The average transmittance of the DLC/BP (film thickness: 11 μm) composite film within the wavelength range of 8–10.5 μm is nearly 90 %, which satisfies the requirements for antireflective and protective coatings. Waddell et al. used a single impact jet apparatus (SIJA) to measure the DTV of Si substrate coated with a 10-μm DLC/BP composite film, and they showed that the DTV increased significantly from 265 m/s when uncoated to 300 m/s after coating. Furthermore, in whirling arm experiments, the average transmittance of a Ge substrate coated with a 10-μm DLC/BP composite film decreased by less than 1 %, indicating that the extent of damage to the sample was relatively low. By characterising the parameters of rain-erosion resistance using DTV and loss of infrared transmission, we have verified that the DLC/BP composite film can provide good protection for optical window materials. According to the report by Wilson et al., DLC/BP composite films have been successfully applied in F-14 fighter aircrafts as antireflective and protective coatings for infrared search and track Ge domes.

7.3 Structural analysis of boron phosphide thin films by reactive magnetron sputtering

7.3.1 Process design of reactive magnetron sputtering

As boron phosphide is a compound thin film, reactive magnetron sputtering was selected as the deposition method owing to considerations of process controllability and the flammability of elemental phosphorus. A stable high-density boron target with a purity of 99.999 %, diameter of 49 mm, and thickness of 3 mm was used. The less toxic phosphine gas was selected as the reactive gas, and high-purity argon gas was selected as the inert gas, which had a purity level of 99.9999 %.

During the process of reactive magnetron sputtering, processing conditions will affect the microstructure of deposited thin films, which in turn will determine their macroscopic performance. Therefore, by adjusting the processing parameters of thin-film deposition, it is possible to control thin-film structure and hence optimize its performance. The main controllable processing parameters include the sputtering power, substrate temperature, and PH_3/Ar flow ratio. Among them, the most crucial factor influencing thin-film quality and compositional structure is the gas flow ratio. Although the sputtering power and substrate temperature have a certain extent of influence on deposition speed and crystallization quality, they do not have a predominant effect on thin-film structure and composition. Therefore, given the prerequisite of good preparations prior to coating, the primary area of research to obtain high-quality boron phosphide thin films and optimized processes should be the effects of PH_3/Ar flow ratio on the structure and performance of boron phosphide thin films fabricated by reactive magnetron sputtering.

Changes in the gas flow ratio were achieved by keeping the flow rate of Ar gas constant while varying the flow rate of PH_3. The flow rate of sputtering gas Ar was set at 50 sccm. Excessively low PH_3 flow rates will limit the deposition speed owing to the lack of precursors in the reaction atmosphere, whereas excessively high PH_3 flow rates will cause target poisoning and difficulties in sputtering; thus, the range of PH_3 flow rate was set as 3–15 sccm. The selection of sputtering power was determined by the size of the target material. Since the size of the sputtering target is fixed, an excessively low sputtering power will cause sputtering difficulties, as the insufficient kinetic energy of Ar^+ ions will cause boron atoms to escape from the target material, and the deposition speed will be too low; in contrast, an excessively high sputtering power will penetrate the target material and cause damage. Therefore, for a boron target with a diameter of 49 mm and thickness of 3 mm, the sputtering power selected was 140 W. The selection of deposition temperature influences thin-film quality. It was found in previous experiments that, when the deposition temperature was relatively low, the adhesion between the boron phosphide thin film and substrate was poor, and the thin film was susceptible to delamination; in contrast, when the deposition temperature was relatively high, the thin film was intact and not prone to delamination, but excessively high temperatures will produce large thermal stress and lead to thin-film fracture. Hence, to remain within the constraints of the deposition system, a deposition temperature of 600 °C was selected. The processing parameter settings are listed in Tab. 7.2.

Tab. 7.2: Processing parameters of boron phosphide films fabricated by reactive magnetron sputtering.

Deposition power (W)	Deposition temperature (°C)	Ar flow rate (sccm)	PH_3 flow rate (sccm)	Target substrate distance (mm)	Radiofrequency power (MHz)
140	600	50	3	80	13.56
140	600	50	6	80	13.56
140	600	50	9	80	13.56
140	600	50	12	80	13.56
140	600	50	15	80	13.56

7.3.2 Composition analysis

Fig. 7.6 shows the full-spectrum XPS of a boron phosphide film surface fabricated using reactive sputtering before and after sputtering by an Ar^+ beam. In addition to the expected B and P elements, the figure also shows high C and O peaks. The intensity of C 1s and O 1s characteristic peaks were relatively high before sputtering, which indicates that the thin-film surface contained substantial proportions of C and O elements. In order to eliminate the influence of impurities in air on the measurements, sputter-

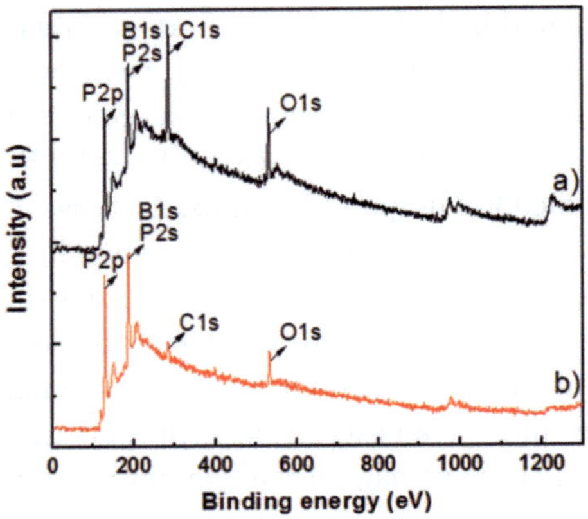

Fig. 7.6: Full-spectrum XPS of boron phosphide thin films: (a) before sputtering; (b) after sputtering.

ing was performed on the thin film for a further 2 min. The characteristic peaks of C 1s and O 1s were found to be significantly weakened as a result. This indicates that C and O elements mainly exist on the thin-film surface because of surface adsorption.

However, we can observe from Fig. 7.6 that the P 2p maximum peak occurred at 130 eV, while the P 2s and B 1s characteristic peaks overlapped near 185 eV and could not be distinguished. This can be solved using the ratio method. It is known that the ratio of P 2p/2s peak area is 1.179. Hence, the area of the P 2s peak can be obtained from the P 2p peak at 185 eV, and the B 1s peak area can be obtained after subtracting the P 2s peak area at 185 eV. Based on the ratio of B and P peak areas and corresponding sensitivity factors, we can obtain the relative composition ratio of B and P.

Fig. 7.7 shows the variations in P/B atomic composition ratio in boron phosphide thin films with different PH_3 flow rates, for which the P/B atomic composition ratios were 1.055, 1.179, 1.180, 1.163, and 1.515, respectively. The curve in the figure shows that the relative concentration of P increased significantly with increasing PH_3 flow rate. At low flow rates, the P/B ratio approached a stoichiometric ratio of 1 : 1; whereas at high flow rates, the thin film gradually transformed to a P-rich state.

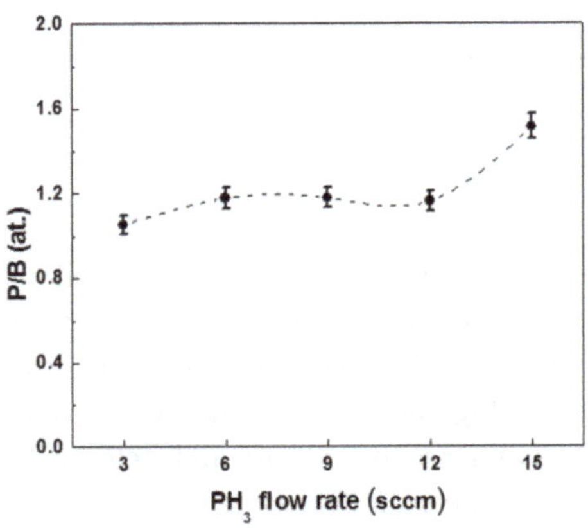

Fig. 7.7: Variations in P/B composition ratio of boron phosphide thin films with PH_3 flow rates.

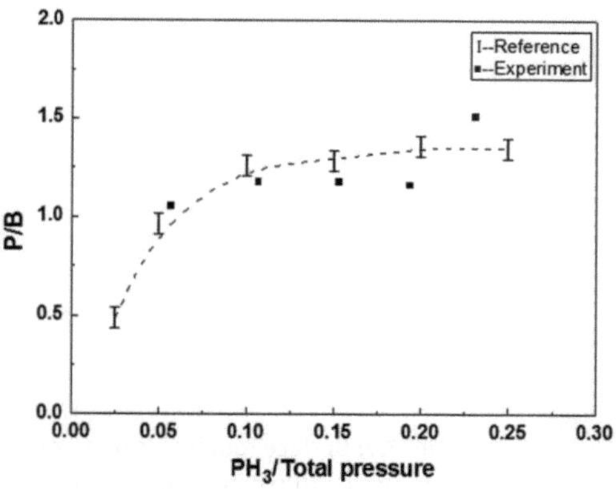

Fig. 7.8: Composition variations of boron phosphide thin films based on other reference and experimental data.

Mackowski et al. fabricated boron phosphide films using reactive radio frequency sputtering (RRFS), and they analyzed the variations in composition. The experimental data obtained in this book and the data from the aforementioned article are both plotted in Fig. 7.8; the two sets of data showed a good fit.

As the PH_3 flow rate continued to increase, the P/B ratio reached a stable value, which is indicative of target poisoning. Two different sputtering modes will occur during reactive sputtering: the metal sputtering mode, which only involves the sputtering of atoms of the target material, and compound sputtering mode, which involves the sputtering of the surface compounds. When the adsorption speed of reactive gas on the target material is greater than the sputtering speed, corresponding chemical reactions will occur between the target material and the reactive gas, leading to the continuous formation of the compound layer and changes in the sputtering mode. This phenomenon is known target poisoning. After the decomposition of excess PH_3 and coverage of the boron target surface, the target material will be contaminated by P elements. This will cause the thin-film composition to remain relatively stable, but will also lead to the sharp decrease in the deposition speed of sputtering. In this book, the P/B ratio of boron phosphide thin films continued to increase and did not tend toward stabilization. Therefore, the sputtering process was in the metal sputtering mode and did not exhibit signs of target poisoning. If the PH_3 flow rate continues to increase, the compound sputtering mode may occur.

7.3.3 Surface morphology

Fig. 7.9 shows the surface roughness of boron phosphide thin films fabricated under different PH_3 flow rates. The root mean square (RMS) surface roughness of the thin films was relatively low overall, and all roughness values were within the range of 1–3 nm. Our experimental results were compared with the results of boron phosphide thin films obtained using the coevaporation technique reported in the literature,

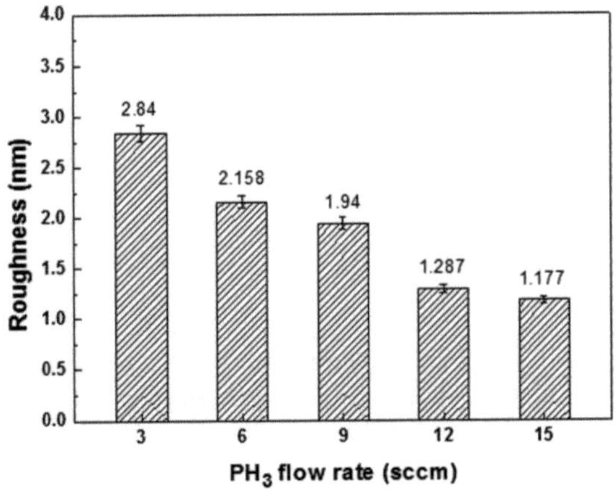

Fig. 7.9: Surface roughness of boron phosphide thin films deposited under different PH_3 flow rates.

which revealed that the surface roughness was less by 1–2 nm in our experiment. Furthermore, when compared to PECVD, the surface roughness was reduced by nearly 10 nm. This indicates that the thin-film surface obtained in this book was smoother. When the flow rate of PH_3 was 3 sccm, maximum surface roughness was 2.84 nm. As the flow rate of PH_3 increased, thin-film roughness decreased gradually and was only 1.177 nm at 15 sccm. Fig. 7.10 shows the surface morphology of boron phosphide thin films fabricated under different PH_3 flow rates, as measured using atomic force microscopy (AFM).

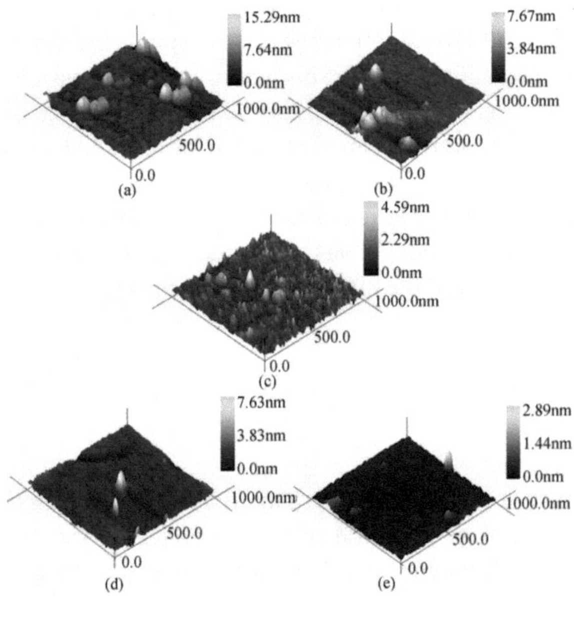

Fig. 7.10: Surface morphology of boron phosphide thin films fabricated under different PH_3 flow rates: (a) 3 sccm; (b) 6 sccm; (c) 9 sccm; (d) 12 sccm; (e) 15 sccm.

Overall, the thin-film surface was smooth and dense. Aside from a few island-like clusters on the surface, there were no obvious defects, and the surface roughness was low. When the PH_3 flow rate was 3 sccm, the thin-film surface showed obvious clustering with numerous clusters that were nonuniformly distributed; the particle size in thin-film growth was also the highest at 3 sccm. Hence, the surface roughness at 3 sccm was the greatest (at 2.84 nm).

As the PH_3 flow rate increased, the number of atomic clusters observed began to decrease, and the volume of these clusters showed a corresponding decrease accompanied by the refinement of surface particles in the thin film. This resulted in a trend of gradual decrease in surface roughness. When the PH_3 flow rate continued to increase to 15 sccm, the thin-film surface only had one prominent 2.89-nm protrusion, whereas the others were extremely fine particles distributed uniformly on the surface. At this point, the coating layer was the densest and most continuous, and hence had the lowest level of surface roughness.

Fig. 7.11 shows the surface morphology of boron phosphide thin films deposited at different PH_3 flow rates observed using scanning electron microscopy (SEM). When the PH_3 flow rate was at a low level of 3 sccm (a), the presence of prominent particles could be observed. When the PH_3 flow rate was increased to 9 sccm (b), crystal grains could not be observed on the thin-film surface, the overall particle size had been significantly reduced, and the continuity and surface density of the coating layer increased. This indicates that, as the PH_3 flow rate increased, the thin films became more refined and uniform, particle sizes of the coating layer gradually decreased, and surface density increased. These observations are consistent with the results of AFM analysis.

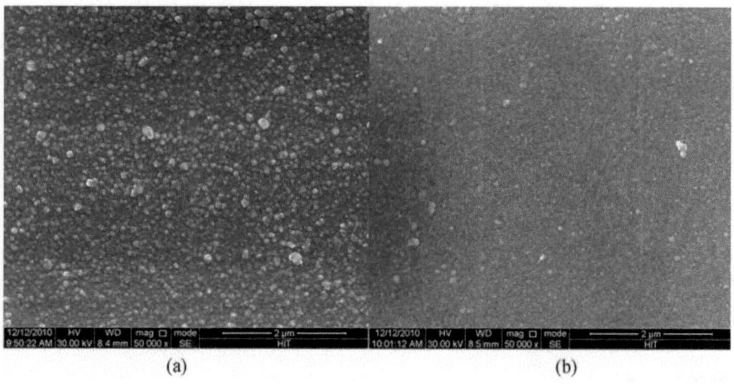

Fig. 7.11: Surface morphology of boron phosphide thin films under different PH_3 flow rates: (a) 3 sccm, (b) 9 sccm.

7.3.4 Crystal structure and bonding states

7.3.4.1 X-ray diffraction (XRD) analysis

Fig. 7.12 shows the XRD patterns of boron phosphide thin films deposited under different PH_3 flow rates. There are two key types of information in the analysis of crystal structures. The first type of key information that can be obtained from the diffraction pattern is the position of diffraction peaks, i.e. the direction of diffraction (angle θ), which is determined by the interplanar spacing d at a given λ. The diffraction direction reflects the lattice constant, and includes information on the sizes and shapes of unit cells. It can be described using Bragg's law. When the PH_3 flow rate was 3 sccm and 6 sccm, diffraction peaks emerged in the thin film when $2\theta = 56.1°$, which correspond to the (220) lattice plane of cubic BP. It should be noted that, when the diffraction angle corresponding to the diffraction peak in the figure was compared to that of the standard sample, the former showed a shift towards smaller angles. By observing the bending direction of the Si substrate using a surface profiler, we determined that the thin film was undergoing compressive stress. Bragg's law shows that, if compressive stress is present in the thin film, and interplanar spacing decreases, then the diffraction angle should shift towards higher angles. However, the experimental results showed precisely the opposite trend. This might have been due to the small grain sizes of boron phosphide, which led to a greater proportion of grain boundaries and greater interatomic distances within the grain boundaries. This may have also been due to the penetration of H and other atoms within the lattice, which led to increased interplanar spacing, causing the shift toward smaller diffraction angles. When the PH_3 flow rate increased further, no significant diffraction peaks were observed in the XRD pattern, implying that the thin film was amorphous. These results show that changes in PH_3 flow rate will cause boron phosphide thin films to transform from a crystalline state to an amorphous state.

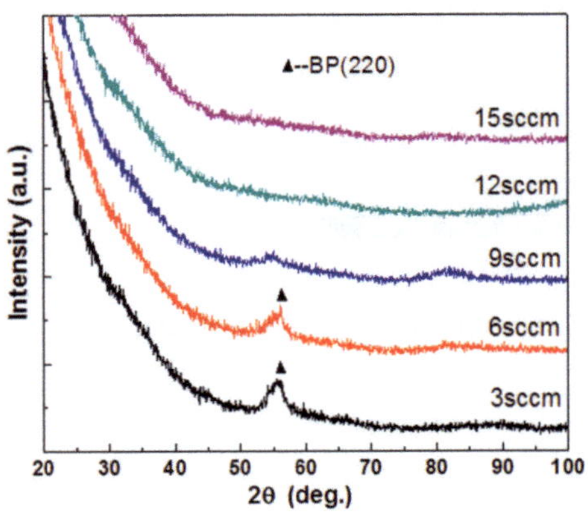

Fig. 7.12: X-ray diffraction patterns of boron phosphide thin films deposited under different PH_3 flow rates.

The second type of key information is the diffraction intensity and the line shape of diffraction peaks, upon which the degree of crystallinity has the most significant impact. Different grain sizes will have different diffraction-peak widths. The diffraction peaks of the fine-grained phase are low and broad, whereas those of large-grained phase are tall and narrow. In addition, the presence of amorphous structures and stress might broaden the diffraction peaks. It can be observed from Fig. 7.12 that, when the PH_3 flow rate increased from 3 sccm to 6 sccm, the peak height slightly decreased, and the full width at half maximum (FWHM) increased from 2.44° to 2.85°. The Scherrer equation can be used to calculate the grain size based on FWHM. The estimated grain sizes are 5.70 nm and 4.95 nm, respectively, for flow rates of 3 and 6 sccm. The grain size decreased with increasing PH_3 flow rate, indicating that the degree of crystallinity in the thin film decreased.

Thin-film structure is determined by its growth mechanism, and is influenced by the processing conditions during deposition. The only variable among the processing parameters in the experiment was the PH_3 flow rate, and the change in PH_3 flow rate affected thin-film composition and deposition speed. On one hand, the stoichiometric ratio has a crucial influence on the crystal growth of III–V group compounds, and it is critical to maintain the ideal stoichiometric ratio. XPS analysis revealed that the relative concentration of P increased significantly with increasing PH_3 flow rate. Thin-film composition gradually transformed from a nearly 1 : 1 stoichiometric ratio at low flow rate to a P-rich state at high flow rate. Hence, nonstoichiometric thin films grown under high flow rates were not conducive to crystal growth and were amorphous. On the other hand, deposition speed plays a crucial role in the growth and migration of thin films. Fig. 7.13 shows the deposition speed of boron phosphide thin films. It can be seen that the deposition speed increased as the PH_3 flow rate increased, and it was in the range of 0.83–1.62 μm/h. The increase in the PH_3 flow rate led to a substantial increase in reactive gas particles; hence, the chemical reactions between the sputtered atoms that reached the substrate and the reactive gas were accelerated to a certain extent, which is reflected as the increase in thin-film deposition speed. Thin-film crystalization is mainly influenced by the diffusion of particles adsorbed on the thin-film surface, and the migration ability of deposited particles on the surface is also affected by deposition speed.

The increase in deposition speed inhibited the diffusion of particles and reduced the surface migration rate. As the particles did not have sufficient time to migrate to the optimal positions, a regular arrangement of the components could not be achieved. Hence, the adsorbed particles could only aggregate near the site of collision with the substrate, and they exhibit a random arrangement. Therefore, a high flow rate will lead to the formation of an amorphous structure with long-range disorder and short-range order.

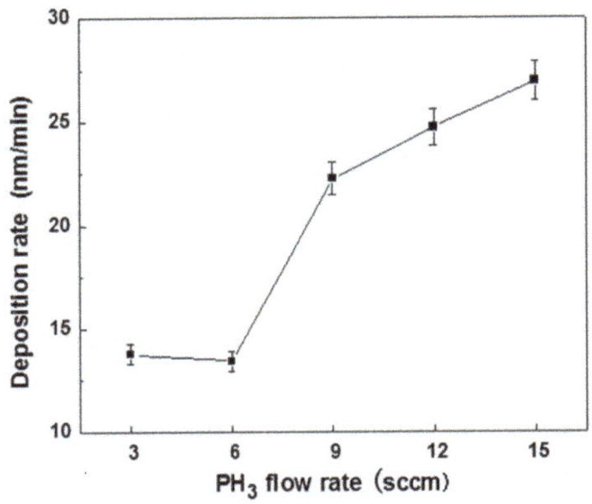

Fig. 7.13: Deposition speed of boron phosphide thin films deposited under different PH_3 flow rates.

7.3.4.2 Fourier transform infrared spectroscopy (FTIR) analysis

Fig. 7.14 shows the FTIR curves of boron phosphide deposited under different PH_3 flow rates. In the figure, the absorption peak near 900 cm^{-1} is the TO vibrational mode of the ZnS substrate, and the small absorption peak at 1457 cm^{-1} was verified through comparison as the B–O bond vibration. A prominent absorption peak was observed near 1260 cm^{-1}. As we know that the σ bond vibration of B–OH is located at 1246 cm^{-1}, while the linear stretching vibration of O–B–O is at 1276 cm^{-1}, it can be inferred that this peak was caused by B–OH and O–B–O.

The main factors influencing the level of infrared absorption include the transition probability of the vibrational energy level and the magnitude of change in dipole moment during molecular vibrations. The molecular structure and symmetry determine the extent of change in dipole moment. When the difference in the atomic electronegativity between both ends of a chemical bond is greater, molecular symmetry will be poorer; hence, the changes in dipole moment during vibration will be greater, and absorption will be stronger. In this book, the absorption peak of B–P bonds did not appear in the FTIR spectra, because the electronegativities of B and P are extremely similar, as a result of which the bond is almost a nonpolar covalent bond; it has a

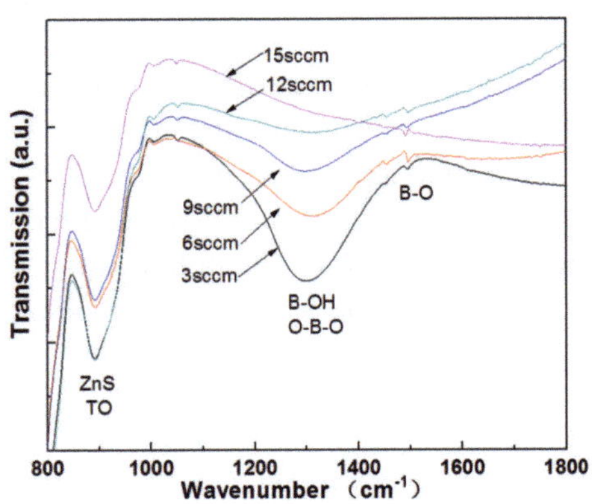

Fig. 7.14: FTIR spectra of boron phosphide thin films deposited under different PH_3 flow rates.

structure with good symmetry and is not infrared active. The intensity of the absorption peak at 1260 cm^{-1} was far greater than that at 1457 cm^{-1}, which may have been caused by the different magnitude of change in dipole moment. As shown in Fig. 7.14, the absorption peak at 1260 cm^{-1} decreased as the PH_3 flow rate increased. This is because the increase in PH_3 flow rate led to a corresponding increase in P in the thin film, which elevated the probability of bonding between B and P, thereby reducing the bonding of B with H and O.

7.3.4.3 Raman analysis

After removing the vibration peak of the monocrystalline Si substrate, the Raman spectra of boron phosphide thin films deposited under different PH_3 flow rates are as shown in Fig. 7.15; all thin films showed the broadening of spectral peaks. It is known that if the thin films are monocrystalline, their lattice vibrational energy will be uniform, and their spectral peak will be very narrow. However, the majority of thin films fabricated in this book were amorphous. When the thin-film lattice is damaged or when the degree of crystallinity is poor, the vibrational energy after excitation will be distributed over a wide range and be reflected as the broadening of spectral peaks.

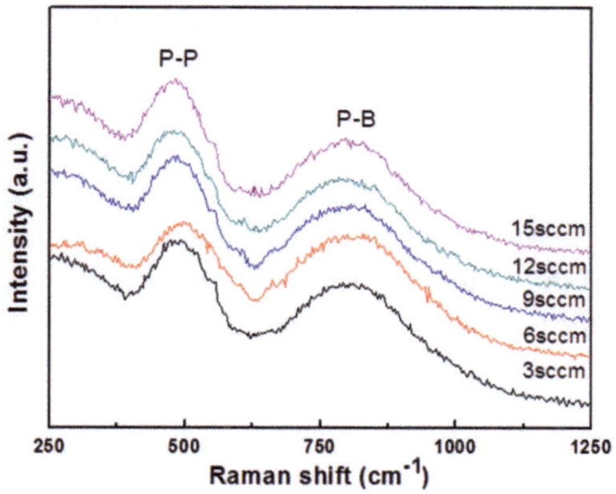

Fig. 7.15: Raman spectra of boron phosphide thin films deposited under different PH_3 flow rates.

Overall, the variations in the Raman spectra are consistent. The vibrational peaks mainly appear near 478 cm^{-1} and 810 cm^{-1}. After comparative analysis, we confirmed that the vibrational peak near 478 cm^{-1} is the symmetrical stretching vibration of P–P bonds. In addition to photon absorption, the occurrence of Raman scattering in the thin film originates from the phonon vibration. The vibrational peak near 810 cm^{-1} belongs to the phonon vibrations of c-BP, which can be split into the TO mode at 800 cm^{-1} and LO mode at 827 cm^{-1}. Based on this, we can infer the main chemical bonding states present in boron phosphide thin films. A schematic diagram of the chemical bonding states in boron phosphide thin films can be found in Fig. 7.16.

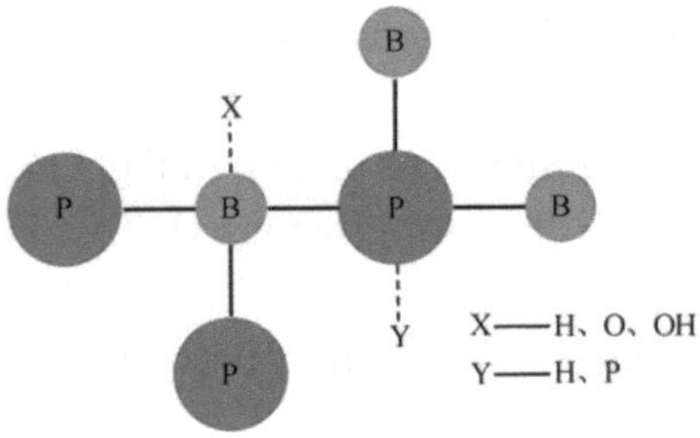

Fig. 7.16: Schematic diagram of chemical bonding states in boron phosphide thin films.

We know that, in BP crystals with a zinc-blende structure, the P atoms are distributed at the vertices of a regular tetrahedron, and the B atom is located at the center. The P and B atoms contribute one valence electron each to form a covalent bond. The covalent bonds are composed of sp^3 hybrid orbitals based on the linear combination of s- and p-state wave functions. By applying Raman spectroscopy, we identified the main chemical bonds in boron phosphide thin films – B–P bonds. If the thin film is completely monocrystalline, then the B atom should only form four covalent B–P bonds with its neighboring P atoms. However, owing to the influence of the deposition processes, the degree of crystallinity was decreased. The thin films fabricated in this book mainly had polycrystalline and amorphous structures. They had large quantities of defects compared to the monocrystalline thin films, which implied that there was a lack of four neighboring atoms surrounding the B and P atoms, thereby producing a number of dangling bonds. Owing to the influence of the precursor, the thin film contained the impurities H and O, which formed bonds with the lone electron pairs in B and P atoms to fill the dangling bonds. B atoms will bond with H and O to form B–OH, B–O, and B–H bonds, while the dangling bonds of P atoms will form P–P and P–H bonds. As can be seen from the analysis above, cubic boron phosphide was present in the thin film, which is consistent with the results of XRD analysis. Furthermore, the Raman spectrum analysis also indicated that, aside from B–P bonds, the thin film contained P–P bonds.

7.3.4.4 XPS analysis

The valence of phosphides is usually labelled using the binding energy of P 2p electrons. The P 2p core-level spectrum of the boron phosphide thin film is located in the range of 129.6–129.7 eV, which is relatively close to the peak position of elemental P (130.0 eV) with a deviation of 0.3–0.4 eV. It is known that the electronegativity of P is 2.19, which is close to that of B (2.04). Hence, the P–B bond formed is virtually a nonpolar covalent bond. Therefore, the electron binding energy of P–B is similar to that of P–P, and the two binding energies are difficult to differentiate. It can be observed from Fig. 7.17 that the FWHM of the P 2p core-level spectrum of the boron phosphide thin film is approximately 2 eV, which is far greater than that of elemental P (approximately 1.3 eV). This indicates that the spectral peaks contain the distributions of two or more types of chemical bonding states.

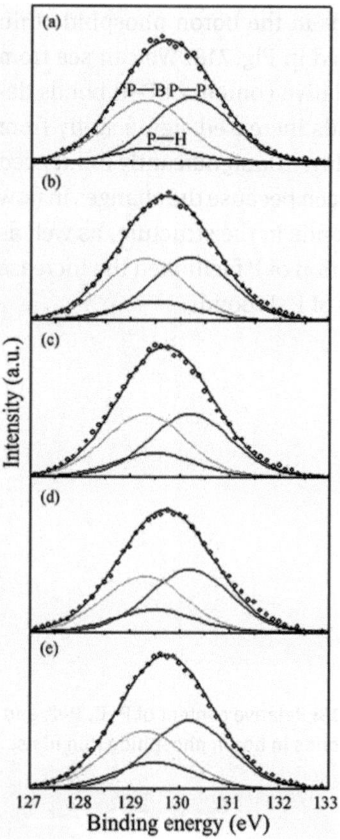

Fig. 7.17: P 2p fitted spectra of boron phosphide thin films deposited under different PH_3 flow rates: (a) 3 sccm; (b) 6 sccm; (c) 9 sccm; (d) 12 sccm; (e) 15 sccm.

With the results above, we can confirm that the element P in the boron phosphide thin film exists in P–B and P–P bonds. Considering the fact that the thin films are mostly amorphous, the structure contains various types of defects. Among them, the predominant defect at an atomic scale involves dangling bonds, which refer to unpaired electrons in P and B atoms, i.e., unsaturated bonds. The reactive gas employed in the experiment was PH_3, which contains a large quantity of H atoms. The H atoms have the tendency to bind with these dangling bonds and form P–H bonds with P atoms. Therefore, we believe that the thin films also contain P–H bonds. The binding energies of the mutual bonding among P, B, and H atoms are very similar; hence, the corresponding peak positions of the resulting bonds will overlap in XPS and cannot be completely separated. Therefore, deconvolution was performed, and three peaks were fitted to the P 2p core-level spectrum of the boron phosphide thin film, i.e., the P–B peak at 129.3 eV, P–H peak at 129.5 eV, and P–P peak at 130.2 eV. During curve fitting, the line shape of the spectral lines was 90 % Gaussian and 10 % Lorentzian, and the FWHM of the spectral peak was 1.9 eV.

The relative contents of P–B, P–P, and P–H bonds in the boron phosphide thin film, obtained through peak deconvolution, are plotted in Fig. 7.18. We can see from the figure that, as the PH_3 flow rate increased, the relative content of P–B bonds decreased from 43.9 % to 37.7 %, while that of P–H bonds increased significantly from 10.3 % to 17.7 %. This indicates that the change in PH_3 flow rate significantly influenced the bonding states in the thin film, which might have been because the changes in flow rate caused the increase in the number of dangling bonds in the structure, as well as the increase in H content. Hence, the continuous addition of P facilitated the increase in P–H bonds and the decrease in the relative content of P–B bonds.

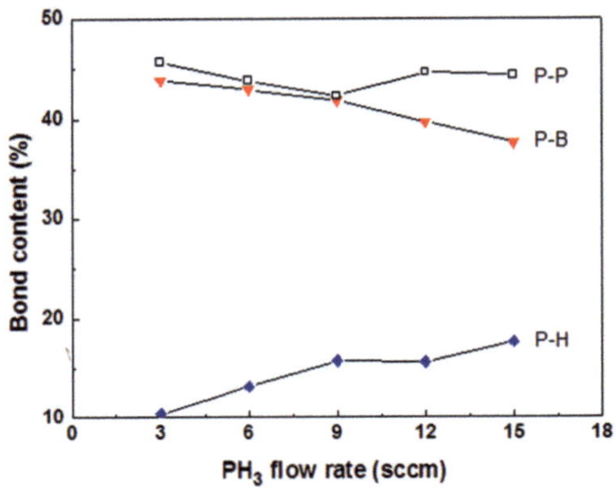

Fig. 7.18: Relative content of P–B, P–P, and P–H bonds in boron phosphide thin films.

7.3.5 Growth mechanism

The thin-film surface showed island-like protruding clusters. Under low flow rates, the protruding particles were large and nonuniformly distributed. With the increase in PH_3 flow rate and deposition speed, the degree of clustering on the thin film decreased, the particles became smaller, and the surface became smoother and more refined. During particle motion near the substrate for thin-film deposition, particle migration is an important dynamic process in thin-film growth, which has a crucial impact on the microstructure of the resulting thin film. When particles such as atoms, molecules, and clusters reach the film surface, they rapidly lose a large proportion of their energy. Thus, their residual energy and deposition speed determine the migration rate of deposited particles on the substrate surface.

If the particles have high mobility, then particles in a metastable position with the second-lowest potential field will overcome the energy barrier and occupy a stable position with the lowest potential field. Under these circumstances, thin-film crystal growth will show good integrity. In the experiment, the PH_3 flow rate had a significant impact on thin-film deposition speed, whereby the increase in the PH_3 flow rate led to a higher thin-film deposition speed. Although this was beneficial to the kinetic energy of the particles, high deposition speeds also led to shorter migration times of the particles on the surface. Thus, when one layer of particles has just reached the surface and has yet to complete the diffusion process, it will have been covered by the subsequent layer, thereby leading to the significant reduction in the migration rates of the deposited particles. Hence, under low PH_3 flow rates, lower deposition speeds and greater migration will imply that the probability of particles packing into nuclei and clusters will be increased dramatically. The resulting thin film will have a high degree of crystallinity and large grain sizes. In contrast, under high PH_3 flow rates, the deposition speed will increase, but the migration rate of deposited particles will decrease. The decrease in aggregation will imply the formation of fewer large clusters. Thus, the thin film will not show grain growth, eventually leading to the formation of amorphous boron phosphide thin films.

When using reactive magnetron sputtering processes to fabricate boron phosphide, among the controllable parameters, including sputtering power, substrate temperature, and gas flow ratio, the most important factor that will determine the quality, structure, and composition of thin films is the gas flow ratio. By changing the PH_3 flow rates when fabricating boron phosphide thin films, the resulting thin films showed differences in terms of chemical composition, bonding states, surface morphology, crystal structure, and other aspects. Under low flow rates, thin-film composition was nearly stoichiometric. The content of B–P bonds was high, and the thin films showed the c-BP nanocrystalline structure. Recrystallization occurred after high-temperature annealing, and the c-BP polycrystalline structure emerged. When the PH_3 flow rate was increased, the deposition speed increased somewhat, but the thin films showed a P-rich state and were amorphous. Although the c-BP polycrystalline structure also appeared at high temperatures, phase transitions from c-BP to $B_{12}P_2$ occurred as well. Hence, the thin-film structures under these conditions had poor thermal stability. Compared to thin films grown under low flow rates, those grown under high flow rates did not exhibit the expected c-BP phase, and had poorer structural thermal stability. Therefore, based on the experimental results above, low PH_3 flow rates should be selected when depositing boron phosphide thin films using reactive magnetron sputtering.

7.4 Thermal stability of boron phosphide thin films

In order to ensure that the thin films were not contaminated by foreign objects during annealing, vacuum annealing was performed in this experiment. If we consider that boron phosphide thin films will be used on the infrared optical elements of high-speed aircraft, the temperatures could reach 500–900 °C. At the same time, we should also consider that the phase transition temperature of boron phosphide is 1050 °C, while the thin-film deposition temperature is 600 °C. Hence, the range of annealing temperature for boron phosphide thin films should be set as 750–1150 °C. Boron phosphide thin films deposited on Si substrates were sealed in a vacuum quartz tube. The degree of vacuum in the tube was maintained at 10^{-2} Pa. The temperature in the furnace was increased to 750 °C, 950 °C, and 1150 °C, respectively, maintained for 1h, and then air cooled to room temperature.

7.4.1 XRD analysis

Fig. 7.19 shows the XRD patterns of boron phosphide thin films during deposition and after annealing at 750–1150 °C. It can be seen from the figure that the boron phosphide films showed a different microstructure during deposition at 600 °C. Specifically, broadened cubic BP diffraction peaks appeared under low flow rates, whereas the thin films were amorphous under high flow rates.

After annealing at 750 °C, except for the (220) plane, thin films fabricated under flow rates of 3 sscm–12 sccm began to show weak diffraction peaks of the cubic BP (111) plane. This indicates that the amorphous thin film had slowly begun to crystallise at this temperature. After annealing at 950 °C, all thin films showed a micro-

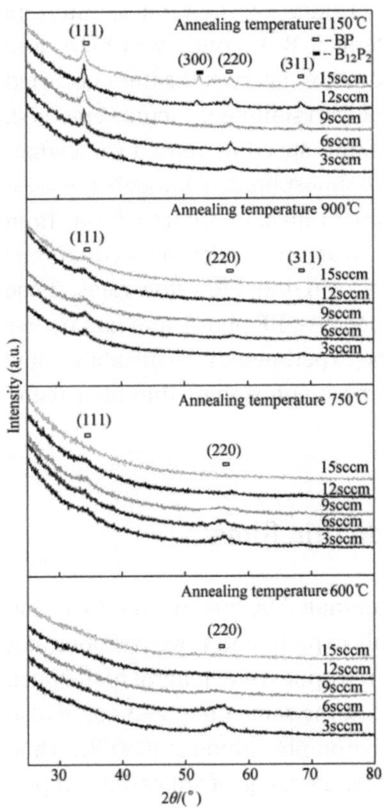

Fig. 7.19: XRD patterns of boron phosphide thin films after vacuum annealing.

crystalline structure with $2\theta = 34.2°$, $57.4°$, and $68.5°$ corresponding to the (111), (220), and (311) planes of cubic BP. However, there were changes in the preferred orientation, wherein the diffraction peak of BP (111) was significantly enhanced, and that of BP (220) was slightly weaker. This indicates that the thin film showed changes in its preferred orientation as the process of crystallization continued, and the system energy reduced. After annealing at 1150 °C, thin films formed under 3–9 sccm still existed in the form of c-BP polycrystals, and the diffraction peaks were all enhanced, especially that of the preferred orientation of BP (111). This indicates that the grains were undergoing continuous growth. However, at high flow rates of 12 sccm and 15 sccm, a new diffraction peak emerged at $2\theta = 52.5°$, which corresponds to the (300) plane of the rhombohedral $B_{12}P_2$. This indicates that the phase transition from c-BP to $B_{12}P_2$ occurred within the temperature range of 950–1150 °C.

When explained from the perspective of thermodynamics, the level of Gibbs free energy at low temperatures is mainly determined by the enthalpy (H) of substances, $H_{\mathrm{loose}} > H_{\mathrm{dense}}$. Hence, at low temperatures, the densely packed c-BP will be the stable phase. At high temperatures, the predominant effect on the Gibbs free energy is the contribution of entropy (S), $S_{\mathrm{loose}} > S_{\mathrm{dense}}$. Hence, at high temperatures, the loosely arranged $B_{12}P_2$ will be the stable phase. Thus, we can conclude that, as temperature increases, the phase transition that occurs in boron phosphide thin films is the transition from the dense structure of c-BP to the loose structure of $B_{12}P_2$. This is consistent with the research by Zhao et al. on the decomposition characteristics of boron phosphide. Their study showed that, within the temperature range of 1000–1050 °C, heated phosphides will undergo decomposition from high-P compounds to more thermally stable low-P compounds and P steam, i.e., $B_{12}P_2$ or $B_{13}P_2$.

7.4.2 Raman analysis

The Raman signals of crystals are often used to characterise the degree of crystallinity and stress states. Therefore, Raman analysis was performed on annealed boron phosphide thin films. Fig. 7.20 shows the Raman spectra of boron phosphide thin films before and after annealing at 1150 °C. For monocrystals with perfect crystalinity, the lattice vibrational energy is uniform and the Raman peak is very narrow. If the lattice is damaged or the degree of crystallinity is insufficient, the vibrational energy after excitation will be distributed over a wider range, and hence, the broadening of spectral peaks will occur. It can be seen from the figure that the thin film deposited at 600 °C only showed a broadened scattering peak, and its peak position was near 810 cm^{-1}.

This is because the boron phosphide thin film mainly exists in an amorphous form. After annealing at 1150 °C, the scattering peak of B–P bonds was significantly enhanced and narrowed, while the increase in PH_3 flow rate was accompanied by the gradual decrease in FWHM from 44 cm^{-1} to 14 cm^{-1}. This indicates that, after annealing, the boron phosphide thin films transitioned from an amorphous state to boron

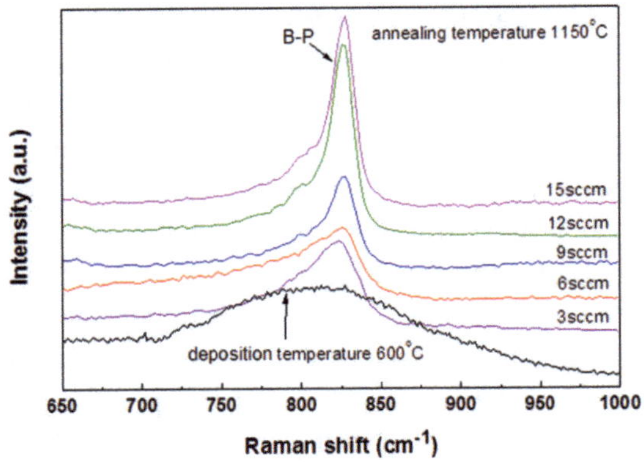

Fig. 7.20: Raman spectra of boron phosphide thin films before and after annealing.

phosphide grains. Furthermore, the average grain sizes of thin films produced under high PH_3 flow rates during the annealing process were larger, which is consistent with the XRD results.

Details on the spectral peaks in Raman spectra are listed in Tab. 7.3. An analysis revealed that the LO mode of c-BP showed a blueshift. When compared to thin films deposited at 600 °C, the scattering peaks of boron phosphide annealed at 1150 °C all showed a right shift by wavenumbers greater than ten, and were already very close to the peak of BP crystals. As the PH_3 flow rate increased, the peak position of B–P bonds shifted from 823 cm^{-1} to 828 cm^{-1}, which is consistent with the BP/Si scattering peaks measured using CSVT. The main cause for the shift in the peak position of Raman scattering is the stress changes in the thin films. During the deposition process at 600 °C, the difference in the thermal expansion coefficients between the boron phosphide thin film and Si substrate produces thermal stress. Furthermore, the defects caused by grain boundary and coating–substrate lattice mismatch produce intrinsic stress. The combined effect of these two factors leads to a shift in the Raman scattering peak. Annealing at 1150 °C removes the thermal stress present in the thin film. The larger grains and reduction in grain boundaries also enable the partial release of intrinsic stress. Thus, the scattering peaks of the thin films tend toward the peak position of boron phosphide crystals.

Tab. 7.3: Spectral peak information of boron phosphide thin films before and after annealing.

Sample	Peak position (cm^{-1})	FWHM (cm^{-1})
600 °C	810	108
3 sccm	823	44
6 sccm	825	42
9 sccm	826	22
12 sccm	827	15
15 sccm	828	14

7.5 Mechanical and optical properties of boron phosphide thin films

7.5.1 Mechanical properties

7.5.1.1 Hardness and elastic modulus

Fig. 7.21 shows the hardness and elastic modulus of boron phosphide thin films measured using the continuous stiffness measurement technique. We can see from the figure that the hardness curve increased rapidly to its peak value in the early stage of indentation and then decreased gradually as indentation depth increased. This is indicative of a hard coating and soft substrate. The hardness reached its maximum level at an indentation depth of 75 nm. The substrate will begin to undergo deformation as the indentation depth increases further, and the hardness will decrease owing to the influence of the soft substrate. The value measured at this point is the composite hardness of the thin film and substrate. The elastic modulus also showed a similar trend of variation. The maximum value was attained at 40 nm, which decreased and reached a stable plateau at approximately 300 nm. We believe that, at this point, the indenter has already penetrated the thin film and is approaching the substrate; thus, the measured values are close to those of the substrate material. It can be inferred from this result that the film thickness is approximately 300 nm, which is consistent with the film thickness data obtained using a surface profiler.

Fig. 7.22 and 7.23 show the hardness and elastic modulus, respectively, of boron phosphide thin films deposited under different PH_3 flow rates. The hardness and elastic modulus both attained their maximum values when the PH_3 flow rate was 6 sccm, and the values were 25.625 GPa and 259.389 GPa, respectively. These values are far greater than those of GaP thin films, as well as Ge, ZnS, and other substrate materials. Compared to the thin film prepared using PECVD by Ogwn et al., which had an average hardness of 15 GPa, the boron phosphide thin film prepared using reactive magnetron sputtering in this book had a significantly higher level of hardness.

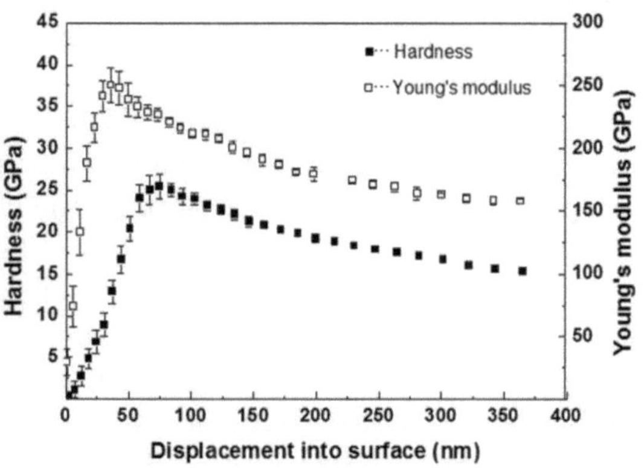

Fig. 7.21: Hardness and elastic modulus of boron phosphide thin films measured using the continuous stiffness measurement technique.

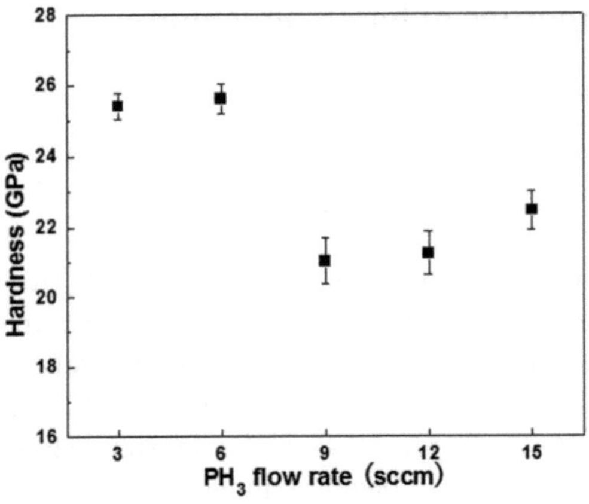

Fig. 7.22: Hardness of boron phosphide thin films deposited under different PH_3 flow rates.

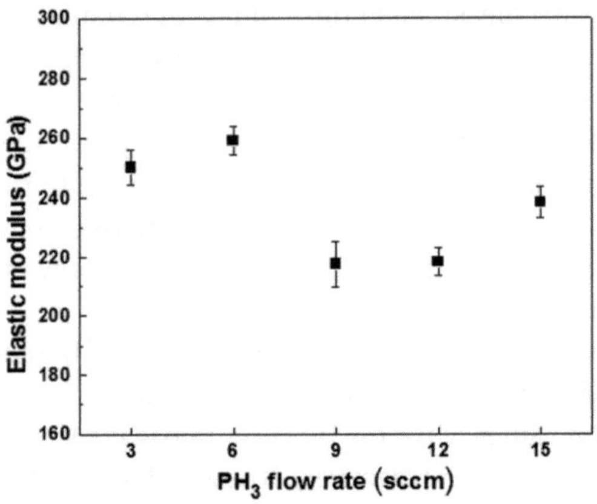

Fig. 7.23: Elastic modulus of boron phosphide thin films deposited under different PH_3 flow rates.

In general terms, the thin films obtained in this book and those deposited using PECVD showed consistent trends of variations under conditions where only the PH_3 flow rate was modified. In other words, the hardness and elastic modulus were high under low PH_3 flow rates, and both showed decreasing trends with increasing PH_3 flow rate. Changes in hardness and elastic modulus are mainly influenced by the chemical bonds and crystal structure of the thin films.

On one hand, it is known that B and P have similar electronegativities, and the B–P bonds in boron phosphide thin films are virtually nonpolar covalent bonds with a high binding energy. The XPS analysis in Chapter 6 has shown that the relative content of B–P bonds in the thin film will decrease with increasing PH_3 flow rate. Hence, the hardness and elastic modulus will also show a corresponding decrease. On the other hand, c-BP has a regular tetrahedral structure similar to that of diamond; its hardness is second only to c-BN, and this structure has the highest hardness level among boron phosphide compounds. The XRD analysis above indicates that the c-BP structure was present when the PH_3 flow rate was 3 and 6 sccm. Since the presence of c-BP has a crucial impact on thin-film hardness and elastic modulus, thin films produced at low PH_3 flow rates exhibited higher levels of hardness. As the PH_3 flow rate increased, the thin film gradually transitioned to an amorphous state with long-range disorder and short-range order, without the presence of c-BP microcrystals. Hence, its hardness level was less than that of thin films produced at lower PH_3 flow rates. An analysis of surface morphology indicated that, for amorphous thin films produced under high flow rates, as the PH_3 flow rate increased, thin-film particles became smaller and surface density increased. Hence, the thin-film hardness showed an increasing trend within a small range.

7.5.1.2 Adhesion performance

Thin-film adhesion is one of the key indicators of mechanical performance, and it has a direct impact on the feasibility and reliability of thin-film applications. However, a quantitative description of thin-film adhesion is relatively difficult, as even by using the same method, different testers might obtain different results. Therefore, adhesion test results generally only have qualitative implications. Currently, there is no unified adhesion testing method for various types of thin films. Methods for testing thin-film adhesion mainly include the scratch test, stretch test, tape test, friction test, ultrasonic test, and centrifuge test.

In order to evaluate the adhesion performance of thin films conveniently and effectively, qualitative assessment was performed using the tape test and abrasion test. These tests were conducted in accordance with the general specifications and requirements of optical thin-film tests found in the Chinese National Standard GJB-2485-95 and US Military Standard MIL-48616 (Tab. 7.4). The thin films underwent the adhesion test, moderate abrasion test, and severe abrasion test. At the end of the tests, laser scanning confocal microscopy was performed to observe the damage on the thin-film surface. The results revealed that all thin films were undamaged. Fig. 7.24 shows the optical micrograph of a boron phosphide thin film after the adhesion test, moderate abrasion test, and severe abrasion test. The coating layer was smooth and showed no signs of delamination. This indicates that the thin film has satisfied the standard requirements, and has good adhesion with the substrate.

Fig. 7.24: Optical micrograph of boron phosphide thin film after tape test and abrasion tests.

Tab. 7.4: Experimental standards to assess the environmental durability of optical thin films.

Experimental item	Experimental description
Adhesion	A 2-cm-wide tape with peel strength greater than 2.74 N/cm is pressed firmly against the thin-film surface and rapidly removed at an angle normal to the surface.
Cleaning durability	The thin film is wiped using a degreased cheese cloth dipped in a mixture of ethanol and ethyl ether.
Moderate abrasion	A degreased cheese cloth is wrapped around the eraser portion of the abrasion tester and rubbed across the thin-film surface 25 times at a pressure of 4.9 N.
Severe abrasion	A degreased cheese cloth is wrapped around the eraser portion of the abrasion tester and rubbed across the thin-film surface 20 times at a pressure of 9.8 N.
Humidity	The thin film is placed under a temperature of 50 ± 2 °C and relative humidity of 95–100 % for 24 h.
High and low temperatures	The thin film is placed under temperatures of −62 ± 2 °C and 70 ± 2 °C for 2 h each.
Water solubility	The thin film is immersed in distilled water for 24 h at room temperature (16–32 °C).
Solvent resistance	The thin film is immersed for 10 min each in trichloroethylene, acetone, and anhydrous ethanol.
Salt solubility	The thin film is immersed in a 4.5 % sodium chloride solution for 24 h at room temperature (16–32 °C).
Ultra-high temperature	The thin film is exposed to high temperatures of 300–500 °C for 1 h.

7.5.1.3 Residual stress

Fig. 7.25 shows the residual stress and thermal stress of boron phosphide thin films fabricated under different PH_3 flow rates. The residual stress measured in the experiment was manifested entirely as compressive stress, and the stress values were relatively small. It can be seen from the figure that, aside from the stress value at 3 sccm, which was on the order of GPa, the stress values at other flow rates were all on the order of MPa. As the PH_3 flow rate increased, thin-film residual stress gradually decreased from 1209.4 MPa to 389.1 MPa. The thermal stress calculated based on the temperature difference was also very small at approximately 200 MPa, and changes in the PH_3 flow rate did not have a significant impact on the generation of thermal stress.

Therefore, the residual stress produced in boron phosphide thin films is mainly caused by internal stress. Internal stress, also known as intrinsic stress, is produced by fairly complex factors. It is mainly determined by thin-film microstructure and defects, and is predominantly due to the interactive effects between grain boundaries and coating–substrate lattice mismatch. During the coating process using reactive mag-

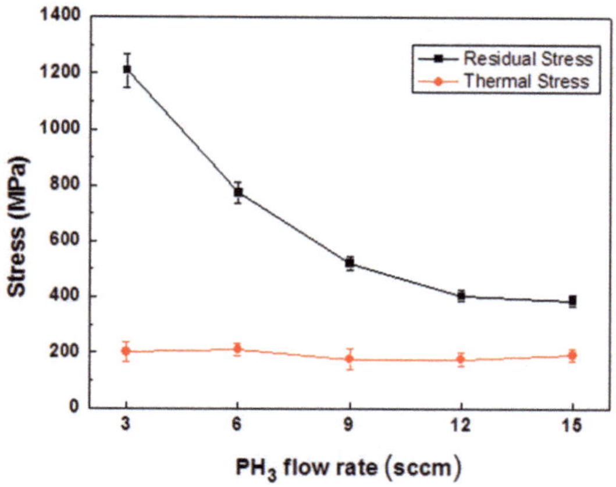

Fig. 7.25: Residual stress and thermal stress of boron phosphide thin films fabricated under different PH_3 flow rates.

netron sputtering, the plasma generated through glow discharge will contain high-speed particles accelerated by the electric field, which will bombard the growing thin-film surface. The energy of the B atoms released from the sputtering target will be far higher than that of the vaporized atoms. The high-speed and high-energy particles that collide with the thin-film surface will penetrate the boron phosphide thin film, and the impact with B or P atoms in the thin film will cause them to deviate from their lattice positions. This will lead to vacancies and displaced atoms, thereby increasing the volume of the thin film. These high-speed and high-energy particles might also enter the gaps within the lattice of the boron phosphide thin film, which will result in the particle "pinning" effect. Therefore, thin films deposited under the processing conditions of reactive magnetron sputtering will have compressive stress. Given that the working pressure remains unchanged, the increase in PH_3 flow rate implies that the relative content of Ar gas will decrease, and the plasma generated through glow discharge will also decrease. Hence, the number of B atoms released because of Ar^+ bombardment will be reduced, and the displacement of lattice atoms and pinning effect caused by the collision of high-energy particles will be weakened. Therefore, compressive stress decreased gradually, which was reflected in the residual stress curve.

7.5.2 Optical properties

7.5.2.1 Infrared absorption

Fig. 7.26 shows the transmission spectra of a ZnS substrate before and after the deposition of a boron phosphide thin film. The figure shows that the ZnS substrate has an absorption peak near 900 cm^{-1}, which is caused by the 3TO vibrational mode of ZnS. After the ZnS substrate was coated with the boron phosphide thin film, a large absorption peak occurred near 1300 cm^{-1}, which might have been caused by the vibration absorption of B–OH and O–B–O bonds.

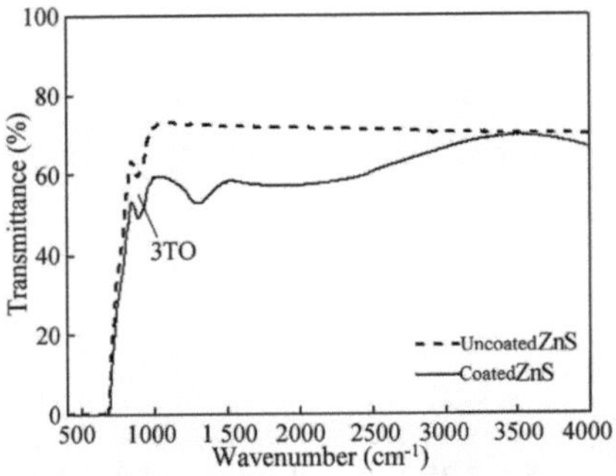

Fig. 7.26: Transmission spectra of Zns substrate before and after deposition of boron phosphide thin film.

The absorption of boron phosphide thin films is mainly determined by extrinsic absorption, and the main impurities are H and O. There were no absorption peaks within the infrared wavelength range of ZnS optical elements (3–5 µm and 8–12 µm), and impurity absorption was relatively low. Within the range of 8–11.5 µm, the average transmittance of the uncoated ZnS substrate was 69.96 %, while the average transmittance after being coated with boron phosphide was 56.49 %. Within the range of 2.5–4 µm, the average transmittance of the uncoated ZnS substrate was 70.32 %, while the average transmittance after being coated with boron phosphide was 66.25 %. The average transmittance had decreased somewhat compared to that before coating because of the high refractive index of boron phosphide ($n \approx 3.0$).

Analysis of the infrared absorption spectra indicated that the boron phosphide thin film did not show significant absorption within the wavelength range of infrared optical elements. However, its relatively high refractive index limited its infrared transmittance. Therefore, boron phosphide thin films cannot be directly applied as an antireflective coating, and will need to be combined with other coating layers with low refractive indices to form a composite coating system.

7.5.2.2 Optical constants

Fig. 7.27 shows the variations in the refractive index and extinction coefficient of boron phosphide thin films with changes in wavelength.

We can see from Fig. 7.27 that the refractive index and extinction coefficient of the thin film showed a decreasing trend with increasing wavelength, which is considered normal dispersion. At a wavelength of 632.8 nm, the refractive index was 3.12 and the extinction coefficient was 0.52, which is similar to the refractive index of CVD thin films (3.0 ± 0.05). Nevertheless, there is still a difference, which might have been because the fabricated thin film was not fully densified and a certain level of porosity was still present. The composition of an ideal BP thin film will have an accurate stoichiometric ratio. The optical and mechanical characteristics of such a thin film will be similar to those of the bulk material. The short-wave absorption limit of crystalline BP is 0.62 µm, whereas the shortwave limit of an amorphous boron phosphide

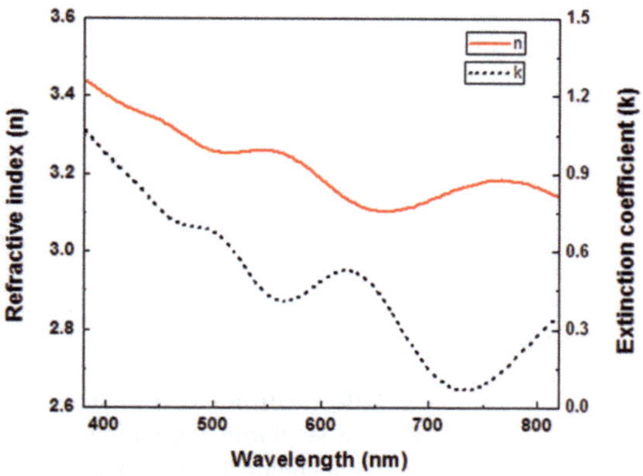

Fig. 7.27: Variations in the refractive index and extinction coefficient of boron phosphide thin films with wavelength.

thin film fabricated using PECVD showed a redshift to 0.95 μm. This indicates that thin-film absorption within the range of 0–950 nm will increase significantly. As the majority of thin films fabricated using magnetron sputtering will be amorphous, the extinction coefficient will increase as they approach the shortwave absorption band. For thinner films, absorption will not have a significant influence on transmittance when the extinction coefficient $k < 10^{-2}$, whereas its impact will be substantial when $k > 10^{-2}$, particularly at the half-wavelength position. We can deduce that, at a film thickness of approximately 300 nm, the extinction coefficient will show an extremely large peak value at 600 nm, which is consistent with the fitted curves of ellipsometric data. According to the classical theory of dispersion, as wavelength increases, the refractive index and extinction coefficient of optical materials both decrease. Therefore, the refractive index of boron phosphide thin films will continue to decrease when the wavelength reaches the infrared range, and the extinction coefficient will approach 0.

7.6 Corrosion resistance and wear resistance of boron phosphide thin films

7.6.1 Corrosion resistance

Fig. 7.28 shows the polarization curves of stainless steel with and without a boron phosphide thin film in 3.5 % NaCl solution. This NaCl concentration was intended to simulate the salt concentration in seawater. The horizontal axis is the corrosion potential, and the vertical axis is the corrosion current. In this polarization curve, better corrosion resistance implies a smaller $\lg(I)$ and more positive corrosion potential, i.e. curves that tend more toward the lower right corner will correspond to better corrosion resistance. It can be seen from the figure that, after the boron phosphide thin film was deposited onto the stainless steel substrate, its polarization curve shifted towards the lower right region with higher potential and lower current density. This indicates

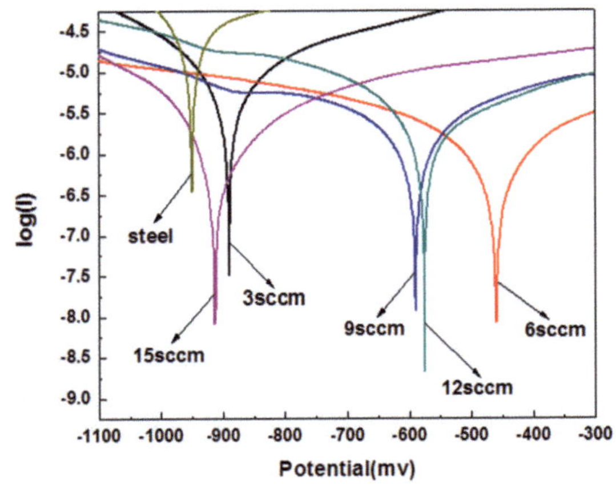

Fig. 7.28: Polarization curves of stainless steel with and without boron phosphide thin film in 3.5 % NaCl solution.

that coating stainless steel substrates with boron phosphide thin films improves their corrosion resistance in seawater.

The main polarization parameters that can be obtained from the polarization curves include the corrosion potential (E_{corr}), corrosion current density (I_{corr}), and breakdown potential (E_{brk}). The polarization parameters of stainless steel with and without boron phosphide thin film are listed in Tab. 7.5.

Tab. 7.5: Polarization parameters of stainless steel with and without boron phosphide thin film.

Sample	E_{corr} (mV)	I_{corr} (A/cm²)	E_{brk} (mV)
Stainless steel	−950	3.47×10^{-7}	−653
3 sccm	−890	3.25×10^{-8}	−480
6 sccm	−460	9.66×10^{-9}	−123
9 sccm	−589	1.17×10^{-8}	−68
12 sccm	−575	2.14×10^{-9}	−58
15 sccm	−913	8.13×10^{-9}	−434

The table above shows that the corrosion potential and breakdown potential of boron phosphide were both higher than those of stainless steel, whereas the corrosion current density was lower than that of stainless steel by 1–2 orders of magnitude. Specifically, the thin film fabricated under a PH_3 flow rate of 6 sccm had the most positive corrosion potential of 460 mV, while the thin film fabricated under 12 sccm had the smallest corrosion current of 2.14×10^{-9} A/cm². During the process of positive polarization, a passivation zone emerged when the electrode potential was greater than the corrosion potential. As the electrode potential increased, the instantaneous current remained stable, and the thin-film surface was in a passivated state, with only a small amount of dissolution and low corrosion rate. This continued until the breakdown potential was reached, which caused pitting or surface erosion of the thin film.

At this point, the positive current density increased rapidly, and there was a sharp increase in the corrosion rate. Compared to the passivation zone of the oxide coating on the stainless steel surface, that of the deposited boron phosphide thin film was significantly wider, and the breakdown potential was substantially higher. These results demonstrate that the deposition of a boron phosphide thin film on stainless steel substrate will increase its corrosion potential and decrease its corrosion current density. Moreover, the presence of the boron phosphide thin film will enlarge its passivation zone while reducing the probability and rate of corrosion. Therefore, boron phosphide thin films can significantly enhance the corrosion resistance of stainless steel in saline solutions.

7.6.2 Wear resistance

Fig. 7.29 shows the friction coefficient of a boron phosphide thin film under a load of 200 g for a test duration of 50 000 s. The occurrence of wear damage can be monitored by observing the changes in the friction coefficient. It can be seen from the figure that the friction coefficient of the thin film is constantly in a fluctuating state but essentially remained at approximately 0.2. However, at approximately 52 000 s, the friction coefficient showed a sudden increase to greater than 0.4. It was confirmed after comparison that this friction coefficient is the same as that of the Si substrate. Hence, we can conclude that the thin film had already been worn through at this point, and the ceramic ball was in contact with the substrate, reflecting the friction coefficient of the substrate.

The wear volume can be calculated based on the cross-sectional area and depth of the wear marks on the thin film, which revealed that the specific wear rate of boron phosphide thin film is $8.95 \times 10^{-10}\ mm^3/(N \cdot m)$. Under similar experimental conditions, the specific wear rate of TiCN, which is a commonly used material with high hardness and wear resistance, is $5 \times 10^{-8}\ mm^3/(N \cdot m)$. Compared to TiCN, the

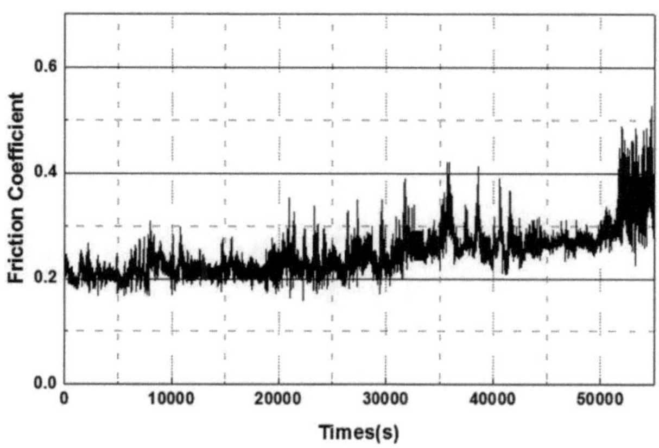

Fig. 7.29: Friction coefficient of boron phosphide thin films.

wear rate of the boron phosphide thin film was less by two orders of magnitude. The hardness level of thin films generally reflects the magnitude of wear resistance. Boron phosphide thin films have a uniform and densified structure, high hardness, high elastic modulus, relatively good toughness, and are not prone to brittle fractures. Hence, they possess relatively long wear lives.

Laser scanning confocal microscopy was used to characterise the morphology of wear marks on the thin-film surface. Fig. 7.30 and 7.31 show the wear mark morphologies of the boron phosphide thin film at 5000 and 52 000 s, respectively.

Based on our analysis, the process of frictional wear on the thin film can be divided into two phases: the transition from the damage mechanism of surface fatigue wear (contact fatigue) to abrasive wear mechanisms. The rotational movement of the silicon carbide ceramic ball causes cyclic loading on the surface of the boron phosphide

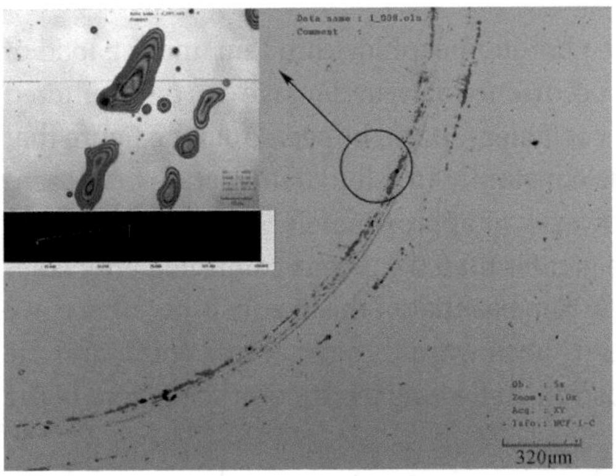

Fig. 7.30: Morphology of wear mark on boron phosphide thin film at 5000 s.

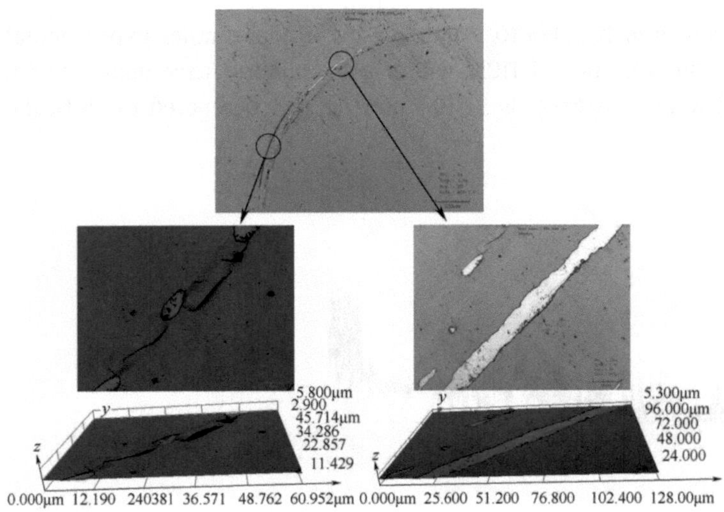

Fig. 7.31: Morphology of wear mark on boron phosphide thin film at 52 000 s.

thin film, producing alternating contact stress. The heat produced during friction increases the temperature of the working surface between the frictional couple. As the thermal expansion coefficients of the thin film and substrate are different, this will result in different thermal deformations. Moreover, owing to the effects of alternating shear stress, the difference in the elastic modulus between the thin film and substrate also leads to different strains. Under the long-term and repeated influence of the two forces, the sample interior, especially the coating–substrate interface, produces a large amount of stress. After a certain number of fatigue cycles, this stress exceeds the coating–substrate adhesion strength, which causes the thin film to separate from the substrate and produce the wrinkled appearance shown in Fig. 7.30.

It can be seen from Fig. 7.30 that at 5000 s the boron phosphide thin film did not exhibit signs of fracture or wear debris, and it only showed localized protrusions, preventing further wear during subsequent friction. The changes in friction coefficient at this point correspond to the fluctuations in Fig. 7.29 within 36 000 s.

After the appearance of wrinkles, the thin film no longer had a smooth and level surface, which now showed localized protrusions. At this point, mutual friction occurs between the ceramic ball and protrusions on the thin-film surface, leading to the repeated plastic deformation of the thin-film surface, which results in fracture and delamination, as shown in Fig. 7.31. This phase involves abrasive wear mechanisms. On the friction coefficient curve, the start of the abrasive wear phase corresponds to the peak in friction coefficient at 36 000 s, and thin-film delamination is reflected in the sudden rise in friction coefficient at approximately 52 000 s.

In order to investigate the influence of PH_3 flow rate on the wear resistance of boron phosphide thin films, comparative analysis was performed to evaluate thin-film wear resistance. In other words, under the same frictional conditions, the greater the degree of wear per cycle or per unit time, the poorer is the wear resistance. Fig. 7.32 shows the friction coefficient of boron phosphide thin films deposited under different PH_3 flow rates. It can be seen from the figure that, as the PH_3 flow rate increased, the friction coefficient of the thin-film surface showed a trend of decrease. With the increase in PH_3 flow rate, the thin film transitioned from a crystalline state to an amorphous state, grain size decreased, and thin-film continuity was enhanced, leading to a gradual decrease in roughness. Surface roughness has a relatively high impact on friction coefficient, as it directly affects the contact state of the surface, distribution of the actual contact area, adhesive contact of asperities, and other factors, thereby modifying the frictional behaviour of thin films. If the surface roughness of the thin film is very large, the surface contact of the frictional couple will be the contact between protrusions instead of overall surface contact. The reduction in contact surface area will lead to large local stress at the points of contact, which is prone to adhesion and shearing during relative motion, producing adhesive wear and leading to the increase in the friction coefficient. A lower thin-film roughness will decrease the probability for the occurrence of adhesive wear mechanisms, and hence, the friction coefficient will show a trend of decrease with decreasing thin-film roughness.

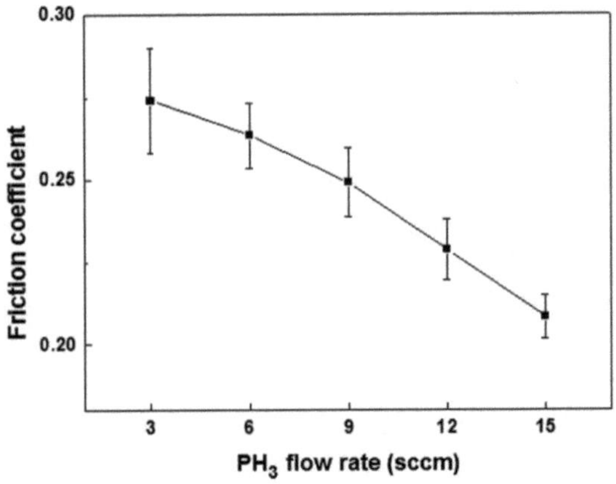

Fig. 7.32: Friction coefficient of boron phosphide thin films deposited under different PH_3 flow rates.

The friction coefficients of boron phosphide thin films at 5000 s were relatively stable and did not undergo great change. This indicates that the thin films were not damaged, and only local wrinkling was observed in the wear-mark morphology. Under these circumstances, the area of wrinkling produced on the thin-film surface can replace wear volume in the calculation of wear rate to evaluate the wear resistance of thin films.

Two factors mainly influence the performance of boron phosphide thin films in the fatigue wear phase. On one hand, the increase in material hardness can enhance its resistance to fatigue wear and reduce the probability of fatigue cracks. On the other hand, surface roughness has a substantial impact on the fatigue strength of the material bearing cyclic loading. Under the effects of alternating loads, the valleys on the surface will lead to stress concentration and produce fatigue cracks. Hence, the reduction of surface roughness can also improve fatigue life. Based on the analysis above, the hardness and surface roughness of thin films will both show a trend of decrease. Fig. 7.33 reveals that, as the PH_3 flow rate increases, the wear rate shows a trend of initial increase followed by a gradual decrease, where the wear rates for 3 and 6 sccm are

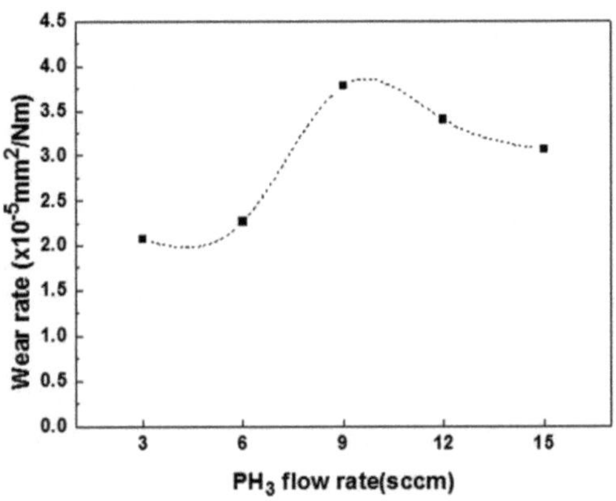

Fig. 7.33: Wear rate of boron phosphide thin films deposited under different PH_3 flow rates.

significantly less than those for high flow rates. This indicates that, in the fatigue wear phase, thin-film hardness has a predominant effect on wear rate. Hence, thin films produced under low flow rates showed lower wear rates. At flow rates of 9–15 sccm, the thin-film hardness showed a slight increase, and the roughness decreased gradually. Hence, under the combined action of both factors, the wear rate showed a small magnitude of decrease.

References

[1] Tropf WJ, Thomas ME, Frazer RK. Windows and domes: past, present, and future. Proceedings of SPIE, 2003: 80–89.

[2] Ondercin RJ, Goldman LM, Harker AB. Development of a repairable IR composite window. Proceedings of SPIE, 1999: 238–249.

[3] Huang ZJ. The trends of development of midwave infrared guided flow dome and optical window materials. Materials Review, 1998, 12(3): 30–33.

[4] Field JE. ELSI conference: invited lecture liquid impact: theory, experiment, applications. Wear, 1999, 233–235: 1–12.

[5] Waddell EM, Monachan BC. Rain erosion protection of materials using boron phosphide coatings. Window and Dome Technologies and Materials II, 1990, 1326: 144–156.

[6] Kelly ES, Ondercin RJ, Detrio JA. Environmental testing of long wave infrared windows. Proceedings of SPIE, 1997, 3060: 68–75.

[7] Gibson DR, Weddell EM, Lewis KL. Advances in ultradurable phosphide-based broadband anti-reflection coatings for sand and rain erosion protection of infrared windows and domes. Optical Engineering, 1994, 2286: 335–346.

[8] Zhu JZ, Liu ZT. Design and materials of infrared anti-reflective and protective coatings. Laser & Infrared, 1996, 26: 275–278.

[9] Wheeler DW, Wood RJK. Solid particle erosion behaviour of CVD boron phosphide coatings. Surface & Coatings Technology, 2006, 200: 4456–4461.

[10] Zhang GF. New infrared anti-reflective and protective coatings. Infrared technology, 1995, 17: 23–28.

[11] Ji SH. Technique and development relating to military optoelectronic equipment infrared window. Journal of Applied Optics, 1996, 17(2): 8–14.

[12] Walter GW. A review of impedance plot methods used for corrosion performance analysis of painted metals. Corrosion Science, 1986, 26(9): 681–703.

[13] Ogwu A, Hellwig T, Doherty S. Electrochemical investigation of the stability and corrosion resistance of boron phosphide films prepared by PECVD. Window and Dome Technologies and Materials IX, 2005, 5786: 365–372.

[14] Mansfield F. Electrochemical impedance spectroscopy (EIS) as a new tool for investigating methods of corrosion protection. Electrochimica Acta, 1990, 35(10): 1533–1544.

[15] Feng X, Shi LY, Hang JZ. Low temperature synthesis of boron phosphide nanocrystals. Materials Letters, 2005, 59: 865–867.

[16] Haddow D, Ogwu A. Corrosion resistance potential of boron phosphide films in a naval environment. The International Society for Optical Engineering, 2003, 5078: 109–120.

[17] Dalui S, Pal AK. Synthesis of n-type boron phosphide films and formation of Schottky diode: Al/n-BP/Sb. Applied Surface Science, 2008, 254: 3540–3547.

[18] Goossens A, Schoonman J. Electrochemical investigations of silicon/boron phosphide heterojunction photoelectrodes. Electrochimica Acta, 1995, 40(10): 1339–1344.
[19] Ginely DS, Baughman RJ, Butler MA. BP-stabilized n-Si and n-GaAs photoanodes. Journal of The Electrochemical Society, 1983, 130(10): 1999–2002.
[20] Takashi U. Boron phosphide-based compound semiconductor device, production method, and light-emitting diode. Patent CN1934717. 2007.
[21] Kumashiro Y, Okada Y. Schottky barrier diodes using thick, well-characterized boron phosphide wafers. Applied Physics Letters, 1985, 47: 64.
[22] Kumashiro Y, Sato K, Chiba S. Preparation of boron and boron phosphide films by photo and thermal chemical vapor deposition processes. Journal of Solid State Chemistry, 2000, 154: 39–44.
[23] Kumashiro K, Hirata K, Sato K. Thermoelectric properties of boron and boron phosphide films. Journal of Solid State Chemistry, 2000, 154: 26–32.
[24] Lewis KL, Savage JA. Phosphide based materials as hard optical coatings. Hard Materials in Optics, 1990, 1275: 46–51.
[25] Dalui S, Hussain S, Varma S. Boron phosphide films prepared by co-evaporation technique: synthesis and characterization. Thin Solid Films, 2008, 516: 4958–4965.
[26] Schmitt JO, Edgar LJH, Liu L. Close-spaced crystal growth and characterization of BP crystals. Physica Status Solidi, 2005, 3: 1077–1080.
[27] Gibson DR, Waddell EM, Kerr JW. Ultra-durable phosphide based anti-reflection coatings for sand and rain erosion protection. Window and Dome Technologies and Materials III, 1992, 1760: 178–200.
[28] Yu HZ. Infrared Optical Materials. Beijing: National Defense Industry Press, 2007: 333–388.
[29] Takenaka T, Takigawa M, Shohno K. Dielectric constant and refractive index of boron monophosphide. Japanese Journal of Applied Physics, 1976, 15: 2021–2235.
[30] Kumashiro Y, Yoshizawa H, Yokoyama T. Epitaxial growth of rhombohedral boron phosphide single crystalline films by chemical vapor deposition. Journal of Solid State Chemistry, 1997, 133: 104–112.
[31] Gibson DR, Waddell EM, Kerr JW. Ultra-durable phosphide based anti-reflection coatings for sand and rain erosion. Window and Dome Technologies and Materials III, 1992, 1760: 178–200.
[32] Nakamoto K. Infrared Spectra of Inorganic and Coordination Compound. Wiley, 1963.
[33] Harris DC. Frontiers in infrared window and dome materials. Infrared Technology XXI, 1995, 2552: 325–335.
[34] Waddell EM, Gibson DR, Willson M. Broadband IR transparent rain and sand erosion protective coating for the F14 aircraft infrared search and track germanium dome. Window and Dome Technologies and Materials, 1994, 2286: 376–385.
[35] Waddell EM, Clark C. Boron phosphide films on new substrate materials. SPIE Conference on Window and Dome Technologies and Materials VI, 1999, 3705: 152–162.
[36] Mackowski JM, Cimma B. Optical and mechanical behaviour of GeC and BP antireflection coatings under rain erosion tests. The International Society for Optical Engineering, 1994, 3: 552–560.
[37] Dalui S, Hussain S, Varma S. Boron phosphide films prepared by co-evaporation technique: synthesis and characterization. Thin Solid Films, 2008, 516: 4958–4965.
[38] Ogwu, T. Hellwig T, Doherty S. Surface characterization and surface energy measurements on boron phosphide films prepared by PECVD. Window and Dome Technologies and Materials IX. 2005, 5786: 130–136.
[39] Harris DC. Materials for Infrared Windows and Domes: Properties and Performance. SPIE Optical Engineering Press, 1999: 3–20.

[40] Zhao PN, Mo JJ. Decomposition temperature of boron phosphide. Chinese Science Bulletin, 1963(11): 53–55.
[41] Ogwu A, Hellwig T, Doherty S. Surface characterization and surface energy measurements on boron phosphide films prepared by PECVD. Window and Dome Technologies and Materials IX, 2005, 5786: 130–136.
[42] Tang WZ. Principles, Technology and Application of Thin-film Material Preparation, 2nd ed. Beijing: Metallurgical Industry Press, 2003, p. 277–279.
[43] Wang QZ. Environmental tribological properties of carbonitride thin films. Masters Thesis, Nanjing: Nanjing University of Aeronautics and Astronautics, 2010, p. 37.
[44] Li XF. Preparation of TiN thin-films and their corrosion and wear properties. PhD Thesis, Taiyuan: Tai Yuan University of Technology, 2010, p. 11–13.

8 Alumina thin films

8.1 Overview

Alumina (Al_2O_3) has excellent mechanical properties and broadband optical transparency, and it is an extremely important mid-infrared transparent material. Hence, alumina thin films can be used as antireflective and protective coatings for infrared windows. Unlike the fabrication of bulk materials, there are special requirements in the preparation process of thin films owing to their adhesion to the substrate. In order to reduce the heat capacity required in the substrate and to decrease the influence of the difference in thermal expansion coefficients between the thin film and substrate materials on thin-film performance, the temperature of thin-film fabrication should not be too high. Therefore, thin-film fabrication is generally a typical nonequilibrium process. Amorphous alumina thin films can be prepared easily, but their mechanical properties are inferior to those of crystalline alumina thin films and they do not satisfy the necessary requirements. Current research on crystalline alumina thin films faces problems involving high-temperature fabrication with expensive and complicated processes, as well as a lack of understanding of the mechanisms of low-temperature fabrication of alumina thin films, which have severely limited the application range of alumina thin films. In view of this, we investigate the structure, low-temperature crystallization mechanisms, and mechanical and optical properties of alumina thin films fabricated using magnetron sputtering. We will also explore low-cost and low-temperature crystallization techniques of alumina thin films in order to produce antireflective and protective coatings of calcium aluminate glass and other mid-infrared transmission materials.

8.2 Research status of alumina thin films

8.2.1 Basic properties

Alumina (or aluminium oxide) has the chemical symbol Al_2O_3 and molecular weight of 102. Al_2O_3 has at least eight types of polymorphs (α-, γ-, δ-, η-, θ-, κ-, χ-, ρ-Al_2O_3), among which, α- is hexagonal, γ- is cubic, δ- is orthogonal, η- is a cubic spinel, θ- is monoclinic, κ- is orthogonal, χ- is cubic, and ρ- is amorphous. Among these Al_2O_3 polymorphs, only α-Al_2O_3 (corundum) is stable, and its oxygen atoms are the most closely packed; the other polymorphs are metastable phases of alumina and will transform to α-Al_2O_3 when heated. For example, when γ-Al_2O_3 is heated to a high temperature, it will transform to corundum; the rate of this transformation will be relatively high only at temperatures above 1000 °C, and the alumina will almost completely transform to α-Al_2O_3 at high temperatures above 1300 °C. Alumina is an amphoteric oxide

https://doi.org/10.1515/9783110489514-011

Tab. 8.1: Basic properties of α-Al_2O_3 and γ-Al_2O_3.

Properties	α-Al_2O_3	γ-Al_2O_3
Density (g/cm^3)	3.96–3.99	3.2, 3.7
Elastic modulus (GPa)	409–441	
Hardness (GPa)	28	
Optical bandgap (eV)	8.8	
Thermal conductivity ($W \cdot m^{-1} \cdot K^{-1}$)	46	
Melting point (°C)	≈ 2050	$\gamma \rightarrow \delta$: 700–800 °C

with good chemical stability. It is insoluble in water, and soluble in inorganic acid and alkaline solutions. Tab. 8.1 lists a few basic properties of α-Al_2O_3 and γ-Al_2O_3.

The metastable polymorphs can all be produced between room temperature and 1000 °C, and the control of the fabrication process to obtain a specific form of crystalline alumina is extremely challenging; as a result, the research on the preparation of alumina is relatively complicated. However, the diversity of alumina polymorphs has also provided an opportunity for its widespread application. Alumina thin films have numerous excellent characteristics, such as high mechanical strength and hardness, good optical properties, good transparency and insulation, high wear resistance, and good chemical inertness. Therefore, it has received substantial attention and is widely applied in multiple fields, including machinery, optics, microelectronics, medicine, and chemical engineering.

8.2.2 Research status of crystalline alumina

Alumina has excellent mechanical, optical and other physical properties, and it is an extremely important mid-infrared transparent protective material. Hence, alumina thin films have crucial applications in the field of infrared window protective coatings. Currently, the mechanical properties of amorphous alumina thin films are unable to satisfy requirements, and the fabrication methods of high-quality crystalline alumina thin films are faced with problems such as high deposition temperature, complex processes, and high costs. This has severely limited the scope of application of alumina as a mid-infrared transparent and protective material.

α-Al_2O_3 and κ-Al_2O_3 fabricated by chemical vapor deposition (CVD) have already been used as protective coatings for cemented carbide tools. α-Al_2O_3 has better thermal stability than all the other crystal shapes of Al_2O_3, and hence could be the preferred option for infrared transparent and protective coatings. However, κ-Al_2O_3 can be more easily obtained during the CVD process, and is more widely applied. κ-Al_2O_3 has relatively good mechanical properties and a high phase transition temperature (900–1050 °C). Thus, it can be a substitute for α-Al_2O_3 in certain applications. However, when the working temperature is sufficiently high, the phase transformation of

κ-Al_2O_3 to α-Al_2O_3 is inevitable. The volume shrinkage that accompanies this phase transition will lead to severe fractures in the thin film, causing a significant increase in the wear rate. Research has shown that the nucleation surface is a key factor influencing alumina polymorphs during thin-film deposition. Therefore, an in-depth understanding of the impact of the nucleation surface on the phase formation and preferred orientation of the thin film would form the foundation for the development of CVD alumina thin films. Exercising precise control over the nucleation stage during thin-film growth will promote α-Al_2O_3 nucleation, which will ensure the production of high-quality thin films. Under appropriate nucleation conditions, alumina thin films with a specific texture can be obtained, which will further improve its cutting performance. In addition, γ-Al_2O_3 will be produced during the CVD process, and research has shown that the introduction of hydrogen sulphide as a catalyst or dopant during deposition will promote the generation of the γ phase.

Although alumina thin films prepared by CVD currently occupy a dominant position in the market for protective alumina coatings of cemented carbide cutting tools, they possess certain shortcomings. These shortcomings are mainly due to the deposition temperature (generally around 1000 °C) required during the fabrication of alumina thin films by CVD. For example, a high deposition temperature could cause harmful chemical reactions between the thin film and substrate. Furthermore, the difference in the thermal expansion coefficient between the substrate and thin film will lead to crack formation during sample cooling. Moreover, a high deposition temperature will significantly limit the range of possible substrate materials and necessitate the use of heat-resistant materials, such as cemented carbide. In order to reduce the deposition temperature of CVD, Kyrylov et al. employed plasma-assisted CVD to fabricate alumina thin films. Their study showed that, when the electrical power density at the cathode is up to 6.6 W/cm^2, α-Al_2O_3 thin films could be produced at 580 °C. This could have resulted from the increase in the migration ability of adatoms on the growth surface due to the energy bombardment.

Brill et al. employed filtered arc deposition to fabricate alumina thin films, and found that at a substrate temperature of 600–700 °C and rf substrate bias of −200 V, the produced alumina thin films contained α-Al_2O_3. When the substrate bias was less than −50 V, alumina thin films containing α-Al_2O_3 were obtained at substrate temperatures above 800 °C, whereas alumina thin films obtained at substrate temperatures below 400 °C were amorphous. Alumina thin films produced at substrate temperatures in the range 400–800 °C contained multiple polymorphs of Al_2O_3, including the α-phase and γ-phase. Yamada-Takamura et al. investigated the growth of crystalline alumina deposited using the filtered arc method within a similar temperature range. Observations under transmission electron microscopy (TEM) revealed that in alumina thin films composed predominantly of α-Al_2O_3, thin-film growth initially showed the γ-Al_2O_3 polymorph structure. When the film thickness reached a few hundred nanometres, the nucleation and growth of α-Al_2O_3 began to occur on the γ-Al_2O_3 layer. Rosén et al. also used the filtered arc technique to deposit alumina thin films. However, when the substrate bias was −100 V and deposition temperature was below 800 °C, alumina thin films containing only α-Al_2O_3 could not be produced. When the substrate bias was −150 V and deposition temperature was 200 °C, alumina thin films composed predominantly of γ-Al_2O_3 were obtained. The discrepancies that have emerged among these studies might have been due to the differences in deposition conditions and the different influences of substrates on polymorph growth. However, there are several issues with the filtered arc deposition of thin films, including poor process reproducibility, poor thickness uniformity, and susceptibility to target poisoning during the deposition process.

Conventional sputtering has been employed to fabricate alumina thin films, and no special treatment was performed on the nucleation surface. In general, crystalline alumina can only be produced when the substrate temperature is greater than 400–500 °C, whereas alumina thin films produced at substrate temperatures below 400–500 °C are amorphous. Crystalline alumina obtained through magnetron sputtering is usually in the γ-phase, and α-Al_2O_3 cannot be obtained directly. However, γ-Al_2O_3 thin films obtained using sputtering shows excellent performance when applied in high-speed metal cutting.

In order to fabricate crystalline alumina thin films at a low deposition temperature, researchers have developed several processing techniques to increase the ion content of deposited particles. These processing techniques are collectively referred to as ionized physical vapor deposition (IPVD). For instance, Schneider et al. used an rf coil to enhance the ionization degree of the deposition ion flux, which enabled the low-temperature preparation of κ-Al_2O_3 and θ-Al_2O_3 thin films at 320 °C and 180 °C, respectively. Khanna and Bhat obtained γ-Al_2O_3 using hollow cathode magnetron sputtering with no substrate heating. Zywitzki et al. employed conventional pulsed direct current sputtering to fabricate alumina thin films. The study showed that, by applying a suitable substrate bias and cathode power, pure α-phase alumina thin films can be fabricated at a deposition temperature of 760 °C, whereas γ-Al_2O_3 can be obtained at temperatures as low as 350 °C. Zywitzki et al. also performed TEM analysis in order to reveal the growth process of alumina thin films composed mainly of α-Al_2O_3. At the initial stage of deposition, the alumina layer formed on the substrate is composed of γ-Al_2O_3; α-Al_2O_3 only began to nucleate and grow when the thin film reached a certain thickness. These findings are similar to the observations by Yamada–Takamura et al. on alumina thin films deposited by the filtered vacuum arc technique. Li et al. used a method similar to that of Zywitzki et al. to fabricate alumina thin films, but they added an external solenoid to form a magnetic trap near the substrate and used a higher substrate bias (up to 400 V). This enabled them to produce crystalline alumina films (unknown polymorph) at a low temperature of 250 °C. However, partial amorphization of the alumina thin films occurred within two days after deposition, which might have been caused by the relatively high compressive stress. Based on the analyses above, it can be seen that the current magnetron sputtering methods for the production of alumina thin films are fairly complex and costly.

As with CVD methods for producing alumina thin films, the nucleation stage of thin-film growth by sputtering is extremely important. For example, a study by Jin et al. showed that by pre-depositing a thin Cr_2O_3 layer on the substrate as a nucleation layer, α-Al_2O_3 could be produced at 400 °C by using rf sputtering with ceramic targets. Both Cr_2O_3 and α-Al_2O_3 have a corundum structure, and the lattice mismatch is small. Hence, pre-depositing Cr_2O_3 will promote the nucleation of α-Al_2O_3. This method is suitable for processing templates, and the deposition temperature of α-Al_2O_3 can further be reduced to 280 °C. This indicates that the main barrier for α-Al_2O_3 growth under low temperatures is in the formation of α-Al_2O_3 nuclei during the nucleation stage. Furthermore, the stability of γ-Al_2O_3 surface energy could be a possible reason for the difficulty in α-Al_2O_3 nucleation in the initial stage. TEM analysis indicates that α-Al_2O_3 showed localized epitaxial growth on Cr_2O_3 grains. Kohara et al. employed a similar method to deposit alumina thin films on a cemented carbide substrate. The difference was that they predeposited a CrN layer on the substrate and performed oxidization to obtain a Cr_2O_3 nucleation layer. It can be seen from the analysis above that, although predepositing a Cr_2O_3 coating as a nucleation layer enabled the low-temperature deposition of crystalline alumina thin films, it caused the coating system to be more complicated and increased the processing complexity.

8.3 Preparation methods of alumina thin films

There are two possible approaches to achieve the low-temperature deposition of crystalline alumina thin films. On the one hand, conventional magnetron sputtering was performed by modifying the processing parameters for the deposition of alumina thin films. On the other hand, a resputtering assisted magnetron sputtering method was designed on the basis of conventional magnetron sputtering for the deposition of alumina thin films.

8.3.1 Deposition process of conventional magnetron sputtering

Firstly, conventional RF magnetron sputtering was performed to deposit alumina thin films. The steps are as follows:

(1) The substrate is ultrasonically cleaned using acetone for 15–30 min, following which it is cleaned using alcohol for 15–30 min and finally using deionized water for 15–30 min.

(2) The substrate treated in step (1) is placed on the sample holder in the magnetron sputtering vacuum chamber, following which the chamber is evacuated. When the pressure in the chamber reaches from 1.0×10^{-4} to 9.9×10^{-4} Pa, heating is initiated. The temperature is adjusted to 25–600 °C and maintained for 0–120 min.

(3) Argon gas is introduced, and when the pressure in the vacuum chamber is 3–5 Pa, resputtering cleaning is performed on the surface of the substrate treated in step (2) for 10–20 min.
(4) A sputtering power of 160 W is applied on the aluminium target for initiation, and the Ar flow rate is controlled at 100–200 mL/min. After pre-sputtering for 3–5 min, O_2 is introduced, and the flow rate is controlled at 3–10 mL/min. When the gas pressure in the vacuum chamber drops to 0.5 Pa, the substrate surface prepared in step (3) is coated.
(5) After sputtering is completed, the temperature is adjusted to a suitable level and maintained for 0–8 h. Then, the temperature of the vacuum chamber is decreased to room temperature, and the alumina thin film sample is retrieved from the chamber.

During the process of reactive sputtering, the reactive gas reacts not only with the target material but also with the coating layer deposited on the chamber wall and substrate. This causes the process to be abnormally complicated, and researchers have dedicated much effort to achieve the effective control of reactive sputtering processes. Berg et al. successfully modeled the reactive sputtering process. The complexity of reactive sputtering processes is closely related to the actual demands of industrial applications. Industrial applications not only require the deposition of thin films that fulfil certain stoichiometric ratios, but also aim to avoid the formation of a compound phase on the metallic target surface (target poisoning). If the reactive gas flow is too low, the compound thin films could not satisfy the stoichiometric ratios. If the reactive gas flow is too high, a compound phase will be formed on the metallic target surface, which will reduce the deposition rate. Therefore, the experiment in this book will require the determination of a suitable O_2/Ar flow ratio to obtain alumina thin films that fulfil the stoichiometric ratios at a high deposition rate.

The deposition temperature and substrate bias are processing parameters with crucial influences on the thin-film crystalline structure and crystallization properties. In order to investigate the effects of processing parameters in conventional magnetron sputtering on the crystalline structure and crystallization properties of alumina thin films, a series of alumina thin films was produced by modifying the deposition temperature and substrate bias in the experiment.

Details on the processing parameters of conventional magnetron sputtering for the fabrication of alumina thin films are listed in Tab. 8.2. To eliminate accidental errors and unexpected factors that might occur during the deposition process, all thin-film samples were prepared in a random order during the deposition process. In addition, in order to verify the stability and repeatability of the magnetron sputtering process, five groups of samples were prepared in a random order under common processing conditions, and independent characterization was performed for these five groups of samples.

Tab. 8.2: Processing parameters of conventional magnetron sputtering for the deposition of alumina thin films.

O_2/Ar flow ratio	Deposition temperature (°C)	Substrate bias (V)	Substrate material
0.02	Room temperature	0	Si (100)
0.025	Room temperature	0	Si (100)
0.03	Room temperature	0	Si (100)
0.04	Room temperature	0	Si (100)
0.05	Room temperature	0	Si (100)
0.04	Room temperature	−40/−80/−120	Si (100)
0.04	150	0/−40/−80/−120	Si (100)
0.04	300	0/−40/−80/−120	Si (100)/Quartz/TCM209
0.04	450	0/−40/−80/−120	Si (100)
0.04	600	0/−40/−80/−120	Si (100)

8.3.2 Deposition process of resputtering assisted magnetron sputtering

In general, alumina thin films produced using conventional magnetron sputtering are usually amorphous. The fabrication of crystalline alumina thin films usually involves processes at deposition temperatures above 500 °C or high-temperature heat treatment after deposition. However, the high-temperature environment of processing conditions will limit the use of heat-sensitive substrate materials. Furthermore, the difference in the thermal expansion coefficients of the substrate material and alumina thin film, as well as the phase changes under high temperatures, will cause extremely high stress between the thin film and substrate, easily leading to thin-film fracture or delamination. Predepositing a layer of chromic oxide thin film on the substrate before alumina deposition will enable the production of crystalline alumina thin films at lower temperatures. However, the predeposited chromic oxide layer will increase the structural complexity of the coating–substrate system, which could lead to adverse effects on the reliability of thin-film application. Ion/plasma-assisted deposition is also an effective technique for the fabrication of crystalline alumina thin films at low temperatures using, for example, bias sputtering and an ion/plasma source to assist deposition. In the process of ion/plasma-assisted deposition, ion bombardment on the thin films will increase the kinetic energy of the atoms around the film surface, promoting the formation of crystalline alumina thin films. However, excessive ion bombardment will also raise the growing velocity of defects in the thin films, which is harmful to the formation of crystalline alumina thin films. Therefore, ion bombardment needs to be controlled carefully. During ion/plasma-assisted deposition, the deposition processes of bias sputtering will impose a large constraint on the bias value applied on the substrate, and high biases cannot be applied. Moreover, the ion/plasma source used during ion/plasma-assisted deposition (e.g. solenoid) increases production costs and the complexity of the deposition system. Furthermore, high-power pulsed magnetron sputtering usually increases the ionic composition of the sputtered particles, which also promotes the deposition of crystalline alumina thin films. However, compared to conventional magnetron sputtering processes, it places a higher demand on the electrical power unit and significantly increases the production costs.

In this book, we designed a resputtering technique that will assist magnetron sputtering in thin film deposition in order to achieve the low-temperature fabrication of crystalline alumina thin films. Resputtering is a type of ion/plasma-assisted deposition technique that also uses the ion bombardment of thin films to increase the kinetic energy of atoms near the thin-film surface, thereby promoting the formation of crystalline alumina thin films. The resputtering technique allows for the application of a high substrate bias and does not involve any additional ion/plasma source. Therefore, this technique not only reduces the constraints of using heat-sensitive materials but also decreases the production costs and complexity of deposition equipment needed to fabricate crystalline alumina thin films compared to ion/plasma-assisted deposition techniques.

In the application of resputtering assisted magnetron sputtering for the fabrication of alumina thin films, the deposition process is divided into two steps: sputtering and resputtering. A deposition cycle is composed of one sputtering step and one resputtering step. During the sputtering step, the total pressure of Ar and O_2 in the deposition chamber is maintained at 0.5 Pa, sputtering power is 160 W, and the sample holder is earthed. Measures are not taken for the ion bombardment of the substrate in this step. During the resputtering step, the electrical supply to the aluminium target is switched off, and a shield is inserted between the aluminium target and substrate. At the same time, the total pressure of Ar and O_2 in the deposition chamber is adjusted to 5.0 Pa, and the O_2/Ar flow ratio remains unchanged. Subsequently, a pulsed direct-current negative voltage is applied on the substrate, whereby the peak voltage is −800 V, duty cycle is 50 %, and frequency is 50 kHz. The negative voltage at this point will excite the thin-film surface to generate plasma. The interval between the resputtering step and sputtering step as well as the total cycle can be adjusted based on the circumstances. In this study, the time ratio was controlled at 0.5, and the total cycle was 30 min. Details of the processing parameters can be found in Tab. 8.3.

The steps for depositing alumina thin films using resputtering assisted magnetron sputtering are as follows:

(1) The substrate is ultrasonically cleaned using acetone for 15–30 min, following which it is cleaned using alcohol for 15–30 min and finally using deionized water for 15–30 min.
(2) The substrate treated in step (1) is placed on the sample holder in the magnetron sputtering vacuum chamber, and the chamber is evacuated. When the degree of vacuum in the chamber reaches from 1.0×10^{-4} to 9.9×10^{-4} Pa, heating is initiated, and the temperature is adjusted to 25–600 °C and maintained for 0–120 min.

Tab. 8.3: Processing parameters of resputtering assisted magnetron sputtering for the deposition of alumina thin films

O_2/Ar flow ratio	Resputtering step/ sputtering step time ratio	Deposition temperature (°C)	Substrate material
0.04	0.5	Room temperature	Si (100)
0.04	0.5	150	Si (100)
0.04	0.5	300	Si (100)/Quartz/TCM209
0.04	0.5	450	Si (100)
0.04	0.5	600	Si (100)

(3) Argon gas is introduced, and when the pressure in the vacuum chamber is 3–5 Pa, resputtering cleaning is performed on the surface of the substrate treated in step (2) for 10–20 min.
(4) A sputtering power of 160 W is applied on the aluminium target for initiation, and the Ar flow rate is controlled at 100–200 mL/min. After pre-sputtering for 3–5 min, O_2 is introduced, and the flow rate is controlled at 3–10 mL/min. When the gas pressure in the vacuum chamber drops to 0.5 Pa, the substrate surface prepared in step (3) is coated. The deposition time is controlled at 10–30 min.
(5) For resputtering assistance, the shield is replaced, and the electrical supply to the aluminium target is switched off. The pressure in the vacuum chamber is adjusted to 5 Pa. Then, the sample holder is connected to the cathode of the pulsed resputtering electrical supply, with a peak voltage of −800 V, and the coating surface undergoes resputtering cleaning for 8–20 min. The pulsed electrical supply is switched off.
(6) Steps (4) and (5) are repeated 1–80 times. After sputtering is completed, the temperature is adjusted to a suitable level and maintained for 0–8 h. Then, the temperature of the vacuum chamber is decreased to room temperature, and the alumina thin film sample is retrieved from the chamber.

8.4 Growth rate, composition, and chemical bonding

8.4.1 Growth rate

When employing reactive magnetron sputtering to fabricate alumina thin films, O_2 is used as the reactive gas. The O_2/Ar flow ratio is a key factor influencing the properties of the resulting alumina thin film, and different O_2/Ar flow ratios will have a substantial impact on the rate of thin-film deposition.

It can be seen from Fig. 8.1 that, as the O_2/Ar flow ratio during sputtering increases, the growth rate of alumina thin film decreased continuously. Specifically, when the O_2/Ar flow ratio increased from 2% to 2.5%, the thin-film growth rate rapidly decreased; when the flow ratio continued to increase to 3%, the decrease in thin-film growth rate slowed down; when the flow ratio increased from 3% to 5%, the growth rate showed an approximately constant decrease.

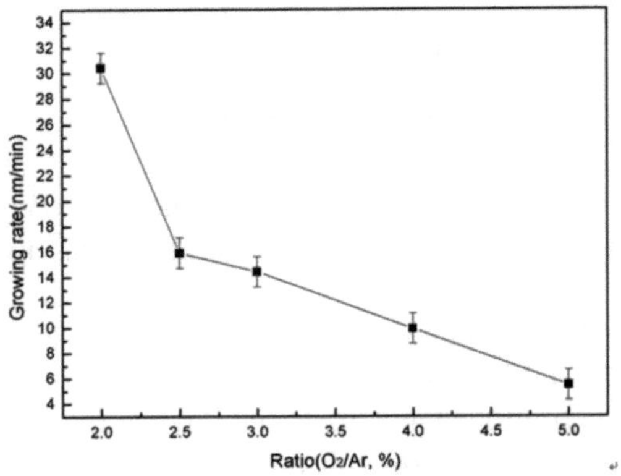

Fig. 8.1: Growth rate of alumina thin films under different O_2/Ar flow ratios.

The variations in thin-film growth rate with O_2/Ar flow ratio reflect the changes in the sputtering mode of the target material and the difference between the sputtering yield of O_2 and Ar gas. With the increase in the O_2/Ar flow ratio from the low initial level, the surface of the target material transitioned from purely metallic to a mixture of metal and oxides. The sputtering mode of the target material changed from the metal sputtering mode to metal-compound mixed sputtering mode. At a certain point when the flow ratio increased from 2.5% to 3%, the surface of the target material was completely covered by oxides, and the sputtering mode of the target material changed to the compound sputtering mode. As the flow ratio continued to increase, because the sputtering yield of O_2 is less than that of Ar, the thin-film growth rate approximated to a constant decline with increasing O_2 ratio.

8.4.2 O/Al composition ratio

Reactive magnetron sputtering was performed to prepare alumina thin films in a mixed atmosphere of O_2 and Ar, and the O_2/Ar flow rate was controlled at 2–5%. The thin-film composition was quantitatively analyzed using XPS and sample EDX. Tab. 8.4 lists the composition test results of alumina thin films fabricated under different O_2/Ar flow ratios (R).

Tab. 8.4: Stoichiometric ratio of O and Al elements in thin films fabricated under different O_2/Ar flow ratios.

Measurement method	$R = 0.02$	$R = 0.025$	$R = 0.03$	$R = 0.04$
EDX	1.01	1.27	1.52	1.54
XPS	1.07	1.31	1.55	1.51

The following conclusion can be drawn based on the test results. The results of quantitative analysis by XPS and sample EDX were in good agreement. When the O_2/Ar flow ratio was 2 % or 2.5 %, the atomic ratios of O and Al elements were less than 1.5, indicating that the alumina thin film is nonstoichiometric. When the O_2/Ar flow ratio continued to increase to 0.03 and beyond, the atomic ratios of O and Al elements in the thin film remained stable at approximately 1.5; at this point, the alumina thin film satisfied the stoichiometric ratio.

8.4.3 Chemical bonding

The spectral peak position of Al 2p in Fig. 8.2 had already been corrected with reference to the position of C 1s (284.6 eV). When the O_2/Ar flow ratio was 2 %, aside from the peak of the ionic state Al^{3+} (71.3 eV), the thin-film XPS also showed the peak of the metallic state Al^0 (74.0 eV), as shown in Fig. 8.2 (a). This indicates that the alumina thin film produced was rich in metallic-state Al atoms. When the O_2/Ar flow ratio was 2.5 %, the peak intensity of the metallic state Al^0 in the thin-film XPS was weaker than that at a flow rate of 2 %, as shown in Fig. 8.2 (b). This indicates that the content of metallic-state Al atoms in the thin film was drastically reduced. When the O_2/Ar flow ratio continued to increase to 3 % or 4 %, the peak of metallic-state Al^0 disappeared

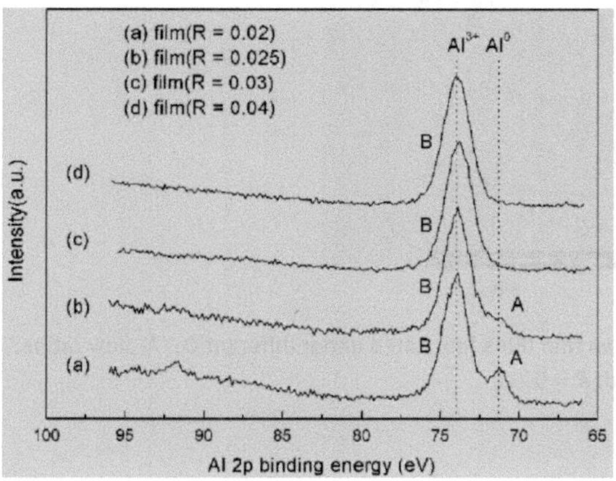

Fig. 8.2: Variation of Al 2p in alumina thin films fabricated under different O_2/Ar flow ratios.

completely, as shown in Fig. 8.2 (c) and (d). This indicates that metallic-state Al atoms were not present in the thin film. The peak intensity of metallic-state Al^0 in Fig. 8.2 gradually decreased and then disappeared as the O_2/Ar flow ratio increased. This pattern of variation implies that the content of metallic Al atoms in the thin film gradually decreased and ultimately reached zero. It also indicates that the fabricated thin films gradually transformed from a mixture of metallic Al and Al_2O_3 to pure Al_2O_3.

8.5 Morphology and structure analysis

8.5.1 Surface morphology

It can be seen from Fig. 8.3 that the impact of O_2/Ar flow ratio on the surface morphology of alumina thin film is obvious. When the O_2/Ar flow ratio was low, relatively large clusters existed on the thin-film surface, as shown in Fig. 8.3(a). However, Fig. 8.3 (b)–(d) show that, as the O_2/Ar flow ratio increased, cluster formation became inhibited. When the O_2/Ar flow ratio was greater than 0.03, the size of the clusters on the thin-film surface was relatively stable and did not decrease further with the increase in the O_2/Ar flow ratio.

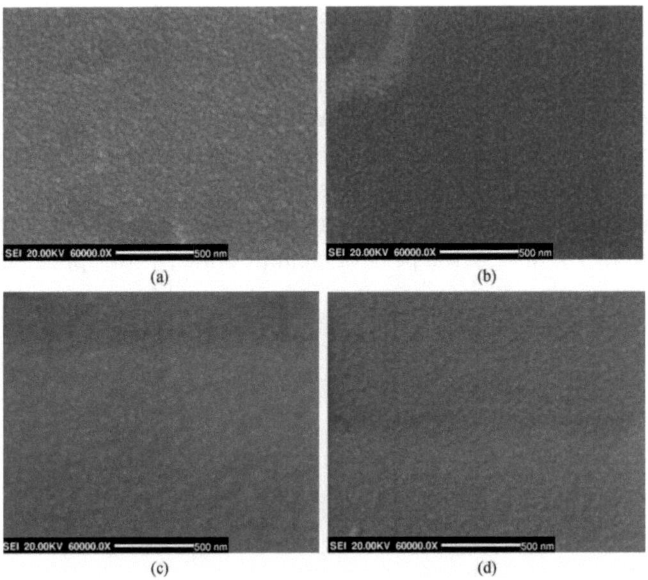

Fig. 8.3: SEM surface morphology of alumina thin films fabricated under different O_2/Ar flow ratios: (a) $R = 0.02$; (b) $R = 0.025$; (c) $R = 0.03$; (d) $R = 0.04$.

8.5.2 Crystal structure

In general, alumina thin films fabricated using magnetron sputtering under room temperature are amorphous. Thin-film amorphization is usually caused by film growth conditions that are far from the energy conditions required for crystallization. Amorphous alumina thin films have a lower level of hardness and smaller refractive index compared to crystalline alumina thin films, and they cannot satisfy the demands of industrial applications. The growth of crystalline thin films must be performed under a certain temperature in order to ensure that the atoms in the thin film have the kinetic energy required to move to the lattice sites. Furthermore, during the thin-film deposition process, the application of a substrate bias will attract energetic ions in the plasma, which collide with the thin-film surface. This will promote the increase in the mobility of atoms at the film surface, facilitating the formation of crystalline alumina thin films. Therefore, this section will investigate the effects of substrate temperature and substrate bias on the crystallization process of alumina thin films.

Fig. 8.4 (a) shows that alumina thin films grown at room temperature are amorphous, and that no significant crystallization occurred as the substrate temperature increased. As shown in Fig. 8.4 (b)–(e), when the substrate temperature was less than 600 °C, the deposited alumina thin film was amorphous. This indicates that deposition temperatures less than 600 °C are not able to provide atoms on the thin-film surface with sufficient energy, which in turn do not have sufficient kinetic energy to form a crystalline alumina thin film.

Typically, during thin-film deposition, the application of a substrate bias will attract energetic ions in the plasma to collide with the thin-film surface, which will promote the increase in the mobility of atoms at the film surface, facilitating the formation of crystalline alumina thin films. Therefore, different substrate biases were applied to prepare alumina thin films at a substrate temperature of 600 °C. The sub-

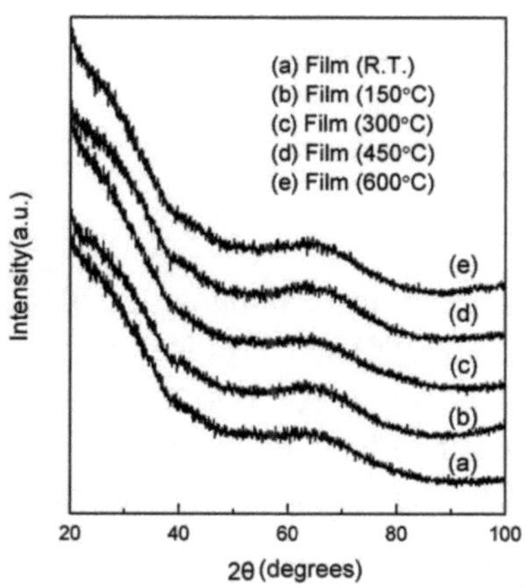

Fig. 8.4: GIXRD diffraction spectra of alumina thin films deposited under different substrate temperatures.

strate biases were controlled at −40 V, −80 V, and −120 V, and the effects of substrate bias on thin-film crystallization were observed.

As shown in Fig. 8.5, alumina thin films fabricated at a substrate temperature of 600 °C and substrate bias of −40 V were amorphous. At the same substrate temperature, when the substrate bias was increased to −80 V or −120 V, the alumina thin films produced were still amorphous. Therefore, although the increase of substrate bias (≤ −120 V) enhanced the collision of energetic ions in the plasma on the growing thin-film surface, it could not produce crystalline alumina thin films.

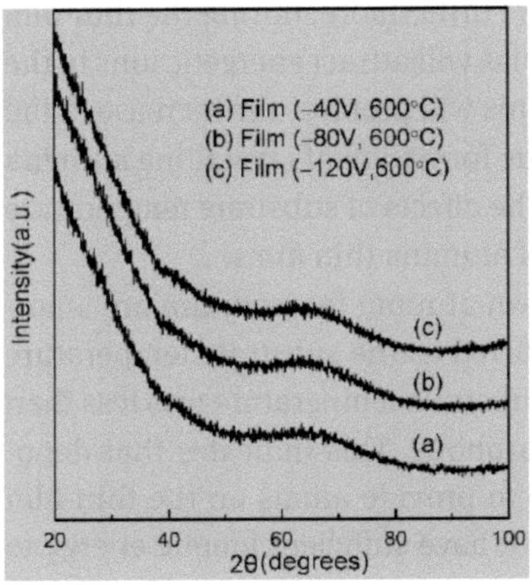

Fig. 8.5: GIXRD diffraction spectra of alumina thin films fabricated under different substrate biases at a substrate temperature of 600 °C.

This indicates that, even at a relatively high deposition temperature (600 °C) and substrate biases of up to −120 V, the surface atoms of alumina thin films did not receive sufficient energy or have enough mobility to move toward the normal lattice sites; hence, the surface atoms were unable to form crystalline alumina thin films.

8.6 Low-temperature crystallization by resputtering assisted deposition

8.6.1 Crystal structure

When reactive magnetron sputtering was employed to fabricate alumina thin films, increases in both the deposition temperature T_S ($T_S \leq 600$ °C) and negative substrate bias ($|V_S| \leq 120$ V) at a relatively high deposition temperature ($T_S = 600$ °C) were unable to yield crystalline alumina thin films. Therefore, in order to produce crystalline alumina thin films at a relatively low temperature, a low-cost assistance method called the resputtering technique was adopted.

Fig. 8.6 shows the GIXRD diffraction spectrum of an alumina thin film produced using resputtering mode magnetron sputtering at a deposition temperature of 300 °C.

To facilitate comparison, the diffraction spectra of alumina thin films fabricated at a deposition temperature of 600 °C and substrate bias of 0 or −120 V are also included in Fig. 8.6. When the deposition temperature was 600 °C and the substrate bias was 0 or −120 V, the alumina thin films produced were amorphous (Fig. 8.6 (a) and (b)). However, Fig. 8.6 (c) shows that the GIXRD diffraction spectrum of the alumina thin film fabricated using resputtering assisted techniques contained crystalline peaks that can be attributed to γ-phase alumina. This indicates that the fabricated thin film at this point contained γ-phase alumina.

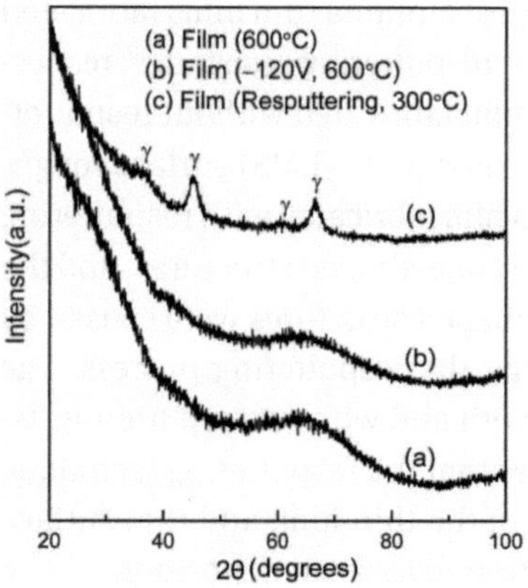

Fig. 8.6: GIXRD diffraction spectra of alumina thin films deposited at 300 °C using resputtering assisted technique and at 600 °C under a substrate bias of 0 or −120 V.

Fig. 8.7 shows the TEM cross section of an alumina thin film deposited using resputtering assisted magnetron sputtering at a deposition temperature of 300 °C.

It can be seen from Fig. 8.7 that the alumina thin film deposited using resputtering assisted magnetron sputtering showed a layered structure, and was composed of coating layers with different microstructures. The amorphous alumina thin film in Fig. 8.4 (c), which was deposited by reactive magnetron sputtering without resputtering assistance, was compared to the crystalline and periodic layered alumina thin film in Fig. 8.6 (c), which was deposited by resputtering assisted reactive magnetron sput-

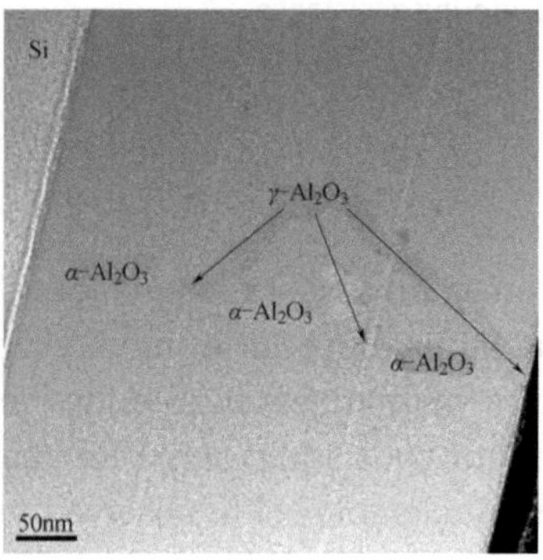

Fig. 8.7: TEM cross section of alumina thin film deposited using resputtering assisted magnetron sputtering at a deposition temperature of 300 °C.

tering. The comparison indicates that the latter is composed of amorphous and crystalline alumina layers. However, the thickness of the crystalline alumina layer is very low (about 7 nm), which has led to difficulties in determining its crystalline characteristics using electron diffraction. Therefore, resputtering assisted reactive magnetron sputtering can be used to produce crystalline alumina thin films. It is not difficult to predict that by adjusting the time ratio between the resputtering and sputtering steps during deposition, as well as the deposition cycle, it is possible to fabricate layered alumina thin films with the required thickness ratio of crystalline/amorphous layers, or even pure crystalline alumina thin films.

Fig. 8.8 (a) and (b) shows the surface morphology of alumina thin films fabricated using reactive magnetron sputtering with and without resputtering assistance, respectively. We can see that the amorphous alumina thin film fabricated without resputtering assistance had a smooth surface, and its root-mean-square (RMS) surface roughness was 0.8 nm. On the other hand, the alumina thin film fabricated with resputtering assistance had a rough surface, which had island and cone-shaped structures, and the RMS roughness was 5.8 nm. These island and cone-shaped structures were caused by the ion bombardment of the thin-film surface during the resputtering process. The surface roughness of alumina thin films fabricated with and without resputtering assistance showed a significant disparity. This indicates that the resputtering technique had an important impact on the surface roughness of the thin film, and that surface roughness can be effectively controlled using resputtering assisted deposition.

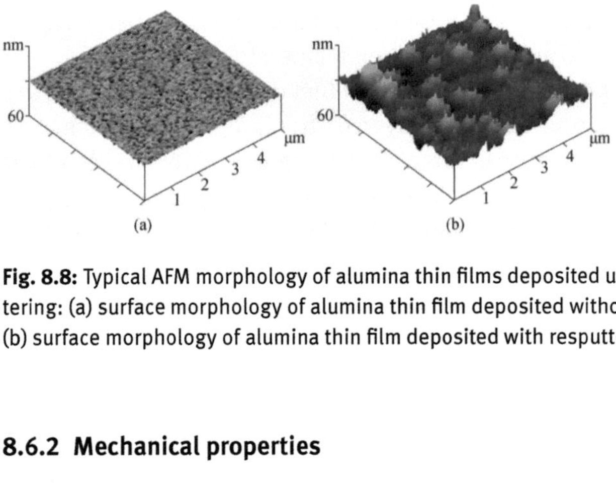

Fig. 8.8: Typical AFM morphology of alumina thin films deposited using reactive magnetron sputtering: (a) surface morphology of alumina thin film deposited without resputtering assistance; (b) surface morphology of alumina thin film deposited with resputtering assistance.

8.6.2 Mechanical properties

Alumina thin films fabricated using reactive magnetron sputtering at room temperature were amorphous. Therefore, this study prepared alumina thin films by modifying the deposition temperature and changing the deposition environment by altering the substrate bias. In addition, resputtering assisted magnetron sputtering was performed to produce alumina thin films. Nanoindentation tests were employed to observe the hardness of alumina thin films produced under different processing conditions using reactive magnetron sputtering.

Fig. 8.9 shows the hardness of alumina thin films produced using reactive magnetron sputtering under different deposition temperatures, substrate biases, and with resputtering assistance. The hardness of the alumina thin film produced at room temperature and without substrate bias was 13.6 GPa. The hardness remained essentially unchanged at approximately 13.6 GPa for alumina thin films produced without substrate bias and at substrate temperatures of 150 °C, 300 °C, 450 °C, and 600 °C. This indicates that below 600 °C, the deposition temperature did not significantly influence the hardness of alumina thin films. Even for alumina thin films fabricated at room temperature, 150 °C, 300 °C, 450 °C, and 600 °C under substrate biases of −40 V, −80 V, and −120 V, the hardness levels remained unchanged at approximately 13.6 GPa. When resputtering assisted magnetron sputtering was employed, the thin-film hardness increased with increasing deposition temperature, and the only exception was at 600 °C. The hardness of alumina thin films produced by resputtering assisted magnetron sputtering at room temperature was 13.6 GPa; as the deposition temperature increased to 150 °C, 300 °C, and 450 °C, the thin-film hardness increased to 14.2, 15.6, and 15.7 GPa, respectively. Nevertheless, when the deposition temperature increased to 600 °C, the thin-film hardness decreased to 14.5 GPa. This indicates that, by employing resputtering assisted magnetron sputtering to fabricate alumina thin films, the content of crystalline alumina layers in the thin film increased gradually as deposition temperature increased from room temperature to 450 °C. However, when the deposition temperature increased further to 600 °C, the content of crystalline alumina layers decreased. This is because, when the deposition temperature increases to a certain critical value with the resputtering technique, the damage caused by ion collision during resputtering will be the predominant effect, which will cause difficulties in the nucleation of crystalline alumina.

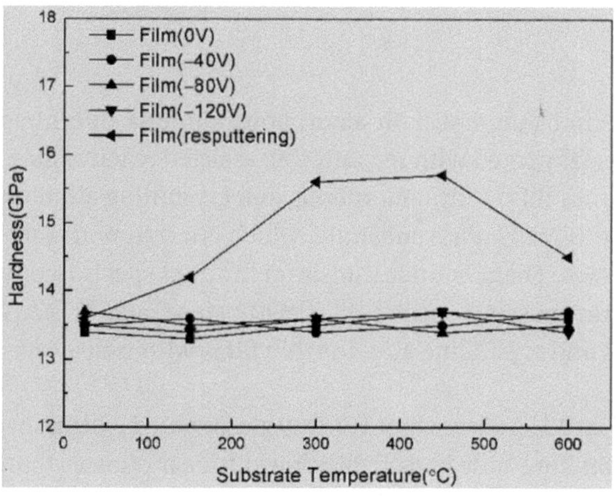

Fig. 8.9: Hardness of alumina thin films fabricated using reactive magnetron sputtering under different deposition temperatures, substrate biases, and with resputtering assistance.

In order to observe the effects of heat treatment on the mechanical properties of alumina thin films deposited using reactive magnetron sputtering, vacuum heat treatment was performed on alumina thin-film samples in this study. The samples were vacuum sealed and insulated at 950 °C for 8 h, followed by cooling in the furnace.

As shown in Fig. 8.10, the hardness values of alumina thin films fabricated using reactive magnetron sputtering without (A) and with (B) resputtering assistance before heat treatment (i.e., the as-deposited state) were 13.6 and 15.6 GPa, respectively. After samples A and B underwent vacuum sealing, insulation at 950 °C for 8 h, and cooling in the furnace, they were labelled as samples C and D, respectively. The hardness values of samples C and D were virtually identical at approximately 16.6 GPa, and was slightly higher than that of the sample prepared with resputtering assistance (B). This is because the degree of crystallization in heat-treated samples C and D was higher than in sample B, and the degree of crystallization in samples C and D was essentially the same.

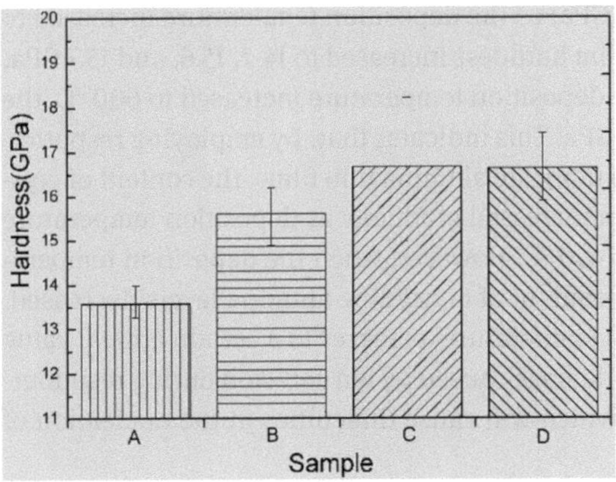

Fig. 8.10: Nanohardness of alumina thin films.

8.6.3 Optical properties

Observations with the naked eye indicate that both amorphous alumina thin films and crystalline alumina thin films fabricated with resputtering assisted reactive magnetron sputtering are transparent. In this study, amorphous and crystalline alumina thin films were deposited onto the TCM209 glass substrate, which is a common material for infrared windows and domes. Then, Fourier transform infrared spectroscopy was performed to measure the transmission spectra of TCM209 glass, and TCM209 glass was coated with amorphous and crystalline alumina thin films within the wavelength range of 2.5–5.0 μm.

The transmission spectra in Fig. 8.11 indicate that TCM209 glass coated with amorphous and crystalline alumina thin films only had slightly lower transmittances than uncoated TCM209 glass. The transmittance of TCM209 glass coated with crystalline

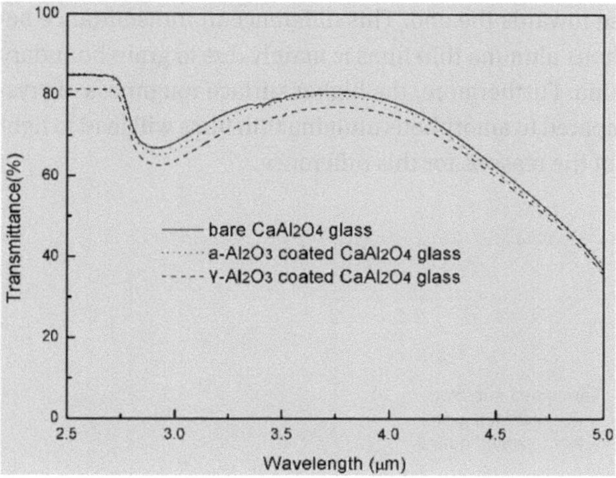

Fig. 8.11: Fourier infrared transmission spectra of TMC209 glass and TCM209 glass coated with amorphous and crystalline alumina thin films.

alumina thin film was the lowest because of the high surface roughness of crystalline alumina thin films produced using resputtering assisted reactive magnetron sputtering. Another possible reason is that alumina and TCM209 glass have similar refractive indices, and that of alumina is slightly higher.

Nevertheless, according to the results of nanoindentation hardness testing of TCM209 glass before and after coating, the hardness of TCM209 glass coated with crystalline alumina thin film was approximately 12 GPa, whereas those of uncoated TCM209 glass and TCM209 glass coated with amorphous alumina thin film were 8 and 10 GPa, respectively. Therefore, TCM209 glass coated with crystalline alumina thin film had better mechanical properties compared to the other two, and it only showed a slight decrease in transmission properties within the wavelength range of 2.5–5.0 μm. Therefore, resputtering assisted reactive magnetron sputtering is an effective method for the deposition of protective crystalline alumina coatings on TCM209 glass products.

In order to measure the refractive indices of amorphous alumina thin films fabricated using reactive magnetron sputtering and of crystalline alumina thin films fabricated using resputtering assisted reactive magnetron sputtering, in this study, we deposited amorphous and crystalline alumina thin films on quartz glass substrates, and measured the transmittance using UV-Vis spectroscopy. Then, the Swanepoel method was employed to calculate the refractive indices of amorphous and crystalline alumina thin films.

It can be seen from Fig. 8.12 that both amorphous and crystalline alumina thin films are transparent within the wavelength range of visible light. As the wavelength decreased, the transmittance of amorphous and crystalline alumina thin films began to decrease. However, when amorphous and crystalline alumina thin films were compared, the transmittance of crystalline alumina began to decrease at a longer wavelength; it also showed a wider wavelength range of transmittance decrease, and a greater magnitude of decrease towards the end. This difference in transmittance between crystalline and amorphous alumina thin films is mainly due to grain boundary scattering in crystalline alumina. Furthermore, the higher surface roughness of crystalline alumina thin films compared to amorphous alumina thin films will lead to light scattering, which is also one of the reasons for this difference.

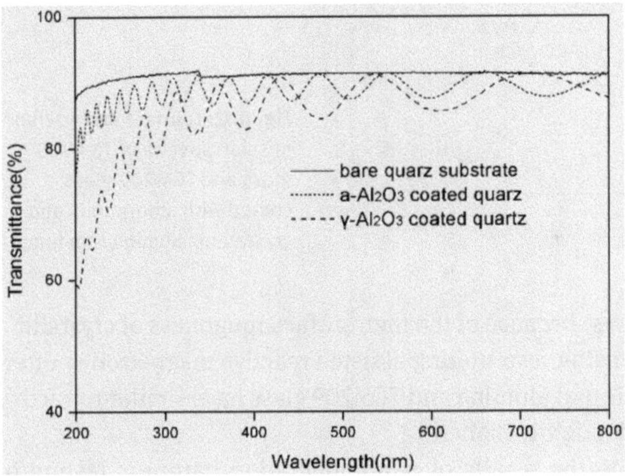

Fig. 8.12: UV-Vis transmission spectra of quartz glass and quartz glass coated with amorphous and crystalline alumina thin films.

Based on the interference effects found in the UV-Vis transmission spectra in Fig. 8.12, the Swanepoel method can be used to calculate the refractive indices of amorphous and crystalline alumina thin films and the quartz glass substrate, as shown in Fig. 8.13. The refractive index of the amorphous alumina thin film is less than that of crystalline alumina thin film because amorphous alumina can be viewed as a lattice in an extreme state of disorder, where the lattice defect density is extremely high and the densification of atomic arrangements is lower than that of corresponding crystalline substances. Therefore, the refractive index of amorphous substances is less than that of corresponding crystalline substances. This has already been reported.

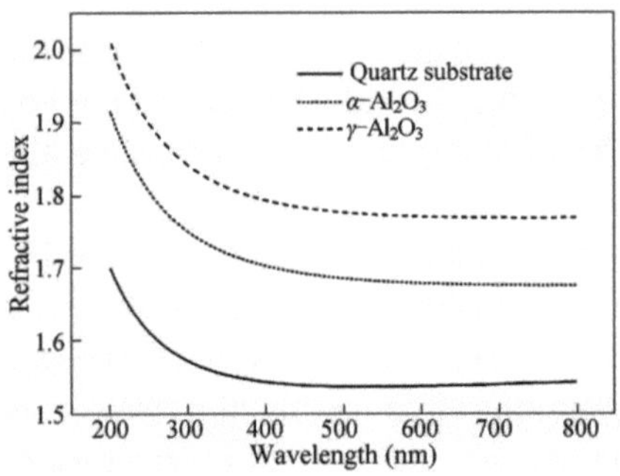

Fig. 8.13: Refractive indices of amorphous and crystalline alumina and quartz glass obtained using the Swanepoel method.

References

[1] Zhao ZW, Tay BK, Lau SP, et al. Microstructural and optical properties of aluminum oxide thin films prepared by off-plane filtered cathodic vacuum arc system. Journal of Vacuum Science & Technology A: Vacuum, Surfaces, and Films, 2003, 21(4): 906–910.

[2] Gitzen WH (ed.). Alumina as a ceramic material. The American Ceramic Society. Westerville, Wiley, 1970.

[3] Levin I, Brandon D. Metastable alumina polymorphs: Crystal structures and transition sequences. Journal of the American Ceramic Society, 1998, 81(8): 1995–2012.

[4] Oliver WC, Pharr GM. An improved technique for determining hardness and elastic-modulus using load and displacement sensing indentation experiments. Journal of Materials Research, 1992, 7(6): 1564-1583.

[5] Wefers K, Misra C. Oxides and hydroxides of aluminium. Alcoa Technical Paper No.19, Revised. Alcoa Laboratories, Pittsburgh, 1987.

[6] Chiritescu C, Cahill DG, Nguyen N, et al. Ultralow thermal conductivity in disordered, layered WSe_2 crystals. Science, 2007, 315(5810): 351–353.

[7] French RH. Electronic band structure of Al_2O_3, with comparison to AlON and AlN. Journal of the American Ceramic Society, 1990, 73(3): 477–489.

[8] Funk R, Schachner H, Triquet C, et al. Coating of cemented carbide cutting tools with alumina by chemical vapor deposition. Journal of the Electrochemical Society, 1976, 123(2): 285–289.

[9] Ruppi S. Deposition, microstructure and properties of texture-controlled CVD α-Al_2O_3 coatings. International Journal of Refractory Metals and Hard Materials, 2005, 23(4): 306–316.

[10] Larsson A, Halvarsson M, Ruppi S. Microstructural changes in CVD κ-Al_2O_3 coated cutting tools during turning operations. Surface & Coatings Technology, 1999, 111(2–3): 191–198.

[11] Halvarsson M, Vuorinen S. Microstructure and performance of CVD κ-Al_2O_3 multilayers. Materials Science and Engineering: A, 1996, 209(1–2): 337–344.

[12] Vuorinen S, Karlsson L. Phase transformation in chemically vapor-deposited κ-alumina. Thin Solid Films, 1992, 214(2): 132-143.

[13] Osada A, Nakamura E, Homma H, et al. Wear mechanism of thermally transformed CVD Al_2O_3 layer. International Journal of Refractory Metals and Hard Materials, 2006, 24(5): 387–391.

[14] Halvarsson M, Vuorinen S. The influence of the nucleation surface on the growth of CVD α-Al_2O_3 and κ-Al_2O_3. Surface & Coatings Technology, 1995, 76–77, Part 1(0): 287–296.

[15] Ruppi S, Larsson A. Chemical vapor deposition of κ-Al_2O_3. Thin Solid Films, 2001, 388(1–2): 50–61.

[16] Larsson A, Ruppi S. Microstructure and properties of CVD Γ-Al_2O_3 coatings. International Journal of Refractory Metals and Hard Materials, 2001, 19(4–6): 515–522.

[17] Müller J, Schierling M, Zimmermann E, et al. Chemical vapor deposition of smooth α-Al_2O_3 films on nickel base superalloys as diffusion barriers. Surface & Coatings Technology, 1999, 120–121(0): 16-21.

[18] Skogsmo J, Halvarsson M, Vuorinen S. Microstructural Study of the κ-$Al_2O_3 \rightarrow \alpha$-$Al_2O_3$ transformation in CVD κ-Al_2O_3. Surface & Coatings Technology, 1992, 54–55, Part 1(0): 186–192.

[19] Kyrylov O, Kurapov D, Schneider JM. Effect of ion irradiation during deposition on the structure of alumina thin films grown by plasma assisted chemical vapor deposition. Applied Physics A, 2005, 80(8): 1657–1660.

[20] Brill R, Koch F, Mazurelle J, et al., Crystal structure characterization of filtered arc deposited alumina coatings: Temperature and bias voltage. Surface & Coatings Technology, 2003, 174–175(0): 606–610.

[21] Yamada-Takamura Y, Koch F, Maier H, et al. Characterization of α-phase aluminum oxide films deposited by filtered vacuum arc. Surface & Coatings Technology, 2001, 142–144(0): 260–264.

[22] Rosén J, Mráz S, Kreissig U, et al. Effect of ion energy on structure and composition of cathodic arc deposited alumina Thin Films. Plasma Chemistry and Plasma Pressing, 2005, 25(4): 303–317.
[23] Cremer R, Witthaut M, Neuschütz D, et al. Comparative characterization of alumina coatings deposited by RF, DC and pulsed reactive Magnetron Sputtering. Surface & Coatings Technology, 1999, 120–121(0): 213–218.
[24] Schütze A, Quinto DT. Pulsed plasma-assisted PVD sputter-deposited alumina thin films. Surface & Coatings Technology, 2003, 162(2–3): 174–182.
[25] Schneider JM, Sproul WD, Matthews A. Phase formation and mechanical properties of alumina coatings prepared at substrate temperatures less than 500 °C by ionized and conventional sputtering. Surface & Coatings Technology, 1997, 94–95(0): 179–183.
[26] Åstrand M, Selinder TI, Fietzke F, et al. PVD-Al_2O_3-coated cemented carbide cutting tools. Surface & Coatings Technology, 2004, 188–189(0): 186–192.
[27] Schneider JM, Sproul WD, Voevodin AA, et al. Crystalline alumina deposited at low temperatures by ionized magnetron sputtering. Journal of Vacuum Science & Technology A, 1997, 15(3): 1084–1088.
[28] Khanna A, Bhat D G. Nanocrystalline gamma alumina coatings by inverted cylindrical magnetron sputtering. Surface & Coatings Technology, 2006, 201(1–2): 168-173.
[29] Zywitzki O, Hoetzsch G. Influence of coating parameters on the structure and properties of Al_2O_3 layers reactively deposited by means of pulsed magnetron sputtering. Surface & Coatings Technology, 1996, 86–87 Part 2(0): 640–647.
[30] Zywitzki O, Hoetzsch G, Fietzke F, et al. Effect of the substrate temperature on the structure and properties of Al_2O_3 layers reactively deposited by pulsed magnetron sputtering. Surface & Coatings Technology, 1996, 82(1–2): 169–175.
[31] Zywitzki O, Hoetzsch G. Correlation between structure and properties of reactively deposited Al_2O_3 coatings by pulsed magnetron sputtering. Surface and Coatings Technology, 1997, 94–95(0): 303–308.
[32] Li Q, Yu YH, Bhatia CS, et al. Low-temperature magnetron sputter-deposition, hardness, and electrical resistivity of amorphous and crystalline alumina thin films. Journal of Vacuum Science & Technology A Vacuum, Surfaces, and Films, 2000, 18(5): 2333–2338.
[33] Jin P, Xu G, Tazawa M, et al. Low temperature deposition of α-Al_2O_3 thin films by sputtering using a Cr_2O_3 template. Journal of Vacuum Science Technology, 2002, 20: 2134–2136.
[34] Andersson JM, Czigany Z, Jin P, et al. Microstructure of α-alumina thin films deposited at low temperatures on chromia template layers. Journal of Vacuum Science & Technology A: Vacuum, Surfaces, and Films, 2004, 22(1): 117–121.
[35] Kohara T, Tamagaki H, Ikari Y, et al. Deposition of α-Al_2O_3 hard coatings by reactive magnetron sputtering. Surface & Coatings Technology, 2004, 185(2–3): 166–171.
[36] Sproul WD, Christie DJ, Carter DC. Control of reactive sputtering processes. Thin Solid Films, 2005, 491(1): 1–17.
[37] Berg S, Nyberg T. Fundamental understanding and modeling of reactive sputtering processes. Thin Solid Films, 2005, 476(2): 215–230.
[38] Swanepoel R. Determination of the thickness and optical constants of amorphous silicon. Journal of Physics E: Scientific Instruments, 1983, 16(12): 1214.
[39] Wang J, Yu YH, Lee S, et al. Tribological and optical properties of crystalline and amorphous alumina thin films grown by low-temperature reactive magnetron sputter-deposition. Surface & Coatings Technology, 2001, 146: 189–194.

9 Yttrium oxide thin films

9.1 Overview

In the 1950s and 1960s, yttrium oxide was successfully used as an additive for heat-resistant oxides, such as yttria-stabilized zirconia (YSZ) ceramics. Owing to its excellent physical and chemical properties, yttrium oxide, a rare-earth ceramic material, has attracted the attention of numerous researchers, and a large amount of detailed research on yttrium oxide ceramics has been conducted. Yttrium oxide has excellent physical and chemical stability at high temperatures, as well as strong mechanical properties (hardness, elastic modulus, and corrosion resistance). It also has adjustable optical properties (broad transmission wave band, optical matching) and dielectric properties. These desirable properties have enabled the widespread application of yttrium oxide in science and engineering, and it has become a milestone in the field of ceramics.

Owing to the thermal stability as well as the good mechanical and optical stability of yttrium oxide, it has been innovatively applied in infrared windows and domes and as their coating material. Compared to other applications, those in the field of infrared optics have more stringent additional requirements, such as resistance to aerothermodynamic failure, good mechanical and optical performance under high temperatures, and protection against prolonged sand and rain erosion. Among the numerous infrared materials, yttrium oxide is resistant to aerothermodynamic failure and shows good mechanical and optical performance under high temperatures. Hence, it is one of the preferred materials for infrared windows and domes. Previous studies have demonstrated the importance of yttrium oxide, and its successful application is regarded as a landmark in the research community. Yttrium oxide has already been successfully applied as an infrared window material with immense success.

The properties of materials are determined by their dimensions and fabrication methods. The physical properties of thin films not only depend on their microstructure, but are also closely related to their preparation method. Based on the principles of thin-film growth, thin-film preparation methods can mainly be classified into physical vapor deposition and chemical vapor deposition. Among these, reactive magnetron sputtering has several advantages, including simple equipment, rapid deposition rate, high thin-film densification, and controllable performance. It has been widely applied in industrial and technological fields. This chapter will mainly focus on the use of reactive magnetron sputtering as a basis to investigate the properties of yttrium oxide, as well as its application as an infrared antireflective and protective coating material.

https://doi.org/10.1515/9783110489514-012

9.2 Crystal structure control of yttrium oxide thin films

Yttrium oxide (Y_2O_3) crystals predominantly consist of three structural types, namely A-, B-, and C-types. Their stability is determined by the temperature and pressure the crystals are subjected to. Fig. 9.1 shows a schematic diagram of the three structural types. A-type Y_2O_3 crystal has hexagonal symmetry with space group P3ml (D^3_{3d}). B-type Y_2O_3 crystal is monoclinic with space group C2/m (C^3_{2k}). C-type Y_2O_3 crystal is cubic with space group Iα3 (T^7_n), lattice constant $a = 1.0604$ nm, and $Z = 16$.

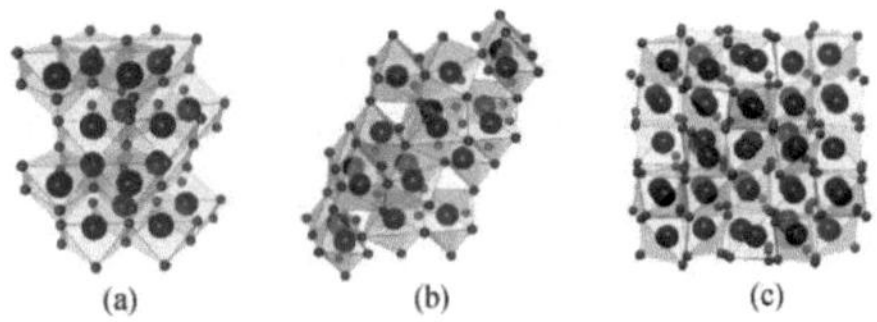

Fig. 9.1: Three structural types of Y_2O_3 crystals: (a) A-type; (b) B-type; (c) C-type.

Under room temperature and atmospheric pressure, Y_2O_3 crystals mainly exist in the C-type structure, whereas the metastable A-type and B-type crystal structures exist under high temperature and high pressure. Among them, the C-type Y_2O_3 crystal is a three-dimensional structure formed by top- or edge-connected oxygen-yttrium (O–Y) octahedra. The O–Y octahedron is equivalent to a fluorite structure, wherein the Y atom is in a body-centred position, and removing two of the eight neighboring O atoms will result in the Y_2O_3 crystal structure. The formation mechanism of this type of crystal configuration can be divided into two. The first is to remove the two O atoms on the face diagonal, which results in a structure with C_2 symmetry. The second is to remove the two O atoms on the body diagonal, which results in a structure with C_{3i} point group symmetry; see Fig. 9.2.

The unit cell of C-type cubic Y_2O_3 crystals includes 16 Y_2O_3 molecular units, i.e. 32 Y atoms and 48 O atoms. Among these, 8 Y atoms have C_{3i} point group symmetry, and 24 Y atoms have C_2 point group symmetry. These two types of O–Y octahedra are connected as shown in Fig. 9.2 (a) in order to form the spatial structure shown in Fig. 9.1 (c). In addition to Y_2O_3, which satisfies the stoichiometric ratio, there are other compounds with low oxygen contents, including $YO_{1.335}$, $YO_{1.401}$, and $YO_{1.458}$.

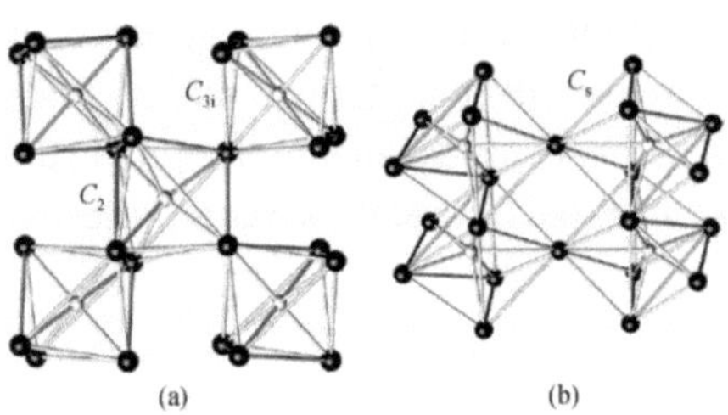

Fig. 9.2: Connection method of oxygen-yttrium octahedron: (a) cubic structure with space group Iα3 (T^7_n); (b) monoclinic structure with space group C2/m.

Among them, $YO_{1.335}$ is the most common under room temperature, and its crystal structure shows hexagonal symmetry with space group P63/mmc and lattice constants $a = 3.810$ Å and $c = 6.034$ Å.

The crystal structure of thin films relies heavily on their preparation method and processing parameters. Different processing parameters can affect the energy and pathway of deposited particles, influencing the dynamics and thermodynamics of thin-film growth and, in turn, determining the thin-film structure and performance. For reactive magnetron sputtering, the adjustable parameters mainly include oxygen partial pressure, substrate temperature, substrate bias, and sputtering power. Under different processing conditions, the thin films will be exposed to different growth environments and will thus form different crystal structures. For yttrium oxide thin films (Fig. 9.3), thermodynamic growth processes tend to form the stable cubic phase, whereas the metastable monoclinic phase or hexagonal phase is formed with deviation from thermodynamic processes. Therefore, this section will investigate the control of thin-film crystal structure based on the processing parameters of reactive sputtering, as well as the pattern of influence of the processing parameters.

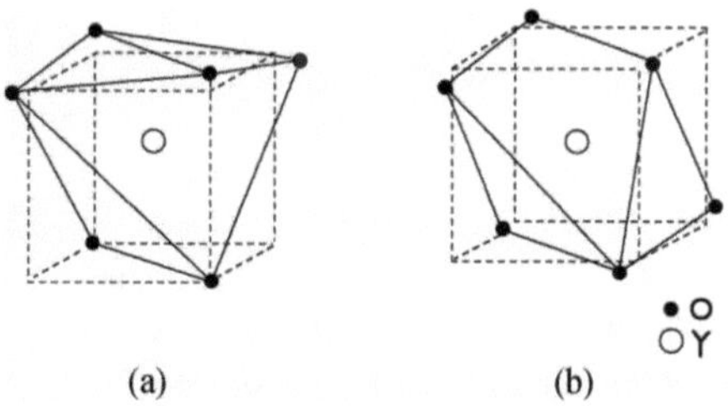

Fig. 9.3: Oxygen–yttrium octahedral structure: (a) C_2; (b) C_{3i}.

9.2.1 Effect of oxygen partial pressure

The fabrication of yttrium oxide thin films using reactive magnetron sputtering generally involves an yttrium metal target and the injection of oxygen gas during sputtering, which causes the sputtered yttrium atoms to react with the oxygen gas and form a thin film on the substrate. Nevertheless, when oxygen gas is introduced, it reacts with not only the sputtered atoms of the target material that reach the substrate to form oxides, but also the surface of the target material. Depending on the amount of oxygen introduced during the sputtering process, three sputtering modes occur: the metallic state, transition state, and poisoned state. Fig. 9.4 shows the corresponding oxygen partial pressure and deposition rate under different oxygen flow rates. As can be seen, the highest deposition rate was achieved during sputtering in the metallic state, whereas sputtering in the poisoned state showed the lowest deposition rate owing to the low sputtering yield caused by the formation of oxides on the target surface.

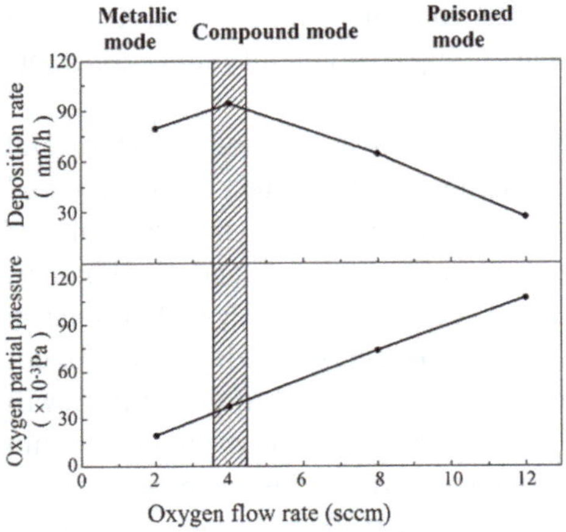

Fig. 9.4: Variations in oxygen partial pressure and deposition rate with oxygen flow rate.

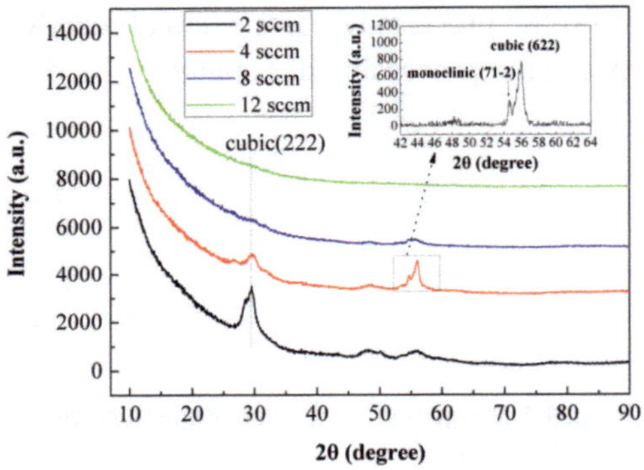

Fig. 9.5: Thin-film crystal structures under different oxygen partial pressures.

Thin films fabricated under different oxygen partial pressures exhibited different crystal structures, as shown in Fig. 9.5. When the oxygen flow rate was 2 sccm, the thin film was polycrystalline. Grazing-incidence x-ray diffraction patterns (GIXRD) showed an obvious diffraction peak for the lattice plane of the cubic phase (222), as well as a few broad dispersive diffraction peak envelopes. When the oxygen flow rate increased to 4 sccm, the cubic (222) diffraction peak was observed, as well as the diffraction peaks of the cubic (622) and weak monoclinic phase ($71\overline{2}$). The preferred orientation of the cubic (222) diffraction peak occurred because the cubic (111) lattice plane had the smallest surface energy. As the oxygen flow rate increased further and entered the poisoned state, the intensity of the diffraction peaks decreased rapidly. When the oxygen flow rate was 12 sccm, no significant diffraction peaks could be observed, and the thin film was amorphous.

Fig. 9.6 shows the 2D images of the surface roughness of yttrium oxide thin films under different oxygen flow rates. Under low oxygen partial pressure, the thin film showed greater surface roughness, surface islands with larger sizes and heights, and lower density of surface islands.

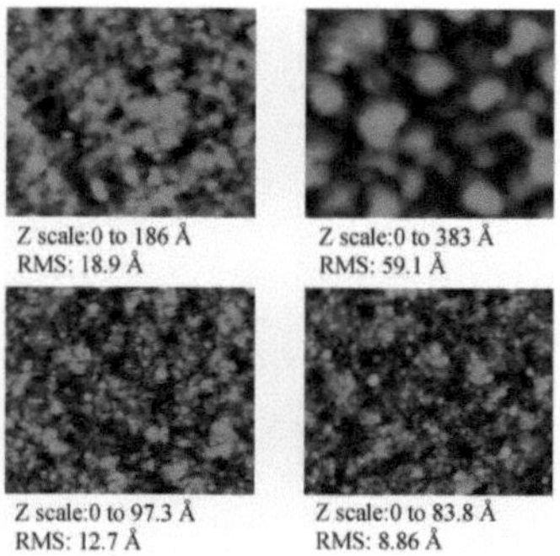

Fig. 9.6: Variations in thin-film surface roughness with different oxygen flow rates.

When the oxygen flow rate reached 4 sccm, surface islands coalesced further and expanded, causing an increase in surface roughness and reduced island density. As the oxygen flow rate increased further, the thin-film surface roughness decreased; furthermore, the thin-film surface was composed of compact and small islands with heights reduced to less than 9 nm, as shown in Fig. 9.6 (c) and (d). Therefore, as the oxygen flow rate increased, the surface roughness decreased, which was determined by the amorphous nature of the thin film.

9.2.2 Effect of substrate temperature

Substrate temperature directly determines the energy of substrate atoms reaching the thin-film surface; hence, it controls the quality of the crystals fabricated. Fig. 9.7 shows the GIXRD patterns of thin films deposited under different substrate temperatures. It can be seen from the figure that thin-film temperature had a direct influence on crystal quality. Under a low substrate temperature, the thin film showed a diffraction peak for the cubic phase, but the peak intensity was relatively weak and surface crystallinity was poor. When the temperature reached 200 °C and 400 °C, in addition to the cubic-phase diffraction peak, there were diffraction peaks for the monoclinic phase corresponding to (202), $(22\bar{2})$, and (020), indicating that the thin film was mainly in a two-phase mixed state of the cubic and monoclinic phases. When the temperature was further increased to 600 °C, the thin film had a single cubic phase and a very strong (222) preferred orientation because the (222) lattice plane in the cubic phase has the lowest surface energy. This indicates that high temperatures could facilitate the formation of the cubic phase, as higher temperatures enable thin-film atoms to obtain greater migration energy, which will facilitate their migration to lower energy for nucleation. Owing to the relatively low energy of the cubic phase, the formation of cubic-phase nuclei will occur easily, which will almost reach thermodynamic equilibrium, followed by growth and nucleation due to the increase of atoms. On the other hand, higher temperatures can also promote the elimination and compensation of various defects. This can facilitate the transition of the monoclinic phase to the cubic

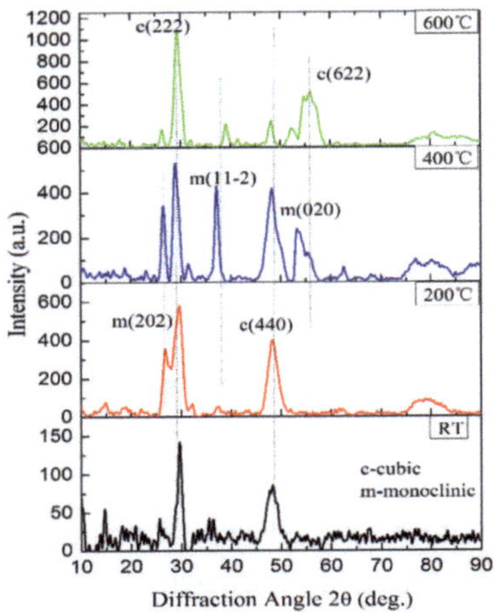

Fig. 9.7: Variations in crystal structure with substrate temperature.

phase to a certain extent, causing the clear appearance of a single cubic phase under high temperatures.

Fig. 9.8 shows the full width at half maximum (FWHM) of thin films fabricated under different substrate temperatures, as well as the grain size of the thin films calculated using the Scherrer equation. It can be observed from the figure that the increase of temperature had a positive effect on thin-film crystallinity. As the temperature increased, the kinetic energy of thin-film atoms increased, facilitating the movement of atoms to the optimal crystallization position and thereby increasing crystallinity. The decrease in FWHM implies that the thin-film grains grew larger under high temperatures, increasing from 6 nm at room temperature to approximately 8 nm at high temperature of 600 °C.

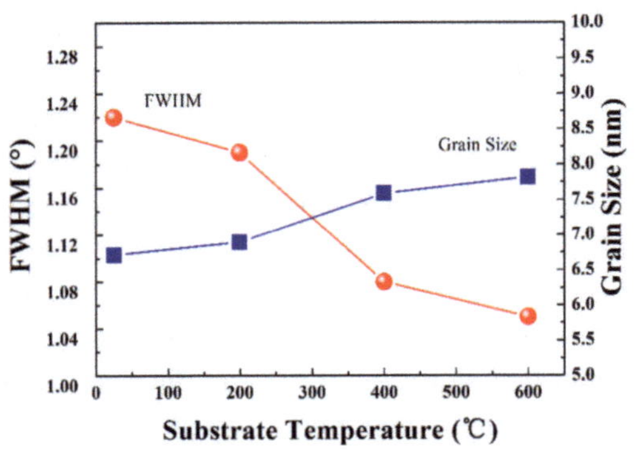

Fig. 9.8: Influence of substrate temperature on FWHM and grain size.

9.2.3 Effect of substrate bias

Bias voltage refers to the application of a certain positive or negative voltage on the substrate during magnetron sputtering. This causes the electrons or ions in the plasma to accelerate toward the thin-film surface under the effects of the electric field and to interact with the thin-film surface. The energy of incident ions can be regulated based on the level of bias voltage, which in turn controls the thin-film structure and properties. Therefore, substrate bias is a key parameter in magnetron sputtering. Hence, research on the influence of substrate bias on the structure of yttrium oxide thin films, as well as its performance control, has important implications.

Fig. 9.9 shows the variations of thin-film deposition rate under different negative substrate biases at room temperature. Under the absence of negative substrate bias, the thin film had a relatively high deposition rate of approximately 150 nm/h. As substrate bias increased, the thin-film deposition rate decreased gradually. A minimum deposition rate of approximately 20 nm/h was reached when the substrate bias was −300 V. As can be seen from the figure, the deposition rate varied with the negative substrate bias. The deposition rate was relatively high when the substrate bias was low, whereas the minimum deposition rate was reached when the substrate bias was very high. The appropriate substrate bias lies within the transition zone. This is because, given that the other processing parameters remain unchanged, the number of sputtered atoms that reached the substrate thin-film surface was fixed. Hence, when the substrate bias was small, the impact of the ions did not result in significant changes in thin-film deposition. However, when the substrate bias increased to a certain value (−160 V in this book), the greater energy of ion bombardment contributed to the displacement of ions deposited on the thin film and led to etching. This caused the collision of deposited atoms with high-energy ions, which released them from adsorption back to vacuum. Hence, the deposition rate at this substrate bias was less than that under normal conditions, and this region is known as the etching zone. The region of substrate bias between these two points is the transition zone, at which the impact of plasma on the deposition rate is small; however, the plasma plays a crucial role in improving the thin-film morphology and crystal structure (see the section below).

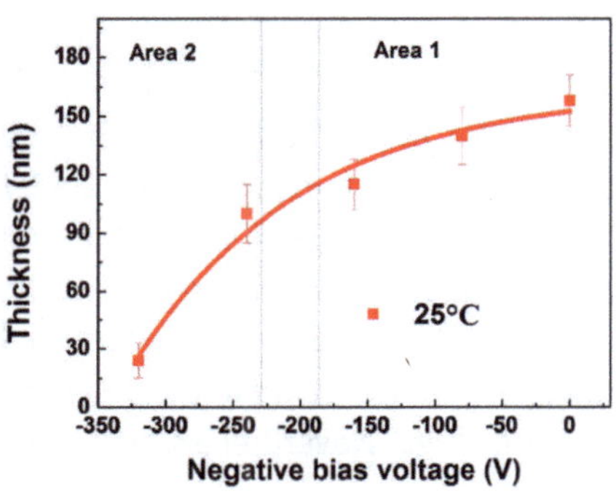

Fig. 9.9: Variation of deposition rate with substrate bias at room temperature.

Fig. 9.10 shows the XRD pattern of thin films under different substrate biases at room temperature. It can clearly be seen that the thin film was polycrystalline. As the substrate bias increased, a significant change occurred in the thin-film crystal structure. Without substrate bias, the thin film had a cubic (222) preferred orientation. As the substrate bias increased (−80 V), a weak monoclinic (202) diffraction peak appeared in addition to the cubic phase. As the substrate bias continued to increase from −160 V to −240 V, a sharp cubic-phase (222) diffraction peak emerged. Furthermore, the monoclinic-phase (003) diffraction peak was very sharp, and weak monoclinic $(51\bar{1})$ and $(71\bar{2})$ diffraction peaks appeared as well. When the substrate bias exceeded −240 V, the intensity of the diffraction peak was severely weakened. Based on the above, the cubic phase can be formed more easily under a lower substrate bias, as the cubic phase is the most stable and shows preferred nucleation and expansion during the growth process. Conversely, the monoclinic phase is metastable and can only be formed under high temperature and high pressure. Hence, the monoclinic phase is more difficult to synthesize compared to the cubic phase. Under a relatively large substrate bias (−160 V), the high-speed collision of energetic argon ions will occur on the thin-film surface. Through the energy and momentum transfer between ions and thin-film atoms, the atoms deposited on the thin film received sufficient energy and formed a high-voltage local field, which can promote the formation and growth of the monoclinic phase. This is the reason why an obvious diffraction peak for the monoclinic phase can be observed under a substrate bias of −160 V, which is similar to the Er_2O_3 phase.

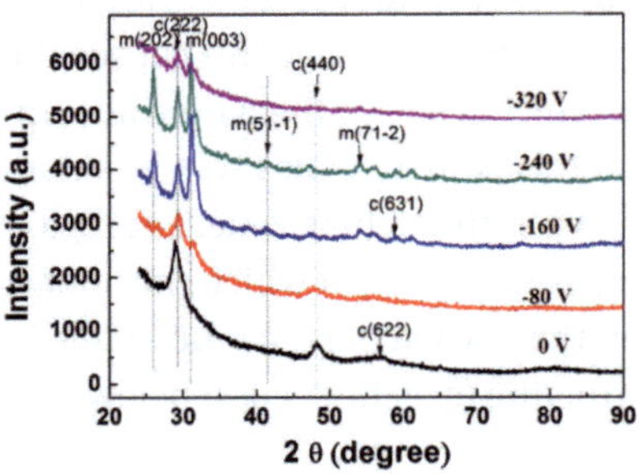

Fig. 9.10: GIXRD of thin films under different substrate biases at room temperature.

Substrate bias not only affects the crystal structure of thin films, but also has significant modifying effects on surface morphology. Fig. 9.10 shows the surface morphology of thin films fabricated under different substrate biases at room temperature. Under room temperature, the cone-shaped islands on the thin-film surface were obvious and relatively high, resulting in a relatively high surface roughness, as shown in Fig. 9.11. Under a substrate bias of −160 V, the modification caused by ion bombardment led to the flatness of thin-film islands. However, this effect did not seem to be apparent, and thin-film roughness did not show a significant change. Nevertheless, when the

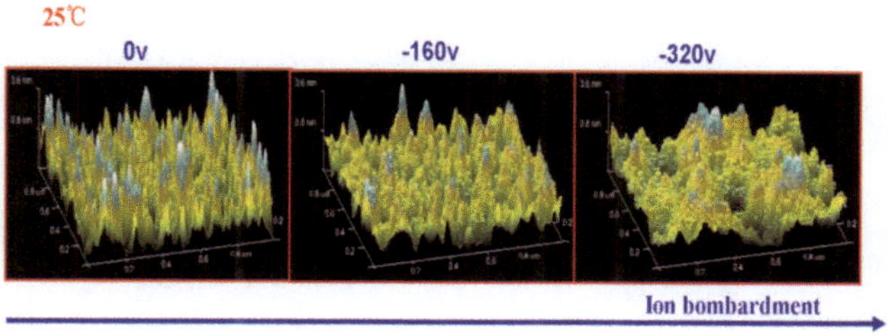

Fig. 9.11: AFM of thin films under different substrate biases at room temperature.

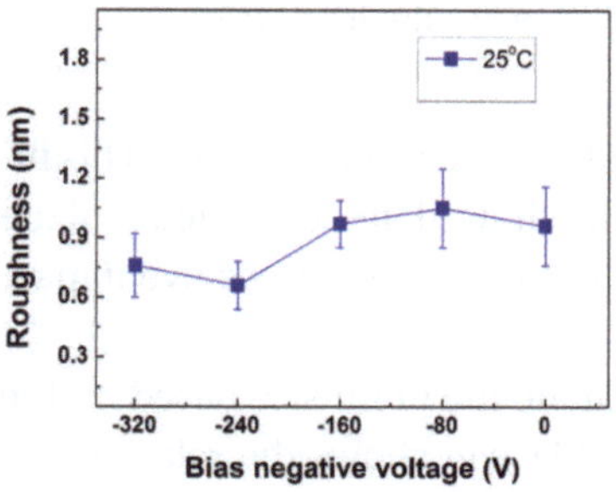

Fig. 9.12: Roughness of thin films under different substrate biases at room temperature.

substrate bias exceeded −320 V and entered the etching zone, most of the tips of the islands were levelled. Simultaneously, the density of thin-film islands increased, island size decreased, and roughness decreased, as shown in Fig. 9.12.

9.2.4 Effect of sputtering power

Sputtering power is a crucial parameter in thin-film deposition by magnetron sputtering. It is determined by the sputtering voltage and plasma current. The sputtering power not only influences deposition rate, but also determines the energy and mass of sputtered ions. Therefore, the sputtering power has an extremely important influence on the growth and microstructure of thin films. Hence, research on sputtering power is an extremely important topic in the control of thin-film structure. This section will investigate the changes in the microscopic crystal structure of thin films fabricated under different radiofrequency powers.

Fig. 9.13 shows the crystal structure of yttrium oxide thin films fabricated under different sputtering powers. The figure indicates that the power had a very large impact on thin-film crystal structure. When the power was low (approximately 60 W), the thin film showed a weak amorphous peak envelope, which included the diffraction peaks of cubic-phase (222) and (440) lattice planes, and monoclinic-phase (202) and (310) lattice planes. When the power increased to 100 W and 140 W, the diffraction peaks of the cubic and monoclinic phases gradually became more obvious, and their intensity increased somewhat; no other diffraction peaks emerged. However, when the power increased to 180 W, there was an extremely clear monoclinic-phase (202)

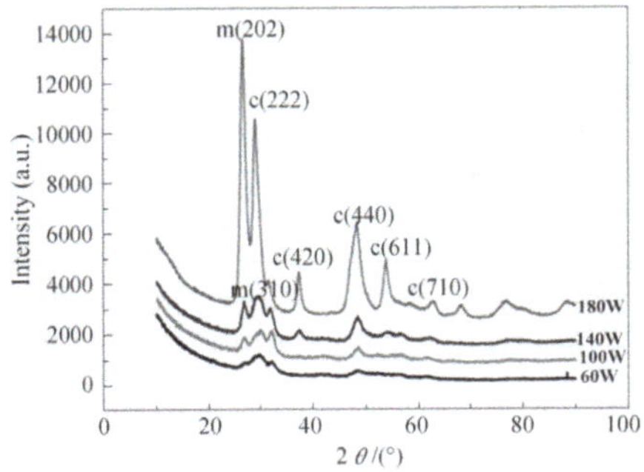

Fig. 9.13: Crystal structure of yttrium oxide thin film under different sputtering powers.

diffraction peak, and the cubic-phase diffraction peak was also very obvious. This indicates that, under a relatively high power, the thin film showed the co-existence of the monoclinic and cubic phases, implying that power has a significant positive impact on the production of the monoclinic phase.

According to the Scherrer equation, we can estimate the grain size based on the FWHM of the diffraction peaks in the XRD pattern. Fig. 9.14 shows the relationship between the diffraction-peak FWHM and grain size of the monoclinic (202) and cubic (222) lattice planes under different sputtering powers. Fig. 9.14 (a) shows that, when the power was relatively low (less than 100 W), the FWHM of the monoclinic-phase (202) diffraction peak was relatively high, which indicates that the thin film has small monoclinic crystals (approximately 10 nm). However, when the power increased to 140 W and 180 W, the FWHM of the monoclinic (202) diffraction peak decreased rapidly; hence, the thin films produced under this sputtering power had relatively large grain sizes (approximately 20 nm). The cubic phase also showed a similar pat-

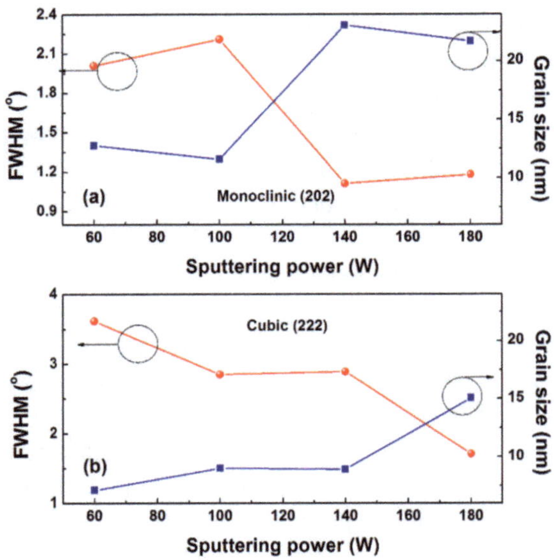

Fig. 9.14: FWHM and grain size of (a) monoclinic (202) and (b) cubic (222) lattice planes under different sputtering powers.

tern. When the power was relatively low, FWHM of the diffraction peak was relativey large and the crystals were small. With the growth of power, FWHM of the cubic (222) diffraction peak gradually declined and the crystal size increased. At the power of 180 W, FWHM of the cubic (222) diffraction peak reached the minimal point and the crystals grew to the maximal size (15 nm). According to the discussion above on the XRD patterns of yttrium oxide thin films fabricated under different sputtering powers, thin-film crystallinity will improve with increasing power while cubic-phase and monoclinic-phase grains will grow simultaneously. However, the optimal power for the rapid growth of the monoclinic phase is 140 W, while that of the cubic phase is 180 W. Therefore, it can be seen that, when both phases exist during the thin-film growth process, the driving force of the monoclinic -phase grain growth is lower than that of the cubic phase.

In summary, this section discussed the changes in thin-film crystal structure with changes in processing parameters in order to reveal the controllability of thin-film structure, as well as transformations between amorphous—crystalline, nanocrystalline—microcrystalline, and cubic—monoclinic phases. In general, temperature and oxygen partial pressure promoted the growth of the cubic phase, whereas sputtering power and substrate bias promoted the growth of the monoclinic phase. This is because different processing parameters control different aspects of the thin-film growth process. Temperature and oxygen partial pressure mainly control the migration kinetic energy of atoms that reach the thin-film surface. A higher level of energy fulfilled the requirements for the thermodynamics and dynamics of the cubic phase; hence the thin film had a greater tendency toward cubic-phase growth. Substrate bias and sputtering power caused plasma-energy acceleration, while energy transfer and collision led to a tendency towards monoclinic-phase growth. In addition, these processing conditions influenced the surface morphology of the thin film and controlled the evolution of surface islands to a smooth thin-film surface.

9.3 Performance control of yttrium oxide film

Thin-film microstructure directly determines the physical and chemical properties of thin films, and these structures are closely related to the fabrication method and deposition processing parameters. Section 9.2 analyzed thin-film microstructure and surface morphology under different processing conditions. Based on the foundation laid above, the present section will investigate the performance control of thin films fabricated under different processing conditions. This will clarify the crucial theoretical and practical significance of performance control in the application of thin films.

Yttrium oxide thin films have excellent physical and chemical properties, and one of the most important properties is its high thermal stability, which can guarantee its application in extreme environments. In particular, yttrium oxide has good optical and mechanical properties. It has a broad transmission wave range (from ultraviolet to far-infrared), large bandgap, and controllable optical constants. It also has superior mechanical properties, including high hardness, strength, and fracture toughness, as well as surface hydrophobicity. The polymorphic characteristics of yttrium oxide enable it to be easily matched with other materials in order to achieve complicated functions. The optical and mechanical properties of thin films are important for infrared windows and can directly determine the failure and fracture characteristics of infrared windows. Therefore, this section will mainly investigate the influences of various processing parameters on the optical, mechanical, and wetting properties of yttrium oxide thin films.

9.3.1 Optical properties of yttrium oxide thin films

Yttrium oxide has superior optical properties and can be used as an optical antireflective thin film for infrared windows. The thickness and optical constants (refractive index and extinction coefficient) of optical thin films are key optical parameters that can determine the transmission, reflection, radiation, and other optical characteristics of thin films. The accurate measurement of thin-film parameters is a prerequisite of optical design. The most common methods used to determine thin-film thickness and optical constants are spectrophotometry and ellipsometry. Spectrophotometry involves the use of a spectrophotometer to measure the transmittance and reflectance curves, which are in turn used to calculate the optical constants. This method can be further divided into the envelope method and full-spectrum fitting. Ellipsometry is another common method used to measure thin-film optical constants. In contrast to spectrophotometry, ellipsometry is nondestructive, noncontact, highly sensitive, and highly accurate. Ellipsometry does not require special sample pretreatment and is suitable for the precise measurements of nanoscale ultra-thin films. Ellipsometry is currently the predominant method for the accurate measurement of the optical constants of optical thin films, and it has been extensively applied in optical designs. This chapter will mainly describe the optical properties of yttrium oxide thin films measured using ellipsometry.

Ellipsometry measures the changes in polarization of elliptically polarized light upon its reflection from a thin-film surface. Then, the optical constants are obtained by fitting the measured data to a reasonable model. A detailed description of the testing principles of ellipsometry has been given in Chapter 5 and will not be elaborated here.

Establishing an optical model is extremely important for the analysis of ellipsometric data. It is only by establishing an accurate and realistic physical model that we can accurately measure the optical constants and thickness of the measured samples. The construction of a model of an optical thin-film system involves two aspects:

(1) establishing a physical structure model of thin-film samples that is as realistic as possible, and
(2) selecting the dispersion relations that can describe the characteristics of a particular thin film.

(1) Establishing a physical structure model

Under general circumstances, a physical model of the substrate-dispersion layers can be established based on the known number of layers. If the discrepancies between the fitted results and experimental results are too large, the dispersion layers will need to be corrected. For an optical thin film with an unknown number of layers and materials, model construction generally begins with a monolayered structure. The optical constants and thin-film thickness are obtained through the inversion of the model. If the discrepancy with the experimental values is substantial, then the model will need to be corrected. We should first consider the problem of heterogeneity, followed by the possibility that the thin film is not a strictly smooth plane; hence, a virtual layer of surface roughness can be established between the thin film and air for fitting. If adjustments to the model and film thickness still do not produce good fitting results, we should consider that the coating is not monolayered. Hence, the modeling of a double layer coating system should be attempted. When modeling multilayered coating systems, we need to consider the possible combinations of layers in the coating system, and the corresponding coating system should be constructed for fitting. If the attempts still do not produce satisfactory results, then a trilayered model should be established for fitting until satisfactory fitting results have been obtained.

(2) Establishing a dispersion model

When fitting the spectral ellipsometric data of the thin film, each film layer needs to be assigned appropriate dispersion relations based on the characteristics of different materials. Transparent thin films are mainly modelled using Cauchy and Sellmeier models. As the yttrium oxide thin films fabricated in this chapter are transparent, the two types of models above were used to fit the measured ellipsometric data in order to obtain the required optical parameters.

This experiment involves the fabrication of a monolayered Y_2O_3 thin film on a P-Si (100) substrate using magnetron sputtering, following which ellipsometry was performed to analyse the optical constants of thin films fabricated under different processing conditions. The correct physical model should first be established before we can fit the ellipsometric data to obtain the results needed.

Based on the discussion above, model establishment includes a physical model and dispersion model. As the sample is a Si substrate coated with a Y_2O_3 thin film, the physical model can be regarded as a Si substrate with a monolayered coating. As Y_2O_3 thin films are transparent from the visible to far-infrared region, the dispersion model selected for the Y_2O_3 coating system was the Cauchy model, and fitting was performed. The fitting between the model and experimental data was not ideal, and the mean-squared error (MSE) was relatively large (16.5). Therefore, this model was corrected to achieve better fitting. In practice, the injection of oxygen gas is required during the fabrication process of yttrium oxide thin films using reactive magnetron sputtering; hence, a SiO_2 reactive layer could be produced on the Si substrate. This could lead to further reactions with the yttrium oxide thin film to produce Y–O-Si, which is a compound with a high refractive index. This has been reported in the literature, which confirms our inference. Hence, we applied a model that includes Si substrate/dielectric layer/Y_2O_3/effective medium (roughness layer). The model of the dielectric layer was INTR_JAW (dielectric layer with high refractive index). The Sellmeier model was used for the Y_2O_3 thin film (suitable for transparent thin films and applies a broader wave range compared to the Cauchy model; the fitted results obtained are more in line with physical implications). The surface-roughness layer was regarded as an effective medium approximation (EMA) of the mixture between Y_2O_3 and air (50 % each). The model is shown in Fig. 9.15. The results of model fitting after corrections are shown in Fig. 9.16. The MSE was 9.319, which is closer to the measured results and hence more realistic. Fig. 9.17 shows the MSE curve of model fitting after correction. It can be observed that the MSE of this model is a global minimum, rather than a local minimum. Thus, it has good reliability, and this model was used to fit the optical constants of Y_2O_3 thin films.

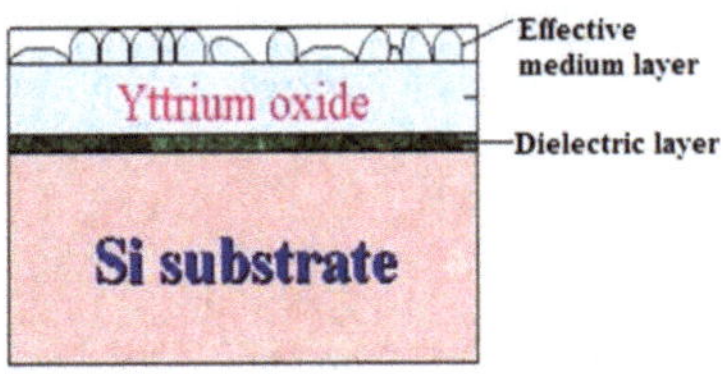

Fig. 9.15: Corrected model.

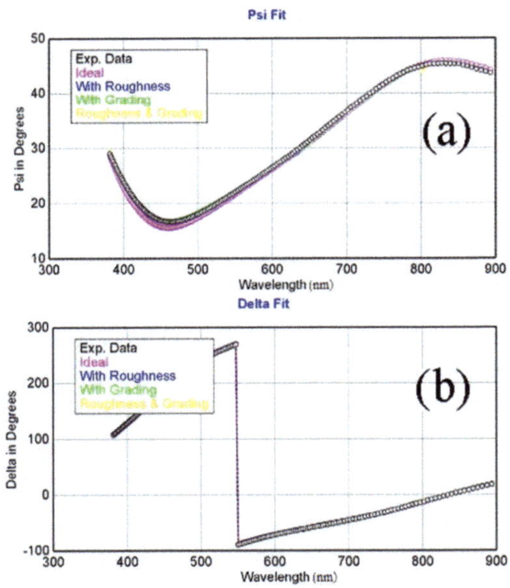

Fig. 9.16: Fitting results of the corrected model.

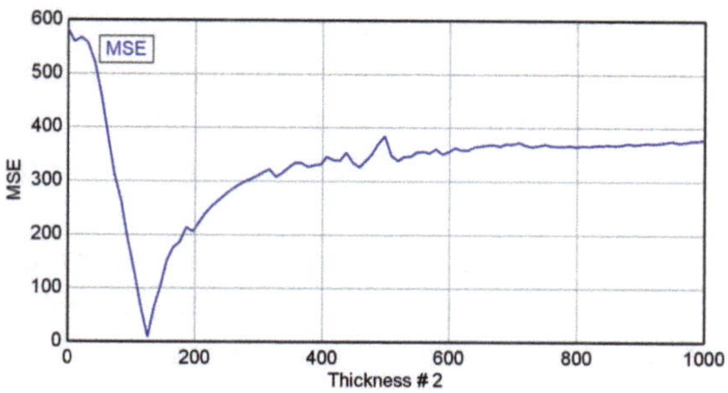

Fig. 9.17: MSE plot after fitting.

9.3.1.1 Optical constants

Using the thin films fabricated under different processing conditions, the model above was employed to fit the obtained optical parameters in order to investigate the influences of different growth conditions on thin-film performance, thereby providing theoretical guidance for practical applications. We have previously discussed the impact of different processing parameters on thin-film crystal structure. These processing parameters included oxygen partial pressure, substrate temperature, substrate bias, and sputtering power. This section will mainly focus on the changes in performance caused by the influence of these parameters on crystal structure, and the parameters include oxygen partial pressure, substrate temperature, substrate bias, and sputtering power.

The refractive indices of yttrium oxide thin films fabricated under varying oxygen partial pressures are shown in Fig. 9.18.

Fig. 9.18 shows that the oxygen partial pressure can control the refractive index in a relatively wide range (1.6–1.9). This can be attributed to the different crystal structures of thin films fabricated under different oxygen partial pressures. The thin-film refractive index reflects the propagation speed of light waves in the material, and refractive index is closely related to material density. Based on the crystal structure

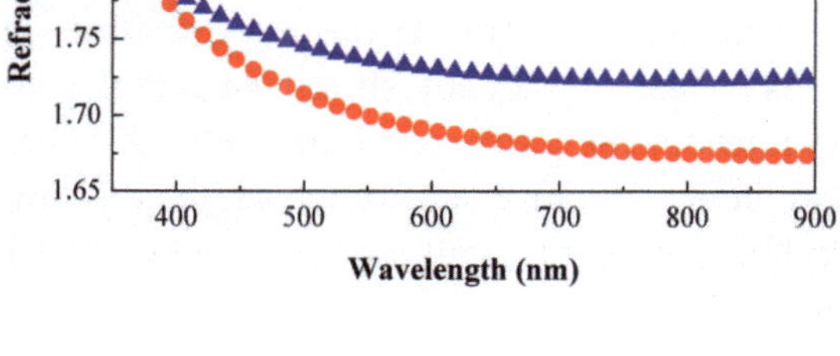

Fig. 9.18: Variations in the refractive index of yttrium oxide thin films with oxygen flow rate.

described above, we know that, when the oxygen partial pressure was relatively low, the thin film had good crystallinity. As the oxygen partial pressure increased gradually and reached 12 sccm, the thin film was already in an amorphous state. Amorphous thin films have looser organization, more holes, and other defects, which decrease the refractive index of the thin film.

The influence of temperature on refractive index showed similar characteristics. The refractive indices of yttrium oxide thin films fabricated under different temperatures are shown in Fig. 9.19. It can be observed that the temperature can control the refractive index in a relatively wide range (1.72–1.95). Similarly, this can be attributed to the influence of temperature on the thin-film crystal structure. Based on the analysis of crystal structure above, when the substrate temperature was relatively low, the thin-film diffraction peak was weaker, crystallinity was poorer, the thin film still contained a large number of small grains, and grain boundaries had numerous vacancies and other defects. This caused the thin film to exhibit a lower density and refractive index. When the substrate temperature was gradually increased, atoms on the thin-film surface received a high level of migration energy, which allowed these atoms to migrate across longer distances on the thin-film surface. Hence, some of the vacancy defects were filled, and the grain sizes increased. This led to an increase in the density and refractive index, with the maximum refractive index occurring at 600 °C (1.95 at 500 nm).

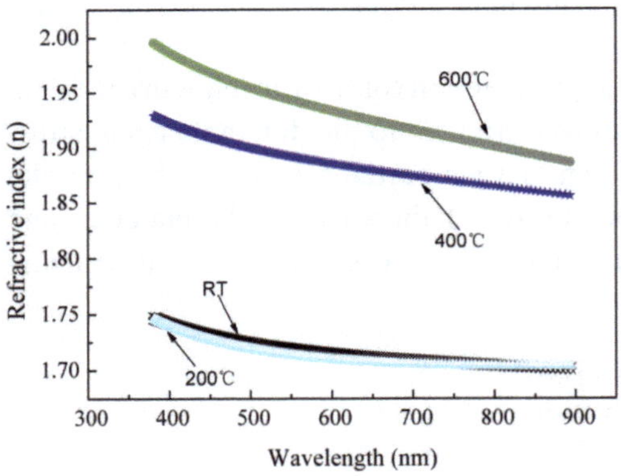

Fig. 9.19: Variations in the refractive index of yttrium oxide thin films with substrate temperature.

The application of negative substrate bias can also control to the refractive index of thin films. The refractive indices of yttrium oxide thin films under different negative substrate biases are shown in Fig. 9.20. The figure shows that as the negative substrate bias increased, thin-film refractive index increased gradually, showing a pattern consistent with the other parameters. This phenomenon can be attributed to the thin-film crystal structure. Without the application of a negative substrate bias (i.e. the substrate is grounded), the resulting thin film had poor crystallinity and a high number of vacancy defects. Hence, the thin film had relatively low density and refractive in-

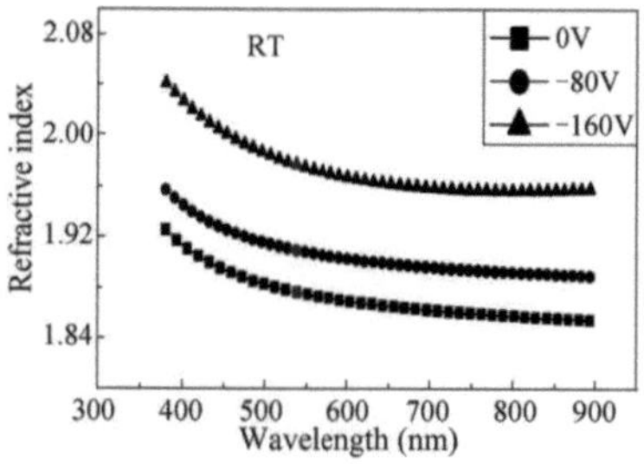

Fig. 9.20: Variations in the refractive index of yttrium oxide thin films with substrate bias.

dex. As shown in the XRD pattern, when the substrate bias was increased, the peak at –80 V became thinner and longer, and crystallinity improved. At a substrate bias of –80 V, the thin film was bombarded by Ar^+ ions of a certain energy during the growth process. On one hand, this effect promoted the energy and momentum transfer in the thin film, thereby increasing the migration energy of thin-film atoms. On the other hand, this effect caused the elimination of weakened atoms on the thin-film surface. The combination of both effects led to greater density. Unlike the effects of temperature, the increase in migration energy of atoms deposited under substrate bias is achieved through momentum transfer. Hence, this process is a deviation from thermodynamic equilibrium. The atomic momentum increases during film growth, which will necessarily lead to the formation of defective structures, such as monoclinic yttrium oxide. When the substrate bias increased further to reach –160 V, momentum transfer was more significant; hence, the XRD pattern revealed a strong monoclinic diffraction peak.

As the monoclinic phase is metastable, it can only be formed under high temperature and pressure. The monoclinic phase has higher density compared to the cubic phase, and a high refractive index (2.0 at 500 nm) can be observed under –160 V. Furthermore, the higher refractive index of the monoclinic phase compared to the cubic phase grown under high temperatures is a manifestation of the high density of the monoclinic phase.

The impact of sputtering power on thin-film structure and refractive index also has significant research implications. The increase in sputtering power will lead to the increase in voltage and breakdown current. An increased voltage implies that the ionized Ar^+ ions are bombarding the target surface at a higher energy. On the other hand, an increased current implies that the number of ions and electrons in the plasma contributing to the current is higher. These changes in the sputtering process will cause thin-film growth to exhibit different physical processes and mechanisms, resulting in thin films with different microstructures and properties. As shown in Fig. 9.21, when the sputtering power was low, thin-film crystallinity was poor, grain size was small, the thin film showed numerous vacancies and other defects, and the thin film had a low refractive index. As the sputtering power increased, the energetic particles bombarding the target material had relatively high levels of energy. This caused the sputtered particles to have high energy as well. In addition, owing to the momentum

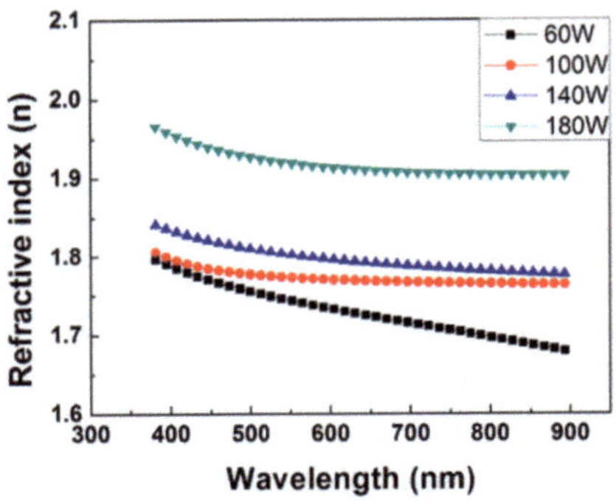

Fig. 9.21: Variations in the refractive index of yttrium oxide thin films with sputtering power

transfer and charge exchange of Ar^+ ions on the target surface, the ions returned to the target surface as neutral atoms. A higher sputtering power led to a greater energy in the reflected atoms. Thus, the thin-film growth process was also affected by particle bombardment, which is similar to the effect of substrate bias. The effect of sputtering power lies in the ionic state of the particles being bombarded. After momentum and energy transfer, as well as the migration and nucleation of thin-film atoms on the surface, a high sputtering power eventually resulted in a microstructure with good crystallinity and the coexistence of the cubic and monoclinic phases. It also improved the density and refractive index of the thin film.

According to the discussion above, different processing parameters can affect the physical nature of reactive magnetron sputtering, and the energy and type of sputtered particles can be modified based on the physical nature. Nevertheless, the modification of the deposited particles using different conditions will involve different pathways; thus, different processing conditions will lead to varying thin-film structure and performance.

9.3.1.2 Density

Density is a key property of thin-film materials and is directly related to porosity. Several methods are available for the measurement of thin-film density, one of which is x-ray reflectometry (XRR). XRR is extensively applied because it is nondestructive and accurate. However, there are strict requirements on thin-film thickness when using XRR to measure thin-film density. In general, a thickness of the order of only tens of nanometres is needed. Furthermore, human error is usually present when data-based structural models are fitted to the data, which could cause a different density morphology. As mentioned previously, the thin-film refractive index is a measure that is closely related to and reflects the density. Hence, refractive index can be used for the qualitative evaluation of thin-film density. According to the comparison of the refractive index of yttrium oxide thin films with that of bulk yttrium oxide, equation (9.1) can be employed to obtain a rough estimate of the relative density P of thin films:

$$P = \left(\frac{n_f^2 - 1}{n_f^2 + 2}\right) \cdot \left(\frac{n_b^2 + 2}{n_b^2 - 1}\right), \tag{9.1}$$

where n_f is the refractive index of the thin film and n_b is the refractive index of the bulk material.

The refractive index of bulk yttrium oxide at 500 nm was used as the criterion (1.97) to measure thin-film density.

Based on the equation above, the refractive indices of thin films grown under different conditions were used to evaluate the impact of different processing parameters on thin-film density, as shown in Fig. 9.22. As with the variations in refractive index, aside from the decrease in density caused by the increase in oxygen partial pressure, the increases in substrate temperature, substrate bias, and sputtering power all led to higher refractive indices and densities. However, it can also be observed that substrate bias had a predominant influence on the increase in refractive index. This is because the substrate bias could provide additional energy to promote thin-film growth, resulting in higher density.

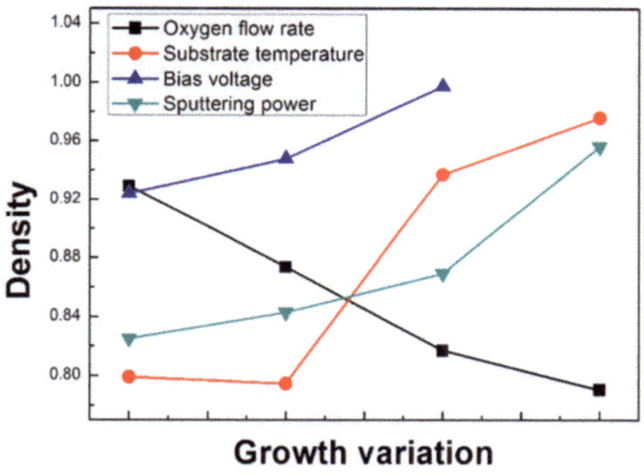

Fig. 9.22: Thin-film density under different growth conditions.

9.3.2 Mechanical properties of yttrium oxide thin films

The structural stability of thin films determines whether they can have practical applications as well as their service duration; the structural stability, in turn, is closely related to mechanical properties. For infrared antireflective and protective coatings, the thin films will need to maintain structural stability within an aerothermodynamic service environment, without being delaminated, while also possessing strong resistance to sand and rain erosion. This implies that the thin films need to have not only strong chemical adhesion, but also superior mechanical properties. Therefore, research on thin-film mechanical properties and the process control under different processing parameters have crucial theoretical and practical implications.

The measurement methods of mechanical properties mainly include microhardness and nanoindentation tests, of which the nanoindentation test has been widely applied in studies on thin-film mechanical properties owing to its excellent sensitivity and accuracy. Nanoindentation testing is the ideal method to measure the hardness (H) and elastic modulus (E) of materials. Material hardness and elastic modulus can be tested using load–displacement (P–h) curves and calculated using the Oliver–Pharr method. However, this method only gives single values for hardness and elastic modulus from one indentation. The hardness and elastic-modulus values of general substrate-coating systems are not constants and will vary gradually with increasing indentation depth. A dynamic testing method (continuous stiffness measurement, CSM) can be applied by testing the contact stiffness–displacement (S–h) relationship during the indentation process, which provides the hardness–displacement (H–h) relationship in one measurement. When the indenter is applied on a sample surface, the surface will undergo plastic deformation to form a pyramidal indentation, followed by the partial recovery of elastic displacement after unloading. Fig. 9.23 (a) shows a typical loading-unloading curve, which indicates that, when the load increased to $P_{\max}$, the indentation depth increased gradually until the maximum value $h_{\max}$. This is followed by the unloading phase. As the load decreased, the indentation depth gradually decreased, but the elastic displacement recovered only partially after complete unloading. The remaining portion of displacement did not recover to its original state owing to plastic deformation, and is defined as h_f, that is, the displacement after complete unloading.

Using the Oliver–Pharr method, the contact depth h_c is given by

$$h_c = h - \varepsilon \frac{P(h)}{S}, \tag{9.2}$$

where h is indentation depth, P is the corresponding load, ε is a constant related to indenter shape, and $\varepsilon = 0.75$ for a Berkovich indenter, and S is the measured con-

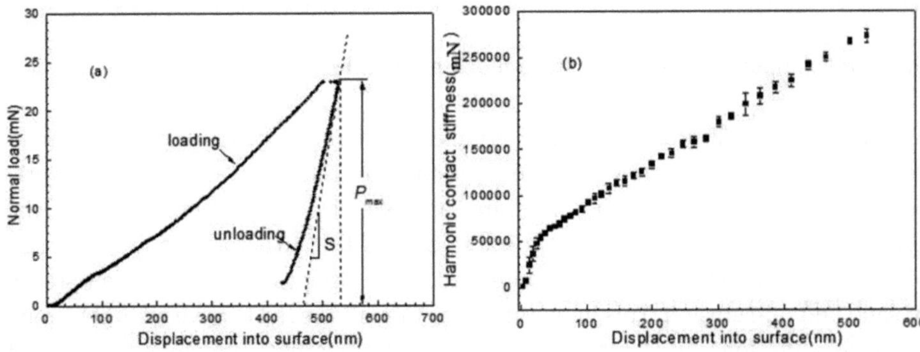

Fig. 9.23: (a) Typical loading curve and (b) contact stiffness–displacement (S–h) curve during the nanoindentation test.

tact stiffness. The hardness (H) and simple elastic modulus (E_r) of the material can be calculated using the following contact area function:

$$A_c = 24.56h_c^2 + C_1h_c + C_2h_c^{1/2} + \cdots + C_8h_c^{1/128}, \tag{9.3}$$

where C_1—C_8 are constants. In this experiment, $C_1 = 25.3$, $C_2 = 2283.4645$, and $C_3 = 34\,749.3301$. Material hardness (H) and simple elastic modulus (E_r) can be calculated using the following equations:

$$H = \frac{P}{A_c}, \tag{9.4}$$

$$E_r = \frac{S\sqrt{\pi}}{2\sqrt{A_c}}. \tag{9.5}$$

The relation between simple elastic modulus E_r and elastic modulus of the material E is given by

$$\frac{1}{E_r} = \frac{1-\nu^2}{E} + \frac{1-\nu_i^2}{E_i}, \tag{9.6}$$

where E and ν are the Young's modulus and Poisson's ratio of the sample, respectively, and E_i and ν_i are the Young's modulus (1141 GPa) and Poisson's ratio (0.07) of a diamond indenter, respectively.

In this book, continuous stiffness measurement was performed on thin films with different thicknesses. Load was applied for the continuous indentation of the indenter from the thin-film surface to the substrate. The hardness value at 1/3 the depth of the thin film was taken as the hardness of the thin film.

9.3.2.1 Effect of temperature

According to the previous discussion, temperature not only affected thin-film microstructure and increased crystallinity, but it could also improve optical performance. Similarly, the substrate temperature also had significant control over the thin-film hardness and elastic modulus. Fig. 9.24 shows the variations of microhardness with indentation depth at 400 °C and 600 °C. As the indentation depth increased, thin-film hardness increased sharply from zero to the maximum and then decreased gradually. When the indenter entered the substrate, the hardness decreased to the relatively low value of the substrate. Corresponding measurements of hardness were also performed using the nanoindentation method, and the hardness value measured at 1/3 the thin-film thickness was regarded as the thin-film hardness. The hardness–displacement curves measured at 400 °C and 600 °C were similar. However, the thin film had a higher level of hardness at a higher temperature (600 °C) because the thin film had better crystallinity under higher substrate temperatures with smaller dislocations and other defects. Thus, the thin film had better resistance against the indenter, which led to a higher level of hardness. In addition to hardness, the nanoindentation test can provide the elastic modulus of the thin film.

According to Griffith's criterion, the elastic modulus and hardness had similar patterns of evolution. Fig. 9.24 shows that the elastic modulus at 400 °C and 600 °C were similar. Comparatively, the elastic modulus at 600 °C was higher.

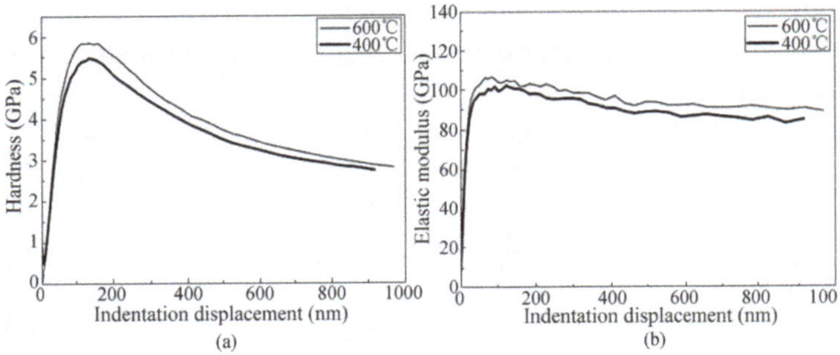

Fig. 9.24: Thin films under different substrate temperatures: (a) hardness; (b) elastic modulus.

9.3.2.2 Effect of substrate bias

As described above, substrate bias has a substantial impact on the properties of thin films, including their microstructure and optical properties. Similarly, substrate bias has a relatively large impact on the mechanical properties of thin films as well.

Fig. 9.25 shows the variations in thin-film hardness and elastic modulus with substrate bias. When the substrate bias was low, the thin-film hardness was relatively low at approximately 10.5 GPa. As the substrate bias increased, the hardness increased rapidly. At −80 V, the hardness increased sharply to 13.5 GPa; at −160 V, the hardness continued to increase to its maximum value of approximately 14.5 GPa. The thin-film hardness began to decrease as the substrate bias increased further because the increased energy of ions caused an etching effect on the thin film and led to damage.

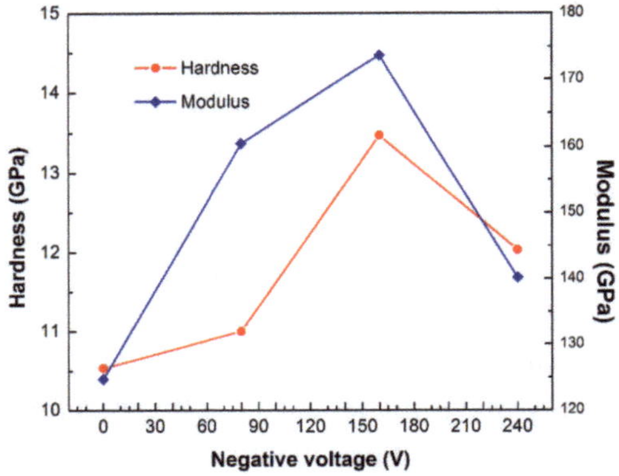

Fig. 9.25: Hardness and elastic modulus of thin films fabricated under different substrate biases.

It can be seen from Fig. 9.25 that the elastic modulus exhibited a variation trend similar to that of hardness. That is, when the substrate bias was low, the elastic modulus was relatively low. When the substrate bias increased to –160 V, the elastic modulus increased to its maximum value 175 GPa). As the substrate bias increased further, the elastic modulus decreased sharply. However, the rate of increase of elastic modulus in the range from –80 V to –160 V was greater than that for hardness. Despite the similar behaviors of the elastic modulus and hardness of thin films, the discussion above indicates that the hardening length and phase were different.

9.3.2.3 Effect of sputtering power

Sputtering power has also shown significant effects on the mechanical properties of thin films. Fig. 9.26 shows that, when the sputtering power was low, thin-film hardness was relatively low at approximately 12.75 GPa. As the sputtering power increased gradually from 60 W to 180 W, the hardness showed an almost linear increase from its minimum value to its maximum value of 13.2 GPa. Nevertheless, the elastic modulus showed a different pattern of variation when compared to the previous processing parameters. When the sputtering power was low, the elastic modulus was relatively small. However, when the sputtering power increased to 100 W, the elastic modulus increased slightly and then decreased drastically as the power increased further. This indicates that the thin-film elastic modulus and hardness did not maintain an entirely consistent trend of variation and showed a certain extent of deviation. A deep-seated reason for this deviation could be attributed to the intrinsic control factors for hardness and elastic modulus.

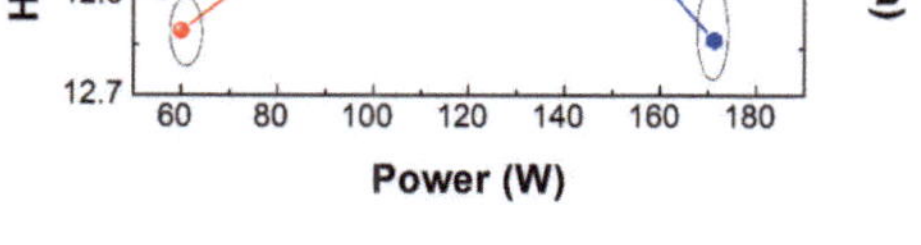

Fig. 9.26: Hardness and elastic modulus of thin films under different substrate biases.

9.3.3 Surface wettability of yttrium oxide thin films

The surface (or interface) of thin films is an extremely important research topic, which spans the entire theoretical and application research on thin films. The surface morphology of a monolayered functional thin film directly determines its application value, whereas the surface morphology of a multilayered coating directly determines the growth of its layered films and the performance of the entire device. Therefore, research on the thin-film surface has significant theoretical and practical implications.

In recent years, yttrium oxide has received widespread attention owing to its excellent physical and chemical properties. Yttrium oxide has a very broad transmission wave rage (0.25–9 μm), high transmittance, high strength, and good thermal stability. Therefore, it has commonly been applied as a material for infrared antireflective and protective coatings. Furthermore, owing to its favorable refractive-index-matching and dielectric properties, yttrium oxide thin films have also been extensively applied as a laser waveguide material and a high-dielectric-constant material in semiconductor devices. The surface morphology of yttrium oxide thin films directly determines the transmittance and environmental stability of yttrium oxide antireflective films. For multilayered coatings, the surface morphology of the initial coating has a direct impact on the growth of subsequent thin films and the performance of the overall device.

In this book, reactive magnetron sputtering was applied to fabricate yttrium oxide thin films under different oxygen flow rates in order to explore their surface morphologies.

9.3.3.1 Effect of oxygen partial pressure

In order to analyze the surface composition and bonding state of the thin films, x-ray photoelectron spectroscopy was performed directly on the thin-film surface without sputtering. Fig. 9.27 shows the full-spectrum scan of the yttrium oxide thin-film surface (2 sccm and 300 °C).

As can be seen from Fig. 9.27, the thin-film surface contained three elements: C, Y, and O. Aside from the intrinsic elements, Y and O, contained in yttrium oxide, C was adsorbed onto the thin film. Tab. 9.1 lists the elemental composition of yttrium oxide thin films fabricated under different oxygen flow rates. The table shows that the surface of thin films fabricated under different oxygen flow rates contained different types and percentage contents of elements. As the oxygen flow rate increased,

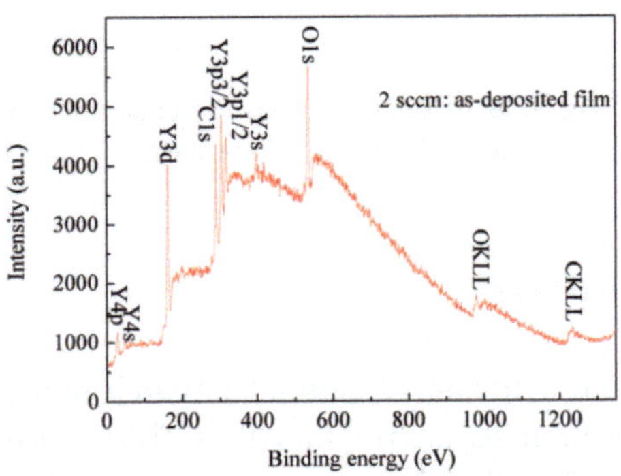

Fig. 9.27: Full-spectrum scan of yttrium oxide thin-film surface (2 sccm, 300 °C).

F and other impurities were introduced into the thin film, which could have been due to the contamination of the crucible; the content of impurities was relatively high when the oxygen flow rate was 12 sccm. The level of adsorbed C fluctuated around 47 % with varying oxygen flow rates. The content of Y also remained relatively stable and was maintained at approximately 15 %. Variations in the content of O were more significant; as the oxygen flow rate increased, the percentage content of O decreased gradually from 36.95 % to 29.92 %. The discussion and analysis above indicate that the thin-film surface contained different elements of varying contents under different processing conditions. This has a significant impact on the physical and chemical properties of the thin-film surface.

Tab. 9.1: Elemental composition of the surface of thin films fabricated under different oxygen flow rates.

Sample	C (%)	O (%)	Y (%)	Other elements (%)
2 sccm	47.9	36.95	15.15	0
4 sccm	49.89	35.74	14.06	0.31
8 sccm	45.22	34.32	16.58	0.83
12 sccm	46.42	29.92	16.08	7.58

In order to determine the bonding state and contents of the elements accurately, narrow-spectrum scans were performed for each element. Fig. 9.28 shows the narrow-spectrum XPS scans of each element on the surface of yttrium-oxide thin films fabricated under an oxygen flow rate of 2 sccm and temperature of 300 °C. It can be seen from Fig. 9.28 (a) that the C 1s peak position was around 285 eV, and the C 1s peak position of the adsorbed, contaminating carbon was used to calibrate the other peak positions. Fig. 9.28 (b) shows the XPS peak for O 1s, which could be deconvoluted into two spectral peaks at 529.0 and 531.3 eV, which corresponded to Y–O bonds in Y_2O_3 and physically adsorbed O_2 or O–H bonds, respectively. All instances of adsorbed oxygen will be referred to as O_β in this section. As can be seen from the figure, the peak area of O_β was relatively large; hence, we can predict that the amount of O_β was greater than that of bonded oxygen. Sensitivity factors were used to accurately analyse the relative contents of the two oxygen states in the thin-film surface under different growth conditions, as shown in Fig. 9.29. The figure shows that different oxygen flow rates had a significant impact on the ratio between adsorbed and bonded oxygen. When the oxygen flow rate was 2 sccm, the thin film contained a high amount of O_β. As the oxygen flow rate decreased, the content of O_β on the thin-film surface decreased gradually, and that of bonded oxygen increased. This indicates that the increase in oxygen flow rate could facilitate the formation of Y–O bonds. Fig. 9.28 (c) shows the XPS peak of Y 3d, which could be deconvoluted into two peaks near 158.3 and 156.3 eV.

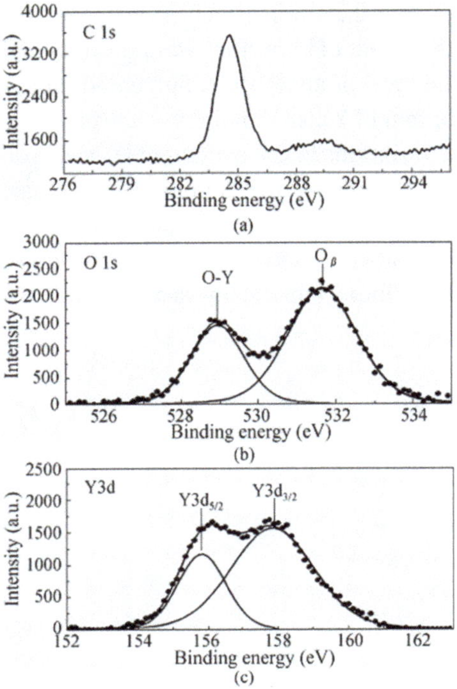

Fig. 9.28: Narrow-spectrum XPS scans of C 1s, O 1s, and Y 3d on the thin-film surface.

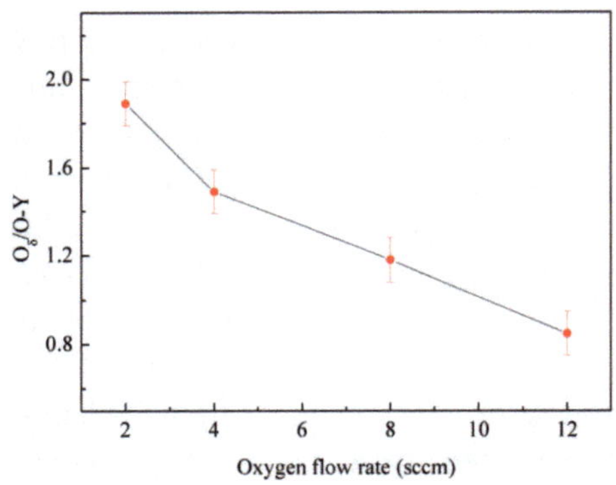

Fig. 9.29: O_δ/Y–O on the surface of thin films fabricated under different oxygen flow rates.

These deconvoluted peaks corresponded to the high-energy state Y $3d_{3/2}$ and low-energy state Y $3d_{5/2}$, and the energy interval between them was 2.02 ± 0.02 eV. The double peak formed by Y 3d was due to spin-orbit splitting. The peak position of Y 3d was higher than the binding energy of pure metallic Y. Hence, the fabricated thin films did not contain pure metallic Y–Y bonds, and the sputtered metallic yttrium had been completely oxidized.

The contact angle between water and the thin-film surface was measured for thin films fabricated under different oxygen flow rates. Fig. 9.30 shows the contact-angle measurement of water on the surface of an yttrium oxide thin film (2 sccm). The inter-

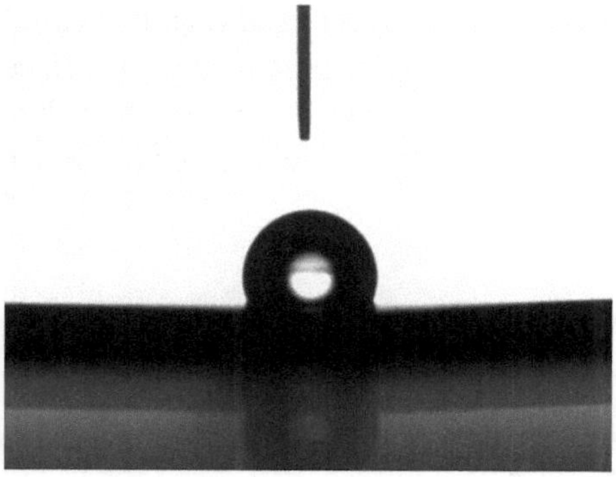

Fig. 9.30: Measurement of wetting angle of water at an oxygen flow rate of 2 sccm.

facial tension between two condensed phases can be calculated using the Young and Van Oss formula as follows:

$$W_a = \gamma_L(\cos\theta + 1) = 2\sqrt{\gamma_S^d \gamma_L^d} + 2\sqrt{\gamma_S^p \gamma_L^p}, \tag{9.7}$$

where W_a is the work function; θ is the contact angle; γ_L^p, γ_L^d are the polarization and dispersion components of surface tension of the liquid phase; and γ_S^p, γ_S^d are the polarization and dispersion components of surface tension of the solid phase, respectively.

In order to determine the surface energy of yttrium oxide thin films, the wetting angles of two different liquids (water and ethylene glycol) were measured, and the surface energy of the yttrium oxide thin film was obtained using equation (9.7). The parameters of surface energy for water and ethylene glycol are listed in Tab. 9.2.

Tab. 9.2: Surface-energy parameters of water and ethylene glycol (10^{-5} N/cm).

Reagents	**Surface energy**	**Dispersion component**	**Polarization component**
Water	72.8	21.8	51.0
Ethylene glycol	48.3	29.3	19.0

The water wettability of yttrium oxide thin films fabricated under different oxygen flow rates can be based on the surface-energy parameters of water and ethylene glycol for different yttrium oxide thin film samples, as listed in Tab. 9.3. The wetting angle of water on the yttrium oxide thin film indicates that the thin-film surface is hydrophobic, and the hydrophobicity of thin films fabricated under different oxygen flow rates were different. At 2 sccm, the water wetting angle was 101.95°. At 4 sccm, the wetting angle increased to 103.1°, which might have been due to the greater roughness of the thin film. When the oxygen flow rate was 12 sccm, the thin film had a relatively large wetting angle because its level of roughness was the lowest. Based on the calculated surface energy of yttrium oxide thin films, it can be seen that, as the oxygen flow rate increased, the surface energy was relatively large at 4 sccm and then decreased gradually. At 4 sccm, the large surface energy was due to the extremely high surface roughness of the thin-film surface. The increase in oxygen flow rate caused the reduction in surface energy, which was due to the decrease in the relative content of surface oxygen.

Tab. 9.3: Wettability and surface energy of yttrium oxide thin films under different oxygen flow rates.

Oxygen flow rate (sccm)	Wetting angle of ethylene glycol (°)	Wetting angle of water (°)	Surface tension	Polarization component	Dispersion component
2	90	101.95	12.1	5.6	6.5
4	86.65	103.1	14	3.3	10.7
8	90.95	101.5	12	6.6	5.4
12	85.9	104.45	10.3	8.1	2.2

9.3.3.2 Effect of substrate bias and temperature

The discussion in the previous section indicates that the wettability of thin films is mainly related to their surface morphology, which in turn is significantly influenced by temperature and voltage bias. These factors can simultaneously lead to surface modification but will produce opposite effects. Therefore, the analysis of the interactive effects between the two factors and their joint modification effects on the thin-film surface will reveal their influence on thin-film wettability.

Firstly, the composition distribution spectra of the surface were analyzed for thin films prepared under a substrate temperature of 600 °C. This was followed by the Lorentzian–Gaussian deconvolution of the O 1s and Y 3d peaks. We can see that the O 1s peak can be split into two peaks at 529 and 531 eV, which corresponded to bonded oxygen (Y–O) and physically adsorbed oxygen, respectively. The composition distribution spectra under room temperature can be seen in Fig. 9.31, where O 1s was clearly split into two peaks, and their relative contents varied continuously with substrate bias. Under unbiased conditions, adsorbed oxygen accounted for a large proportion, which decreased gradually with the increase in substrate bias; in contrast, the relative content of bonded oxygen increased gradually with the increase in substrate bias. Owing to the presence of spin-spin splitting, Y 3d can be divided into two equal pairs of peaks. Each pair of Y $3d^{5/2}$ peaks has peak positions of approximately 156 and 158 eV, which correspond to Y–O and Y–O_δ, respectively. As with the $O_{1\delta}$ phase, under unbiased conditions, the peak area of Y–O_δ indicates that its relative content was fairly high and was similar to the Y–O bond phase. Nevertheless, as the substrate bias increased, the relative content of Y–O_δ decreased gradually. When the substrate bias increased to −320 V, the Y–O_δ content decreased to its minimum.

The results obtained at 600 °C are shown in Fig. 9.32 and are unlike those at low temperature. Without substrate bias, the relative content of adsorbed oxygen for O 1s

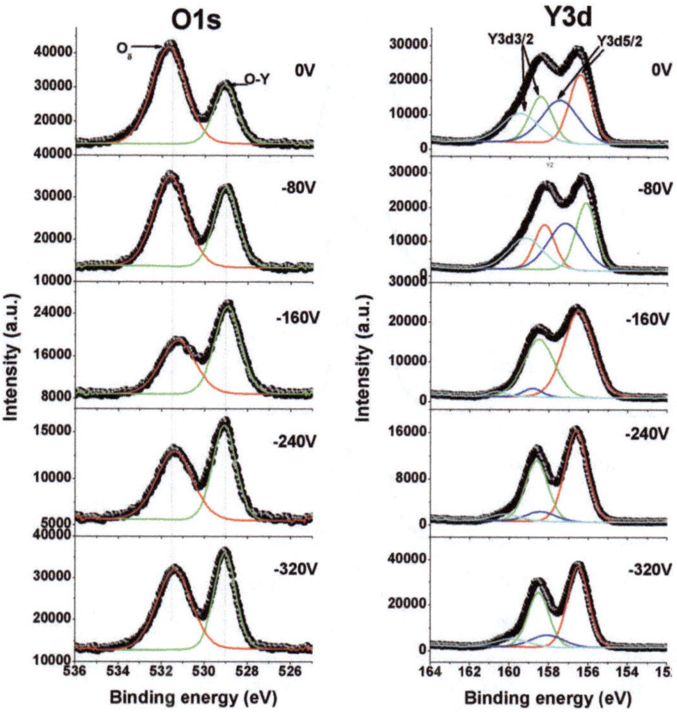

Fig. 9.31: Deconvolution of O 1s and Y 3d energy spectra of thin films fabricated at room temperature under different substrate biases.

was fairly high, but bonded oxygen had the highest proportion in the thin film. As the substrate temperature increased, the content of adsorbed oxygen decreased slightly but increased again at −320 V. For Y 3d, when the substrate bias was low, the content of Y–O_δ was relatively high. As the substrate bias increased, the content of Y–O_δ decreased gradually. At −160 V, the content of Y–O_δ reduced to its minimum value.

The ratio of elements can be obtained using XPS elemental sensitivity analysis, as shown in Fig. 9.33. It can be observed that, under different substrate temperatures, the elemental distribution of the thin-film surface showed different responses to substrate bias. At a low temperature, the O/Y ratio was relatively large, which implies that the thin film contained excess oxygen. As the substrate bias increased, the relative content of the oxygen phase decreased gradually. At −160 V, the O/Y ratio reached its minimum value. Then, as the substrate bias increased further, the O/Y ratio increased once again to a stoichiometric state. On the other hand, at a high temperature and without substrate bias, the O/Y ratio was relatively small. As the substrate bias increased, the O/Y ratio decreased slightly and reached its minimum value of approximately 1.25 at −240 V, following which the O/Y ratio returned to its initial stoichiometric ratio as substrate bias increased further.

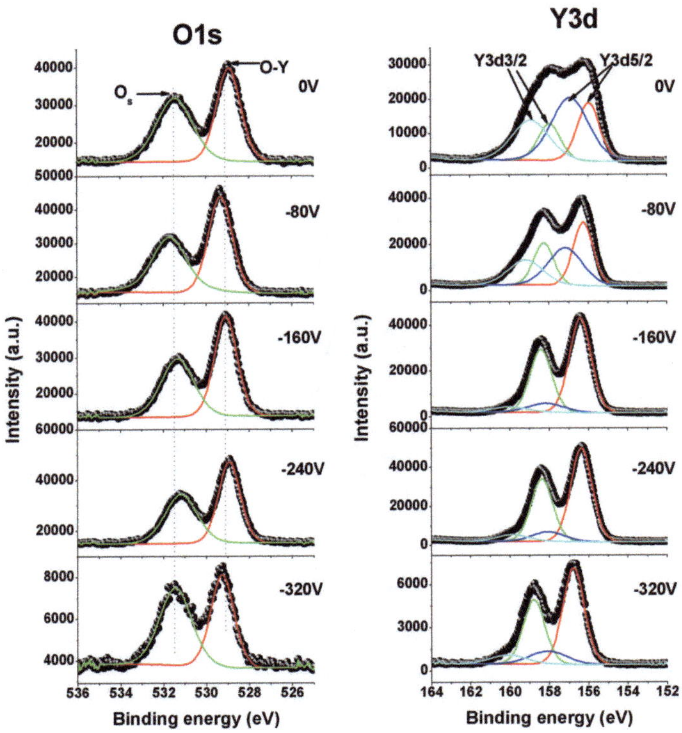

Fig. 9.32: Deconvolution of O 1s and Y 3d energy spectra of thin films fabricated at 600 °C under different substrate biases.

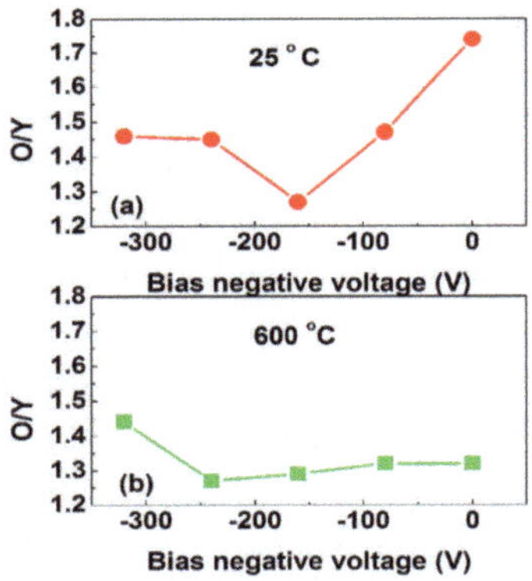

Fig. 9.33: O/Y ratio of thin-film surface under different temperatures: (a) room temperature; (b) 600 °C.

Fig. 9.34 shows the wetting angle of water and ethylene glycol on the thin-film surface under different temperatures and substrate biases. We can observe a small difference in the wettability of thin films between high and low temperatures. When the substrate bias was 0 V, the wetting angle was relatively large; it decreased gradually as the substrate bias increased. When the substrate bias was increased further, the wet-

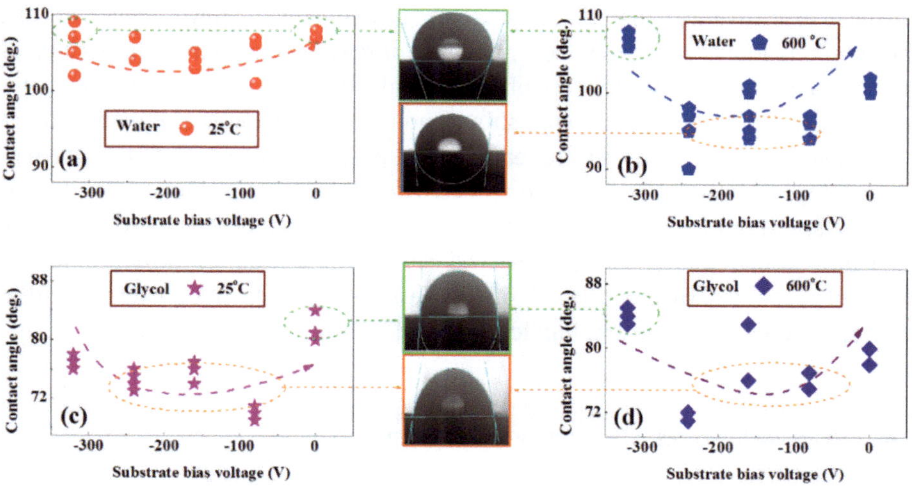

Fig. 9.34: Wetting angles of water and ethylene glycol on the thin-film surface under different temperatures and substrate biases.

ting angle returned to its maximum value. For water, at a high temperature, there were significant differences in the wettability of thin films under different substrate biases. For ethylene glycol, at a high temperature, the wettabilities of thin films were similar to that of water under different substrate biases. In the initial stage, the wetting angle of ethylene glycol was large; subsequently, as the substrate bias increased, the wetting angle showed a trend of decrease followed by an increase to its original level.

Numerous factors could influence the surface wettability of thin films. The predominant factors include surface roughness and surface composition. Based on previous analyses, we know that the surface roughness is low at high substrate biases. Nevertheless, the wetting angle of thin films under a high substrate bias was comparable to that without substrate bias. This indicates that the impact of roughness on the thin-film surface wettability is fairly small. On the other hand, the element ratio and wetting angle of thin films showed consistent variations. At the lowest O/Y ratio, the thin-film surface showed the smallest wetting angles for water and ethylene glycol. However, under high substrate biases, the O/Y ratio returned to its original level, and the wetting angles for water and ethylene glycol were comparable to those under the initial conditions.

9.4 Growth pattern and deposition rate

Reactive magnetron sputtering involves the injection of reactive gases during sputtering to enable reactions that produce the required compound thin films. Owing to its simple setup, strong controllability, and good thin-film structure performance, it has been extensively applied in industrial and scientific fields. Nevertheless, despite the simple methods and strong controllability of reactive magnetron sputtering, we still lack a thorough understanding of its physical nature. Owing to its complex physical nature, it has nonlinear relationships, and a typical example is the appearance of reactive hysteresis loops during reactive sputtering. The presence of reactive hysteresis loops will cause the deposition process to exhibit three obvious zones, i.e. the metallic sputtering mode, transition mode, and poisoned mode. The growth conditions under different sputtering modes will cause the thin films to have significantly different properties. Hence, research on the clarification of reactive hysteresis and its control patterns has important implications and effects.

9.4.1 Reactive hysteresis loop and growth pattern

After injecting the reactive gas during the sputtering process, the reactive gas in the vacuum crucible will react with not only the sputtered particles on the substrate surface to form the required compound thin film, but also the surfaces of the crucible wall and the sputtering target. As the nature of the resulting compounds is different from that of the metallic state, significant changes will occur in the sputtering parameters. The yttrium oxide thin films used in this book were mainly fabricated by reactive magnetron sputtering. Hence, the sputtering target was metallic yttrium, and the reactive gas was high-purity oxygen.

Fig. 9.35 shows typical manifestations of reactive hysteresis loops. Reactive hysteresis occurs not only for the discharge voltage, but also for the reactive gas partial pressure and discharge current of the crucible. The discharge voltage (Fig. 9.35 (a)) was approximately 400 V before the injection of the reactive gas. As oxygen increased gradually (stepwise increments of 0.2 sccm, with an interval of 2 min for each step to detect changes in voltage and other parameters), at point 1, the voltage remained at the sputtering voltage of 400 V. When the oxygen flow rate increased to a value slightly above point 2, the sputtering voltage decreased rapidly from point 2 to 180 V. Hence, point 2 was taken as the critical point. Subsequently, as the oxygen flow rate continued to increase, the discharge voltage remained at 180 V. The zone where the discharge voltage was 400 V is known as the metallic sputtering mode. The zone beyond the critical point of point 2, where the discharge voltage dropped to 180 V, is known as the poisoned mode. The zone between the metallic sputtering and poisoned sputtering modes is known as the transition zone. As the oxygen flow rate decreased to point 4 (almost overlapping with point 2), the sputtering voltage did not return to its initial value (400 V); instead, it continued to remain at a lower state. When the oxygen flow rate continued to decrease to a value slightly lower than point 3, the sputtering voltage immediately returned to its original value of 400 V. The zone between point 4 and point 3 is known as the hysteresis loop zone, which results in reactive hysteresis. The reactive hysteresis loop of oxygen partial pressure is shown in Fig. 9.35 (b).

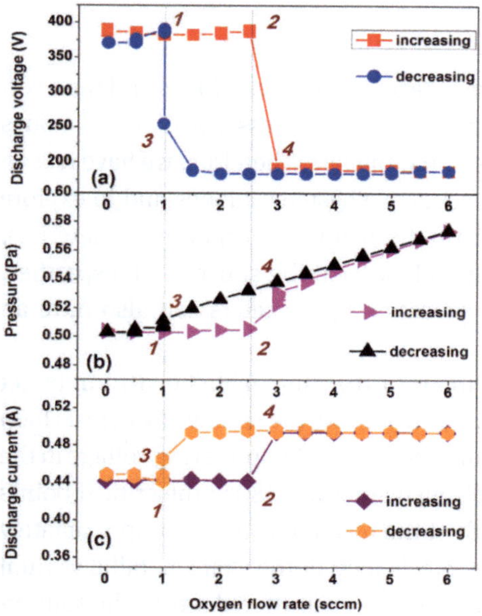

Fig. 9.35: Reactive hysteresis loops: (a) reactive hysteresis loop of discharge voltage; (b) reactive hysteresis loop of oxygen flow rate; (c) reactive hysteresis loop of discharge current.

In the metallic sputtering mode, the oxygen partial pressure was extremely low and was virtually zero. This is because the injected oxygen was completely consumed in the reaction or pumped out of the crucible. In the poisoned sputtering mode, the injected oxygen was not completely consumed, and hence, the oxygen partial pressure showed a trend of linear increase with oxygen flow rate beyond point 2. When the oxygen flow rate was reduced, the oxygen partial pressure decreased linearly from a relatively high level to point 3, following which further reductions in oxygen flow rate caused oxygen partial pressure to return to the level in the poisoned state at zero. Fig. 9.35 (c) shows the hysteresis of sputtering current, which exhibited the opposite trend to that of voltage. In the metallic sputtering mode, the minimum level of sputtering current reached was 0.44 A. Beyond point 2, the sputtering current increased to 0.48 A in the poisoned sputtering mode and continued to remain at this value. The analysis above indicates that changes occurred in all sputtering parameters under different sputtering modes. In the metallic mode, sputtering voltage was relatively high, oxygen partial pressure was very low, and sputtering current was relatively low. The trends were reversed in the poisoned mode. As the transition zone was very narrow, rapid changes occurred in the voltage, oxygen partial pressure, and other parameters in this region. Hence, the clarification of these physical parameters and their patterns of evolution plays an extremely crucial role in gaining a better understanding of controllable reactions and in the sputtering of thin-film samples that can satisfy application requirements.

9.4.2 Control of reactive hysteresis loop

The previous section discussed the presence of reactive hysteresis loops and their corresponding modes in reactive sputtering. The sputtering processes in different modes correspond to different physical processes and parameters. Therefore, we have reason to speculate that it is possible to control the reactive hysteresis loops and to explore the possible factors. Owing to the differences in the attributes of sputtering materials, the reactive hysteresis loops of different materials will exhibit substantial disparities. However, for a specific material, external processing parameters can also have an impact.

Fig. 9.36 shows the reactive hysteresis loops of the same metallic yttrium target tested under different sputtering times. Fig. 9.36 (a) shows the reactive hysteresis loop of a new unsputtered yttrium target. The figure shows that the sputtering voltage in the metallic state was 400 V and that in the poisoned state was 180 V; the critical points were at 2.5 sccm and 1.3 sccm, and the width of the reaction zone was approximately 1.2 sccm. Fig. 9.36 (b) shows the reactive hysteresis loop of the same metallic yttrium target tested after 20 h of sputtering. Compared to the unsputtered target, the sputtering voltage in the metallic state was 400 V, with a slight increase near the critical point. However, the critical point after sputtering was at 1.8 sccm, and the width of the transition zone was 0.4 sccm. These observations indicate that the reactive hysteresis loop is affected by sputtering time. This is because, as the sputtering time increased, the width and depth of the etched groove increased, which led to changes in the magnetic field on the surface of the etched groove, thereby affecting the sputtering hysteresis loop.

In addition to sputtering time, there are several factors that can influence and control the reactive hysteresis loops. In recent years, researchers have employed a series of methods to investigate the control of reactive hysteresis loops. The main methods include spectrometric detection, real-time feedback control, and the change of process

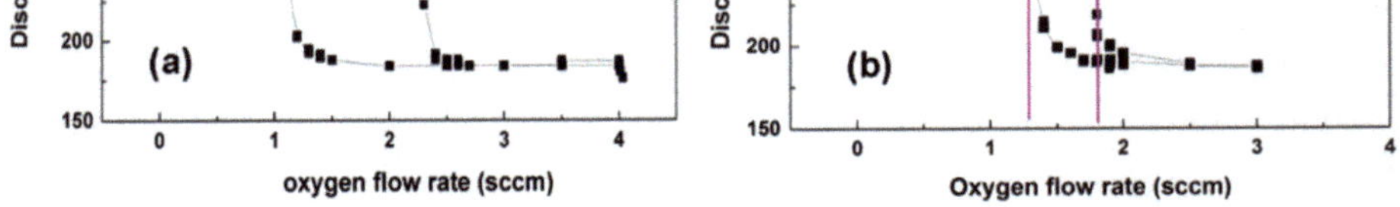

Fig. 9.36: Testing of reactive hysteresis loops: (a) new unsputtered target; (b) after 20 h of sputtering.

parameters, such as nitrogen, pumping rate, and target size, in order to modify and control the reactive hysteresis phenomena. The researchers aimed to obtain satisfactory deposition conditions and deposition window.

Fig. 9.37 shows the reactive hysteresis loops of yttrium targets sputtered under different pumping speeds. It can be seen that the pumping speed had a substantial impact on the results. Under a low pumping speed, the discharge voltage decreased rapidly from 400 V to 180 V when the oxygen content exceeded the critical point, which corresponded to the rapid transition of the metallic sputtering mode to the poisoned mode. Simultaneously, the oxygen partial pressure at the critical point showed drastic changes. Owing to the rapid transition at the critical points between the metallic mode and the poisoned mode with the increase and decrease of oxygen partial pressure, thin-film deposition near the critical points is extremely unstable. Furthermore, the reactive hysteresis loop showed a rapid transition zone under a low pumping speed. Therefore, the pumping speed is one of the solutions for achieving stable deposition conditions near the critical points. Fig. 9.37 (b) and (d) show the reactive hysteresis loops tested under a high pumping speed. As can be seen, the transition zone in the reactive hysteresis loops was removed under a high pumping speed, and a very wide transition zone existed between the metallic sputtering mode and poisoned mode. Hence, the transition zone can be used for stable thin-film deposition. Similarly, the critical points of sudden changes did not occur for oxygen partial pressure under a high pumping speed, and the hysteresis effect had been removed. The removal of the hysteresis effect will enable us to achieve the goal of stable thin-

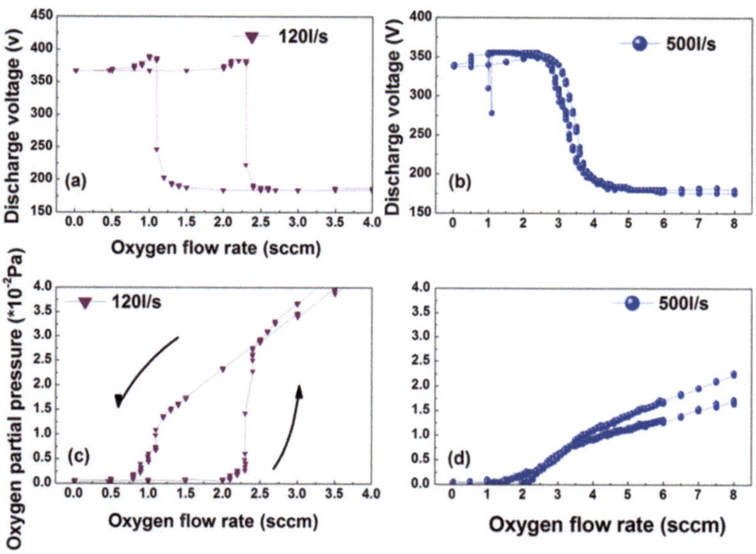

Fig. 9.37: Reactive hysteresis loops of yttrium target tested under pumping speeds of 120 l/s and 500 l/s.

film deposition within the transition zone. As the transition zone is located between the poisoned and metallic states, the fabricated thin films will have rich physical structures, which can satisfy various types of requirements.

9.4.3 Growth pattern and deposition rate

The conditions of thin-film growth differ under different sputtering modes. Hence, the microstructure and performance of the resulting thin films also vary. In practical industrial applications, one of the performance indicators that has received widespread attention is deposition rate. Achieving higher deposition rates is a crucial pathway for increasing production yield. Therefore, clarifying thin-film growth rates under different sputtering modes and determining the relationship between growth rate and sputtering mode are extremely urgent and important issues. The following section will discuss the relationship between deposition rate and sputtering mode.

As can be seen from Fig. 9.38, different sputtering modes corresponded to different deposition rates; that is, thin-film deposition rates are dependent on the sputtering mode. In the metallic sputtering mode, the surface of metallic targets will be partially

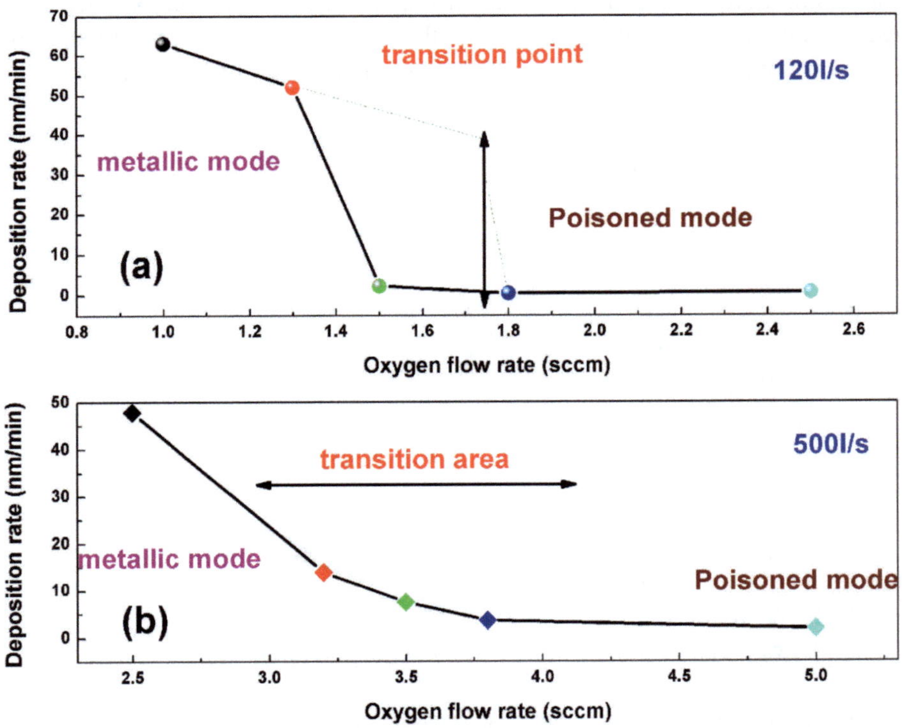

Fig. 9.38: Deposition rates of thin films under different sputtering modes.

oxidized. Under the bombardment of energetic ions (400 eV), the metallic target surface will exhibit a higher sputtering rate, and more atoms will be sputtered to reach the thin-film surface for coating. In the poisoned mode, the target surface will be completely oxidized and will form a dense layer of oxides. Furthermore, the sputtering voltage under this condition is 180 V, which implies that the energy carried by energetic ions bombarding the target surface will be lower (180 eV). Moreover, the sputtering yield of oxides is much lower than that of metals. These factors will lead to a lower number of particles sputtered per unit time and per unit area. Hence, the atoms reaching the substrate surface to form a thin film will be reduced, thereby decreasing the deposition rate. Therefore, to achieve a higher deposition rate for a specific sputtering target, adjustments are generally made to the processing parameters to ensure that the thin film will grow under the metallic sputtering mode. However, owing to the high deposition rate under the metallic mode, there is insufficient time for the reaction because of the lack of oxygen, which will cause the resulting thin film to have stronger metallic properties. Therefore, obtaining thin films with superior performance under high deposition rates will require more complicated techniques, such as high-frequency pulsed laser deposition, high-energy magnetron sputtering, and localized oxygen supply.

9.5 Adhesion of yttrium oxide thin film and zinc-sulphide substrate

According to the previous investigations on optical and mechanical properties, the refractive index of yttrium oxide is approximately 1.8, which can be controlled within a large range, and yttrium oxide is able to the transmit the entire wave range. Furthermore, yttrium oxide has a high level of hardness and elastic modulus. Yttrium oxide is hydrophobic, and thus also has good resistance to rain and sand erosion. More importantly, yttrium oxide has structure and performance stability under high temperature and pressure, and it is resistant to aerothermodynamic failure. These characteristics indicate that yttrium oxide can be applied as infrared antireflective and protective coatings. Zinc sulphide (ZnS) has been widely applied as a material for infrared windows owing to its infrared multi-spectral transmission characteristics. There is a high level of index-matching between yttrium oxide and zinc sulphide, which can produce good antireflective effects. In addition, one of the key factors in assessing the applicability of yttrium oxide is its ability to ensure that coating delamination will not occur even under multiple-stage and repeated heating in aerothermodynamic environments. This property is determined by the interface between the Y_2O_3 thin film and ZnS window, i.e. the interface composition and adhesion strength. Therefore, research on the interface composition and adhesion strength between the Y_2O_3 thin film and ZnS window is extremely important. This section will first explore the composition distribution and bonding state of the Y_2O_3/ZnS interface. Based on the results, we investigate the interfacial adhesion under different conditions in order to determine the conditions for achieving strong interfacial adhesion.

9.5.1 XPS composition analysis of interface region

Fig. 9.39 shows the XPS full spectrum of the Y_2O_3 thin-film surface after sputtering by Ar^+ ions; the sputtering time was 15 min. The figure indicates that the initial thin-film surface before sputtering contained C and F elements in addition to Y and O. However, after Ar^+ sputtering, the characteristic peaks of these impurities disappeared, indicating that CO_2 and other gases in air had contaminated the thin-film surface. The full-spectrum XPS clearly shows the characteristic peaks of Y 3d, Y 3p, and O 1s. Among them, the peak position of Y 3d was near 157 eV, that of Y 3p was near 300 eV, and that of O 1s was around 531 eV.

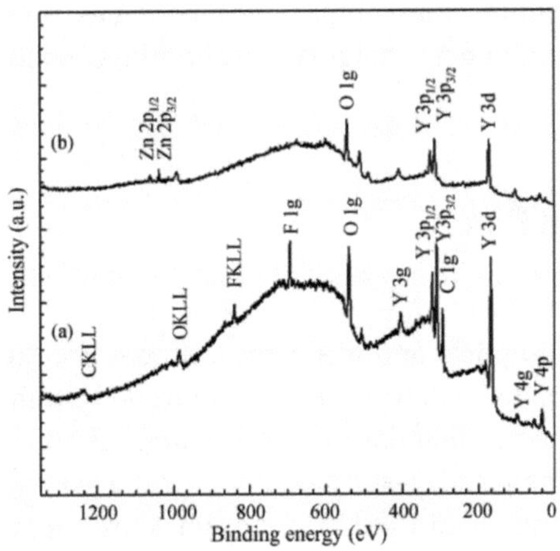

Fig. 9.39: Typical XPS full spectrum of Y_2O_3 film: (a) before Ar^+ sputtering; (b) after Ar^+ sputtering.

In order to explore the adhesion performance of the Y_2O_3 thin film and ZnS window substrate, XPS profile analysis was performed to investigate the composition distribution at the interface region. Variations in the coating–substrate composition with Ar^+ sputtering time are shown in Fig. 9.40. The atomic contents of Zn, S, Y, and O were calculated based on the integrated area under the Zn 2p, S 2p, Y 3d, and O 1s spectral peaks. Initially, only Y and O were present; as the etching time increased, the contents of Zn and S increased gradually beyond a certain period of etching, while the contents of Y and O decreased gradually, exposing the interface region. Subsequently, the contents of the four elements intersected; Y and O gradually disappeared, whereas the ZnS substrate was exposed. It can be seen from the figure that the mixed composition zone was broad, which might have been because the hand-polished ZnS substrate surface was relatively rough.

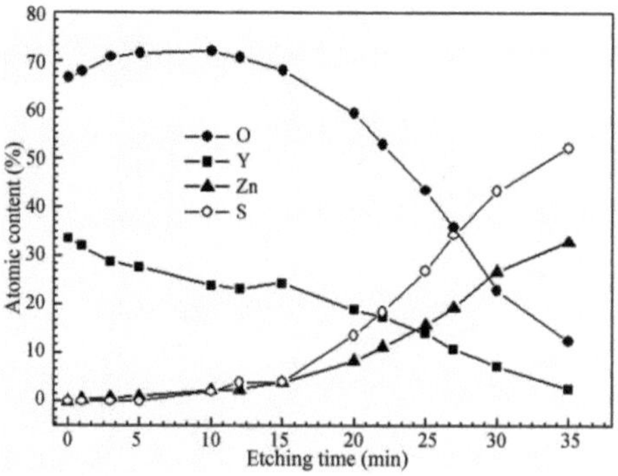

Fig. 9.40: XPS composition analysis of Y_2O_3/ZnS interface.

In order to determine interfacial adhesion, it is necessary to analyse the chemical states of each element at the interface. The analysis of the chemical states of elements is an important analytical function of XPS. Before performing the analysis of elemental chemical valence, we must first perform the accurate calibration of binding energy because, after the sample releases photoelectrons after excitation by x-ray irradiation, positive holes will remain on the sample. If the sample chamber conductor has good point contact with the spectrometer, then positive charge will be immediately neutralized by the conduction current in the sample holder. If the sample is a semiconductor or an insulator, then the conduction current from the spectrometer can completely neutralize the positive charge on the sample surface. This will cause the accumulation of positive charge on the sample surface and reduce the kinetic energy of emitted photoelectrons. Hence, the measured binding energy will be higher than the actual value. The external standard method is generally employed for peak correction.

Samples used for XPS analysis are prone to contamination from external hydrocarbons. When the sample is loaded into the spectrometer chamber, it will be further contaminated by residual diffusion pump oil in the sample chamber. The carbon in such contaminants can be used for charge correction, and the C 1s binding energy is generally taken as 284.6 eV. However, since this experiment involves XPS profile analysis, external carbon will only be present on the thin-film surface and the coating–substrate interface, but not inside the thin film (Fig. 9.41). Hence, it is difficult to use C 1s to position the entire process. Hence, the double internal standard method was employed for positioning. In other words, the external standard method was used to position the O 1s peak of Y_2O_3 thin film sputtered for 1 min. The peak position was 529.53 eV, which is consistent with the results in the literature. This is because, when the Ar^+ sputtering time was shorter than 10 min, the spectral peaks of S and Zn did not appear, as the interface region had not yet been reached. The main component was Y_2O_3, and the O 1s peak position did not show significant displacement. Hence, the O 1s peak position at 529.53 eV can be used for spectral peak correction of XPS spectra after 5 min of sputtering. The narrow O 1s peak obtained after 5 min of Ar^+ sputtering is shown in Fig. 9.42. The figure shows that, aside from the O 1s peak of

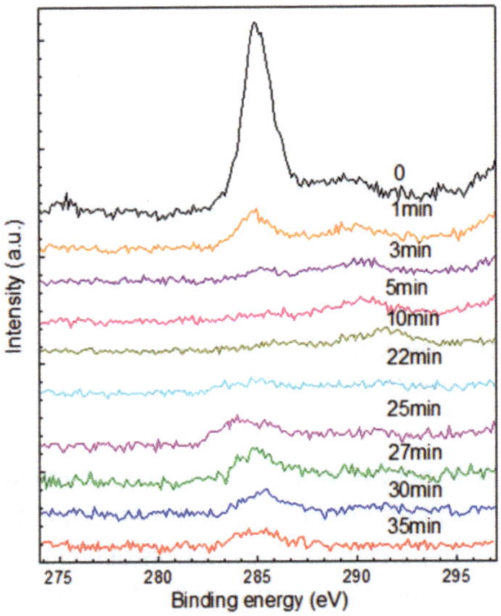

Fig. 9.41: C 1s spectra under different sputtering times.

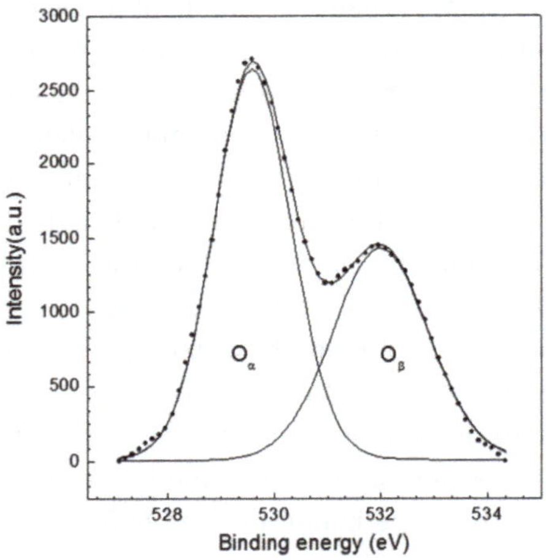

Fig. 9.42: O 1s spectrum after 5 min of sputtering.

Y_2O_3, the peak for O_β existed at 532 eV. This peak resulted from the physical adsorption of oxygen in the thin film, which is common in oxide thin films. Its binding energy will not change with the differences in the series of samples, and hence, it can be used to correct the other spectral peaks.

XPS profile analysis revealed that the overall coating system is composed of three parts:

(1) surface layer (Ar^+ sputtering time less than 5 min), which contains organic contaminating carbon and oxygen commonly found in air;
(2) pure oxide layer (Ar^+ sputtering time 5–12 min);
(3) interface transition layer (Ar^+ sputtering time 15—35 min), which is a mixed region that contains components from the thin film and the substrate.

In order to analyse the composition and bonding state of the coating system, deconvolution was performed on the narrow spectrum of each element. Splitting might occur in the inner energy levels of the three energy levels, Y 3d, Zn 2p, and S 2p, owing to the interactions between electron orbital motion and spin. Hence, doublets will be presented in all photoelectron spectra recorded, and the energy interval in the doublet will increase with increasing atomic number. For the same element, the chemical shift of the spin-orbit coupling doublet is extremely consistent, and the energy interval between the doublet rarely exceeds 0.2 eV.

Fig. 9.43 and 9.44 show the deconvoluted peaks of O and Y, respectively, under different sputtering times.

Fig. 9.43 shows that O 1s mainly had two XPS peaks, which were located near 529.5 eV and 532 eV. Among them, 529.5 eV is the characteristic peak of oxygen in Y–O, whereas 532 eV is caused by the physically adsorbed oxygen in the thin-film or surface-adsorbed O–H.

Fig. 9.43 also shows that the relative content of O_β located near 532 eV showed a significant decrease from 47 % before sputtering to 40 % after 5 min of sputtering. However, as the Ar^+ sputtering time increased, the O_β relative content remained within

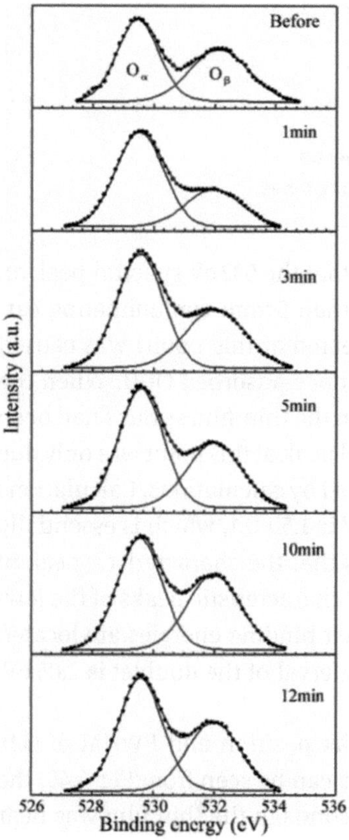

Fig. 9.43: O 1s fitted spectra under different sputtering times.

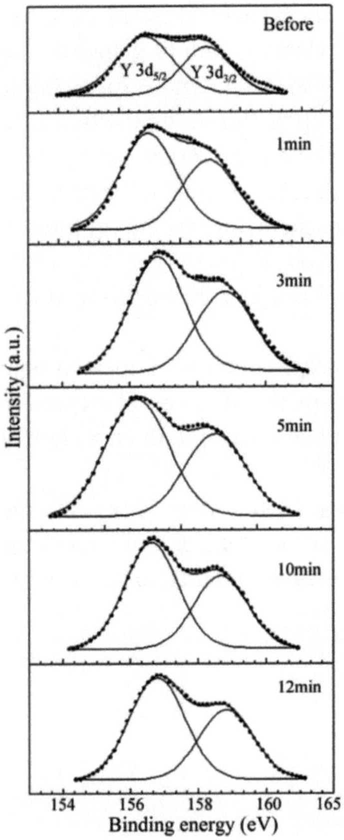

Fig. 9.44: Y 3d fitted spectra under different sputtering times.

a certain range and did not disappear. This indicates that the 532 eV spectral peak on the thin-film surface layer (Ar^+ sputtering time less than 5 min; contaminating carbon on the thin-film surface had completely disappeared at this point) was caused by physically adsorbed oxygen in the thin film or surface-adsorbed O–H. When the Ar^+ sputtering time was greater than 5 min, the O–H on the thin-film surface had been completely etched away by Ar^+, and the 532 eV spectral peak at this point was only due to physically adsorbed oxygen. This point can be verified by calculations. Calculations indicate that the ratio of the O and Y atoms at 529.5 eV is 1.5 ± 0.1, which is essentially consistent with the ideal value of 1.5. Fig. 9.44 shows that the characteristic peak of Y 3d is composed of two spectral peaks, which are the characteristic peaks of the low-energy state Y $3d_{5/2}$ and high-energy state Y $3d_{3/2}$. Their binding energies are located near 156.8 eV and 158.8 eV, respectively; the energy interval of the doublet is 2.05 eV, and the intensity ratio is 1.5.

Fig. 9.45 and 9.46 show the variations in the peak position and FWHM of O 1s and Y 3d characteristic peaks with sputtering time. As can be seen from Fig. 9.45, the peak position of the O 1s characteristic peak of Y–O bonds in the thin film was near

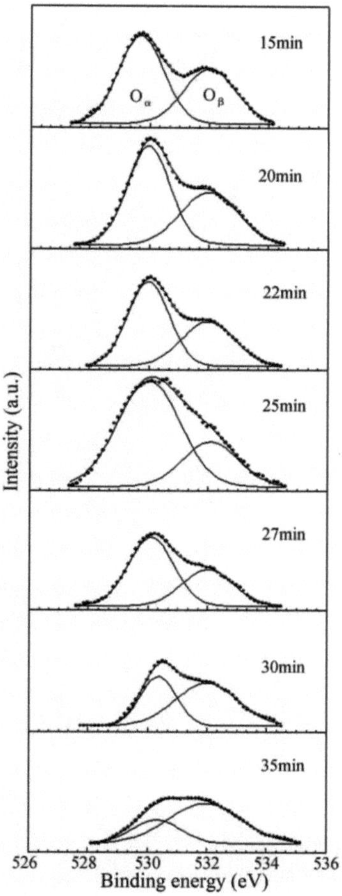

Fig. 9.45: Interface O 1s fitted spectra under different sputtering times.

529.5 eV. However, as the sputtering time and depth increased to the interface region, the FWHM of the O 1s characteristic peak was maintained at approximately 1.7 eV and did not undergo significant changes. However, the O 1s peak position gradually shifted toward higher binding energies, and its final position was 530.31 eV, which is 0.8 eV higher than the binding energy of the O 1s characteristic peak for Y–O bonds. This indicates that chemical reactions had occurred between O and the ZnS substrate in the interface region, which caused a shift in the peak position of O 1s. O in the interface region can react with Zn to form Zn–O bonds, with S to form S–O bonds, and with Zn and S to form $ZnSO_4$. However, if O and S form S–O bonds, since the electronegativity of O (3.65) is greater than that of Zn (1.6), O will have a much greater electron affinity than Zn. Hence, the S 2p binding energy in S–O bonds will be much greater than that in Zn–S bonds. However, Fig. 9.46 shows that the maximum binding energy of S $2p_{3/2}$ in the interface region is 161.58 eV, which is less than that of S $2p_{3/2}$ in Zn–S bonds (161.9 eV). Hence, the presence of S–O bonds can be excluded. Furthermore, since the binding energy of S $2p_{3/2}$ in $ZnSO_4$ is 169.5 eV, this also indicates that $ZnSO_4$ was not

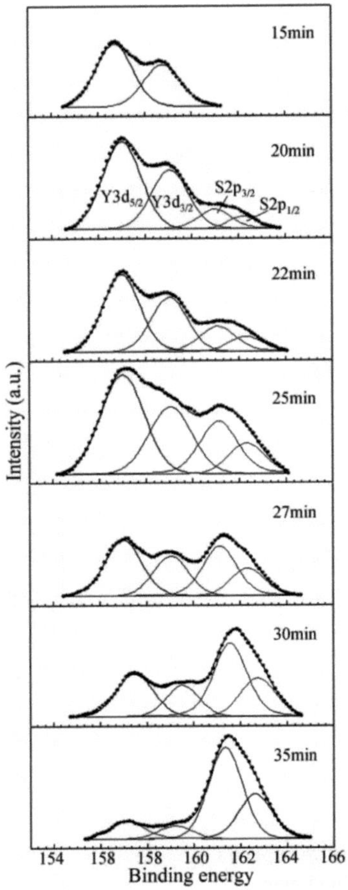

Fig. 9.46: Interface Y 3d fitted spectra under different sputtering times.

formed in the interface region. Based on the analysis above, we know that O did not form chemical bonds with S; instead, it reacted with Zn to form Zn–O bonds.

Fig. 9.47 shows the variations in the position and FWHM of the O 1s characteristic peak with sputtering time. As can be seen, the position of the O 1s peak in the interface region increased gradually as the sputtering depth increased, and the O 1s peak was finally located at 530.31 eV. This is consistent with the binding energy of O 1s in Zn–O bonds (530.1–530.4 eV). As the sputtering time increased, Y atoms decreased gradually, and Zn atoms increased. Hence, the chances for bonding between O and Y were reduced, whereas those between O and Zn increased. This caused the binding energy of O 1s to shift gradually toward higher binding energies. The Zn–O bonds formed between O and Zn can also be verified by the chemical shift in the Zn 2p spectral peak.

Fig. 9.48 shows the fitted spectra of interface Zn 2p under different sputtering times.

We can see from Fig. 9.48 that, as the sputtering depth increased, the peak position of the Zn $2p_{3/2}$ characteristic peak showed a gradual chemical shift toward higher

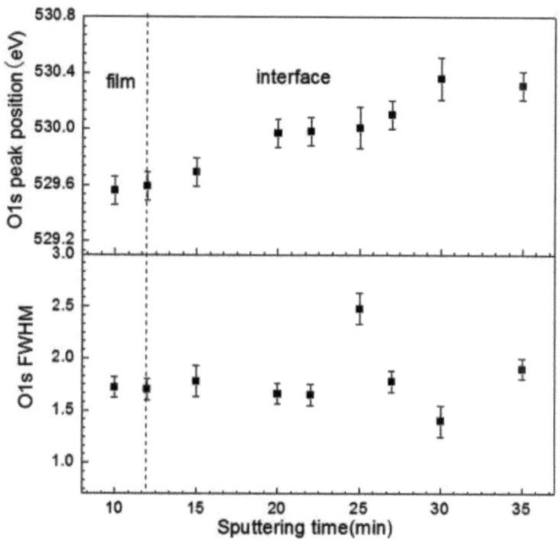

Fig. 9.47: Peak position and FWHM of interface O 1s peak under different sputtering times.

binding energies. The maximum binding energy of interface Zn $2p_{3/2}$ (1022.55 eV) was higher by 0.55 eV than that of the ZnS substrate. This coincides with the binding energy of Zn 2p in Zn–O (1022.5 eV).

Fig. 9.49 shows the variations in the position and FWHM of the Y 3d characteristic peak with sputtering time. As can be seen, the positions of Y $3d_{5/2}$ and Y $3d_{3/2}$ characteristic peaks of Y–O bonds in the thin film were 156.8 eV and 158.8 eV, respectively; the FWHM values of peaks were maintained around 1.9 eV. However, as the sputtering time and depth increased to the interface region, the FWHM of the Y 3d characteristic peak decreased slightly and fluctuated around 1.8 eV. This indicates that changes were occurring in the bonding environment surrounding the Y atoms. Furthermore, the positions of the Y 3d doublet gradually shifted toward higher binding energies with increasing sputtering depth, and the final positions were 157.15 eV and 159.20 eV, respectively. This was an increase of 0.35 eV compared to the binding energy of Y 3d in Y–O bonds, which indicates that chemical reactions had occurred between the Y atoms and ZnS substrate, resulting in the shift of the Y 3d peak position.

Y atoms in the interface region can react with S atoms to form Y_2O_3, with Zn atoms to form intermetallic compounds (YZn, YZn_3, YZn_5, YZn_{12} etc.), and with Zn and S to form compounds with all three elements. However, according to the principle of electronegativity, the electronegativity of Zn (1.6) is smaller than that of O (3.65); hence, the electron affinity of Zn is poorer than that of O. Therefore, when binding occurs between Y and Zn, the binding energy of Y 3d needs to be lower than that between Y and O. This is not in agreement with the experimental results, which indicates that intermetallic compounds of Y and Zn were not formed. Furthermore, the electronegativity of Y (1.2) is smaller than that of Zn (1.6). Hence, when compounds containing Y, Zn, and S are formed, the binding energy of Zn 2p should be lower than that in ZnS, which does not conform to our experimental results, and compounds containing Y, Zn, and S are yet to be discovered. Based on the analysis above, we know that Y reacted with S to

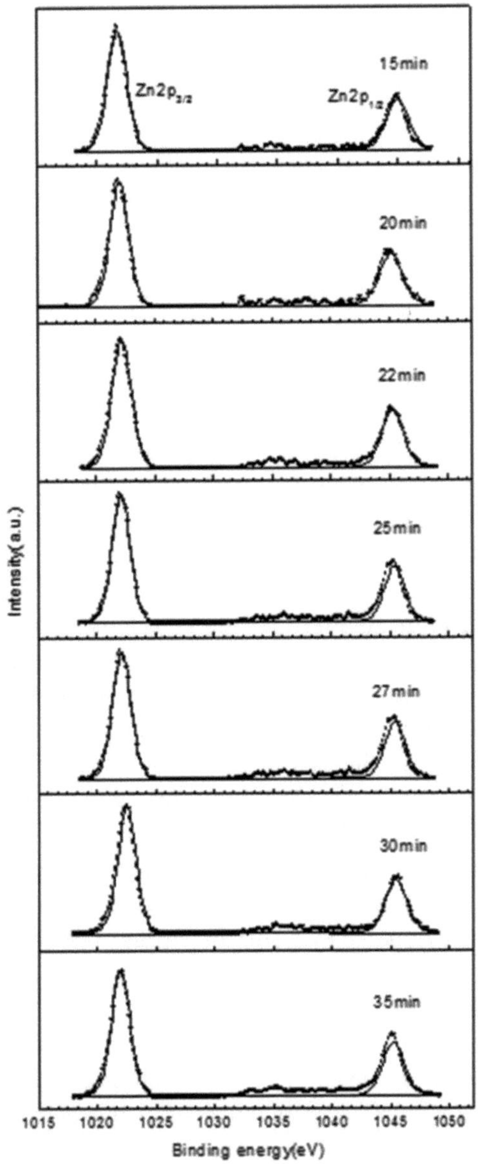

Fig. 9.48: Interface Zn 2p fitted spectra under different sputtering times.

form Y_2S_3. Fig. 9.49 shows that the peak positions of interface Y $3d_{5/2}$ and Y $3d_{3/2}$ characteristic peaks showed gradual chemical shifts toward higher binding energies as the sputtering depth increased, and the peaks were finally located at 157.15 eV and 159.20 eV, respectively. This result is consistent with the Y $3d_{5/2}$ peak position in Y_2S_3 (157.1–157.4 eV). Furthermore, the FWHM coincided with that of Y $3d_{5/2}$ in Y_2S_3 (1.8 eV). The reaction between Y and S to form Y_2S_3 can be verified by S_{2p}.

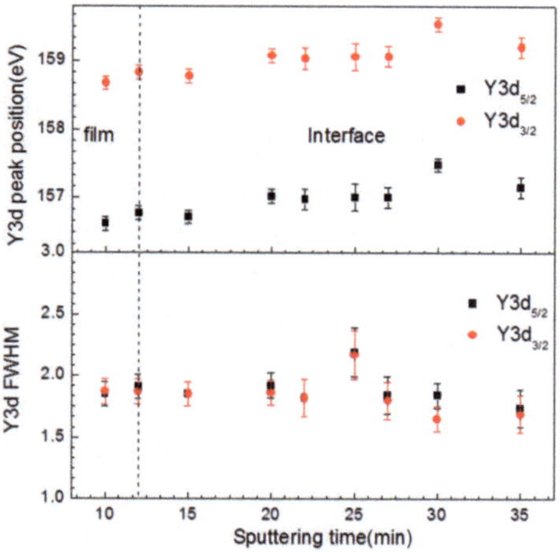

Fig. 9.49: Peak position and FWHM of interface Y 3d under different sputtering times.

Fig. 9.50 shows the variations in the position and FWHM of the S 2p characteristic peak with sputtering time. It can be seen that, as the sputtering time increased, the doublet peak positions of S 2p showed gradual a chemical shift toward higher binding energies. The peak positions of S $2p_{3/2}$ and S $2p_{1/2}$ increased from 161.01 eV and 162.19 eV to 161.58 eV and 162.76 eV, respectively, while their FWHM values fluctuated around 1.7 eV. This result is in agreement with the binding energy of S $2p_{3/2}$ in Y_2S_3 (160.7–161.3 eV) and its FWHM (1.6 eV). Moreover, as the sputtering energy continued to increase, the binding energy of S $2p_{3/2}$ increased further until it approached the binding energy of S $2p_{3/2}$ (161.9 eV) in the ZnS substrate.

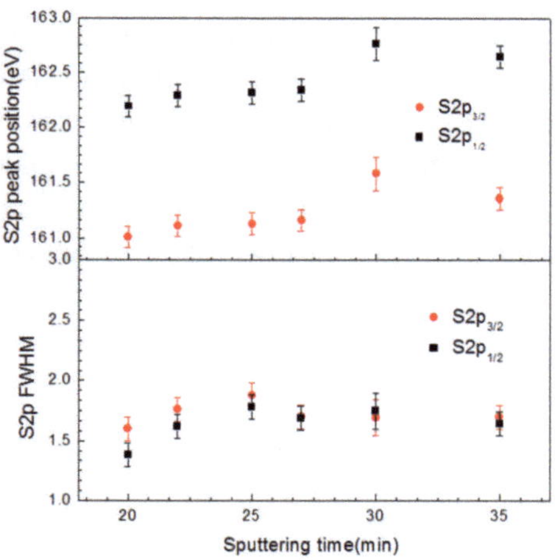

Fig. 9.50: Peak position and FWHM of interface S 2p under different sputtering times.

9.5.2 Interfacial adhesive strength

Adhesion is one of the most important mechanical properties of coating–substrate systems, which directly determines whether the thin film is suitable for practical applications. Although it is difficult to use the nanoscratch test to obtain a precize measurement of the coating–substrate interfacial adhesive strength, it is possible to perform highly precise measurements of changes in the scratch surface profile, lateral force, and friction coefficient before, during, and after scratching. These measurements can be used to determine the critical load for fracture and delamination, which gives us a semi-quantitative evaluation of coating–substrate binding and adhesion properties.

Fig. 9.51 shows the variations in the scratch surface profile and friction coefficient of the Y_2O_3 thin film within the scratch distance. The scratch surface profile is composed of three curves: pre-scanning, scratching, and post-scanning. During the experiment, an indenter was placed close to the sample surface. Then, a force of 0.1 mN was applied on the sample, and the indenter was moved over a distance of 700 µm on the sample surface for pre-scanning. The indenter was returned to its original position, and a force of 0.1 mN was applied for a distance of 100 µm. Subsequently, the vertical load applied by the indenter increased linearly and continuously from 0 to 150 mN over a distance of 500 µm, following which a force of 0.1 mN was applied for the last 100 µm. The indenter was returned to the original position again, and a force of 0.1 mN was applied to scratch over the groove to scan the residual depth of the groove. The profile curve at 0 indicates the thin-film surface; negative displacement in the *Y* direction indicates that the indenter had scratched into the thin film, and positive displacement indicates bumps and the accumulation of cracked chippings on the thin-film surface.

The surface profile shown in Fig. 9.51 is divided by two dotted lines into three zones, which correspond to the photograph of the scratch in Fig. 9.52 (a). In the first zone, the thin film only underwent elastic deformation. During this process, friction increased linearly, and the friction coefficient remained constant. As the scratch load increased, the residual scratch depth and width increased, but the thin film did not show fracture or delamination. When the indenter reached the intersection of the sec-

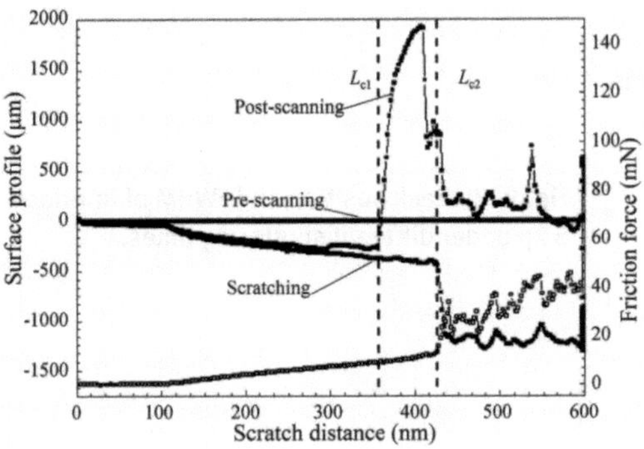

Fig. 9.51: Scratch surface profile and friction curves of Y_2O_3 thin film.

ond zone (at the first dotted line in Fig. 9.5 (a)), the scratching force exceeded the coating adhesive strength. The coating began to fracture and bulge, and the post-scanning curve showed sudden step changes. The SEM micrograph in Fig. 9.52 (b) indicates that parallel cracks in the coating expanded along the direction of scratching. When they reached a certain distance, angular cracks began to emerge; the angular cracks expanded at a certain angle from the scratching direction. The corresponding normal load at this point is known as the critical load L_{c1}, which indicates the critical point of coating fracture. As the scratch load increased further and reached the intersection between the second and third zones (at the second dotted line in the figure), friction increased sharply and drastic changes appeared in the scratching and post-scanning curves. This result indicates that the coating had completely fractured and was delaminated from the substrate.

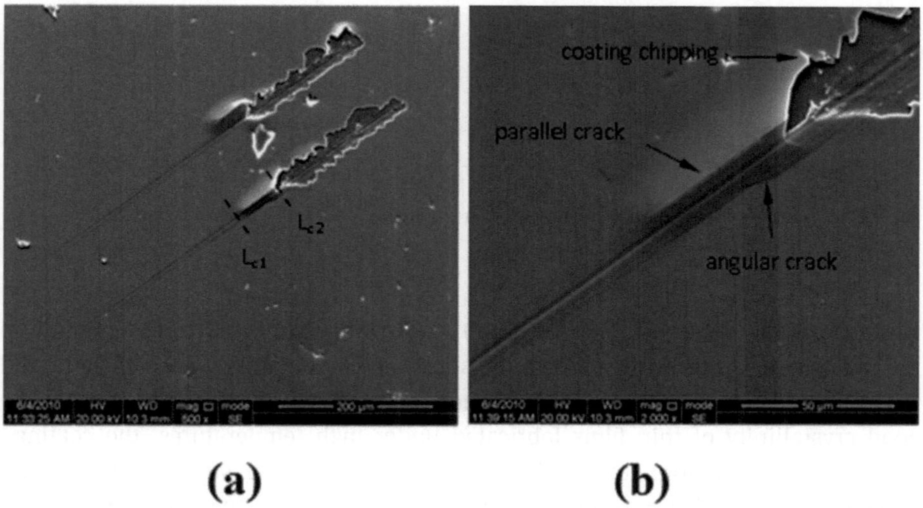

Fig. 9.52: SEM morphology of scratch surface: (a) overall scratch morphology; (b) local scratch morphology.

Fig. 9.53 shows the friction coefficients and maximum friction coefficients of Y_2O_3 thin films grown on ZnS substrates under different substrate biases and temperatures. Fig. 9.53 (a) and (b) show the adhesion coefficients obtained from nanoscratch tests of yttrium oxide thin films fabricated under different substrate biases at room temperature. It can be seen that substrate bias had a significant impact on the coating–substrate adhesive strength. When the substrate bias was low (0 V or −80 V), thin-film adhesion coefficients were comparable. When substrate was −160 V, the adhesion coefficient was increased.

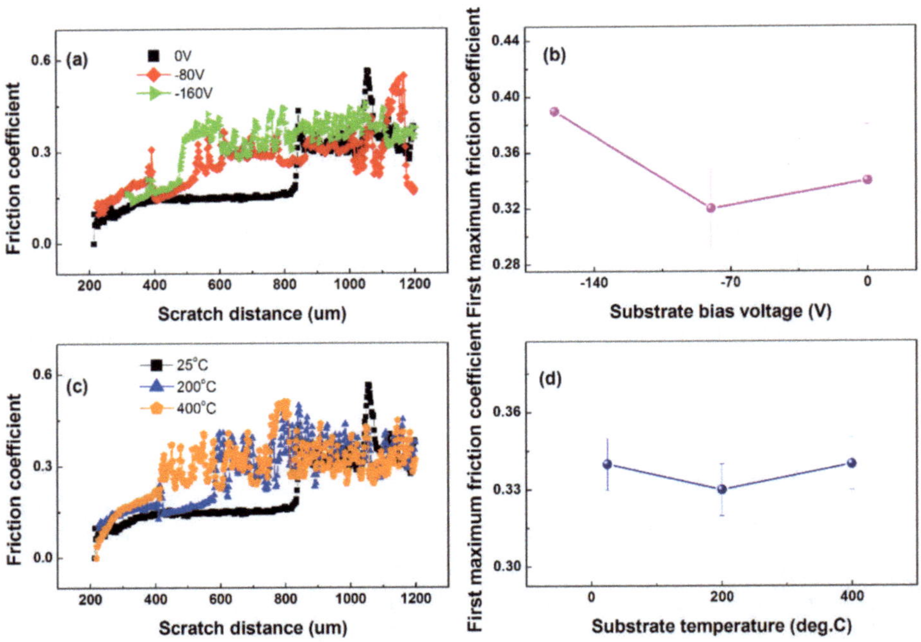

Fig. 9.53: Fiction coefficient and maximum friction coefficient under different substrate biases (a), (b), and temperatures (c), (d).

However, it can be seen from Fig. 9.53 (c) and (d) that the increase in temperature did not have a significant impact on the substrate-coating adhesive coefficient, and the values remained within a certain error range. This indicates that, despite the good crystallinity of thin films fabricated under high temperatures, the coating–substrate adhesive strength is relatively poor, whereas the substrate bias could enhance coating–substrate adhesive strength. Therefore, thin films with ultra-strong adhesion and superior optical and mechanical properties can be fabricated under the appropriate bias-assisted deposition conditions.

References

[1] Curtis C. Properties of yttrium oxide ceramics. Journal of the American Ceramic Society, 1957, 40: 274–378.

[2] Lefever R, Matsko J. Transparent yttrium oxide ceramics. Materials Research Bulletin, 1967, 2: 865–869.

[3] Nigara Y. Measurement of the optical constants of yttrium oxide. Japanese Journal of Applied Physics, 1968, 7: 404-408.

[4] Nigara Y, Ishigame M, Sakurai T. Infrared Properties of Yttrium Oxide. Journal of the Physical Society of Japan, 1971, 30: 453–458.

[5] Klemm KA, Patterson HS, Johnson LF, Moran MB. Protective optical coatings for diamond infrared windows. SPIE's 1994 International Symposium on Optics, Imaging, and Instrumentation: International Society for Optics and Photonics, 1994: 347–56.
[6] Korenstein R, Cremin P, Varitimos T, Tustison RW. Optical properties of durable oxide coatings for infrared applications [C]. AeroSense 2003: International Society for Optics and Photonics, 2003: 169–78.
[7] Zinkevich M. Thermodynamics of rare earth sesquioxides. Progress in Materials Science, 2007, 52: 597–647.
[8] Repelin Y, Proust C, Husson E, Beny J. Vibrational spectroscopy of the C-form of yttrium sesquioxide. Journal of Solid State Chemistry, 1995, 118: 163–169.
[9] Saraiva M, Georgieva V, Mahieu S, Van Aeken K, Bogaerts A, Depla D. Compositional effects on the growth of Mg(M)O films. Journal of Applied Physics, 2010, 107: 034902-10.
[10] Strijckmans K, Leroy W, Degryse R, Depla D. Modeling reactive magnetron sputtering: Fixing the parameter set. Surface & Coatings Technology, 2012, 206: 3666–3675.
[11] Cheng XR, Qi ZM, Zhang GB, Zhou HJ, Zhang WP, Yin M. Growth and characterization of Y_2O_3 thin films. Physica B, 2009, 404: 146–149.
[12] de Rouffignac P, Park JS, Gordon RG. Atomic layer deposition of Y(2)O(3) thin films from yttrium tris(N, N-diisopropylacetamidinate) and water. Chemical Materials, 2005, 17: 4808–4814.
[13] Zhang SQ, Xiao RF. Yttrium oxide films prepared by pulsed laser deposition. Journal of Applied Physics, 1998, 83: 3842–3848.
[14] Cho MH, Ko DH, Jeong K, Whangbo SW, Whang CN, Choi S, et al. Structural transition of crystalline $Y_2 O_3$ film on Si (111) with substrate temperature. Thin Solid Films, 1999, 349: 266–269.
[15] Anders A. Discharge physics of high power impulse magnetron sputtering. Surface & Coatings Technology, 2011, 205: S1–S9.
[16] Adelhelm C, Pickert T, Balden M, Rasinski M, Plocinski T, Ziebert C, et al. Monoclinic B-phase erbium sesquioxide (Er_2O_3) thin films by filtered cathodic arc deposition. Scripta Materialia, 2009, 61: 789–792.
[17] Ramana CV, Mudavakkat VH, Bharathi KK, Atuchin VV, Pokrovsky LD, Kruchinin VN. Enhanced optical constants of nanocrystalline yttrium oxide thin films. Applied Physics Letters, 2011, 98: 031905.
[18] Pearce SJ, Parker G J, Charlton M D B, Wilkinson JS. Structural and optical properties of yttrium oxide thin films for planar wave guiding applications. Journal of Vacuum Science & Technology A, 2010, 28: 1388–1392.
[19] Nishide T, Shibata M. Orientation and surface properties of sol-gel derived Y_2O_3 films. Journal of Sol-Gel Science and Technology, 2001, 21: 189–193.
[20] Pulker H. Characterization of optical thin films. Applied Optics, 1979, 18: 1969–1977.
[21] Oliver WC, Pharr GM. An improved technique for determining hardness and elastic modulus using load and displacement sensing indentation experiments. Journal of Materials Research, 1992, 7: 1564–1583.
[22] Zhu JQ, Zhu YK, Shen WX, Wang YJ, Han JC, Tian G, et al. Growth and characterization of yttrium oxide films by reactive magnetron sputtering. Thin Solid Films, 2011, 519: 4894–4898.
[23] Barve SA, Jagannath, Mithal N, Deo MN, Chand N, Bhanage BM, et al. Microwave ECR plasma CVD of cubic Y_2O_3 coatings and their characterization. Surface & Coatings Technology, 2010, 204: 3167–3172.
[24] Chambers J, Parsons G. Physical and electrical characterization of ultrathin yttrium silicate insulators on silicon. Journal of Applied Physics, 2001, 90: 918–933.
[25] Wang Z, Wu J, Fang Q, Zhang JY. Photoemission study of interfacial reactions during annealing of ultrathin yttrium on SiO_2/Si(1 0 0). Applied Surface Science, 2005, 239: 464–469.

[26] Craciun V, Howard J, Lambers E, Singh R, Craciun D, Perriere J. Low-temperature growth of Y_2O_3 thin films by ultraviolet-assisted pulsed laser deposition. Applied Physics A, 1999, 69: S535–S538.

[27] Safi I. Recent aspects concerning DC reactive magnetron sputtering of thin films: a review. Surface & Coatings Technology, 2000, 127: 203–218.

[28] Berg S, Nyberg T. Fundamental understanding and modeling of reactive sputtering processes. Thin Solid Films, 2005, 476: 215–230.

[29] Depla D, Strijckmans K, Degryse R. The role of the erosion groove during reactive sputter deposition. Surface & Coatings Technology, 2014, 258: 1011–1015.

[30] Severin D, Kappertz O, Kubart T, Nyberg T, Berg S, Pflug A, et al. Process stabilization and increase of the deposition rate in reactive sputtering of metal oxides and oxynitrides. Applied Physics Letters, 2006, 88: 161504.

[31] Strijckmans K, Depla D. A time-dependent model for reactive sputter deposition. Journal of Physics D Applied Physics, 2014, 47: 235302.

[32] Niu D, Ashcraft RW, Chen Z, Stemmer S, Parsons GN. Chemical, physical, and electrical characterizations of oxygen plasma assisted chemical vapor deposited yttrium oxide on silicon. Journal Electrochemical Society, 2003, 150: 102–109.

[33] Durand C, Dubourdieu C, Vallee C, Loup V, Bonvalot M, Joubert O, et al. Microstructure and electrical characterizations of yttrium oxide and yttrium silicate thin films deposited by pulsed liquid-injection plasma-enhanced metal-organic chemical vapor deposition. Journal of Applied Physics, 2004, 96: 1719–1729.

[34] Moulder JF, Stickle WF, Sobol PE, Bomben KD. Handbook of X-ray Photoelectron Spectroscopy. Eden Prairie MN, Perkin Elmer; 1992.

[35] Ji W, Lin J, Tang S, Du YW. Preparation of ZnS nanoparticles by ultrasonic radiation method. Applied Physics A, 1998, 66: 639–641.

[36] Chen M, Wang X, Yu Y, Pei Z, Bai X, Sun C, et al. X-ray photoelectron spectroscopy and auger electron spectroscopy studies of Al-doped ZnO films. Applied Surface Science, 2000, 158: 134–140.

[37] Islam MN, Ghosh T, Chopra K, Acharya H. XPS and X-ray diffraction studies of aluminum-doped zinc oxide transparent conducting films. Thin Solid Films, 1996, 280: 20–25.

[38] Quinn J, Lemoine P, Maguire P, McLaughlin J. Ultra-thin tetrahedral amorphous carbon films with strong adhesion, as measured by nanoscratch testing. Diamond and Related Materials, 2004, 13: 1385–1390.

10 Infrared transparent conductive oxide thin film

10.1 Overview

In the development of electro-optical systems, transparent conducting materials (TCM) have become an indispensable part of aerospace electro-optical systems, serving as the defrosting/demisting heater material and radar-cross-section (RCS) blocker material of the electro-optical windows. In order to reduce the weight of aerospace vehicles, the trend of function integration of components has exerted increasingly strict requirements on the electro-optical detection system. Not only must the highly integrated multi-spectral electro-optical detection system realize the highly efficient transmission of wide waveband, but the window is also required to have a relatively low sheet resistance (i.e. high carrier concentration and mobility) to shield electromagnetic waves and thereby achieve stealth. Ideal infrared electro-optical systems detect infrared signals using the front large diameter window. However, electromagnetic radiation can also easily traverse the window and affect the sensitive infrared detector with electromagnetic interference. Moreover, according to Drude's free electron theory, when the vibrational frequencies of photons and free electrons are close in the infrared range, collisional scattering is likely to occur, leading to obvious absorption. Therefore, we know that, intrinsically, it is difficult to balance the requirements of broadband, multispectral, high-efficiency transmission (especially high efficiency transmission in the infrared range) and the performance of electromagnetic shielding and stealth.

Currently, there are mainly two conventional approaches for preparing infrared transparent conductive films: one is the fabrication of a metal/metal-mesh thin film on the external window surface, while the other is the fabrication of n-type or p-type doped oxide semiconductor thin films. Both have their advantages and disadvantages. Specifically, metal or metal-mesh films possess the merit of excellent conductivity but shortcomings of soft texture and vulnerability to rain erosion and sand erosion. In addition, the superior conductivity of metal results in high carrier concentration and low mobility, and it is also the reason for high reflection and absorption in the required band.

In contrast, the coefficients of thermal expansion of n-type and p-type doped oxide semiconductor coatings are close to those of the substrates, and their physical and chemical properties are more stable than those of metals. However, it should be noted that outstanding electro-optical properties can only be obtained by heating the substrates, making it difficult for these coatings to meet the application requirements of heat-sensitive substrates.

To address the issues existing in the research of infrared transparent conductive films described above, in this chapter, we report the following work, which was conducted previously. First, we fabricated mid-infrared transparent conductive crys-

https://doi.org/10.1515/9783110489514-013

talline indium oxide (In_2O_3) and Sn-doped indium oxide films (In_2O_3:Sn) with preferential orientation using plasma bombardment magnetron sputtering at room temperature and explored the relationship between the negative bias voltage and material structure and electro-optical performance. Subsequently, based on the principle of element doping, n-type Ru-doped yttrium oxide coatings with different ruthenium doping amounts were prepared at different substrate temperatures using the method of double-target magnetron co-sputtering; in this manner, far-infrared transparent conductive oxide coatings were obtained.

10.2 Films prepared by plasma bombardment magnetron sputtering

10.2.1 Microstructure of In_2O_3 and In_2O_3:Sn films

10.2.1.1 Morphology

The morphology of In_2O_3 and In_2O_3:Sn films is closely related to their electro-optical performance; therefore, it is necessary to characterize the morphology of these thin films. In this work, we adopt the Veeco atomic force microscope (AFM) to characterize the morphology of individual sections of an area of 2 µm × 2 µm on the fabricated sample films.

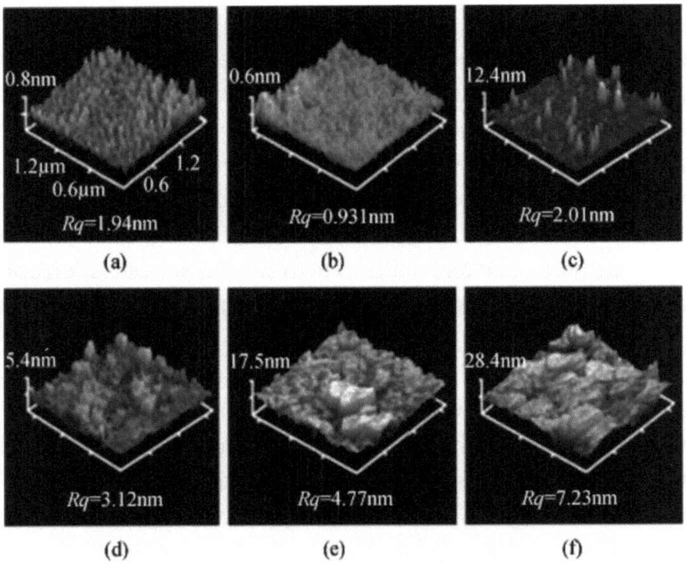

Fig. 10.1: Morphology of In_2O_3 films fabricated using the plasma bombardment magnetron sputtering method with different plasma energy values: (a) magnetron sputtering; (b) |−500 V|; (c) |−600 V|; (d) |−700 V|; (e) |−800 V|; (f) |−900 V|.

Fig. 10.1 shows the morphology of In_2O_3 films deposited directly using magnetron sputtering as well as using plasma bombardment magnetron sputtering involving different negative bias voltages. It can be observed that, for the In_2O_3 film fabricated under a relatively low negative bias voltage ($|V_p| < |-600\ V|$), the $2\ \mu m \times 2\ \mu m$ area exhibits a compact and uniform surface without cavities or island structures. However, with the increase of negative bias voltage ($|V_p| > |-600\ V|$), cones and spatial island structures occur and gradually spread over the entire surface. When $|V_p|$ further increases to $|-800\ V|$, the island morphology becomes dominant on the entire sample surface with an obvious occurrence of gullies, whereas the cones gradually disappear.

It can be seen from Fig. 10.2 that, with the increase of negative bias voltage, the size of particles on the surface steadily increases. Particle morphology can hardly be observed on the surface in Fig. 10.2 (a) while small homogeneous granular particles occur on the surface when the negative bias voltage increases to $|-600\ V|$ (shown in Fig. 10.2c). At $|V_p| = |-800\ V|$, particle growth is rather noticeable, showing good agreement with the surface morphological characteristics observed using AFM in Fig. 10.1.

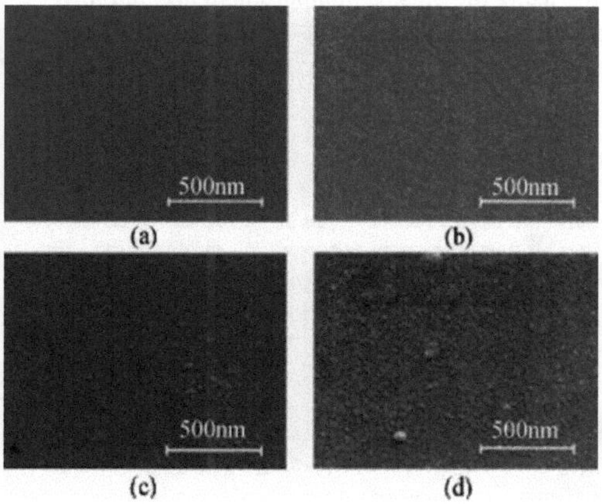

Fig. 10.2: SEM of In_2O_3 films fabricated using the plasma bombardment magnetron sputtering method with different levels of plasma energy: (a) magnetron sputtering; (b) $|-400\ V|$; (c) $|-600\ V|$; (d) $|-800\ V|$.

Fig. 10.3 demonstrates the morphology of In_2O_3:Sn films deposited using the plasma bombardment magnetron sputtering method with different values of plasma energy under different negative bias voltages, which exhibits phenomena similar to those observed in the In_2O_3 sample films. However, the morphology mainly features cone structures without a distinct occurrence of island structures.

We can see from the scanned morphology shown in Fig. 10.4 that, with the increase of negative bias voltage, the size of particles on the surface shows an increasing trend, and the distribution of these particles changes from a dispersed pattern to an aggregated pattern. These two variations are similar to the changes in surface morphology observed in the In_2O_3 films. Compared to In_2O_3 films, In_2O_3:Sn films experience a more significant change in particle size on the coating surface. Therefore, it can be concluded that the surface morphology has a close relationship with the magnitude of

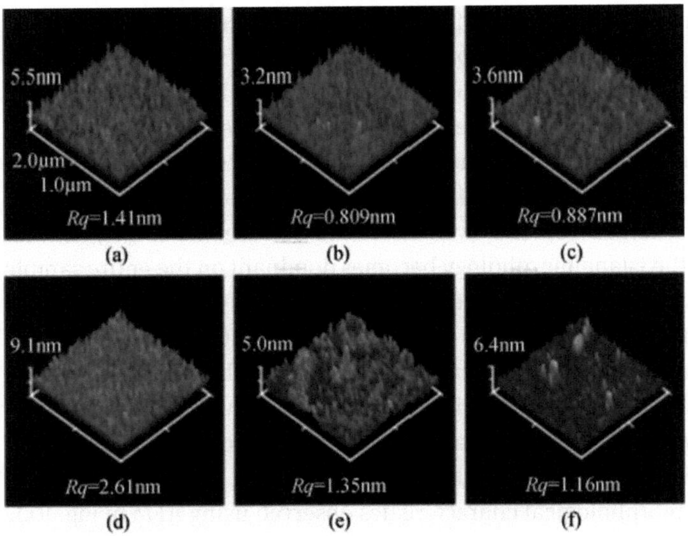

Fig. 10.3: Morphology of In_2O_3:Sn films fabricated using the plasma bombardment method with different values of plasma energy: (a) without plasma bombardment; (b) |−400 V|; (c) |−500 V|; (d) |−600 V|; (e) |−700 V|; (f) |−800 V|.

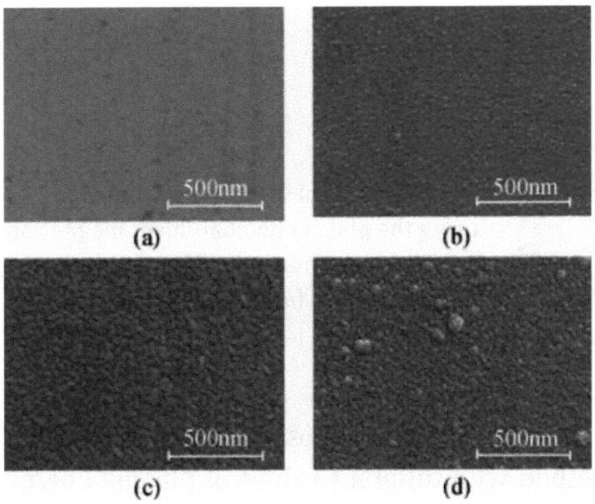

Fig. 10.4: Morphology of In_2O_3:Sn films fabricated using the plasma bombardment magnetron sputtering method with different values of plasma energy: (a) without plasma bombardment; (b) |−400 V|; (c) |−600 V|; (d) |−800 V|.

argon-ion energy in the process of plasma bombardment. Under a relatively low negative bias voltage, the primary function of argon-ion bombardment is to enhance the atomic mobility along the surface and eliminate the surface defects for generating a uniform and compact surface. Under the bombardment of a high argon-ion energy, the primary function is to exert irritation damage to produce the island morphology with a great scale of gullies on the surface of In_2O_3 films, causing a significant difference in surface morphology between In_2O_3 and In_2O_3:Sn films.

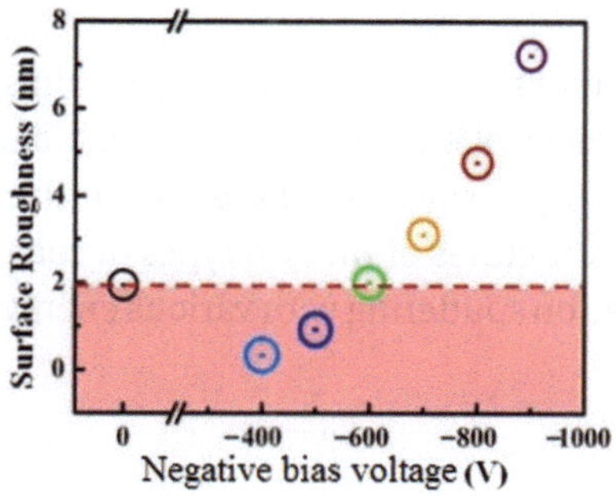

Fig. 10.5: Relationship between surface roughness and negative bias voltage of In_2O_3 film.

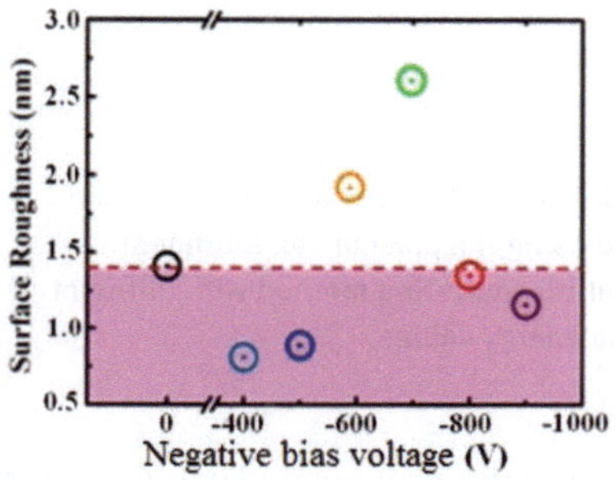

Fig. 10.6: Relationship between surface roughness and negative bias voltage of In_2O_3:Sn film.

Fig. 10.5 shows the relationship between the surface roughness of the In_2O_3 film and negative bias voltage. It can be observed that, with the increase of negative bias voltage, the roughness of the In_2O_3 film monotonously increases from 0.314 nm to 7.23 nm. When the negative bias voltage is less than $|-600\ \mathrm{V}|$, the surface roughness is less than that of the sample film fabricated using magnetron sputtering. However, when the negative bias voltage exceeds $|-700\ \mathrm{V}|$, the surface roughness of the thin film dramatically increases.

In the contrast, as the $|V_p|$ increases, the surface roughness of the In_2O_3:Sn film first decreases, then increases, and finally decreases again. At $|V_p| = |-600\ \mathrm{V}|$, the surface roughness reaches a peak of ≈ 2.6 nm. This is most likely caused by the grain growth generated during the process of high-energy argon plasma bombardment. Based on the variation in surface morphology of these two materials under plasma bombardment, we can conclude that

(1) plasma bombardment magnetron sputtering is a technique that can accurately and efficiently monitor the surface roughness of coating materials, and
(2) a significant variation in surface roughness is likely caused by the grain growth under the impact of high-energy plasma bombardment.

10.2.1.2 Crystalline structure

It can be learned from the surface morphological analysis of In_2O_3 and In_2O_3:Sn films that, under plasma bombardment with different values of energy, both types of films show remarkable variations in the internal crystalline structure. Therefore, in this section, we systematically analyze the crystalline structures of the two types of coatings fabricated using plasma bombardment magnetron sputtering with various energy levels.

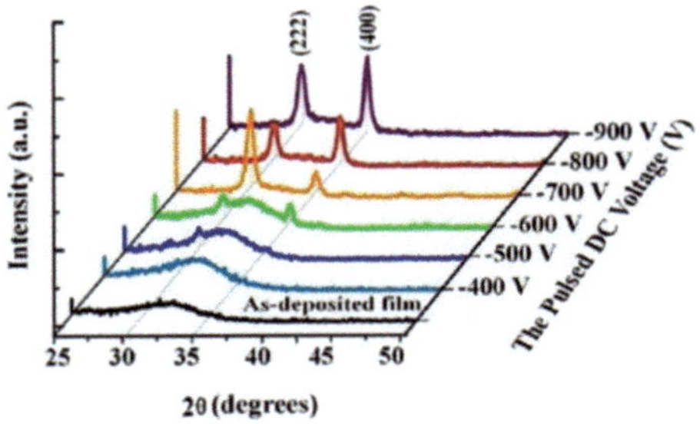

Fig. 10.7: XRD structures of In_2O_3 films fabricated using the plasma bombardment magnetron sputtering method with different plasma energy values.

Fig. 10.7 illustrates the XRD crystalline structure of In_2O_3 films deposited using plasma bombardment magnetron sputtering with various energy values. Under a relatively low negative bias voltage ($|V_p| < |-500\ V|$), the crystalline structure of the In_2O_3 film is similar to that of the In_2O_3 film deposited directly using magnetron sputtering; both films have amorphous structures. When $|-500\ V| < |V_p| < |-700\ V|$, argon plasma with a certain amount of kinetic energy provides energy for enhancing the atomic shift along the coating surface during the fabrication of the In_2O_3 film, which helps the In_2O_3 atoms near the surface to move into the lattice positions of crystalline In_2O_3 under the bombardment. In the meantime, this amount of energy also plays a role in driving more free oxygen atoms to participate in the construction of In_2O_3 film structures, leading to the (222) preferential orientation. When $|V_p| > |-700\ V|$, the bombardment of argon ions with a relatively high energy on the thin film gradually changes the preferential orientation of the In_2O_3 film from (222) to (400). Irradiation damage is the major process occurring during the bombardment of high-energy argon ions on In_2O_3 films. Compared to the (222) crystallographic plane, (400) shows a stronger bombardment-resistance effect. Therefore, after the bombardment of high-energy argon ions, the In_2O_3 film exhibits a preferential orientation of ⟨400⟩. A similar phenomenon is observed in the high-energy radio-frequency (HERF) fabrication of In_2O_3:Sn films.

Fig. 10.8 shows the XRD crystalline structures of In_2O_3:Sn films fabricated using the plasma bombardment magnetron sputtering method with different values of plasma energy. It is found that, at a relatively low negative bias voltage ($|V_p| < |-500\ V|$), the In_2O_3:Sn film is amorphous, has and its structure is not significantly different from that of the In_2O_3 film fabricated directly using magnetron sputtering. When $|-500\ V| < |V_p| < |-700\ V|$, positively charged argon ions are accelerated to impact the coating surface under the action of the negative bias voltage. The kinetic energy of the positively charged argon ions is converted partially into internal energy, as well as energy for atomic motion along the surface. Therefore, the atomic mobility along the surface is enhanced, and the surface crystallization of In_2O_3:Sn indicates the (222) crystallographic plane. Meanwhile, Sn atoms in the film perfectly enter the doping positions, with no obvious occurrence of the SnO_2 peak in Fig. 10.8. When $|V_p| > |-700\ V|$, under the bombardment of argon ions, there is no transition of crystal orientation, but the decrease of the (222) peak intensity can be observed. This indicates that the doping causes an increase of energy required for the transition of crystal orientation. However, the energy provided by Ar^+ at $|V_p| = |-900\ V|$ can hardly satisfy the requirement of preferential orientation transition, and instead, the irradiation damage of the soft crystal orientation ⟨222⟩ is induced. The peak drop is shown in Fig. 10.8.

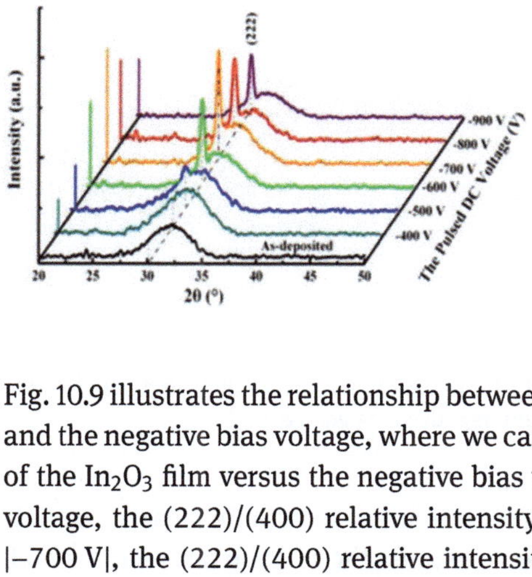

Fig. 10.8: XRD structures of In_2O_3:Sn films fabricated using the plasma bombardment magnetron sputtering method with different values of plasma energy.

Fig. 10.9 illustrates the relationship between the relative peak intensity of (222)/(400) and the negative bias voltage, where we can see the change of preferential orientation of the In_2O_3 film versus the negative bias voltage. With the increase of negative bias voltage, the (222)/(400) relative intensity shows an increasing trend. When $|V_p| = |-700\ V|$, the (222)/(400) relative intensity reaches its peak value with the highest degree of preferential orientation of ⟨222⟩. With a further increase of negative bias voltage ($|V_p| > |-700\ V|$), the preferential orientation steadily changes from ⟨222⟩ to ⟨400⟩. At $|V_p| = |-900\ V|$, the (222)/(400) relative intensity falls to the minimum, i.e., the preferential orientation along ⟨400⟩ reaches the highest degree.

The relationship between the grain size and negative bias voltage is studied further. The grain size on the (400) crystallographic plane can be calculated using the Scherrer equation, which is written as

$$D = \frac{K\lambda}{B\cos(\theta)}, \tag{10.1}$$

where D is the grain size (nm), K a constant (usually, $K = 0.89$ is considered for the calculation of In_2O_3/In_2O_3:Sn films according to the literature), λ the $K\alpha$ incident wave-

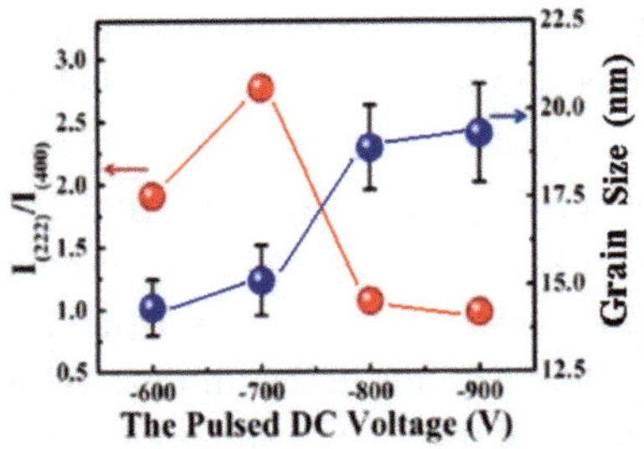

Fig. 10.9: Variations of the relative intensity ratio (222)/(400) of diffraction peaks in In_2O_3 film and the grain size calculated based on the (400) crystallographic plane versus the negative bias voltages.

length of Cu (1.54 Å), θ the Bragg angle of diffraction (°), and B the full width at half maximum (FWHM) of the diffraction peak.

It can be observed from Fig. 10.10 that the grain size of the In_2O_3:Sn film gradually increases with the increase of negative bias voltage. When $|V_p| = |-600\ V|$, the grain size calculated based on the (400) crystallographic plane is approximately 13.5 nm; when $|V_p| = |-900\ V|$, the grain size increases to almost 20 nm, which is in agreement with the pattern of the XRD image.

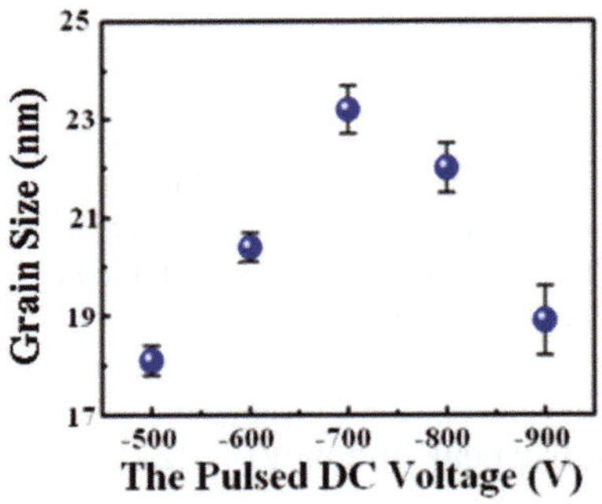

Fig. 10.10: Grain size of In_2O_3:Sn films fabricated under different negative bias voltages.

In order to characterize the pattern of preferential orientation more clearly, the pole figures (Fig. 10.11 and 10.12) are adopted here to express the relation between the preferential orientations of In_2O_3 and In_2O_3:Sn films under various negative bias voltages and to obtain the angle between the two orientations.

Fig. 10.11 demonstrates the crystallization orientations of (222) and (400) of In_2O_3 films fabricated under negative bias voltages of $|V_p| = |-500\ V|$ and $|V_p| = |-800\ V|$. It can be seen from the tests that, at $|V_p| = |-500\ V|$, the diffraction intensity is rather concentrated at the pole of the (400) pole figure, indicating that the (400) crystallographic plane is parallel to the substrate surface to a certain extent. Meanwhile, it is known from the (222) pole figure that the maximum values of diffraction intensity occur at the pole, suggesting that the (222) crystallographic plane is parallel with the substrate surface as well. For another group of diffraction rings on the (222) pole figure, a relatively wide diffraction ring can be observed at a diffraction angle of approxi-

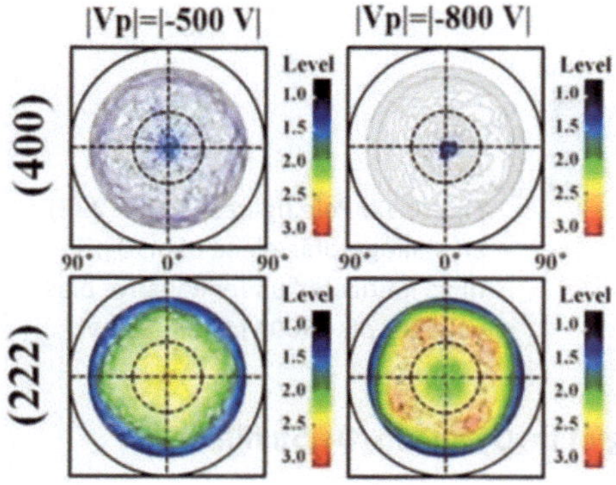

Fig. 10.11: Pole figures on the (222) and (400) crystallographic planes of In_2O_3 films fabricated under negative bias voltages of |−500 V| and |−800 V|.

mately 60°. This diffraction ring corresponds to the {110} intraplate texture. Compared with the situation at $|V_p|$ = |−500 V|, when the negative bias voltage is |−800 V|, most of the diffraction-intensity points are concentrated at the pole in the (400) pole figure with only a few points distributed in the proximity of the pole. This shows that, under the bombardment of high-energy plasma, the ⟨400⟩ crystal orientation is strictly perpendicular to the substrate surface. As shown in the (222) pole figure, except for the relatively strong diffraction spots at the pole, four obliquely symmetric diffraction spots appear on the ring around the 60° diffraction angle. Therefore, it is proved that, apart from being parallel with the substrate surface, the (222) crystallographic plane also has an increased preferential orientation of intraplate texture along the {110} crystallographic plane family, which is in agreement with the previous XRD results.

Fig. 10.12 shows the (222) orientation of In_2O_3:Sn films fabricated under negative bias voltages of $|V_p|$ = |−500 V| and $|V_p|$ = |−800 V|. Tests show that there only exists the texture of (222) orientation in the In_2O_3 films fabricated under the two negative bias voltages. When $|V_p|$ = |−500 V|, the diffraction intensity is relatively concentrated at the pole of the (222) pole figure, suggesting that the (222) crystallographic plane is parallel with the substrate surface to a certain degree. In addition, compared with the (222) pole figure of the In_2O_3 film under identical negative bias voltage, this pole figure lacks the other group of diffraction rings at a diffraction angle of approximately 60°. This tells us that the doped Sn atoms can restrain the shift of oxygen and indium atoms during the process of argon plasma bombardment, making it difficult to form the intraplate texture.

Compared with the situation of $|V_p|$ = |−500 V|, when the negative bias voltage is |−800 V|, it can be seen from the (222) pole figure that most of the diffraction intensity points are concentrated on the diffraction ring around 60°. This is because under the bombardment of plasma with a relatively high energy the particles in motion overcome the energy required for forming the intraplate texture and, therefore the preferential orientation of texture in the {110} crystallographic plane family degree is enhanced.

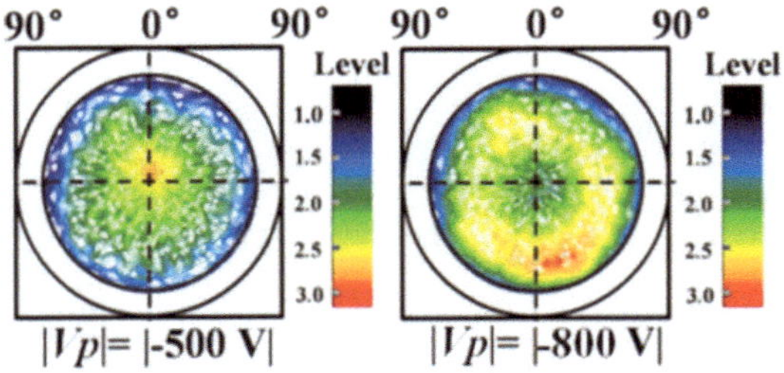

Fig. 10.12: Pole figures on the (222) crystallographic plane of In_2O_3:Sn films fabricated under negative bias voltages of |−500 V| and |−800 V|.

10.2.2 Composition and chemical bonding of In_2O_3 and In_2O_3:Sn films

10.2.2.1 Composition characterization

We can adjust the In/Sn/O content ratio in In_2O_3 and In_2O_3:Sn films by changing the negative bias voltage. Energy-dispersive x-ray spectrometry (EDS) was used to demarcate the Sn/In content ratio in the In_2O_3:Sn films. However, for In_2O_3 films, EDS cannot efficiently separate the oxygen content in the film and substrate owing to the interference from oxygen existing in the substrate. Therefore, we can only adopt x-ray photoelectron spectrometry (XPS) to implement the associated analysis. Fig. 10.13 shows the SEM surface morphology of a selected area of an In_2O_3:Sn film fabricated on a glass substrate under a negative bias voltage of |−600 V| and the typical elemental-energy spectral peaks of Sn and In of the film-surface EDS. It can be seen from the EDS anal-

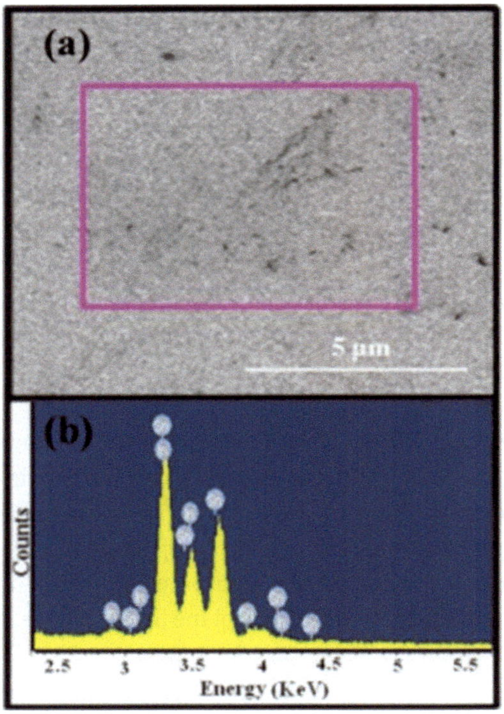

Fig. 10.13: (a) SEM morphology of EDS area on the In_2O_3:Sn film surface; (b) EDS analysis of In and Sn spectral peaks of the In_2O_3:Sn film.

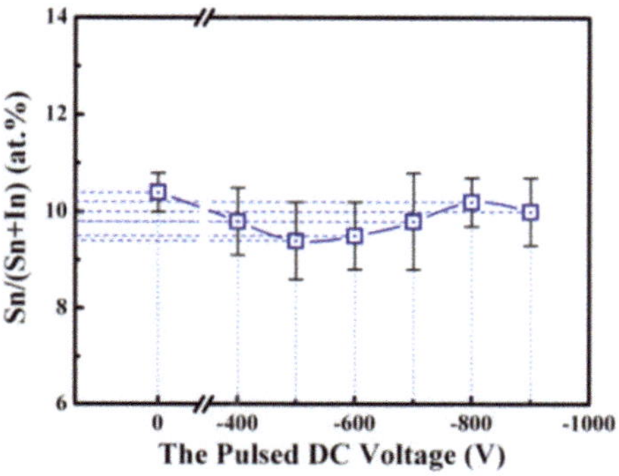

Fig. 10.14: Sn/(Sn + In) atomic percentage of In_2O_3:Sn films fabricated under different negative bias voltages.

ysis that, as the negative bias voltage changes ($|-400\ V| < |V_p| < |-900\ V|$), there is no significant variation in the Sn:In atomic ratio in the In_2O_3:Sn film fabricated using plasma bombardment magnetron sputtering, and the variation that exists can be considered to be within the error range (Fig. 10.14).

10.2.2.2 Chemical bonding of elements

From structure and composition analysis, we can obtain the In/Sn/O atomic concentration ratio of In_2O_3 and In_2O_3:Sn films deposited using plasma bombardment magnetron sputtering with various energy levels. Corresponding chemical bonding can be affected by the plasma bombardment as well. In this section, we study the chemical bonding of major elements in In_2O_3 and In_2O_3:Sn films by controlling the plasma bombardment energy with different negative bias voltages. The variation in chemical state of major elements in the In_2O_3 and In_2O_3:Sn films will be measured and demarcated using the XPS method.

Fig. 10.15 illustrates typical XPS full-scan spectra of In_2O_3 and In_2O_3:Sn films fabricated using conventional magnetron sputtering. Characteristic peaks of In/Sn/O/C elements can be clearly observed in the In_2O_3 and In_2O_3:Sn film specimens, whereas there are no characteristic peaks for other elements. Therefore, it can be preliminarily determined that the content of impurity elements in the specimens is rather low. Meanwhile, the occurrence of the weak C peak is mainly due to the fact that the exposure of the specimen in air can cause the coating surface to absorb the C impurity to a certain degree. In the In_2O_3 and In_2O_3:Sn films, the 3d spectral peaks of In, Sn, and O have the strongest intensities. Therefore, when conducting further narrow-area scanning and analysis of chemical bonding, we mainly scan the 3d-peak binding energy zones of In, O, and Sn.

First, we analyze the indium element in the In_2O_3 and In_2O_3:Sn films. Fig. 10.16 shows the In 3d core-level spectra of In_2O_3 and In_2O_3:Sn films fabricated under different negative bias voltages ($|-500\ V|$ and $|-800\ V|$). The In 3d spectra of In_2O_3 and In_2O_3:Sn films include two spectral peaks, i.e., the $3d_{5/2}$ peak located at ≈ 444.4 eV and the $3d_{3/2}$ peak at ≈ 452.0 eV.

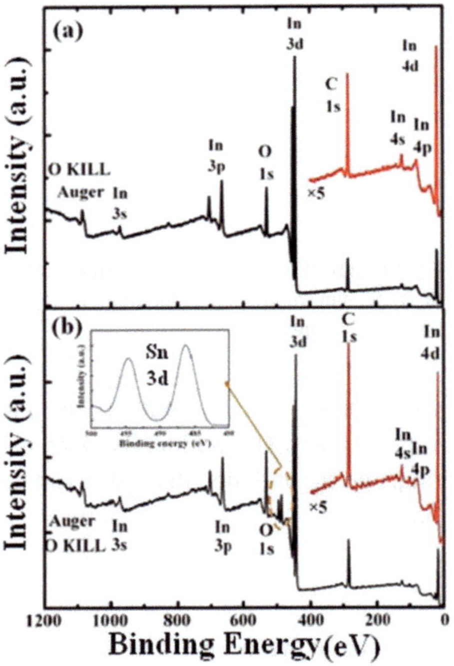

Fig. 10.15: Typical XPS full-scan spectra of (a) In_2O_3 film and (b) In_2O_3:Sn film fabricated using magnetron sputtering.

The existence of the In double peak ($3d_{5/2}$ and $3d_{3/2}$ peaks) can be observed in Fig. 10.16, which is the result of orbital electron splitting subsequent to the spin process. Among the two peaks, the one with smaller azimuthal quantum number generally has higher binding energy after the spin splitting, whereas the one with larger azimuthal quantum number has lower binding energy. Moreover, the intensity of the latter peak is greater than that of the former. Similar phenomena can

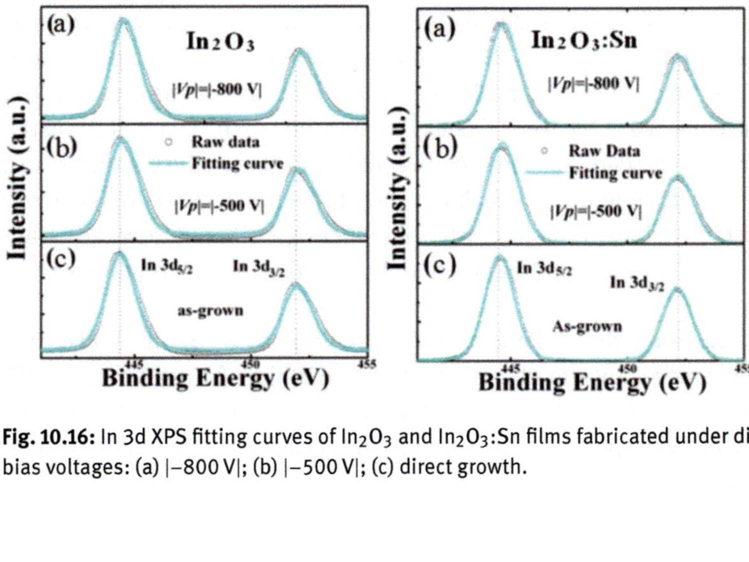

Fig. 10.16: In 3d XPS fitting curves of In_2O_3 and In_2O_3:Sn films fabricated under different negative bias voltages: (a) |−800 V|; (b) |−500 V|; (c) direct growth.

be observed in transition metals such as Cd and Y. As shown in Fig. 10.16, with the increase of negative bias voltage, the position of the In spectral peak exhibits a slight shift (approximately 0.3 eV), possibly because the high-energy plasma bombardment promotes the gradual binding of unsaturated In–O bonds and the new free oxygen enhances the binding of In–O bonds, making the binding bonds shift towards the high-energy side. In order to verify the conjecture, we will further analyze the binding energy of oxygen below. In the meantime, we also observed that the relative position of $3d_{5/2}$ and $3d_{3/2}$ peaks essentially remains constant, which is consistent with the results in the literature.

Fig. 10.17 and Tab. 10.1 show the relative intensity ratio of the 3d double peak and the FWHM of the XPS fitting curve for In in In_2O_3 and In_2O_3:Sn films, respectively, fabricated under different negative bias voltages. With the increase of negative bias voltage, the FWHM of the In 3d spectral peak remained constant at approximately 1.5, and the relative intensity ratio of the $3d_{5/2}$ and $3d_{3/2}$ peaks also essentially stayed constant. Therefore, the fitting of the In 3d peak can be considered reasonable. According to Fig. 10.16 and 10.17, for In_2O_3 and In_2O_3:Sn films fabricated under plasma bombardment with various values of energy, the In 3d binding-energy variation can be preliminarily regarded as the result of further binding between indium ions and free oxygen inside the coating under the action of plasma bombardment.

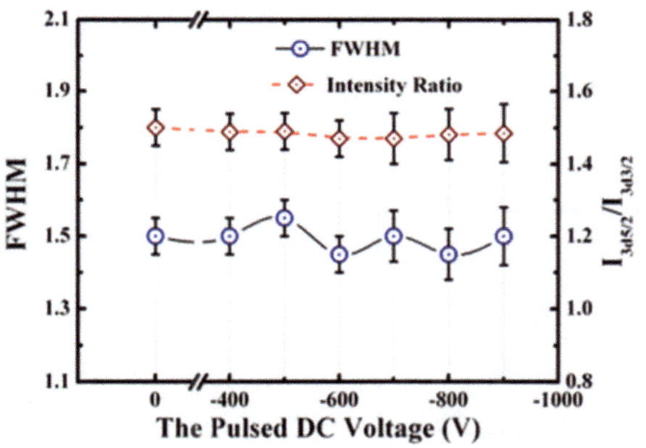

Fig. 10.17: Variation of relative intensity ratio and FWHM of $3d_{5/2}$ and $3d_{3/2}$ peaks for In_2O_3 films over V_p.

Tab. 10.1: Relative intensity ratio and FWHM of $3d_{5/2}$ and $3d_{3/2}$ peaks for In_2O_3:Sn films deposited under different V_p values.

V_p (V)	0 V	−400 V	−500 V	−600 V	−700 V	−800 V	−900 V
$I_{3d\,5/2}/I_{3d\,3/2}$	1.48	1.50	1.47	1.48	1.52	1.47	1.49
FWHM ($I_{3d\,5/2}$)	1.49	1.50	1.48	1.47	1.51	1.50	1.50

Then, we performed high-resolution XPS analysis in the Sn $3d_{5/2}$ bond energy zone. Fig. 10.18 illustrates the Sn 3d fitting curves of In_2O_3:Sn films fabricated under the conditions of $|V_p| = |0\ V|$, $|V_p| = |-500\ V|$, and $|V_p| = |-800\ V|$, respectively. It can be seen that two peaks exist in the fitting curves at 486.2 eV and 487.1 eV, respectively. These two peaks correspond to two different oxidation states, i.e., Sn^{2+} (around 486.2 eV) and Sn^{4+} (around 487.1 eV). We can acknowledge from the figure that, with the increase of negative bias voltage, the Sn^{4+} concentration gradually increases while the Sn^{2+} peak shows a significant decrease. This is because the energy of Ar^+ plasma bombardment on the specimen surface steadily increases over the increase of negative bias voltage, driving the Sn^{2+} ions doped in the film to lose electrons and transform into the Sn^{4+} upper state under the bombardment of Ar^+ ions. The transformation of Sn^{2+} ions into Sn^{4+} ions can also be regarded as a position-number increase of Sn^{4+} ions replacing In^{3+} ions, which promotes the occurrence of more free electrons in the entire coating.

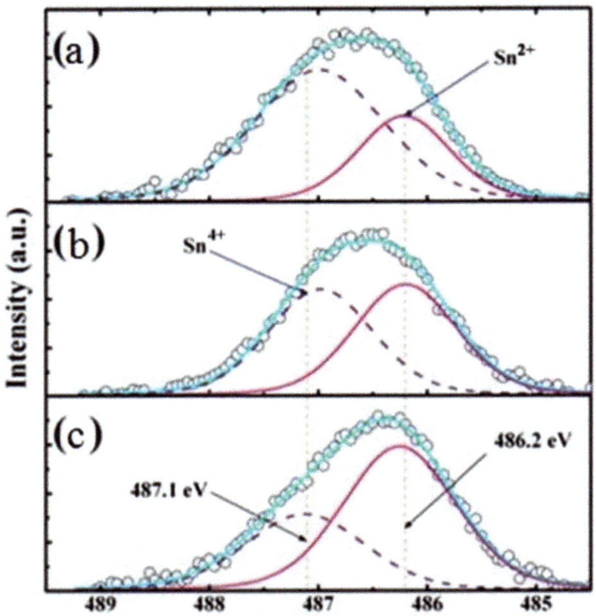

Fig. 10.18: Sn $3d_{5/2}$ XPS fitting curves of In_2O_3:Sn films fabricated under different negative bias voltages: (a) $|V_p| = |-800\ V|$; (b) $|V_p| = |-500\ V|$; (c) fabricated using magnetron sputtering.

Fig. 10.19 shows the I^{2+}_{Sn}/I^{4+}_{Sn} ratio and Sn^{4+} FWHM of In_2O_3:Sn films fabricated under different negative bias voltages. It can be clearly observed that, with the increase of negative bias voltage, the I^{2+}_{Sn}/I^{4+}_{Sn} ratio steadily decreases from 1.75 to 0.35, and the Sn^{4+} FWHM narrows from 1.2 to 1.0. This proves that the plasma bombardment process is beneficial to the transformation of low-valence Sn^{2+} ions into high-valence Sn^{4+} ions.

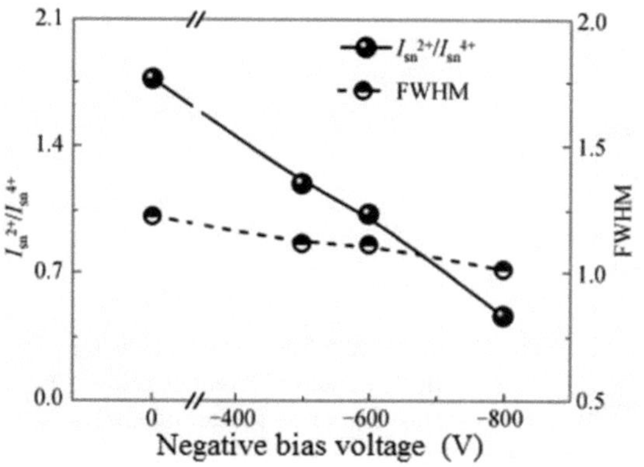

Fig. 10.19: I_{Sn}^{2+}/I_{Sn}^{4+} ratio and Sn^{4+} FWHM of In_2O_3:Sn films fabricated under different negative bias voltages.

The analysis of 3d narrow-band photoelectron spectral peaks of In and Sn indicates that both In and Sn atoms in the In_2O_3 and In_2O_3:Sn films fabricated using plasma bombardment magnetron sputtering chemically react with O atoms, forming the corresponding In–O and Sn–O bonds. Therefore, it is quite necessary for us to analyze the bonding state of oxygen.

Fig. 10.20 shows the O 1s core-level spectra of In_2O_3 films fabricated under different negative bias voltages. In the In_2O_3 films fabricated using plasma bombardment magnetron sputtering, oxygen has two bonding energies at 530.0 ± 0.2 eV and 531.6 ± 0.2 eV, respectively. The O_α peak located around 530.0 eV is an O–In bond (an In atom corresponding to this type of oxygen atom has 6 closest oxygen ions around it). The high-energy O_β peak (around 531.6 eV) is mainly associated with the O–In bonds at the oxygen-vacancy zone. With the increase of $|V_p|$, the intensity of the O_α peak steadily decreases. This is because a higher particle bombardment energy can better enhance the surface atomic mobility and is thus more promotive of the reaction of free oxygen atoms and surface indium atoms. A similar phenomenon can be observed in Al_2O_3 coatings fabricated using plasma bombardment magnetron sputtering, with the formation of saturated O–In bonds. The high-energy O_β peak (around 531.6 eV) exists primarily because of the O–In bonds at the oxygen-vacancy zone. These oxygen vacancies of the O–In bond force the In^{3+} ions to provide extra free electrons to the thin film. More vacancies result in more free electrons. Practically, the superior electrical properties of In_2O_3 film originate from the extra free electrons provided by the oxygen vacancies. A similar result is obtained in In_2O_3:Sn films fabricated using plasma bombardment magnetron sputtering.

An x-ray photoelectron spectrometer can be used not only to analyze the chemical bonding of each element in In_2O_3 and In_2O_3:Sn films, but also to conduct quantitative analysis. The relative content of surface elements can be determined by calculating the areas below the corresponding element peaks in the obtained element spectrogram and substituting the sensitivity factors of each element into equation (10.2). Owing to the adoption of the sensitivity factor in this approach, it is named the sensitivity factor method.

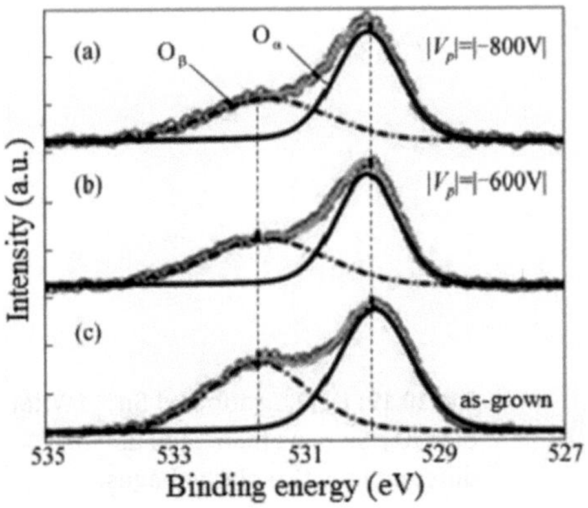

Fig. 10.20: O 1s XPS fitting curves of In_2O_3 films fabricated under different negative bias voltages: (a) magnetron sputtering; (b) $|V_p| = |-600\,V|$; (c) $|V_p| = |-800\,V|$.

According to this method, the relative content ratio of elements in the In_2O_3 film can be expressed by the equation

$$\frac{n_i}{n_j} = \left(\frac{I_i}{I_j}\right)\left(\frac{S_j}{S_i}\right), \tag{10.2}$$

where i and j represent elements In and O, I the photoelectron intensity (major peak area of corresponding element), n_i the relative content of element i, and S_i the sensitivity factor of element i.

By measuring the relative concentration ratio in the In_2O_3 surface atomic layer, we can acquire the relative atomic content of In and O in the In_2O_3 film. Considering that element In belongs to the transition metal series, the 3d peak is used as the analysing peak. Therefore, based on equation (10.2), the relative atomic content ratio of In and O can be written as

$$C_{In} = \frac{I_{In3d}/S_{In}}{I_O/S_O + I_{In3d}/S_{In}}, \tag{10.3}$$

where C_{In} is the relative atomic content concentration of In in the In_2O_3 film.

The relative atomic content of O and In can be obtained using equation (10.2). During the calculation process, the total intensity of In photoelectron spectral peaks can be obtained by calculating the areas of $3d_{5/2}$ and $3d_{3/2}$ peaks, while that of O photoelectron spectral peaks only needs the consideration of the area of the O 1s peak.

Fig. 10.21 shows the atomic ratios O/In and O_α/O_β of In_2O_3 films fabricated under different negative bias voltages. We can see that, with the increase of negative bias voltage, the ratio of O to In gradually increases. When $|V_p| = |-900\,V|$, O/In reaches its maximum of approximately 1.38. In contrast, the atomic ratio O_α/O_β experiences a significant decrease from the initial value of approximately 0.93 to 0.68. This indicates that a high negative bias voltage has a favourable effect on the increase of O/In, making it close to a stoichiometric value. On the other hand, the atomic ratio O_α/O_β of the In_2O_3 film provides a sensitivity factor index for the oxygen-vacancy content level.

Tab. 10.2 lists the In/Sn atomic ratio and O/(1.5 In + 2 Sn) atomic ratio of In_2O_3:Sn films fabricated under different negative bias voltages.

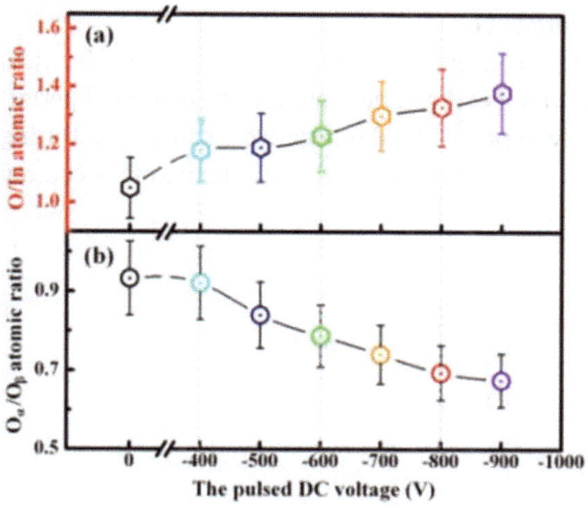

Fig. 10.21: O/In and O_α/O_β of In_2O_3 films fabricated under different negative bias voltages.

Tab. 10.2: In/Sn atomic ratio and O/(1.5 In + 2 Sn) atomic ratio of In_2O_3:Sn films fabricated under different negative bias voltages.

V_p (V)	0	−400	−500	−600	−700	−800	−900
Sn/In (%)	10.12	11.56	10.35	10.83	11.20	10.11	10.27
O/(1.5 In + 2 Sn) (%)	1.41	1.35	1.36	1.21	1.26	1.22	1.19

For a perfect In_2O_3:Sn crystal, the value of O/(1.5 In + 2 Sn) should be unity. A value higher or lower than unity is indicative of oxygen excess or oxygen deficit. General In_2O_3:Sn films can be denoted as $In_{2-x}Sn_xO_{3-\delta}$, where x denotes the number of In ions replaced by Sn ions in the In_2O_3 lattice, and δ is the oxygen vacancy number.

10.2.3 Photoelectric performance of In_2O_3 and In_2O_3:Sn films

Photoelectric performance is a critical index for evaluating the photoelectric components of In_2O_3 and In_2O_3:Sn films. Subsequent to the plasma bombardment with different values of energy, In_2O_3 and In_2O_3:Sn films undergo substantial variations in terms of structure, surface morphology, and composition. These variations have an indivisible intrinsic relation with the photoelectron performance of the In_2O_3 and In_2O_3:Sn films. To evaluate the electrical performance, we primarily test the variation of carrier concentration, mobility, and electrical resistivity versus the negative bias voltage in In_2O_3 and In_2O_3:Sn films. To evaluate the optical performance, we mainly test the transmission of UV-visible light and thereby acquire the relation between optical bandgap and negative bias voltage.

10.2.3.1 Variation of electrical performance

Fig. 10.22 shows the variations of carrier concentration (n), mobility (μ), and electrical resistivity (ρ) with the negative bias voltage. It can be seen that the electrical performance of the In_2O_3 film is intensively impacted by the negative bias voltage. During the increase of $|V_p|$ from $|-400\,V|$ to $|-900\,V|$, the electrical performance of the coating decreases before increasing again.

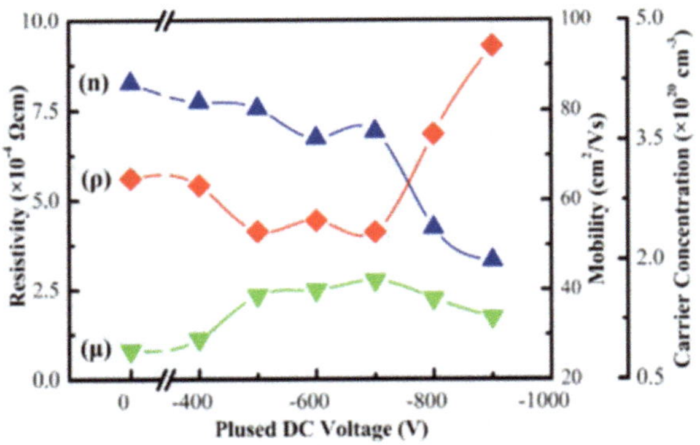

Fig. 10.22: Electrical performance of In_2O_3 films fabricated under different negative bias voltages.

In general, the superior electrical performance of the In_2O_3 film is attributed to the large number of oxygen vacancies inside the coating. Each oxygen vacancy can provide two free electrons to the film. The relationships among the electrical resistivity, carrier concentration, and electron mobility can be expressed by

$$\rho = \frac{1}{ne\mu}, \tag{10.4}$$

where ρ denotes the electrical resistivity of the coating ($\Omega\cdot$cm), μ the Hall mobility of free electrons (cm^2/V$\cdot$s), n the free-electron concentration in the coating (cm^{-3}), and e the charge of an electron ($1.6021892 \times 10^{-19}$ C).

When $|V_p| = |-700\,V|$, we obtain the minimum electrical resistivity of $4.11 \times 10^{-4}\,\Omega\cdot$cm. The Hall mobility of the thin film also gradually increases with the increase of the negative bias voltage and reaches a peak of 42.1 cm^2/V$\cdot$s at $|V_p| = |-700\,V|$. There are two reasons for the Hall mobility increase. One is that the enhancement of the crystallization degree reduces the number of defects in the film, leading to a drastic decrease of the number of scattering kernels caused by the defects. The other is the decrease of grain-boundary scattering, which is due to the increase of grain size. This can be explained by the fact that, with the increase of negative bias voltage, the grain size increases and, thus, the grain-boundary content decreases, which has already been verified by the XRD observation.

Fig. 10.23 shows the electrical performance of In_2O_3:Sn films fabricated under different negative bias voltages. As the negative bias voltage increases, the electrical resistivity, carrier concentration, and Hall mobility all show a decrease-to-increase

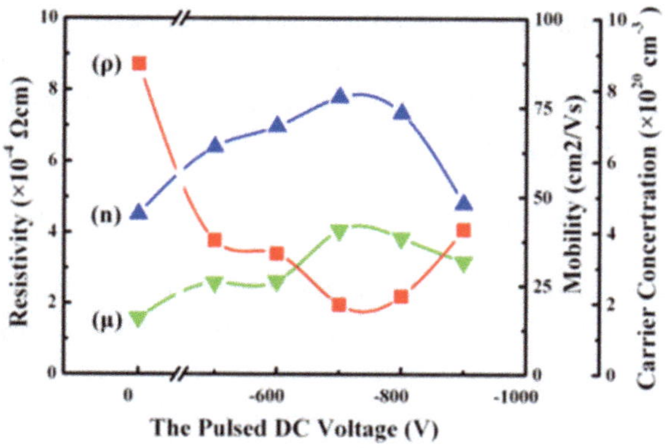

Fig. 10.23: Electrical performance of In_2O_3:Sn films fabricated under different negative bias voltages.

trend. The minimum electrical resistivity is 8×10^{-4} Ω·cm at $|V_p| = |-700\,V|$. This trend is due to the increase of the plasma-bombardment energy with the increase of negative bias voltage, which causes the doped low-valence Sn^{2+} ions that replace the In^{3+} lattice positions to lose further electrons and transform into high-valence Sn^{4+} ions, leading to the increase of carrier concentration. Moreover, the enhancement of the degree of film crystallization increases the mobility of free electrons. The electrical resistivity of the film is, therefore, optimized under the action of both these factors. However, at $|V_p| > |-800\,V|$, the severe plasma bombardment on the In_2O_3:Sn film causes surface radiation damage, causing both carrier concentration and mobility to decrease and eventually leading to the decrease of electrical resistivity.

10.2.3.2 Variation of optical performance

Fig. 10.24 shows the transmission of UV-visible light in In_2O_3 films fabricated under different negative bias voltages. We can see that every In_2O_3 film specimen has a mean transmission greater than 80 % in the visible range. When the wavelength is less than 500 nm, transmission decreases owing to the effect of the absorption edge existing in the In_2O_3 films.

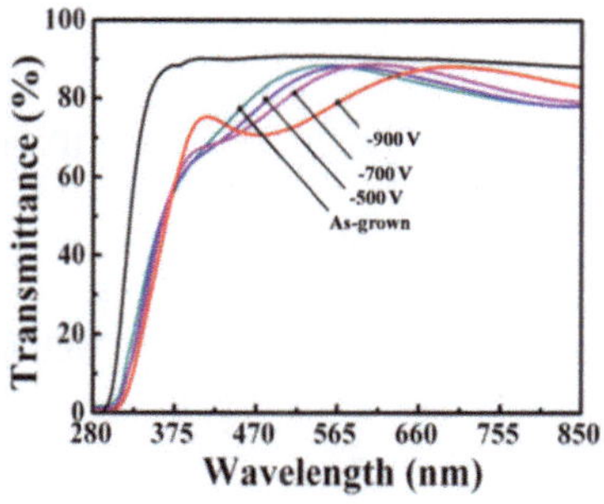

Fig. 10.24: Transmission of UV-visible light in In_2O_3 films fabricated under different negative bias voltages.

The optional bandgap of the In_2O_3 film is one of the critical indexes for determining the application range of the films. According to the Tauc law, the optical bandgap of the film can be obtained easily based on the relationship between bandgap and absorption bandgap, as expressed in the equation

$$(\alpha h\upsilon)^{1/n} = A(h\upsilon - E_g), \tag{10.5}$$

where α denotes the absorption coefficient (m^{-1}), $h\upsilon$ the incident electron energy (eV), h the Planck's constant ($6.62606957 \times 10^{-34}$ J·s), A the band-tail parameter, and E_g the film optical bandgap (eV).

The bandgap of transition-type semiconductor materials such as In_2O_3 can be determined using the index n. When $n = \frac{1}{2}$, the semiconductor material corresponds to the allowed direct transition; when $n = 2$, it corresponds to the allowed indirect transition; when $n = \frac{3}{2}$, it corresponds to the forbidden direct transition; and when $n = 3$, it corresponds to the forbidden indirect transition. The In_2O_3 and In_2O_3:Sn films studied in this work are semiconductors of allowed direct transition, i.e., $n = \frac{1}{2}$. We can obtain the absorption coefficient (α) by using the Fresnel equation of transmission:

$$T = \exp(-\alpha d), \tag{10.6}$$

where T is the material transmission and d is the film thickness (m).

When we calculate the absorption coefficient, because the substrate thickness is far greater than the deposited coating thickness, the effect of substrate glass on the film absorption coefficient needs to be considered. Therefore, an approximate treatment is employed in which the entire absorption coefficient of the film-substrate system is considered to be equal to the product of absorption coefficient and thickness:

$$A = \alpha d = \alpha(d_{\text{film}} + d_{\text{sub}}). \tag{10.7}$$

When the two layers of the film-substrate system, i.e. film and substrate, are separated, the absorption coefficient of the film can be expressed as

$$A_{\text{film}} = \alpha_{\text{film}} d_{\text{film}} \tag{10.8}$$

and the absorption coefficient of the substrate as

$$A_{\text{sub}} = \alpha_{\text{sub}} d_{\text{sub}}. \tag{10.9}$$

Then, the absorption coefficient of the film-substrate system should be

$$A = A_{\text{film}} + A_{\text{sub}} = \alpha_{\text{film}} d_{\text{film}} + \alpha_{\text{sub}} d_{\text{sub}}. \tag{10.10}$$

Therefore, we obtain

$$\alpha_{\text{film}} d_{\text{film}} + \alpha_{\text{sub}} d_{\text{sub}} = \alpha(d_{\text{film}} + d_{\text{sub}}),$$

i.e. the film absorption coefficient,

$$\alpha_{\text{film}} = \alpha + (\alpha - \alpha_{\text{sub}}) \frac{d_{\text{sub}}}{d_{\text{film}}}. \tag{10.11}$$

Fig. 10.25 shows the relationship between $(\alpha h\upsilon)^2$ and $h\upsilon$ of the In_2O_3 films fabricated under different negative bias voltages. The optical bandgap can be acquired from the figure by using the slope method. We can see that the optical bandgap gradually decreases from 3.58 eV to 3.45 eV with the increase of negative bias voltage. The bandgap variation can be explained by the fact that the film surface is crystalized under plasma bombardment with a certain amount of energy, reducing the number of defects and carrier concentration in the crystal. According to the Burstein-Moss theory of band filling, the bandgap filling effect is gradually weakened with the decrease of carrier concentration, leading to a decline of optical bandgap and a redshift of the absorption limit.

Fig. 10.26 (a) shows the UV-visible transmission of In_2O_3:Sn films fabricated under different negative bias voltages. For all the values of negative bias voltage, the optical transmission of the films is always greater than 80 % in the visible range. With the

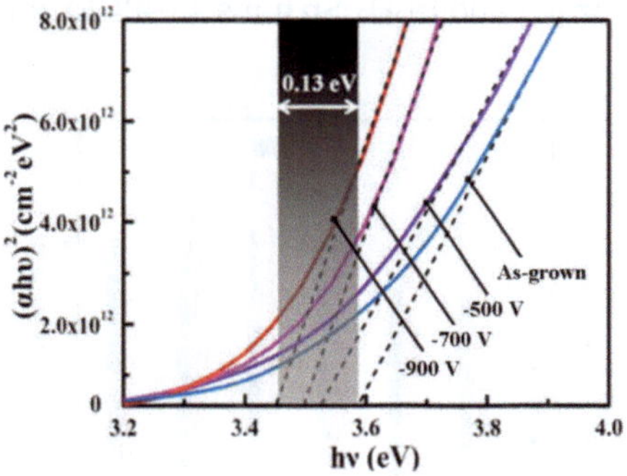

Fig. 10.25: Relationship between $(\alpha h\upsilon)^2$ and $h\upsilon$ of In_2O_3 films fabricated under different negative bias voltages.

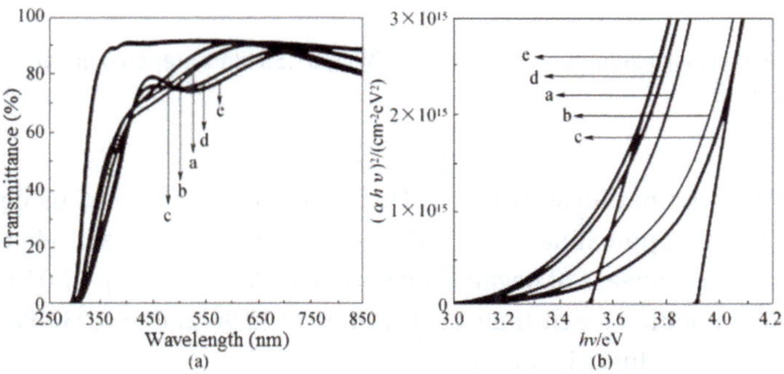

Fig. 10.26: Films fabricated under different negative bias voltages of In_2O_3:Sn: (a) UV-visible transmission, and (b) relation between $(\alpha h\upsilon)^2$ and $h\upsilon$: (a) direct growth; (b) |−500 V|; (c) |−600 V|; (d) |−800 V|; (e) |−900 V|.

increase of negative bias voltage, the carrier concentration of the In_2O_3:Sn films increases, and the optical bandgap is broadened from 3.52 eV to 3.9 eV. Subsequently, under plasma bombardment damage on the coating surface, both the carrier concentration and optical bandgap decrease.

10.2.3.3 Microhardness of In_2O_3 and In_2O_3:Sn film

Generally, In_2O_3 and In_2O_3:Sn films fabricated using magnetron sputtering at room temperature are all in the amorphous state with relatively poor mechanical performance; therefore, it is difficult to apply them as optical window components. Furthermore, adopting a high-temperature fabrication or annealing process can hardly guarantee the sensitivity requirement of the thermosensitive window substrate. Therefore, plasma bombardment magnetron sputtering is used in this work to fabricate crystalline In_2O_3 and In_2O_3:Sn films with superior mechanical performance. We characterize the hardness and elastic modulus of In_2O_3 and In_2O_3:Sn films prepared under different negative bias voltages by using the nanoindentation device.

Fig. 10.27 shows the hardness and elastic modulus of In_2O_3 and In_2O_3:Sn films deposited on glass substrates under different negative bias voltages. The indentation depths of hardness and elastic modulus of the In_2O_3 and In_2O_3:Sn films are all 30 nm.

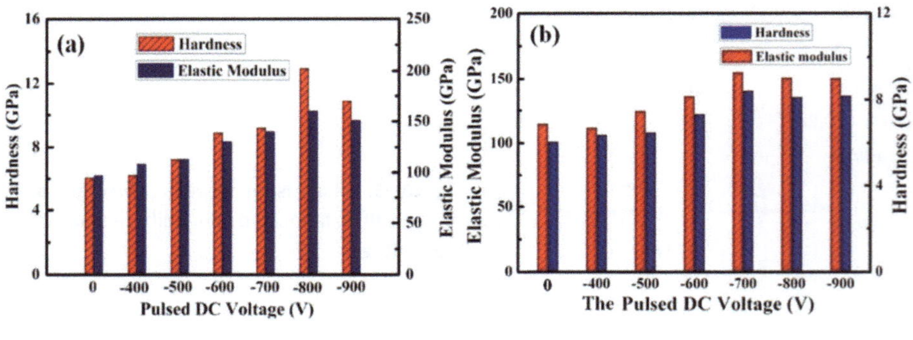

Fig. 10.27: Microhardness and elastic modulus of (a) In_2O_3 and (b) In_2O_3:Sn films prepared under different negative bias voltages.

As shown in Fig. 10.27 (a), the hardness of the In_2O_3 film directly fabricated using magnetron sputtering is only approximately 6.06 GPa. Both the hardness and elastic modulus increase with the increase of negative bias voltage. When $|V_p| = |-800\,V|$, the hardness and elastic modulus reach their peak values of 12.96 and 160.03 GPa, respectively. The increase of hardness is caused by

(1) the transition of the In_2O_3 film from an amorphous state to a crystalline state and
(2) the variation of preferential crystal orientation from the (222) to (400) crystallographic plane under the high-energy plasma bombardment.

It has already been proved that the {400} family of crystallographic planes has high hardness. With the further increase of negative bias voltage ($|V_p| > |-800\,V|$), the material hardness and elastic modulus experience a certain degree of decrease. This is because a certain amount of energy of the high-energy plasma can exert irradiation damage on the film surface, leading to a decrease of the hardness of the specimen films.

In contrast, the variation of hardness and elastic modulus in the In_2O_3:Sn films with the negative bias voltage is shown in Fig. 10.27 (b). It can be observed that the increase of negative bias voltage can improve the film hardness in a certain range; beyond that range, however, the hardness does not significantly increase with the increase of negative bias voltage. This can be explained by the fact that no high-hardness grains are formed in the preferential orientation ⟨400⟩ during the process of high-energy plasma bombardment.

10.3 Structure and performance of Y_2O_3:Ru

10.3.1 Surface morphology of ruthenium-doped yttrium oxide film

Fig. 10.28 shows the AFM surface morphology of Ru-doped Y_2O_3 (RYO) films fabricated using a metal ruthenium target under different values of sputtering power. The sputtering power of the ruthenium target has a significant influence on the surface morphology of the RYO film. We can see that the RYO films exhibit compact surfaces without obvious holes. With the increase of ruthenium-target sputtering power, island structures and cones gradually appear on the surfaces.

When the ruthenium-target sputtering power increases from 10 W to 40 W, the surface roughness of the film steadily increases (Fig. 10.29). This phenomenon is caused by the variation of the kinetic energy of sputtering particles. When the ruthenium-target sputtering power is only 10 W, the film-forming ruthenium atoms with a relatively low amount of kinetic energy are slowly deposited on the substrate surface, producing an extremely low surface roughness of approximately 0.213 nm. When the ruthenium-target sputtering power increases to 40 W, the deposition speed substantially increases, causing the surface roughness of the film to increase to approximately 0.837 nm.

Fig. 10.30 illustrates the surface morphology of Y_2O_3:Ru films fabricated under a ruthenium-target sputtering power of 40 W, yttrium target sputtering power of 60 W, and different substrate temperatures. It can be seen that all the specimens possess a smooth surface without remarkable, distinct holes or island structures. At room temperature, we can observe some cones on the surface. As the temperature increases, the number of cones gradually decreases.

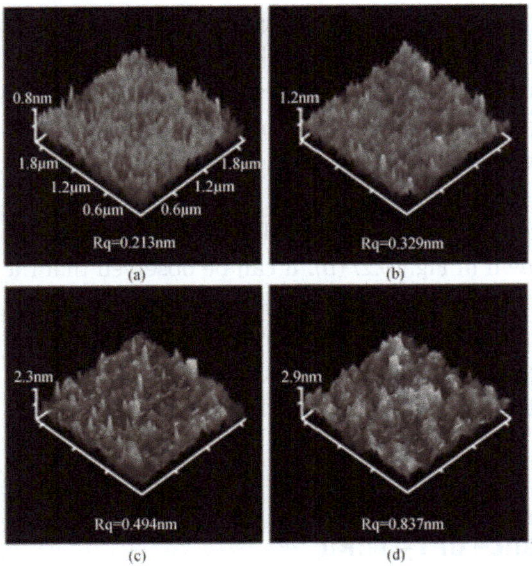

Fig. 10.28: AFM surface morphology of Y_2O_3:Ru films fabricated using a metal ruthenium target under different values of sputtering power: (a) 0 W; (b) 10 W; (c) 20 W; (d) 40 W.

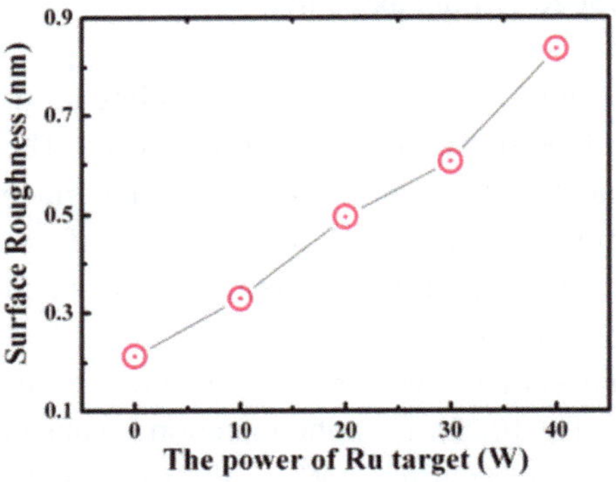

Fig. 10.29: Relation between surface roughness and ruthenium-target sputtering power of Y_2O_3:Ru film.

As Fig. 10.31 shows, with the increase of substrate temperature, the film roughness decreases significantly from the initial value of 0.837 nm down to 0.159 nm. This result indicates that the substrate-temperature increase can efficiently accelerate the diffusion rate and extend the range of atoms along the surface, driving the surface particles to shift into the gullies between nucleation sites and consequently reducing the nucleation-induced surface roughness.

10.3.2 Crystal structure of ruthenium-doped yttrium oxide film

Fig. 10.32 shows the XRD patterns of Y_2O_3:Ru films fabricated under different values of ruthenium-target sputtering power. Under a relatively low ruthenium-target power (10 W), at $2\theta = 30°$, we can find a fairly large, wide peak envelope in the XRD diffrac-

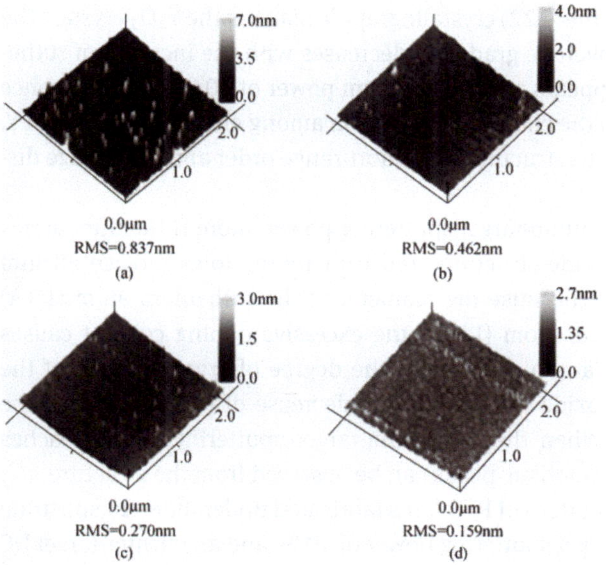

Fig. 10.30: AFM surface morphology of Y_2O_3:Ru films fabricated at different substrate temperatures: (a) room temperature; (b) 200 °C; (c) 400 °C; (d) 600 °C.

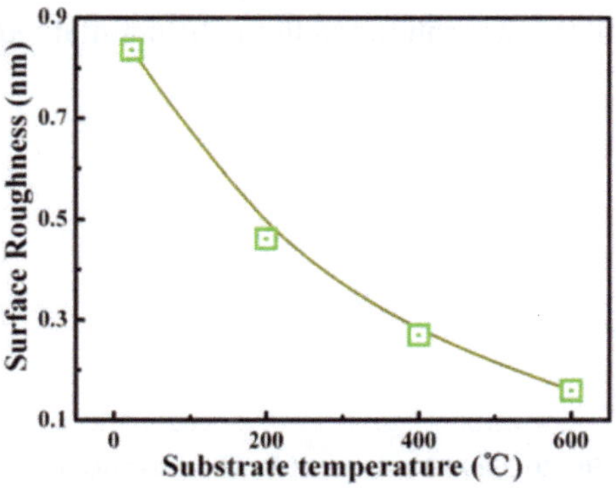

Fig. 10.31: Surface roughness of Y_2O_3:Ru films fabricated at different substrate temperatures.

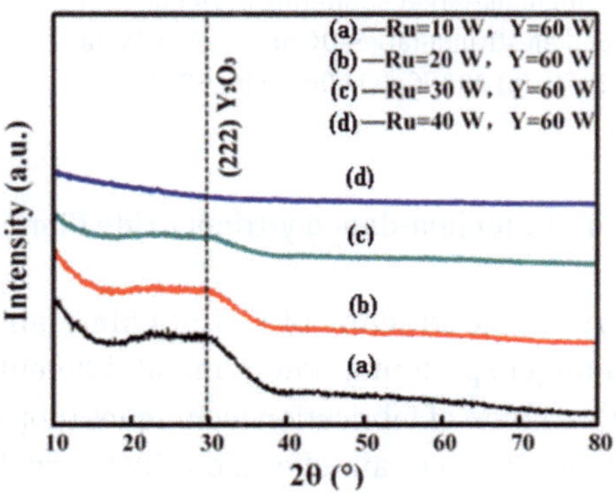

Fig. 10.32: XRD pattern of RYO films fabricated under different values of ruthenium-target sputtering power: (a) 0 W; (b) 10 W; (c) 20 W; (d) 40 W.

tion pattern corresponding to the (222) crystallographic plane in the Y_2O_3 crystal. The intensity of the wide peak envelope gradually decreases with the increase of ruthenium power and almost disappears at the ruthenium power of 40 W. The occurrence of the peak envelope is due to the chemical interaction among elements Ru, O, and Y, which generates the amorphous structure with short-range order and long-range disorder.

With the increase of ruthenium-target sputtering power, more ruthenium atoms are involved in the yttrium oxide structure, and ruthenium atoms replace yttrium atoms in the lattice structure. Because the diameter of the ruthenium atom (1.3 Å) is less than that of the yttrium atom (1.8 Å), the excessive doping content causes a serious structural disorder and a decrease of the degree of crystallization of the original amorphous yttrium oxide films, that is, a decrease of the intensity of the amorphous peak envelope. When the ruthenium-target sputtering power reaches 40 W, no obvious crystal characteristic peaks can be observed from the structure.

Fig. 10.33 shows the XRD pattern of RYO films fabricated under different substrate temperatures, a ruthenium-target sputtering power of 40 W, and an yttrium-target DC power of 60 W. It can be observed that Y_2O_3:Ru films fabricated at either room temperature or 600 °C are all in the amorphous state. From the viewpoint of thermodynamics and film-growth dynamics, although the amorphous state is a metastable state, as long as the metastable state can grow rapidly after nucleation, a great variation of free energy of the system must take place and the amorphous state can be maintained at a high temperature.

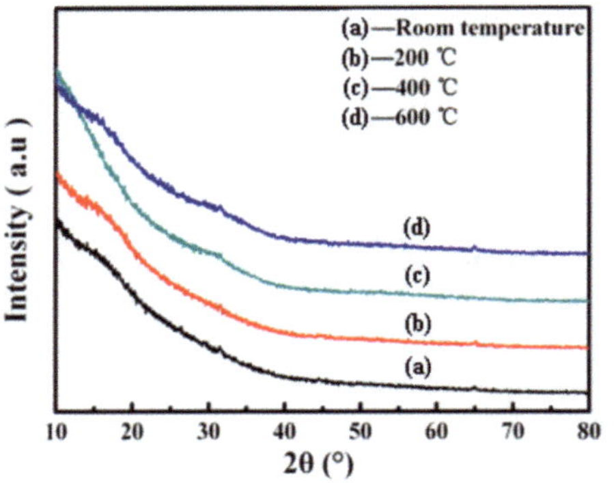

Fig. 10.33: XRD pattern of RYO films fabricated under different substrate temperatures, a ruthenium-target sputtering power of 40 W, and an yttrium-target DC power of 60 W: (a) 23 °C; (b) 200 °C; (c) 400 °C; (d) 600 °C.

10.3.3 Composition and chemical bonding of ruthenium-doped yttrium oxide film

By analysing the surface morphology and crystalline structure of Y_2O_3:Ru films fabricated under different values of ruthenium-target sputtering power and at different substrate temperatures, it is verified that the variation of fabrication techniques (doping amount and substrate temperature) can generate certain effects on the internal structure of Y_2O_3:Ru films. The variation above has a close correlation with the content of film components and chemical bonds between elements. In this section, we will focus on the variations of the Ru/Y/O content and chemical bonding of elements in Y_2O_3:Ru films.

10.3.3.1 Component characterization

In the test, we adjust the Ru-doping content in the Y_2O_3:Ru films by changing the Ru-target sputtering power and demarcate the ratio of elements Ru and Y in the specimens by using the energy-dispersive x-ray spectroscopy (EDS) method.

Fig. 10.34 shows the EDS peaks of Y and Ru and the SEM morphology of the measured EDS zone of the Y_2O_3:Ru films. The EDS test and analysis indicate that, at room temperature, in the Y_2O_3:Ru films fabricated under the Ru-target sputtering power of 10 W, 20 W, 30 W, and 40 W and Y-target DC power of 60 W, the Ru:Y atomic ratios are 0.26, 0.41, 0.57, and 0.94, respectively. It can also be inferred from the test that the Ru-doping content has a relatively small variation range around 0.94 in the Y_2O_3:Ru films when the Ru-target sputtering power is controlled at 40 W, Y-target DC power at 60 W, and substrate temperature within the range of 23–600 °C (see the inset in Fig. 10.35).

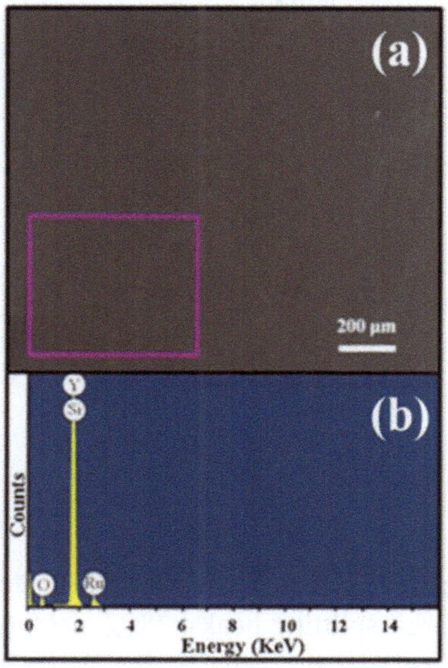

Fig. 10.34: (a) EDS analysis on the spectral peaks of each element in the Y_2O_3:Ru films; (b) SEM morphology of the EDS zone on the film surface.

10.3.3.2 Chemical bonding of elements

From the previous EDS analysis of the specimens, we have obtained the content of elements Ru and Y in the Y_2O_3:Ru films. However, the combined state of the three major elements, Ru, Y, and O, cannot be acquired from this analysis. In order to analyze the

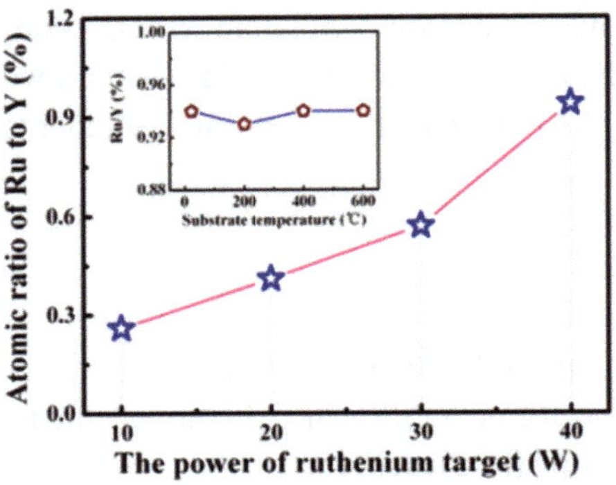

Fig. 10.35: Ru:Y atomic ratio of Y_2O_3:Ru films fabricated under different values of Y-target sputtering power. Inset:Ru:Y atomic scale of Y_2O_3:Ru films fabricated under a Ru target sputtering power of 40 W and different substrate temperatures.

chemical state, we can only adopt XPS to further investigate the effect of Ru-doping amount and substrate temperature on the chemical bonding of each element in the Y_2O_3:Ru films.

Fig. 10.36 illustrates a typical full-scan XPS spectrum of an Y_2O_3:Ru film prepared using double-target co-sputtering. The spectrum indicates the existence of elements Ru, Y, and O in the Y_2O_3:Ru film without peaks for other elements, indicating that the impurity content in the RYO film is quite low. The most remarkable peaks are Ru 3d, Y 3d, and O 1s. Therefore, we will focus on these peak positions during the scanning process when we conduct small-range analysis on the elements.

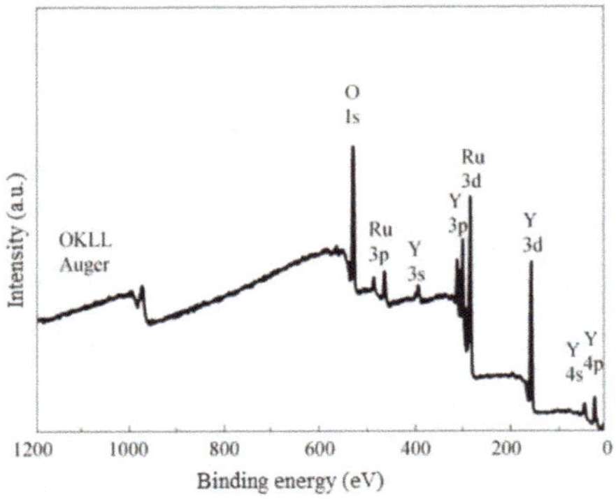

Fig. 10.36: Typical full-scan XPS spectrum of Y_2O_3:Ru film.

Fig. 10.37 shows the Ru 3d core-level spectra of Y_2O_3:Ru films with various amounts of Ru-doping, which include two groups of spin-orbit splitting peaks. $3d_{5/2}$ spectral peaks of RuO_2 and RuO_3 are located at 281.1 eV and 282.6 eV, respectively. The distance between the spin splitting peaks of $3d_{5/2}$ and $3d_{3/2}$ is approximately 4.2 eV. With the increase of Ru-doping content, low-valence Ru^{4+} and high-valence Ru^{6+} corresponding to RuO_2 and RuO_3 undergo a significant change of intensity, while the peak-position distance between $3d_{5/2}$ and $3d_{3/2}$ remains unchanged.

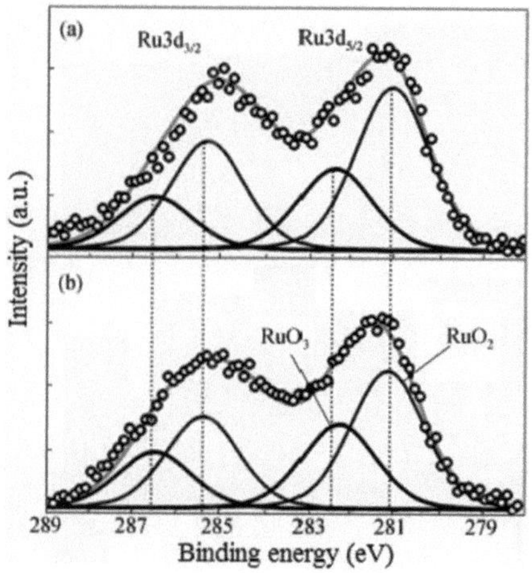

Fig. 10.37: Ru 3d XPS fitting curves of Y_2O_3 films with different Ru-doping amounts: (a) 57 %, (b) 26 %.

Fig. 10.38 demonstrates the relative intensity ratio of Ru 3d peaks under different Ru-doping concentrations, where we can clearly identify changing trend of the Ru^{4+} / Ru^{6+} atomic percentage. Ru^{4+} / Ru^{6+} is 1.3 when the Ru-doping content is 26 %, which means that the content of Ru^{4+} in Y_2O_3:Ru films gradually increases with the increase of doping content. This is because at a relatively low Ru-doping content, a part of low-valence RuO_2 can continue to react with excessive oxygen, forming high-valence RuO_3. However, with the increase of doping amount, the concentration of O atoms surrounding reaction Ru atoms gradually decrease, reducing the probability of reaction and the RuO_3 content.

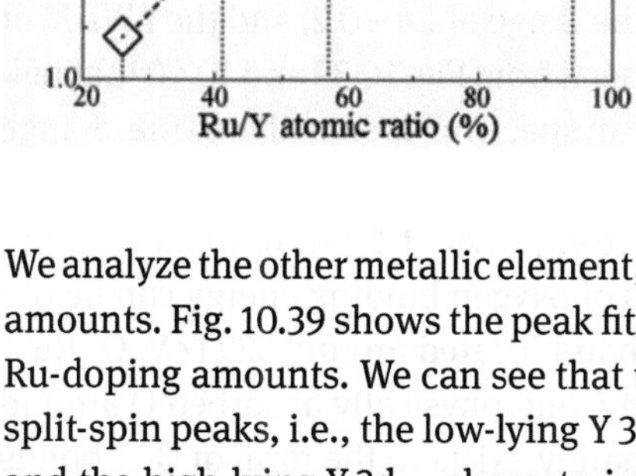

Fig. 10.38: Relative intensity ratio of Ru $3d_{5/2}$ and $3d_{3/2}$ spectral peaks in Y_2O_3:Ru films with different Ru-doping amounts.

We analyze the other metallic element, Y, in the Y_2O_3:Ru films with different Ru-doping amounts. Fig. 10.39 shows the peak fitting of element Y in Y_2O_3:Ru films with different Ru-doping amounts. We can see that the Y 3d characteristic peak is composed of two split-spin peaks, i.e., the low-lying Y $3d_{5/2}$ characteristic peak located around 156.8 eV and the high-lying Y $3d_{3/2}$ characteristic peak around 158.8 eV, with an energy gap of approximately 2.05 eV.

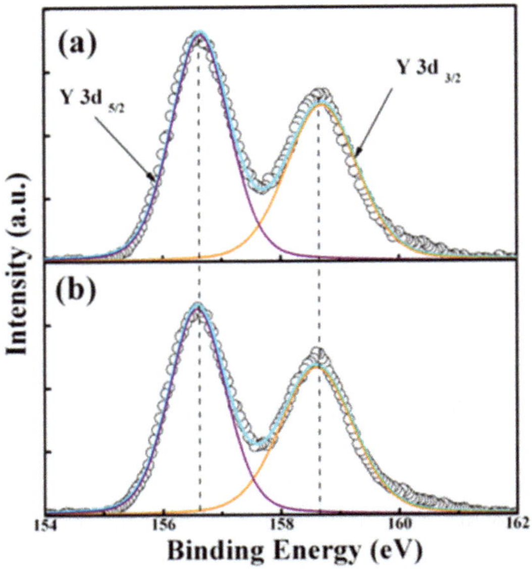

Fig. 10.39: Y 3d XPS fitting curves of Y_2O_3:Ru films with different Ru-doping amounts: (a) 26 %; (b) 57 %.

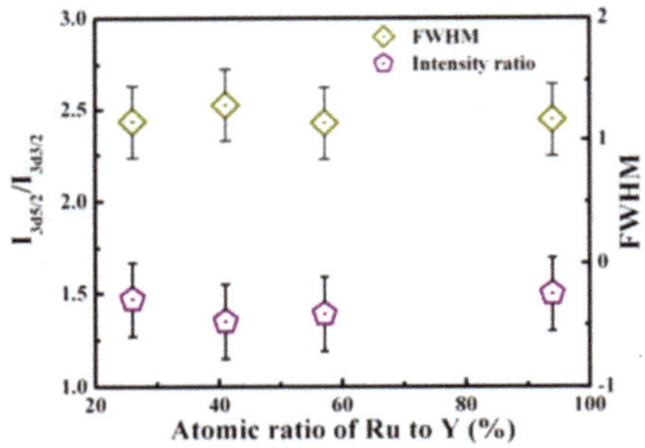

Fig. 10.40: Relative-intensity ratio and FWHM of Y $3d_{5/2}$ and $3d_{3/2}$ spectral peaks with different Ru-doping amounts.

Fig. 10.40 shows the variation patterns of the relative-intensity ratio and fitting-curve FWHM of the Y 3d peak of Y_2O_3:Ru films with different Ru-doping amounts. With the increase of Ru-doping content, we can observe that the relative-intensity ratio of the Y double peak ($3d_{5/2}$ and $3d_{3/2}$) remains within the range of 1.4 ± 0.2, and the FWHM of the Y 3d spectral peaks within 1.2 ± 0.2. It can be seen from Fig. 10.39 and 10.40 that neither the Y 3d binding energy nor the photoelectron spectrum is sensitive to the change of Ru-doping content.

Fig. 10.41 illustrates the O 1s core-level spectral peaks of Y_2O_3:Ru films with different Ru-doping amounts, where all four types of oxygen binding energy can be observed in the Y_2O_3:Ru films. They are the O–Y bond located around 529.1 eV, O–Ru^{4+} bond around 530 eV, O–Ru^{6+} bond around 531 eV, and physically absorbed O around 531.6 eV, respectively. The oxygen in the films mainly exists in the form of O–Y bonds and O–Ru bonds as well as the physically absorbed oxygen on the film surface, which is inevitable because of the exposure of the specimen to air. As shown in Fig. 10.41,

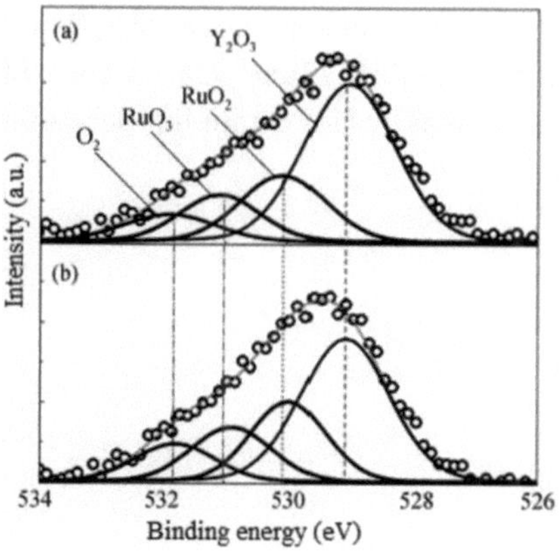

Fig. 10.41: Variation of O 1s XPS fitting curve of Y_2O_3:Ru films versus the Ru-doping amount: (a) 26 %; (b) 57 %.

when the Ru-doping concentration gradually increases with the increase of Ru-target sputtering power, the relative intensity of spectral peaks corresponding to the O–Ru bonds steadily increases. Therefore, it can be proved that the Ru content in the films gradually increases, which shows agreement with the EDS analysis results. We will also conduct an analysis on the chemical bonding of each element in the Y_2O_3:Ru films fabricated under an Ru-target RF power of 40 W, Y-target DC power of 60 W, reaction pressure of 1.0 Pa, and different substrate temperatures.

Fig. 10.42 shows the Ru 3d core-level spectra of Y_2O_3:Ru films fabricated at substrate temperatures of room temperature and 600 °C. Compared to those of Y_2O_3:Ru films with different Ru-doping amounts shown above, the RuO_2, RuO_3, Ru^{4+}, and Ru^{6+} peak positions display no shift in the graph. Compared with the Y_2O_3:Ru film fabricated at room temperature, the one fabricated at 600 °C has an increased relative content of RuO_3 but a decreased relative content of RuO_2. Next, we calculate the relative content of RuO_2 and RuO_3 by using the areas of the two spin peaks of RuO_2 and RuO_3.

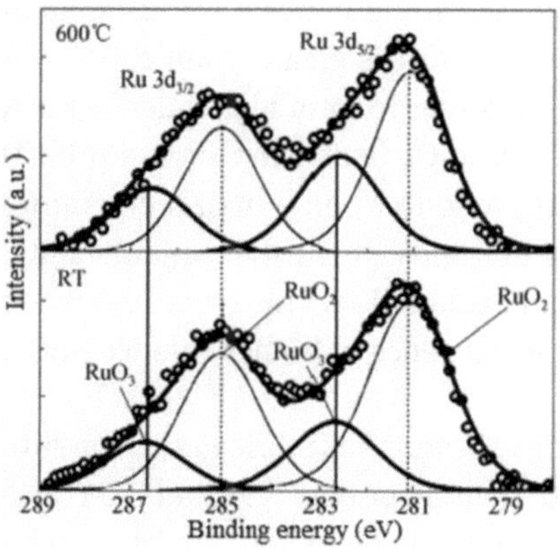

Fig. 10.42: Ru 3d XPS fitting curves of Y_2O_3:Ru films fabricated at the substrate temperatures of room temperature and 600 °C.

Fig. 10.43 shows the relative-intensity ratio of Ru^{4+} and Ru^{6+} spectral peaks of Y_2O_3:Ru films fabricated at different substrate temperatures. The RuO_2/RuO_3 relative content can be obtained using the I^{4+}_{Ru}/I^{6+}_{Ru} ratio. With the increase of substrate temperature, the relative content I_{RuO_2}/I_{RuO_3} decreases from 2.5 to 1.7.

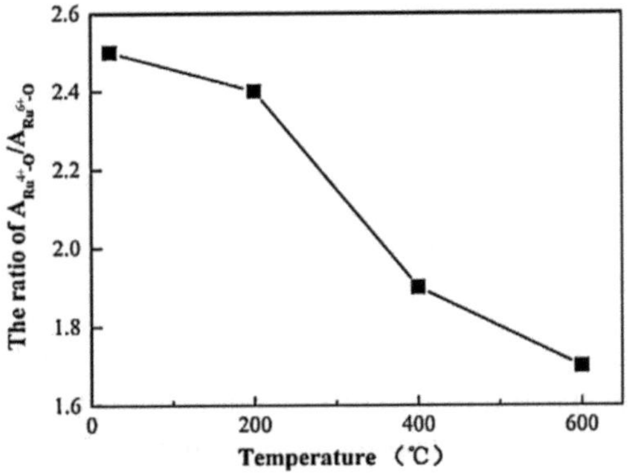

Fig. 10.43: Relative intensity ratio of Ru^{4+} and Ru^{6+} spectral peaks of Y_2O_3:Ru films fabricated at different substrate temperatures.

The content of RuO_x ($x = 2$ or 3) can be related to the following chemical equations:

$$Ru_{(s)} + O_{2(g)} \rightarrow RuO_{2(s)} + RuO_{3(s)} \quad \text{(low temperature)}, \tag{10.12}$$

$$Ru_{(s)} + O_{2(g)} \rightarrow RuO_{2(s)} + RuO_{4(s)} \quad \text{(high temperature)}, \tag{10.13}$$

$$RuO_{2(s)} + O_{2(g)} \rightarrow RuO_{3(s)} \quad \text{(high temperature)}, \tag{10.14}$$

low-temperature decomposition:

$$RuO_{4(s)} \rightarrow RuO_{2(s)} + O_{2(g)} \quad \text{(low temperature)}. \tag{10.15}$$

At a relatively low temperature, RuO_3 is mainly formed through the reaction between RuO_2 and excessive oxygen (equation (10.12)). However, the mechanism for the further generation of RuO_3 on the high-temperature substrate is not fully understood yet. Lin et al. suggest that, when the substrate temperature is greater than 200 °C, Ru metal atoms are likely to be oxidized by the oxygen atoms to form high-valence RuO_3 and RuO_4. During the subsequent cooling process, RuO_4 is decomposed into RuO_2 (equations (10.13) and (10.15)). Some authors proposed that, on a high-temperature substrate, thermal oxidation will occur between RuO_2 and O_2 to further generate RuO_3 (equation (10.14)). In this work, we prefer the latter explanation. This is because when analysing the Ru 3d peaks, we have not found the existence of higher-valence RuO_4, i.e., Ru^{8+}.

Fig. 10.44 shows the Y 3d core-level spectra of Y_2O_3:Ru films fabricated at the substrate temperatures of room temperature and 600 °C. It can be observed that the Y $3d_{5/2}$ bond energy of the Y_2O_3:Ru film deposited at the substrate temperature of room temperature (158.4 eV) is 0.4 eV lower than that of the Y_2O_3 film without Ru (156.85 eV).

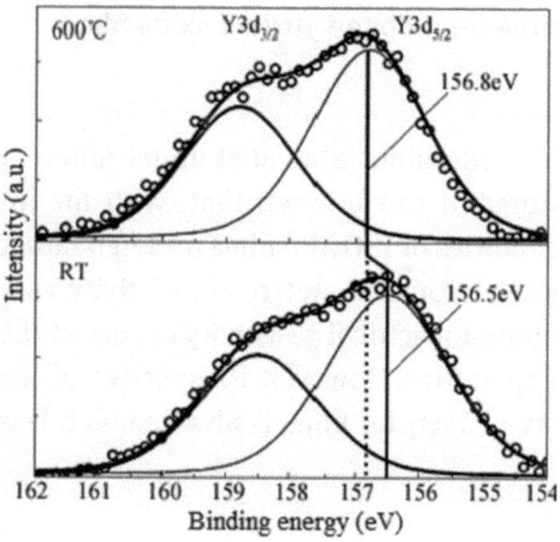

Fig. 10.44: Y 3d XPS fitting curves of Y_2O_3:Ru films fabricated at the substrate temperatures of room temperature and 600 °C.

This also proves that the oxide of nonstoichiometric yttrium exists in the Y_2O_3:Ru films. When the substrate temperature is 600 °C, the Y $3d_{5/2}$ bond energy is significantly greater (156.8 eV). Therefore, we can verify that a high temperature is beneficial to improving the Y/O ratio and gaining a superior oxide of stoichiometric yttrium.

Fig. 10.45 illustrates the O 1s core-level spectra of Y_2O_3:Ru films fabricated at the substrate temperatures of room temperature and 600 °C. We can see that, with the increase of substrate temperature, the peak intensity ratio undergoes significant changes. O–Y bonds at 529.0 eV gradually shift towards the high-energy side, which is in agreement with the previous results of the Y 3d peak. We also observe from the analysis O–Ru^{6+} and O–Ru^{4+} peaks that the relative intensity of the O–Ru^{6+} peak steadily increases, whereas that of the O–Ru^{4+} peak gradually decreases, again showing agreement with the previous results of the Ru 3d peak. Moreover, when the substrate temperature increases, the content of absorbed oxygen on the surface gradually decreases.

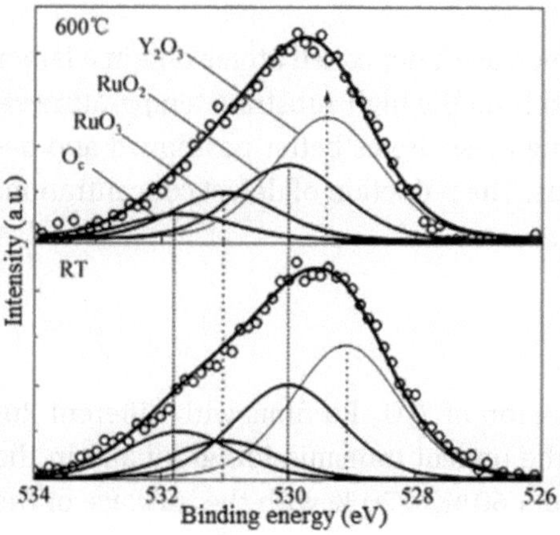

Fig. 10.45: O3d XPS fitting curves of Y_2O_3:Ru films fabricated at the substrate temperatures of room temperature and 600 °C.

10.3.4 Electrical and optical properties of ruthenium-doped yttrium oxide film

10.3.4.1 Electrical properties

Tab. 10.3 shows the electrical properties of Y_2O_3:Ru films fabricated under different Ru-doping amounts and substrate temperatures. It can be seen that, with the increase of Ru-doping amount, the electrical properties of Y_2O_3:Ru films undergo rather significant changes. That is, the carrier concentration and electrical resistivity vary across several orders of magnitude. The minimum electrical resistivity occurs at the Ru-doping content Ru/Y = 0.94 at room temperature. However, irrespective of the amount of Ru doping, the electrical resistivity of Y_2O_3:Ru films is always much less than that of Y_2O_3 films.

Tab. 10.3: Electrical properties of Y_2O_3:Ru films fabricated under different Ru-doping concentrations and substrate temperatures.

No.	Doping concentration Ru/Y (%)	Substrate temperature (°C)	Carrier concentration n (cm^{-3})	Electrical resistivity ρ (Ω·cm)	Sheet resistance (Ω/□)
1	26	Room temperature	1.86×10^{16}	8.24×10^{3}	6.92×10^{8}
2	41	Room temperature	2.73×10^{17}	1.81×10^{3}	1.52×10^{8}
3	57	Room temperature	6.83×10^{20}	7.46×10^{-1}	6.27×10^{4}
4	94	Room temperature	1.68×10^{22}	3.36×10^{-3}	283.4
5	94	200	6.16×10^{21}	7.78×10^{-3}	838.4
6	94	400	7.86×10^{20}	2.27	2.17×10^{5}
7	94	600	4.04×10^{20}	2.29	2.06×10^{5}

There are two reasons for the low electrical resistivity:

(1) doped Ru atoms replacing the Y atoms in the lattice structure provide extra free electrons for the coatings, and
(2) there exist great quantities of vacancy defects and metal interstitial atoms.

However, when the substrate temperature increases, deposited atoms obtain a larger surface mobility, and internal energy gained from the high substrate temperature efficiently increases the atomic energy, making these atoms better positioned and decreasing the defect concentration of the films. The reduction of defect concentration, in turn, decreases the carrier concentration.

10.3.4.2 Optical properties

Fig. 10.46 demonstrates the optical transmission of Y_2O_3:Ru films with different Ru-doping amounts. It can be observed that the optical transmission gradually in the visible range (around 550 nm) decreases from 60 % to 20 % with the increase of the Ru-doping amount.

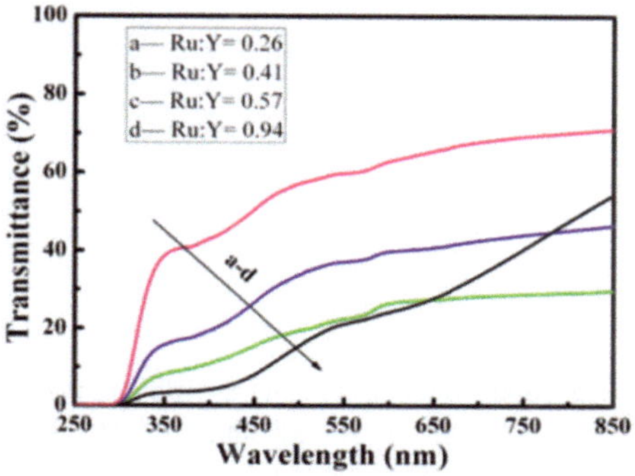

Fig. 10.46: Optical transmission of Y_2O_3:Ru films with different Ru-doping amounts.

Fig. 10.47 illustrates the optical bandgaps of Y_2O_3:Ru films fabricated with different Ru-doping amounts. It can be observed that, when the Ru-doping content increases, the optical bandgap of Y_2O_3:Ru films increases from 1.9 eV to 2.48 eV. This can be explained by the fact that the increase of carrier concentration causes the Fermi energy to increase. The carrier-concentration growth results from the increased doping concentration.

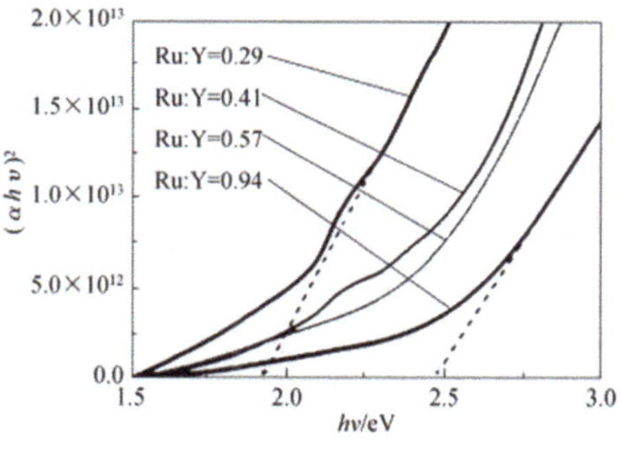

Fig. 10.47: Optical bandgap of Y_2O_3:Ru films fabricated with different Ru-doping amounts.

For comparison, transmission and optical bandgap in the waveband around 550 nm for Y_2O_3:Ru films fabricated at different substrate temperatures when Ru/Y = 0.94 by using the method described above are shown in Fig. 10.48 and 10.49. The transmission gradually increases with increasing substrate temperature. When the substrate temperature increases from room temperature to 600 °C, the optical bandgap exhibits a decreasing trend. This is because the increasing substrate temperature reduces the number of defects in the films as well as the number of oxygen vacancies, leading to a decrease of free electrons, a decrease of Fermi energy, and thereby a reduction of bandgap.

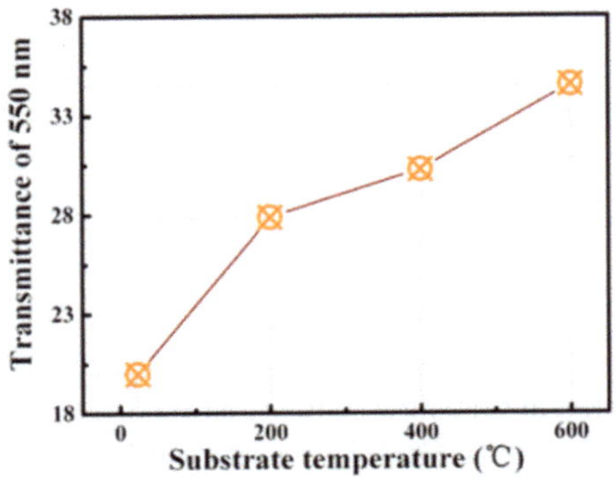

Fig. 10.48: Transmission in the visible range around 550 nm of Y_2O_3:Ru films fabricated at different substrate temperatures.

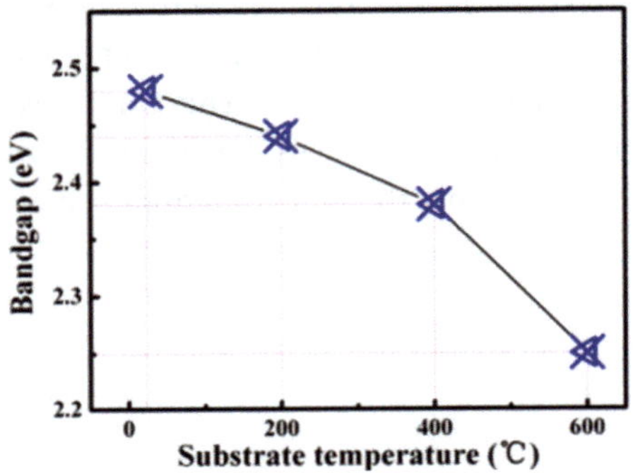

Fig. 10.49: Optical bandgap of Y_2O_3:Ru films fabricated at different substrate temperatures when Ru:Y = 0.94.

10.4 Mechanism analysis and application of infrared transparent conduction

10.4.1 Mechanism analysis of infrared transparent conduction

The theory of photon/electron interaction, constructed based on the Drude free-electron theory, essentially involves the adoption of properties of transmission and reflective index to explain the relationship between the external controllable properties (e.g. carrier concentration, relaxation time, etc.) and internal uncontrollable properties (e.g. effective mass, etc.) for traditional metals and transparent conductive oxide (TCO) materials.

In this theory, electron oscillation is considered to be induced by the electrical-field component of an optical-wave electromagnetic field, which can be described using the Lorentz oscillator equation, where electron displacement, velocity, and acceleration are a function of time and angular frequency of the electromagnetic field. Therefore, the relation between frequency and optical conductivity is obtained as

$$\sigma = \frac{J}{E} = \frac{-\mathrm{i}\omega ne^2/m^*}{(\omega_0^2 - \omega^2) - \mathrm{i}\omega/\tau} = \frac{\omega ne^2}{m^*}\left\{\frac{-\mathrm{i}(\omega_0^2 - \omega^2) + \omega/\tau}{(\omega_0^2 - \omega^2)^2 + (\omega/\tau)^2}\right\}. \tag{10.16}$$

By solving the Lorentz oscillator motion equation and Maxwell's equation simultaneously, we can obtain the dielectric function, $\varepsilon^* = \varepsilon_1 + \varepsilon_2$, the imaginary part and real part of which are

$$\varepsilon_1 = \varepsilon_\infty - \frac{\sigma_0 \tau}{\varepsilon_0}\left\{\frac{1}{1+\omega^2\tau^2}\right\}. \tag{10.17}$$

$$\varepsilon_2 = \frac{\sigma_0}{\varepsilon_0 \omega}\left\{\frac{1}{1+\omega^2\tau^2}\right\}. \tag{10.18}$$

When $\varepsilon_1 = 0$, i.e. $N = K$, the corresponding frequency is named plasma frequency,

$$\omega_p = \sqrt{\frac{\sigma_0}{\varepsilon_0 \varepsilon_\infty \tau}} = \sqrt{\frac{ne^2}{\varepsilon_0 \varepsilon_\infty m^*}}, \tag{10.19}$$

and the corresponding wavelength is named plasma wavelength,

$$\lambda_p = \frac{2\pi c_0}{\omega_p} = 2\pi c_0 \sqrt{\frac{\varepsilon_0 \varepsilon_\infty m^*}{ne^2}}. \tag{10.20}$$

From this equation, we can readily see the relationship between carrier concentration and plasma wavelength. When $\lambda = \lambda_p$, free carriers as a whole resonate with the incident electromagnetic wave and promote the absorption. The dielectric function $\varepsilon_1 = 0$ at this point, and the frequency of this incident electromagnetic wave becomes the mutation frequency of this material, which experiences a dramatic change in optical properties, i.e. the wavelength upper limit of the TCO coating in the transmission zone.

When $\lambda < \lambda_p$, the frequency of incident electromagnetic wave is in the visible range at a relatively high value. Because electrons are not able to keep pace with this frequency, they can hardly absorb the visible photons. Therefore, they exhibit the transmission of electromagnetic waves with frequency greater than ω_p.

When $\lambda > \lambda_p$, the frequency of incident electromagnetic waves is in the near- and mid-infrared regions; ε_1 shows a relatively large negative value, whereas ε_2 has a relatively large positive value. Therefore, they perform an intense reflection on electromagnetic waves with a wavelength higher than λ_p.

According to the literature, we know that, for In_2O_3 films, $\varepsilon_\infty = 4$, $m^* = 0.3m_0$, and for In_2O_3:Sn films, $\varepsilon_\infty = 3.95$, $m^* = 0.4m_0$.

Fig. 10.50 shows the relationship between plasma wavelength and carrier concentration of In_2O_3 and In_2O_3:Sn films. It can be observed that with the decrease of carrier concentration the plasma wavelength gradually shifts towards the infrared waveband with a growing degree of red shift. By comparing relationships between plasma wavelength and carrier concentration in almost twenty publications with the curve in this figure, we can observe that the relation between carrier concentration and plasma wavelength can be perfectly fitted using Drude's free electron theory. Thenwe conduct fitting on the curve between plasma wavelength and carrier concentration in Y_2O_3:Ru films. Owing to the lack of data on efficient electron mass and high-frequency dielectric constant for Y_2O_3:Ru films, according to the averaging theory of efficient electron

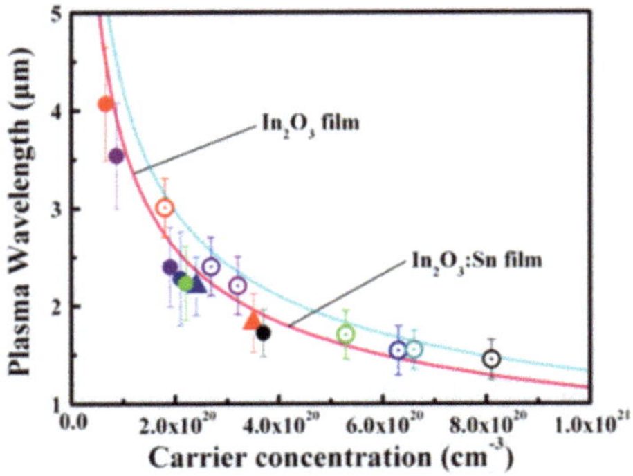

Fig. 10.50: Relationship between plasma wavelength and carrier concentration of In_2O_3 and In_2O_3:Sn films; Triangles represent data for In_2O_3 films found in in the literature and circle for In_2O_3:Sn films in the literature.

mass in amorphous multicomponent transparent conducting oxides, it is known that the mixed conducting orbital energy level that is formed by the s orbital energy level of positive ions and p orbital energy level of oxygen is the major part constituting the efficient electron mass. Moreover, the efficient electron mass has no significant effects on the bond-length and bond-angle variation of multiple positive ions in the doped oxide. Therefore, the efficient electron mass of Y_2O_3:Ru films is taken as $(m^*_{Y_2O_3} + m^*_{RuO_2})/2$, where $m^*_{Y_2O_3} = 0.5m_e$ and $m^*_{RuO_2} = 2.7m_e$. For the high-frequency dielectric constant of Y_2O_3:Ru films, since the conductivity of the Y_2O_3 film ($\varepsilon_\infty = 15$) reduces the dielectric constant, we assume that the high-frequency dielectric constant of Y_2O_3:Ru films are 5, 10, and 15, respectively.

It can be seen from Fig. 10.51 that when the high-frequency dielectric constant is 5 and the carrier concentration is approximately 10^{20} cm^{-3}, the plasma wavelength can reach the far-infrared waveband. With the increase of the high-frequency dielectric constant, a high-efficiency transmission in the far-infrared waveband can also be realized under a fairly large carrier concentration, i.e., superior conductivity.

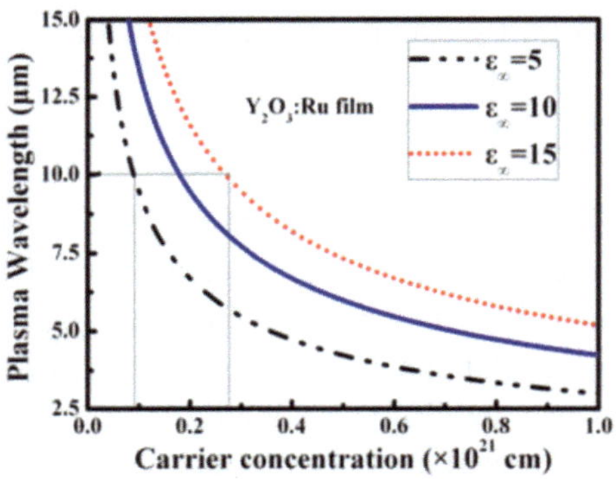

Fig. 10.51: Relationship between plasma wavelength and carrier concentration of Y_2O_3:Ru films under three different values of the high-frequency dielectric constant.

10.4.2 Zinc sulphide window far-infrared transparent conductive oxide thin film

ZnS substrate material is a semiconductor material with a wide bandgap that has been commonly used as optical components in the fields of aeronautics, astronautics, and military industries. Currently, it is mainly applied in windows, lens, and domes of optical systems. Although ZnS has superior optical properties in both mid- and far-infrared wavebands, it cannot shield against the interference of electromagnetic waves owing to the lack of conductivity. In order to realize transparent conductivity in the far-infrared waveband on ZnS substrates, we fabricated Y_2O_3:Ru films at different substrate temperatures.

Fig. 10.52 shows the infrared transmission curves of Y_2O_3:Ru films, ITO films, CNT films, and metal/dielectric multilayer materials at different substrate temperatures.

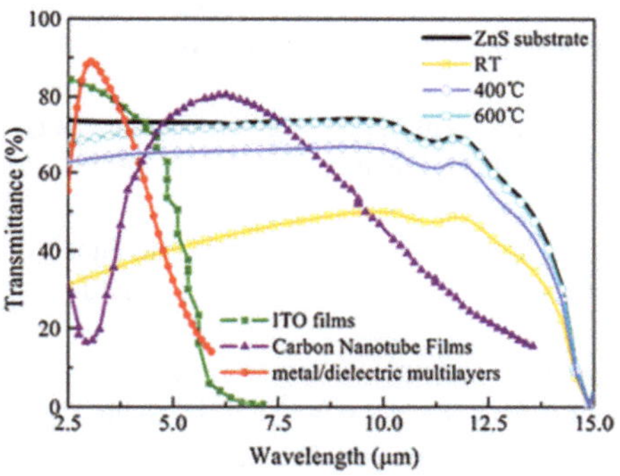

Fig. 10.52: Infrared transmission of different transparent conductive oxide materials: Y_2O_3:Ru films, ITO films, CNT films, and metal/dielectric multilayer materials at different substrate temperatures.

We can see from Fig. 10.52 that the infrared transmission of the Y_2O_3:Ru film gradually increases with the increase of substrate temperature. When the substrate temperature is 600 °C, the transmission in the far-infrared waveband is close to that of the ZnS substrate, and the electrical resistivity is approximately 2.18 Ω·cm. Compared with the Y_2O_3:Ru film in the wide waveband, the ITO film can only realize the high-efficiency transmission within the 3–5 μm waveband when the carrier concentration is approximately 10^{19} cm^{-3}, but it is not transparent in the far-infrared waveband. Metal/dielectric materials can be transparent only in a relatively narrow waveband range. This is quite similar to the situation of CNT films, which have a very high efficiency of transmission only within 3–8 μm and is not transparent in the far-infrared waveband.

References

[1] Barr TL, Liu YL. An X-ray photoelectron spectroscopy study of the valence band structure of indium oxides. Journal of Physics and Chemistry of Solids, 1989, 50(7): 657–664.
[2] Baur J, Silverman E. Challenges and opportunities in multifunctional nanocomposite structures for aerospace applications. MRS Bulletin, 2007, 32(04): 328–334.
[3] Becker L. Influence of IR sensor technology on the military and civil defense. International Society for Optics and Photonics, 2006, 61270S-15.
[4] Bellingham JR, Phillips WA, Adkins CJ. Electrical and optical properties of amorphous indium oxide. Journal of Physics Condensed Matter, 1990, 2(28): 6207.
[5] Bloemer MJ, Scalora M. Transmissive properties of Ag/MgF_2 photonic bandgaps. Applied Physics Letters, 1998, 72(14): 1676–1678.
[6] Brewer SH, Franzen S. Calculation of the electronic and optical properties of indium tin oxide by density functional theory. Chemical Physics, 2004, 300(1): 285–293.
[7] Celozzi S, Lovat G, Araneo R. Electromagnetic Shielding. Wiley Online Library, 2008. p. 1–50.
[8] Chan HYH, Takoudis CG, Weaver MJ. High-pressure oxidation of ruthenium as probed by surface-enhanced Raman and X-ray photoelectron spectroscopies. Journal of Catalysis, 1997, 172(2): 336–345.
[9] Christou V, Etchells M, Renault O, et al. High resolution x-ray photoemission study of plasma oxidation of indium–tin–oxide thin film surfaces. Journal of Applied Physics, 2000, 88(9): 5180–5187.
[10] Craciun V, Craciun D, Wang X, et al. Transparent and conducting indium tin oxide thin films grown by pulsed laser deposition at low temperatures. Journal of Optoelectronics and Advanced Materials, 2003, 5(2): 401–408.
[11] Craciun V, Howard J, Lambers ES, et al. Low-temperature growth of Y_2O_3 thin films by ultraviolet-assisted pulsed laser deposition. Applied Physics A, 1999, 69(1): S535–S538.
[12] Geetha S, Satheesh KKK, Rao CRK, et al. EMI shielding: Methods and materials – A review. Journal of Applied Polymer Science, 2009, 112(4): 2073–2086.
[13] Gerfin T, Grätzel M. Optical properties of tin-doped indium oxide determined by spectroscopic ellipsometry. Journal of Applied Physics, 1996, 79(3): 1722–1729.
[14] Glasgow B, Bell W. The Future of Anti-Aircraft Imaging Infrared Seeker Missile Threats. IEEE, 1999, 4: 457–465.
[15] Gordon RG. Criteria for Choosing Transparent Conductors. MRS Bulletin, 2000, 25(08): 52–57.
[16] Hamberg I, Granqvist CG. Theoretical model for the optical properties of In_2O_3:Sn films in the 0.3-50 μm range. Solar Energy Materials & Solar Cells, 1986, 14: 241–256.
[17] He L, Wu Z, Li Z, et al. X-ray photoelectron spectroscopy studies of indium-tin-oxide treated via oxygen plasma immersion ion implantation. The European Physical Journal Applied Physics, 2013, 62(03): 30301.
[18] Hewitt R, Winograd N. Oxidation of polycrystalline indium studied by x-ray photoelectron spectroscopy and static secondary ion mass spectroscopy. Journal of Applied Physics, 1980, 51(5): 2620–2624.
[19] Hoefnagels JPM, Langereis E, Kessels WMM, et al. Plasma-surface interaction and surface diffusion during silicon-based thin-film growth. Plasma Science. IEEE Transactions on, 2005, 33(2): 234–235.
[20] Johnson LF, Moran MB, Infrared transparent conductive oxides. Proc. of SPIE, 2001: 289–299.
[21] Joseph S, Marcovitch O, Yadin Y, et al. High durability antireflection coatings for silicon and multispectral ZnS. Defense and Security Symposium. International Society for Optics and Photonics, 2007: 65450T-10.

[22] Kamei M, Shigesato Y, Takaki S. Origin of characteristic grain-subgrain structure of tin-doped indium oxide films. Thin Solid Films, 1995, 259(1): 38–45.

[23] Kim YT. Achievement of zero temperature coefficient of resistance with RuO_x thin film resistors. Applied Physics Letters, 1997, 70(2): 209–211.

[24] Li X, Yan Y, Mason A, et al. High mobility CdO films and their dependence on structure. Electrochemical and Solid-State Letters, 2001, 4(9): C66–C68.

[25] Lin YT, Chen CY, Hsiung CP, et al. Growth of RuO_2 nanorods in reactive sputtering. Applied Physics Letters, 2006, 89(6): 063123-3.

[26] Manoj PK, Gopchandran KG, Koshy P, et al. Growth and characterization of indium oxide thin films prepared by spray pyrolysis. Optical Materials, 2006, 28(12): 1405–1411.

[27] Marcovitch O, Zipin H, Klein Z, et al. Infrared transparent conductive coatings deposited by activated reactive evaporation. SPIE, 1990, 1442: 58.

[28] Medvedeva JE. Averaging of the electron effective mass in multicomponent transparent conducting oxides. Europhysics Letters, 2007, 78(5): 57004.

[29] Music D, Zumdick NA, Hallstedt B, et al. On the solubility of yttrium in RuO_2. Journal of Applied Physics, 2011, 110(5): 054317.

[30] Ordal MA, Long, LL Bell RJ, et al. Optical properties of the metals Al, Co, Cu, Au, Fe, Pb, Ni, Pd, Pt, Ag, Ti, and W in the infrared and far infrared. Applied Optics, 1983, 22(7): 1099–1119.

[31] Park Y, Choong V, Gao Y, et al. Work function of indium tin oxide transparent conductor measured by photoelectron spectroscopy. Applied Physics Letters, 1996, 68(19): 2699–2701.

[32] Puddephatt RJ. Low temperature chemical vapour deposition of ruthenium and ruthenium dioxide on polymer surfaces. Journal of Materials Chemistry, 1999, 9(10): 2439–2444.

[33] Qiao Z, Latz R, and Mergel D. Thickness dependence of In_2O_3: Sn film growth. Thin Solid Films, 2004, 466(1): 250–258.

[34] Reeber RR, Powell GW. Thermal expansion of ZnS from 2 to 317 K. Journal of Applied Physics, 1967, 38(4): 1531–1534.

[35] Rhodes C, Franzen S, Maria JP, et al. Surface plasmon resonance in conducting metal oxides. Journal of Applied Physics, 2006, 100(5): 054905.

[36] Saha B, Thapa R, Chattopadhyay K. Bandgap widening in highly conducting CdO thin film by Ti incorporation through radio frequency magnetron sputtering technique. Solid State Communications, 2008, 145(1): 33–37.

[37] Schulz LG. An experimental confirmation of the Drude free electron theory of the optical properties of metals for silver, gold, and copper in the near infrared. JOSA, 1954, 44(7): 540–545.

[38] Schulz LG. The experimental study of the optical properties of metals and the relation of the results to the Drude free electron theory. Advances in Physics, 1957, 6(21): 102–144.

[39] Song W, So SK, Wang D, et al. Angle dependent X-ray photoemission study on UV-ozone treatments of indium tin oxide. Applied Surface Science, 2001, 177(3): 158–164.

[40] Su X, Sun W, Wei Q, et al. Design and fabrication of antireflection coatings on ZnS substrate. International Society for Optics and Photonics, 2006: 614907-5.

[41] Tahar RBH, Ban T, Ohya Y, et al. Tin doped indium oxide thin films: Electrical properties. Journal of Applied Physics, 1998, 83(5): 2631–2645.

[42] Uvarov V, Popov I. Metrological characterization of X-ray diffraction methods for determination of crystallite size in nano-scale materials. Materials Characterization, 2007, 58(10): 883–891.

[43] Waddell EM, Gibson DR, Wilson M. Broadband IR transparent rain and sand erosion protective coating for the F14 aircraft infra-red search & track of germanium dome. Proc. of SPIE, 1994: 376–385.

[44] Weiher R, Ley R. Optical properties of indium oxide. Journal of Applied Physics, 1966, 37(1): 299–302.

[45] Yamaguchi T, Takahashi H, Sudoh A. Optical behavior of a metal island film. JOSA, 1978, 68(8): 1039–1044.
[46] Yu LS, Harper JME, Cuomo JJ, et al. Alignment of thin films by glancing angle ion bombardment during deposition. Applied Physics Letters, 1985, 47(9): 932–933.
[47] Zhang XT, Liu YC, Zhi ZZ, et al. Resonant Raman scattering and photoluminescence from high-quality nanocrystalline ZnO thin films prepared by thermal oxidation of ZnS thin films. Journal of Physics D Applied Physics, 2001, 34(24): 3430.

Index

a-D 168
a-D coating 158
a-D thin film 171, 173, 174, 179, 180, 182, 184, 187, 188, 195, 199, 200, 202, 206, 208, 218, 222
aerodynamic force 55
aerodynamic heating 55
aerodynamic heating environment 66
AlON ceramics 22
AlON transparent ceramics 22
alternating multilayered film 232
alternating multilayered ta-C film 229, 230
alumina thin film 365, 367, 371
– crystal structure 377
– GIXRD diffraction spectrum 379
– growth rate 374
– mechanical properties 380
– optical properties 382
– preparation methods 369
– SEM surface morphology 376
aluminium oxynitride (AlON) 7
amorphous carbon (a-C) 167
amorphous diamond film 167, 179, 185, 202, 220, 320
amorphous diamond thin film 200, 218
antireflective coating 15, 33
arc-spot control technique 155, 156
atmospheric infrared transmittance spectrum 3
atmospheric window 1

bend test 49
blackbody 5
blackbody radiation 2, 3
blunted cone outer flow field 83
boron phosphide thin film 325, 326, 329, 331, 332, 334, 335, 341, 348, 349
– corrosion resistance 355
– friction coefficient 360
– hardness 350
– mechanical properties 349
– optical constants 354
– optical properties 349, 354
– thermal stability 345
– wear resistance 355, 357

chalcogenide glass 30
chemical vapor deposition 152
coating design 126
– basic theories 113
coating separation 57
coating sputter deposition method 160
coating system
– parametric variations 138
coating uniformity 162
coating–substrate adhesion 48
coating–thickness distribution 163
combined thermal/mechanical state 55
compound coating 149
compound sputtering mode 335
conservation of energy 75
conservation of mass 74
conservation of momentum 74
continuous stiffness measurement (CSM) 234
convective heat-transfer coefficient 89, 91
conventional magnetron sputtering 369
conventional sputtering 368
cosputtering 149
coupling coefficient 209
crystalline alumina 366
crystalline alumina thin film 368, 378
curve-fitting analysis 182
Czochralski process 13

D-peak FWHM 215
depolarization design 132
depositing speed 149
diamond crystal 197
diamond-like carbon (DLC) 325
direct numerical simulation (DNS) 76
dispersion 60

elastic modulus 47
Electre blunt cone 81
electromagnetic radiation 1, 2
electron energy loss spectrum analysis 178
ellipsometry modeling 291
emissivity 60
emittance 41
equation of state 75
European Space Agency (ESA) 81
evaporation condition
– deposition process 152
extinction coefficient 197

far-field boundary condition 81
FCVA 154
FCVA deposition coating 155
film density 286
filtered cathodic vacuum arc (FCVA) technique 320
filtered cathodic vacuum arc deposition 154, 155
flux limiter 79
Fourier transform infrared spectroscopy (FTIR) analysis 340
fracture strength 59
FTIR 340
FWHM 172, 209, 339, 342, 347, 392, 396, 430

G-peak FWHM 215
germanium 9
germanium thin film 293
germanium TO model 314
germanium-carbide coating 265, 267, 269–271, 273–278, 280, 282, 288, 290, 295–297, 299, 300, 303, 305, 306, 308–310, 313
germanium-carbide composite film 317, 320
germanium-carbide film 263, 264, 281
germanium-carbide thin film 272
germanium-target sputtering 276, 278
germanium-target sputtering power 283, 287, 290, 307, 312
gradient multilayered film 231
grazing-incidence x-ray reflectometry 188, 229

hardness 45
heat exchanger method (HEM) 19
heat-flux density 85, 89
heat-sensitive substrate material 371
heat-transfer coefficient 85
high-pressure high-temperature method (HPHT) 14
high-refractive-index material 132, 136
hot spot 28
hot-pressed polycrystalline material 20
hydrogenated amorphous carbon (a-C:H) 167
hydrogenated DLC 167
hydrogenated tetrahedral amorphous carbon (ta-C:H) 167
hydrometeor impact 98

In
– 3d narrow-band photoelectron spectral peaks 453
In_2O_3 film 440, 443
– bonding 448
– composition 448
– microhardness 460
– photoelectric performance 455
– XRD crystalline structure 444
In_2O_3:Sn film 441
– photoelectric performance 455
– chemical bonding 448
– composition 448
– grain size 446
– microhardness 460
indentation test 48
infrared antireflective and protective coating 32
– service environment 55
infrared antireflective coating 32
infrared countermeasure 1
infrared guidance 1
infrared imaging 1
infrared optical material 59
infrared transparent conduction 474
infrared transparent conductive oxide thin film 439
infrared window ix, 1
– aerodynamic heat/strength failure of 56
– failure 71
– military applications 11
infrared window functional film 61
infrared window material 217
– development of 57
infrared-transparent substrate material 8
interfacial adhesion 425
ionic crystal 16

large eddy simulation (LES) 76
large-area filtered cathodic vacuum arc uniform deposition 157
long-wavelength infrared window (LWIR) 2
longitudinal expansion wave 93
low-refractive-index material 132, 136

Macleod software 138
magnesium fluoride (MgF_2) 17
magnetron co-sputtering 264, 312, 461
– system 264
magnetron sputter deposition 160
– system 162
magnetron sputtering 147, 164
– system 148, 159

– target 160, 164, 165
– thin film deposition 372
material failure mechanism 57
material refractive index 66
maximum allowable thermal flux of thermal shock 62
metal sputtering mode 150, 335
metallic sputtering mode 418
mid-wavelength infrared window (MWIR) 2
modulation transfer function (MTF) 6
monolayered antireflective coating 128
monolayered dielectric film 113, 114, 118
monolayered film 128, 132
monolayered ta-C film 235, 240
monolayered ta-C thin film 237
monolayered uniform dielectric film 117, 119, 120
moth eye structure 34
multilayer film window
– stress theory 66
multilayered a-C film structure 231
multilayered amorphous diamond film 225, 227, 231
– mechanical properties 233
multilayered dielectric film 132
multilayered film 123, 244
– stress theory 225
multilayered film-substrate system 226
multilayered ta-C film 235, 240, 242, 245, 249, 251
multilayered ta-C thin film 237

nonhydrogenated DLC 167
nylon bead impact 98

optical distortion 57
optical glass 9, 28
optical thickness 140
optical thin film 113, 132
optical thin-film material 32
optical-thickness deviation 140
oxide glass 28

P-polarization 129, 137
P-polarized light 129
P-polarized wave 127
PECVD 154, 336
periodic film thickness 247
periodic multilayered film 124
planar magnetron sputtering techniques 159
plane uniform magnetron sputtering 158
plasma bombardment magnetron sputtering 440
plasma-beam magnetic filtration technique 155, 156
plasma-enhanced chemical vapor deposition (PECVD) 152
pointed cone model 87
pointed cone outer flow field 86
polarization splittin 133
polarization splitting 133, 135
polycrystalline 9
polycrystalline aluminium oxide 24
polycrystalline material 20
precision guidance 57
pull-off test 49

quarter-wave stack depolarization design 133
quarter-wave stack depolarizing coating system 136

radiation disturbance saturation 57
rain erosion 91, 92, 95, 330
– damage behavior 92
– protection against 99
– resistance 98
– whirling-arm 96
Raman analysis 170
Raman spectrum 170
Rayleigh wave 93
reactive hysteresis loop 419, 420
reactive magnetron sputtering 332, 333, 417
reactive magnetron sputtering process 345
reactive sputtering 149, 370
real in-line transmittance (RIT) 24
refractive index and dispersion 60
resputtering assisted deposition 378
resputtering assisted magnetron sputtering 371, 379
– processing parameters 373
Reynolds-averaged Navier–Stokes (RANS) 76
Rfit2000 software 287
rocket sled test 96
Roe-FDS scheme 79
rupture 56
ruthenium-doped yttrium oxide film
– chemical bonding 464
– composition 464

– crystal structure 462
– electrical properties 472
– optical properties 472
– surface morphology 461

S-polarization 129, 137
S-polarized 127
S-polarized light 129
S-polarized wave 127
sand erosion 100, 101
– damage behavior 100
– environment temperature 102
– impact angle 101
– impact velocity 101
– sand shape 102
– test 103
semiconductor crystal 9
shielding effectiveness (SE) 36
short-wavelength limit 60
Silicon (Si) 12
single-crystal 9
single-crystal aluminium oxide 18
single-crystal material 9
single-crystal Si 12
single-layer film window 62
– optical theory 63
Sn
– 3d narrow-band photoelectron spectral peaks 453
sodium chloride (NaCl) 16
sp^2 hybridization 189, 209
sp^3 bond 172
sp^3 hybridization 174–177, 180, 248
spatial frequency 6
sputtering power 395, 409
Stefan–Boltzmann law 4
stress theory
– multilayer film window 66
structural failure test 67
substrate bias 393
subsurface crack model 94
surface superstructure 34
symmetric boundary condition 81
symmetrical coating system 133

ta-C film 233, 235
ta-C thin film 232
TE wave 116
tetracoordinate hybridization 167
tetrahedral amorphous carbon (ta-C) 167
thermal conductivity 49
thermal evaporation deposition 150
thermal expansion coefficient 50
thermal shock 67
thermal shock resistance 59
thermal stability 206
thin film 169, 180, 212
– crystal structure 390, 395
– electrical properties 302
– fabrication 145
– hardness 169
– mechanical properties 305, 406
– optical parameters 192
– optical thickness 140
– physical thickness 146
– refractive index 145
– stress 204
– surface composition 191
– surface morphology 188
– surface roughness 391
– thickness 200
transmission wavelength ranges 44
transparent ceramics 21
transparent conducting thin-film 35
transverse optical mode (TO mode) 328
transverse shear wave 93
turbulence equation 76

vacuum coating 152
vibrational energy level 43
viscous flux
– discrete scheme 80

wall boundary condition 80
water jet impact 97
Wien maximum radiance law 6
window failure
– causes 71
window material
– comprehensive test 68

x-ray diffraction (XRD) analysis 338
x-ray photoelectron spectroscopy (XPS) 179, 218
x-ray photoelectron spectrum 177
x-ray photoelectron spectrum analysis 174
x-ray reflectometry (XRR) 404
x-ray spectroscopy (EDX) 242
XPS analysis 342

XPS depth profiling 220
XRD analysis 346, 350
XRD pattern
- thin film 394

yttrium oxide 387
yttrium oxide film 461
- performance control 397
yttrium oxide thin film 387, 388, 402
- crystal structure 396
- deposition rate 417
- growth pattern 417
- mechanical properties 405
- optical properties 398
- refractive index 404
- surface energy 414
- surface wettability 409
- water wettability 413
yttrium oxide thin-film surface 410

zinc-sulphide substrate 318–320

Bei Fragen zur Produktsicherheit wenden Sie sich bitte an:
If you have any questions regarding product safety,
please contact:

Walter de Gruyter GmbH
Genthiner Straße 13
10785 Berlin
productsafety@degruyterbrill.com